Science and Technology in the Global Cold War

Transformations: Studies in the History of Science and Technology

Jed Z. Buchwald, general editor

Dolores L. Augustine, *Red Prometheus: Engineering and Dictatorship in East Germany, 1945–1990*

Lawrence Badash, *A Nuclear Winter's Tale: Science and Politics in the 1980s*

Lino Camprubí, *Engineers and the Making of the Francoist Regime*

Mordechai Feingold, editor, *Jesuit Science and the Republic of Letters*

Larrie D. Ferreiro, *Ships and Science: The Birth of Naval Architecture in the Scientific Revolution, 1600–1800*

Gabriel Finkelstein, *Emil du Bois-Reymond: Neuroscience, Self, and Society in Nineteenth-Century Germany*

Kostas Gavroglu and Ana Isabel da Silva Araújo Simões, *Neither Physics nor Chemistry: A History of Quantum Chemistry*

Sander Gliboff, *H. G. Bronn, Ernst Haeckel, and the Origins of German Darwinism: A Study in Translation and Transformation*

Niccolò Guicciardini, *Isaac Newton on Mathematical Certainty and Method*

Kristine Harper, *Weather by the Numbers: The Genesis of Modern Meteorology*

Sungook Hong, *Wireless: From Marconi's Black-Box to the Audion*

Jeff Horn, *The Path Not Taken: French Industrialization in the Age of Revolution, 1750–1830*

Alexandra Hui, *The Psychophysical Ear: Musical Experiments, Experimental Sounds, 1840–1910*

Myles W. Jackson, *Harmonious Triads: Physicists, Musicians, and Instrument Makers in Nineteenth-Century Germany*

Myles W. Jackson, *Spectrum of Belief: Joseph von Fraunhofer and the Craft of Precision Optics*

Paul R. Josephson, *Lenin's Laureate: Zhores Alferov's Life in Communist Science*

Mi Gyung Kim, *Affinity, That Elusive Dream: A Genealogy of the Chemical Revolution*

Ursula Klein and Wolfgang Lefèvre, *Materials in Eighteenth-Century Science: A Historical Ontology*

John Krige, *American Hegemony and the Postwar Reconstruction of Science in Europe*

Janis Langins, *Conserving the Enlightenment: French Military Engineering from Vauban to the Revolution*

Wolfgang Lefèvre, editor, *Picturing Machines 1400–1700*

Staffan Müller-Wille and Hans-Jörg Rheinberger, editors, *Heredity Produced: At the Crossroads of Biology, Politics, and Culture, 1500–1870*

William R. Newman and Anthony Grafton, editors, *Secrets of Nature: Astrology and Alchemy in Early Modern Europe*

Naomi Oreskes and John Krige, editors, *Science and Technology in the Global Cold War*

Gianna Pomata and Nancy G. Siraisi, editors, *Historia: Empiricism and Erudition in Early Modern Europe*

Alan J. Rocke, *Nationalizing Science: Adolphe Wurtz and the Battle for French Chemistry*

George Saliba, *Islamic Science and the Making of the European Renaissance*

Suman Seth, *Crafting the Quantum: Arnold Sommerfeld and the Practice of Theory, 1890-1926.*

Leslie Tomory, *Progressive Enlightenment: The Origins of the Gaslight Industry 1780-1820*

Nicolás Wey Gómez, *The Tropics of Empire: Why Columbus Sailed South to the Indies*

Science and Technology in the Global Cold War

edited by Naomi Oreskes and John Krige

The MIT Press
Cambridge, Massachusetts
London, England

© 2014 Massachusetts Institute of Technology

All rights reserved. No part of this book may be reproduced in any form by any electronic or mechanical means (including photocopying, recording, or information storage and retrieval) without permission in writing from the publisher.

Set in Stone Sans and Stone Serif by Toppan Best-set Premedia Limited.

Library of Congress Cataloging-in-Publication Data

  Science and technology in the global cold war / edited by Naomi Oreskes and John Krige.
    pages cm
  Includes bibliographical references and index
  ISBN 978-0-262-02795-3 (hardcover : alk. paper)—ISBN 978-0-262-52653-1 (pbk. : alk. paper)
  1. Technology and state. 2. Science and state. 3. Security, International. I. Oreskes, Naomi, editor of compilation. II. Krige, John, editor of compilation.
  T14.5.S3768 2014
  338.9'2609045—dc23
  2014009830

150377131

to Dan Kevles and Paul Forman, who started the conversation, and to our families, who endured it

Contents

Acknowledgments   ix

Introduction   1
Naomi Oreskes

1   Science in the Origins of the Cold War   11
Naomi Oreskes

2   Atomic Tracings: Radioisotopes in Biology and Medicine   31
Angela N. H. Creager

3   Self-Reliant Science: The Impact of the Cold War on Science in Socialist China   75
Sigrid Schmalzer

4   From the End of the World to the Age of the Earth: The Cold War Development of Isotope Geochemistry at the University of Chicago and Caltech   107
Matthew Shindell

5   Changing the Mission: From the Cold War to Climate Change   141
Naomi Oreskes

6   Fighting Each Other: The N-1, Soviet Big Science, and the Cold War at Home   189
Asif Siddiqi

7   Embedding the National in the Global: US-French Relationships in Space Science and Rocketry in the 1960s   227
John Krige

8   Bringing NASA Back to Earth: A Search for Relevance during the Cold War   251
Erik M. Conway

9 Calculating Times: Radar, Ballistic Missiles, and Einstein's Relativity   273
Benjamin Wilson and David Kaiser

10 Defining (Scientific) Direction: Soviet Nuclear Physics and Reactor Engineering during the Cold War   317
Sonja D. Schmid

11 The Cold War and the Reshaping of Transnational Science in China   343
Zuoyue Wang

12 When *Structure* Met Sputnik: On the Cold War Origins of *The Structure of Scientific Revolutions*   371
George Reisch

13 Big Science and "Big Science Studies" in the United States and the Soviet Union during the Cold War   393
Elena Aronova

Concluding Remarks   431
John Krige

About the Authors   443
Index   447

Acknowledgments

This volume would not have been possible without the generous support of the Francis Bacon Foundation as part of the biennial Francis Bacon Prize in the History and Philosophy of Science, awarded to Naomi Oreskes in 2009. The editors would also like to thank Professor Jed Buchwald and the California Institute of Technology for their support of Naomi Oreskes as the 2009 Francis Bacon Fellow and of John Krige as the Eleanor Searle Visiting Professor in 2009. We are extremely grateful for the hard work and collegiality of all the contributors to the volume, to two anonymous referees, and to the indefatigable Charlotte Goor, who assisted in every way with extraordinary diligence and with even more extraordinary patience.

Introduction

Naomi Oreskes

Historians are committed to studying science and technology in context. We accept, as a matter of principle that is rarely debated, that the context of intellectual work is part of its history, and that any account of knowledge that does not include its cultural origins is at best incomplete, and at worst misleading in intellectually and politically significant ways. Yet the question of exactly how context affects content remains challenging. Even though history of science and history of technology are mature disciplines, the scientists (and to a somewhat lesser extent engineers) that we study still tend to resist our approach, often viewing contextualization as diminishing their claims to objectivity and the universality of scientific knowledge. Although we insist on the importance of context, we are often at pains to demonstrate in a convincing manner just how the broader social, political, economic, or religious environments of knowledge production really matter to the knowledge being produced.

One area in which historians have recently produced a substantial corpus of convincing work addressing this challenge is the history of science and technology in the Cold War. The Cold War presents a rich opportunity for historians because the dramatic expansion of science and engineering supported by national governments and the relation of governmental support for science and engineering to geopolitical conflict and ambition invite appraisal of the relationship between those conflicts and ambitions and the science that, in some way, supported and enabled them. The arms race, most obviously, would not have occurred without the East-West political conflict that is often taken to define the Cold War, but it also *could* not have occurred without the work of scientists and engineers. Much has already been written about the role of scientists in building the nuclear weaponry that defined the Cold War, but the space race, the exploration of the deep oceans and the deep interior of the Earth, the rise of telecommunications and civilian nuclear power, and many other scientific and technological developments were also directly tied to the global conflict that the Cold War entailed.

One person whose thinking about the Cold War has strongly influenced the scholarship presented in this volume is the historian Daniel Kevles. In the 1980s, thinking about how the Cold War did or did not alter the intellectual trajectory of American physics, Kevles famously concluded that physics is what physicists do. On some level that is undeniable, yet Kevles' formulation left unanswered the historical questions of *why* they do what they do rather than what they did before, what they might otherwise have done at the time, and what they may yet do in the future. Work in the 1980s and the 1990s also left largely unaddressed the role of sciences other than physics in the Cold War.

Since the 1980s, when Kevles—along with Paul Forman—defined the debate about Cold War science along certain axes, numerous scholars have joined the discussion, amplified it, and extended it in diverse ways. Chief among these has been the extension of investigations beyond physics to show how numerous other fields—agriculture, biomedicine, computer science, ecology, geology, meteorology, seismology, oceanography—were affected at least as much as physics was by the Cold War's constraints and opportunities.[1] Historians of the social sciences have also tracked how Cold War geopolitical concerns stimulated new interest on the part of various patrons, including the US federal government and private foundations, in new disciplines such as "area studies," communications, and cognitive science, as well as encouraging extended work in conventional fields such as philosophy and psychology.[2]

Virtually all of the scholars who have looked at the question agree that during the Cold War military largesse changed the scope and the capacity of science in many domains. Virtually all scholars also agree that new institutions and new institutional arrangements emerged, some of which played major roles in the shape and structure of science and the conditions of the scientific workplace after World War II. However, scholars disagree as to whether these changes were primarily constraints or primarily opportunities, and many have simply skirted the question of to what degree Cold War conditions changed the content or the character of the scientific knowledge that was produced (or not produced).

But what is the purpose of studying the historical context of science and technology if we don't believe that it does—or at least may—shape their content? Every historical account stands in implicit relief against a plausible counterfactual in which matters could have turned out differently. It is this hovering counterfacticity that gives history its emotional force. For if it were inevitable that things turned out as they did, if history followed deterministic laws, as Marx and others once thought (or hoped), then history could indeed be a science, but it would be far less emotionally and imaginatively fertile than it is.

This volume takes up, with vigor, the questions "What did scientists do in the Cold War?" and "Why did they do those things and not other things?" A conference that explored the topics discussed herein was held at the California Institute of Technology on May 7–9, 2010, with the generous support of the Bacon Foundation, as part of the biennial Francis Bacon Prize in the History and Philosophy of Science, awarded to

Naomi Oreskes in 2009. One conference and one volume could hardly hope to settle the questions we have raised, and we do not suppose in any way to have achieved that here. Yet each of the chapters in this volume, in its own way, delves deeply into what has been, and arguably should still be, a central focus of history of science as a discipline: how the social context of scientific work affects its content. Each chapter attempts to give some indications of how the Cold War context either enabled or disabled certain kinds of investigations and intellectual achievements. Each chapter addresses the question of whether the Cold War context was more constraint or more opportunity. In this sense, the volume may be seen as a contribution not only to Cold War history, but also to the long-standing interest in history of science in the role of patronage, and of social context broadly construed.

With some notable exceptions, scholars who have studied scientific patronage have rarely been willing to claim that that patronage *caused* scientists to work in particular ways, much less to draw certain conclusions about the natural world.[3] On the other hand, the purpose of patronage, military or otherwise, is, in most cases, to adjust the focus of attention and influence the direction of work. Patient-driven patronage may shift biologists' attention toward cures for particular diseases on which their attention had not been focused previously. Artistic patronage may create a demand for portraiture that might not otherwise be of much interest to artists. And military patronage is intended to garner scientific attention to questions of military pertinence and concern. The questions for historians—questions we take up in this volume (in some cases explicitly and in others implicitly)—are the following: In what specific manner did Cold War patronage affect the patterns and priorities of scientific research? What consequences, if any, did those adjustments have? How did these patterns vary in different national contexts? What role did national ambitions play in fostering, enabling, or disabling certain lines of investigation? What happened to scientists who tried to do things other than what their national governments wanted them to do?

Our topics are how the Cold War shaped and altered the trajectories of science and existing technologies, how it created new sciences and technologies, how it affected the relationships between scientists and engineers and their patrons, and how scientists and engineers managed, negotiated, and adjusted those relationships—with varying degrees of success—as they attempted to achieve their own goals in relation to state patrons, fellow scientists (friendly or competitive), and personal aspirations. Science and technology were different after the Cold War than they were before it; that claim is indisputable. In this volume we are concerned with the historical understanding of the character and dimensions of that change, and with the specific ways in which Cold War politics, anxieties, and aspirations were or were not significantly responsible for those changes.

One theme that emerged clearly from the conference that produced these chapters was that, whatever the particular science involved (whether related to weapons and their delivery systems, agriculture, isotopes, the speed of light in a vacuum, or the transmission of sound in the sea), and whatever the political system that science was operating in (capitalist, communist, or hybrid), the knowledge produced bore some significant relation to the goals of the nation-state (or nation-state equivalent, in the case of the Soviet Union) that was helping to procure it. But the specific relations varied significantly from nation to nation, just as the goals of nation-building varied. In the United States and the Soviet Union, weapons—including their testing, hiding, detection and delivery—were of paramount importance. In France and China, issues of independence and self-reliance loomed large.[4] Although our volume is more US-centric that we had originally hoped it would be, an important contribution is the presentation of historical studies of Europe, the Soviet Union, and China, and their close juxtaposition with stories from the to-date-better-studied United States. Thus, the chapters are organized not by national origin, but by discipline, in order to facilitate cross-comparisons of Cold War approaches to knowledge in different national and political contexts within the larger rubric of the Cold War.

We focus not only on what happened to science during the Cold War, but also on what happened to science because of the Cold War. We address these topics in terms of the structure of science, the research agendas, who was setting those agendas, and how and why. We have consciously avoided discussing nuclear weapons and nuclear medicine (already discussed thoroughly by others) and computer science (ably covered by Paul Edwards and Janet Abbate, although there is still much to be done in the domains of artificial intelligence, operations research, linear programming, and the rise of numerical methods and simulation). There is more to be said, as well, about Cold War considerations in biology and ecology. Indeed, if our arguments are correct, almost any area of science could have been a topic for this volume; limitations of time and space precluded comprehensiveness, lest we embark on an encyclopedia of Cold War science.[5]

We admit to the historical presumption that the demands, desires, and expectations—either stated or implied—of both immediate patrons and society at large could hardly *not* have affected the scientific and technological knowledge produced during the Cold War, yet we recognize that the ways in which these effects played out were diverse, culturally situated, and in no way predetermined or predictable. In a sense we are arguing that, although it may be difficult to demonstrate specific consequences, such consequences are to be *expected*, and it is the historian's job to determine what they are and how they unfolded in various particular and diverse instances.

Our focus is not so much on how the Cold War affected individual scientists—that territory is well trod, as we certainly know that Cold War anxieties caused leading scientists to become victims of anti-communist "witch hunts" in the United States and murderous xenophobia in the Soviet Union.[6] Rather, our focus is on the contents of scientific (and engineering) research activities. In some cases, such as that of Harold Urey, we see active agency in an individual scientist: as Matthew Shindell notes, Urey strove to build an ambitious research program outside the shadow of the atomic bomb, but was ultimately unable to find a patron that could support the scale of his ambitions outside of national-security concerns. Likewise, we see individual agency in Benjamin Wilson and David Kaiser's account of how Irwin Shapiro mobilized Cold War resources to test the general theory of relativity, and in George Reisch's consideration of how Cold War politics led Thomas Kuhn to use the word 'paradigm' to describe what he had previously called scientific dogma. But for the most part, these are stories of groups of scientists, communities, and the institutions in which they worked—institutions whose shape, character, and in some cases existence owed much to the aspirations of their host nations to use knowledge to political, social, and economic effect. Angela Creager shows us how scientists working in conventional academic settings made use of the opportunities the Cold War offered to expand investigations using isotopes as a research and medical tool. Erik Conway and John Krige focus on the development of new forms of institutional support and motivation for certain kinds of scientific research in certain kinds of settings.

Questions of agency and causality raise the thorny issue of overdetermination. Many things were happening in the world between 1945 and 1989, and we need to consider whether the category "the Cold War" is apposite when considering science outside the United States and the Soviet Union. For example, how did global geopolitics interact with nationalism, development, and post-revolutionary politics in India and in China? Sigrid Schmalzer and Zuoyue Wang invite us to reconsider the periodization of the Cold War that many scholars of the United States and the Soviet Union have taken for granted. Their chapters, and those of Asif Siddiqi and Sonja Schmid, also invite us to consider more deeply our use of the categories "basic science" and "applied science."

Recurrent themes throughout this volume are tension and debate over what constitutes "pure," "basic," or "fundamental" science, how to characterize the relation of such science to "applied" science, technology, and practical knowledge, and the political and epistemic valence that these categories carried in different cultural and national settings. These topics are highly familiar to historians of US science: the pure/applied distinction and the place of "basic" research in American science is one of the standard tropes of the field. However, the histories told in this volume also illustrate how efforts to patrol the boundaries of *both* pure and applied science are also arguments about which forms of knowledge are most important, valuable, and necessary to the nation. In the past, historians of science often accepted scientists' assertions that basic science was, well, *basic*—that is to say, foundational—and therefore a necessary precursor to "useful" or "applied" knowledge. In China and the Soviet Union during the Cold War—and even sometimes in the United States—that assertion wasn't broadly accepted, much less supported by developments on the ground. The chapters in this volume show that the category "basic science" itself has been contested and disputed—that basic science wasn't necessarily considered obviously more valuable and important than applied science, and wasn't necessarily viewed as a precursor to the work of nation-building through practical knowledge and technologies.

Questions about categories—how historical actors used them as tools of both cognition and persuasion, and how we use them as tools of analysis—invite self-scrutiny into both the Cold War origins of our own concern with the role of scientific and technical knowledge in national goals and its effect on our categories of analysis (a topic ably taken up by Elena Aronova and by George Reisch). The Cold War clearly speaks to us today in part because the question of the autonomy and uses of science (and indeed, of all forms of knowledge) remains sharp for us, as both an epistemic question and a social one. As John Krige notes at the end of this volume, Paul Forman and Dan Kevles' famous historical interventions were not merely interpretive; they were also normative, as they reflected Forman's and Kevles' own views on the value and necessity of autonomous science (or not). They addressed the question of the necessity and desirability of a scientific community functioning largely independently of the larger world that surrounds and sustains it (or not).

As historians, we might argue that the very idea of an independent scientific community is at best quaint, and surely one that no Marxist would have accepted during the Cold War. Yet the fact that so many scientists in the United States insisted that they *were* independent—that they did "basic science" even while being wholly or nearly entirely funded by the US military—and the fact that until recently many historians (again, at least in the United States) accepted this argument suggest, at minimum, that autonomy was an important value to these scientists, one they felt obligated to insist they had protected and not lost or even compromised. In China, however, a different cultural setting led scientists to insist on the reverse: the practical value of their work, and its close connection to the needs of the state and the people the state ostensibly represented. Autonomy as Americans understood it would have been deeply problematic for Chinese scientists.

Dan Kevles was surely right when he argued that American scientists weren't pawns in the Cold War but were active partners who made conscious decisions and helped to shape and inspire the expectations of their patrons and communities. At least this was true of the entrepreneurial leaders of US physics, and Sonja Schmid's and Asif Siddiqi's studies of rocketry and nuclear energy in the Soviet Union make clear that this was as true in the Soviet Union as it was in the United States. Sigrid Schmalzer's and Zuoyue Wang's studies of science in China support a similar claim. But decisions are made in context, so we might also frame our question this way: If science is what scientists do, then what did scientists decide to do in the Cold War, and how were those decisions shaped by the exigencies and opportunities of the period? Clearly, the availability of funds, instruments, research platforms, personnel, and moral and logistical support, along with personal commitments and cultural context, made some decisions more attractive than others. They also made some choices effectively impossible.

Choice is a useful category, but only up to a point. What roles did patronage, patriotism, national ambitions, and Cold War geopolitics play in shaping scientists' beliefs about the direction their science could or should take and in defining the spheres of possibility? Finally, and perhaps most important, how did scientists' choices, decisions, and resistance affect what we learned—or failed to learn—about the natural world?

Notes

1. Chandra Mukerji, *A Fragile Power: Scientists and the State* (Princeton University Press, 1989); Peter Galison, *Image and Logic: A Material Culture of Microphysics* (University of Chicago Press, 1997); Peter Galison and Bruce Hevly, eds., *Big Science: The Growth of Large-Scale Research* (Stanford University Press, 1992); David H. DeVorkin, *Science with a Vengeance: How the Military Created the US Space Sciences After World War II* (Springer-Verlag, 1992); David H. DeVorkin, "The Military Origin of the Space Sciences in the American V-2 Era," in *National Military Establishments and the Advancement of Science and Technology: Studies in the 20th Century History*, ed. Paul Forman and José M. Sánchez-Ron (Kluwer, 1996); Stuart W. Leslie, "Science and Politics in Cold War America," in *The Politics of Western Science 1640–1990*, ed. Margaret Jacob (Humanities Press, 1992); Stuart W. Leslie, *The Cold War and American Science: The Military-Industrial-Academic Complex at MIT and Stanford* (Columbia University Press, 1993); Michael T. Bernstein, "The Cold War and Expert Knowledge: New Essays on the History of the National Security State," *Radical History Review* 63 (1995): 1–6; Michael A. Dennis, "Accounting for Research: New Histories of Corporate Laboratories and the Social History of American Science," *Social Studies of Science* 17 (1987): 479–518; Michael Dennis, A Change of State: The Political Cultures of Technical Practice at the MIT Instrumentation Laboratory and the John Hopkins University Applied Physics Laboratory, 1930–1945, PhD dissertation, Johns Hopkins University, 1991; Michael A. Dennis, "Our First Line of Defense: Two University Laboratories in the Postwar American State," *Isis* 85 (1994): 427–455; Michael A. Dennis, "Historiography of Science: An American Perspective," in *Science in the Twentieth Century*, ed. John Krige and Dominique Pestre (Harwood, 1997); Noam Chomsky et al., *The Cold War and the University: Toward an Intellectual History of the Post-War Years* (New Press, 1997); Ronald E. Doel, "Scientists as Policymakers, Advisors, and Intelligence Agents: Linking Contemporary Diplomatic History with the History of Contemporary Science," in *The Historiography of Contemporary Science and Technology*, ed. Thomas Soderqvist (Harwood, 1997); Ronald E. Doel, "Constituting the Postwar Earth Sciences: The Military's Influence on the Environmental Sciences in the USA after 1945," *Social Studies of Science* 33, no. 5 (2003): 635–666;

Jacob Darwin Hamlin, *Oceanographers and the Cold War: Disciples of Marine Science* (University of Washington Press, 2005); Jacob Darwin Hamlin, *Poison in the Well: Radioactive Waste in the Oceans at the Dawn of the Nuclear Age* (Rutgers University Press, 2008); David Hounshell, "The Cold War, RAND, and the Generation of Knowledge, 1946–1962," *Historical Studies in the Physical and Biological Sciences* 27 (1997): 237–267; Barton Hacker, "Military Patronage and the Geophysical Sciences in the United States: An Introduction," *Historical Studies in the Physical and Biological Sciences* 30 (2000): 309–313; Ronald Rainger, "Constructing a Landscape for Postwar Science Roger Revelle, the Scripps Institution and the University of California, San Diego," *Minerva* 39 (2001): 327–352; Mark Solovey, "Introduction: Science and the State During the Cold War: Blurred Boundaries and a Congested Legacy," *Social Studies of Science* 31, no. 2 (2001): 165–170; Mark Solovey, "Project Camelot and the 1960s Epistemological Revolution: Rethinking the Politics-Patronage-Social Science Nexus," *Social Studies of Science* 31, no. 2 (2001): 171–172; David Kaiser, "Cold War Requisitions, Scientific Manpower, and the Production of American Physicists after World War II," *Historical Studies in the Physiological and Biological Sciences* 33, no. 1 (2002): 131–159; David Kaiser, "The Physics of Spin: Sputnik Politics and American Physicists in the 1950s," *Social Research* 73, no. 4 (2006): 1225–1252; John Krige, Science, Technology and Civil Security in the 1950s, seminar, Johns Hopkins University, 2003; Naomi Oreskes, "A Context of Motivation: US Navy Oceanographic Research and the Discovery of Sea-Floor Hydrothermal Vents," *Social Studies of Science* 33, no. 5 (2003): 697–742; Naomi Oreskes and Ronald E. Doel, "Physics and Chemistry of the Earth," in *The Cambridge History of Science*, volume V: *Modern Physical and Mathematical Sciences*, ed. Mary Jo Nye (Cambridge University Press, 2002), 538–552; Paul N. Edwards, *A Vast Machine: Computer Models, Climate Data, and the Politics of Global Warming* (MIT Press, 2010); Paul N. Edwards, *The Closed World: Computers and the Politics of Discourse in Cold War America* (MIT Press, 1996), chapter 3; Elena Aronova, Karen Baker, and Naomi Oreskes, "From the International Geophysical Year through the International Biological Program to LTER: Big Science and Big Data in Biology, 1957–present," *Historical Studies in the Natural Sciences* 40, no. 2 (2010): 183–224; Rebecca S. Lowen, *Creating the Cold War University: The Transformation of Stanford* (University of California Press, 1997); S. S. Schweber, "Theoretical Physics and the Restructuring of the Physical Science: 1925–1975," in *Big Culture: Intellectual Cooperation in Large-Scale Cultural and Technical System*, ed. Giuliana Gemelli (Editrice CLUEB, 1994); Everett Mendelsohn, Merritt Roe Smith, and Peter Weingart, eds., *Science, Technology, and the Military* (Kluwer, 1988); James R. Fleming, *Fixing the Sky: The Checkered History of Weather and Climate Control* (Columbia University Press, 2010); Angela N. Creager and Maria Jesús Santemases, "Radiobiology in the Atomic Age: Changing Research Practices and Politics in Comparative Perspective," *Journal of the History of Biology* 39, no. 4 (2006): 637–647; Naomi Oreskes and Erik Conway, "Challenging Knowledge: How Climate Science became a Victim of the Cold War," in *Agnotology: The Making and Unmaking of Ignorance*, ed. Robert N. Proctor and Londa Schiebinger (Stanford University Press, 2008); Bruno J. Strasser, *La fabrique d'une nouvelle science: La biologie moléculaire à l'âge atomique (1945–1964)* (Leo S. Olschki, 2006); Soraya de Chadavarian, *Designs for Life: Molecular Biology after World War II* (Cambridge University Press, 2002); Stephen Bocking, "Ecosystems, Ecologists, and the Atom: Environmental Research at Oak Ridge National Laboratory," *Journal of the History of Biology* 28 (1995): 1–4; *American Hegemony and the Post-war Reconstruction of Science in Europe* (MIT Press, 2008). A comprehensive bibliography produced by Cold

War Science & Technology Studies Program is available at http://www.cmu.edu/coldwar/knowl.htm.

2. Irving Lewis Horowitz, *The Rise and Fall of Project Camelot: Studies in the Relationship between Social Science and Practical Politics* (MIT Press, 1967). See also Jamie Cohen-Cole, "The Creative American: Cold War Salons, Social Science, and the cure for Modern Society," *Isis* 100 (2009): 219–262; Slava Gerovitch, "Perestroika of the History of Technology and Science in the USSR: Changes in the Discourse," *Technology and Culture* 37, no. 1 (1996): 102–134; Slava Gerovitch, "Writing History in the Present Tense: Cold War-Era Discursive Strategies of Soviet Historians of Science and Technology," in *Universities and Empire: Money and Politics in the Social Science During the Cold War*, ed., Christopher Simpson (New Press, 1998); Joy Rohde, "Grey Matters: Social Scientists, Military Patronage, and Democracy in the Cold War," *Journal of American History* 96, no. 1 (2010): 99–122; Mark Solovey, "Riding Natural Scientists' Coattails Onto the Endless Frontier: The SSRC and the Quest for Scientific Legitimacy," *Journal of the History of the Behavioral Sciences* 40, no. 4 (2004): 393–422; Solovey, "Science and the State During the Cold War," pp. 165–170; Solovey, "Project Camelot," pp. 171–172; Mark Solovey and Hamilton Cravens, eds., *Cold War Social Science: Knowledge Production, Liberal Democracy, and Human Nature* (Palgrave MacMillan, 2012); Rebecca Lemov, *World as Laboratory: Experiments with Mice, Mazes, and Men* (Hill and Wang, 2005); Janet Martin-Nelson, "This War for Men's Minds: The Birth of a Human Science in Cold War America," *History of the Human Sciences* 23, no. 5 (2010): 131–155.

3. The obvious counterexample is Mario Biagioli, who went further along this line than many historians have found comfortable. See Biagioli, *Galileo, Courtier: The Practice of Science in the Culture of Absolutism* (University of Chicago Press, 1993); Steven Shapin and Simon Schaffer, *Leviathan and the Air-Pump: Hobbes, Boyle and the Experimental Life* (Princeton University Press, 2011); Robert E. Kohler, *Partners in Science: Foundations and Natural Scientists* (University of Chicago Press, 1991).

4. On French self-reliance, see Gabrielle Hecht, *The Radiance of France: Nuclear Power and National Identity after World War II* (MIT Press, 1998).

5. Edwards, *The Closed World*; Janet Abbate, *Inventing the Internet* (MIT Press, 1999).

6. Jessica Wang, *American Science in an Age of Anxiety: Scientists, Anticommunism, and the Cold War* (University of North Carolina Press, 1999); Michael A. Bernstein, *A Perilous Progress: Economists and Public Purpose in Twentieth-Century America* (Princeton University Press, 2001); Ellen Schrecker, *No Ivory Tower: McCarthyism and the Universities* (Oxford University Press, 1986).

1 Science in the Origins of the Cold War

Naomi Oreskes

In *Military and Political Consequences of Atomic Energy*, first published in the United Kingdom in 1948 (and in the United States a year later under the catchier title *Fear, War and the Bomb*), the physicist P. M. S. Blackett declared that the dropping of the atomic bombs on Japan was "not so much the last military act of the Second World War as the first major operation of the cold diplomatic war with Russia now in progress."[1] Blackett was one of many, then and now, who have tried to assess the role of the atomic bomb—and therefore, implicitly, of science—in ending World War II and launching the Cold War, as well as the significance of the Cold War in altering the course of science.[2]

At the end of World War II, many scientists emphasized the bomb's significance—perhaps because the greater the bomb's role, the greater their role. If the bomb were crucial either in ending World War II or in beginning the Cold War, then science and scientists were crucial too, and perhaps had a further role to play in helping to control it. Niels Bohr argued that the unprecedented power of nuclear weapons necessitated new forms of international governance.[3] George Orwell agreed that nuclear weaponry heralded the dawn of a new age—and not a good one. In the essay in which he coined the term "Cold War," Orwell argued that the atomic bomb was so terrifying that it would put an end to conventional warfare, but that, counterintuitively, this was *not* good, because the bomb would put in its stead a hideous peace as "horribly stable as the slave empires of antiquity." The world would find itself in a permanent state of "cold war," and the West would be unable to act decisively when conditions called for it.[4] Blackett (and others) didn't believe that the atomic bomb would make other forms of weaponry obsolete, much less end conventional warfare; they noted dryly that US generals had shown no sign of giving up their conventional forces. The atomic bomb, Blackett suggested, was in some sense an extension of the World War II policy of massive bombardment of civilians, and just as unethical. After all, how was destroying a Japanese city with one bomb much different from destroying a German city with many bombs (an argument that the bomb's defenders would later use, although to opposite effect than Blackett intended)?[5] Blackett noted that a "huge weight of ordinary bombs" had been dropped on Germany "without leading to a decisive failure of either production or civilian morale," suggesting that, as awesome and frightening as atomic weaponry was, it had not been decisive in World War II and it wasn't likely to prove decisive in future wars either:

> Three million tons of ordinary bombs were dropped by British and American aircraft in the European and Pacific Wars. Since one atomic bomb of the 1945 type produces ... about the same material destruction as 2,000 tons of ordinary bombs, it is certain that a very large number of atomic bombs would be needed to defeat a great nation by bombing alone.[6]

Nuclear weapons were powerful, to be sure, but it was a mistake to overestimate their significance, because that might lead to hysteria, which in turn would make it harder to negotiate with the Soviet Union to find a route to a lasting, stable peace. (It would also make it more difficult to develop civilian nuclear power generation.[7]) Overestimation of the bomb's power was leading to "a hysterical search for 100 per cent security," a security that could never be achieved.[8] It was generating pressure for a huge buildup of weaponry. (Blackett cited the logic of the "Irishman" who "on seeing a stove advertised to save half one's fuel, he bought two to save it all!"[9]) Worst of all, hysteria about the power of the bomb in the hands of enemies led to the hideous suggestion that a pre-emptive strike might be justified—a suggestion that would indeed be made at various points during the Cold War and afterward.[10]

In hindsight we can see that both Blackett and Orwell were partly right. The world did plunge into a Cold War—a deep freeze of animosity between the United States and the Soviet Union that chilled much of the rest of the world as well—and Orwell's term took hold to describe it.[11] The Cold War had a range of negative consequences, though perhaps not as dark or as monolithically negative as Orwell feared.[12] And science was central to the Cold War, because it had been scientists who had perceived the possibility of nuclear weapons, scientists who had built them, and scientists who continued to develop the means of testing, hiding, detecting, and delivering them.[13] Blackett, for his part, was correct that the Cold War climate included a significant component of hysteria, leading both sides to demonize the other, and to insist on the necessity of stockpiling tens of thousands weapons that neither side ever used or wanted to use. Nor did these expenditures prevent either side from spending comparable resources on conventional weapons. On neither side were generals prepared to give up their conventional forces—even after the development of the hydrogen bomb, which was thousands of times as powerful as the bombs dropped on Hiroshima and Nagasaki. The premise that nuclear weaponry would be more economical than conventional forces also proved false, as both sides built both massive nuclear forces—tens of thousands of nuclear weapons, thousands of bombers, hundreds of ICBMs (and still more intermediate-range ballistic missiles), and scores of nuclear submarines—and massive conventional forces of troops, tanks, battleships, and aircraft carriers.[14] As

President Dwight Eisenhower put it in his farewell address, both sides were compelled to "create a permanent armaments industry of vast proportions."[15] This industry cast a long shadow, as citizens of not only the United States and the Soviet Union but the rest of the world as well lived under the threat of Mutual Assured Destruction. Indeed, at the start of his presidency, Eisenhower went further, describing life during the Cold War as "not a way of life at all in any true sense." "Under the cloud of threatening war," he continued, "it is humanity hanging from a cross of iron."[16]

In 1997, the historian Walter LaFeber looked back and summarized the American Cold War experience this way: "It has cost Americans $8 trillion in defense expenditures, taken the lives of nearly 100,000 of their young men and women, ruined the careers of many others during the McCarthyite witch hunts, [and] led the nation into the horrors of Southeast Asian conflicts. ... It has not been the most satisfying chapter in American diplomatic history."[17] No doubt one could say something similar from the Soviet perspective. There were costs to other nations as well, as they felt compelled to establish their own nuclear weapons programs, participated (both knowingly and inadvertently) in the nuclear tests of other nations, or became sites of nuclear weapons facilities and thus potential targets in a war.[18]

What did science have to do with all this? The atomic bomb could not, of course, have been built without scientific insight and technical prowess—the discovery of nuclear fission, the detailed determinations of the requirements for critical mass, and the extensive work on materials, electronics, and conventional explosives that made the atomic bomb possible, and so scientists and historians of science have placed great emphasis on the role of the atomic bomb in creating the Cold War world.[19] The bomb, it seemed obvious at the time, at least to the scientists who had helped to build it, had transformed the world. As Martin Sherwin argued in a book that has gone through multiple editions and has been repeatedly described as "definitive," the bomb destroyed the old world—a world where war was generally fought between near neighbors—and replaced it with a new world of global conflict.[20] In the old world, wars were fought by uniformed soldiers, primarily on battlefields; in the new world, warfare would spread everywhere—on the land, in space, and beneath the sea. Civilians in cities would be the primary targets: the threat of megaton nuclear weapons meant that no one was safe. As Nevil Shute made indelibly clear in his novel *On The Beach*, even those who survived nuclear blasts in remote locations would be victims of fallout. And scientists, it seemed, were largely to blame, for they had started the whole thing.[21] But had they really? As Michael Gordin has noted, the Cold War was as much about *knowledge about knowledge*—who had it and who didn't—as it was about the knowledge itself.[22] And as Odd Arne Westad has emphasized, the global Cold War was as much about politics and ideology as it was about advanced weapons and their delivery systems.[23] While historians of science and technology have emphasized the role of scientific and technical knowledge—the role of the bomb in triggering the Cold War and the arms race in sustaining it—political historians have tended to see the matter somewhat differently.

The Political Origins of the Cold War

In his classic work *The United States and the Origins of the Cold War, 1941–1947*, John Lewis Gaddis found the political origins of the Cold War not so much in the use of the bomb at the end of the war as in irreconcilable differences between two opposed political and economic systems.[24] His periodization immediately tells us that the bomb is, at most, one piece of a larger story.

The driving force behind President Franklin Roosevelt's approach to World War II, Gaddis argues, was a desire to end it correctly by paying attention to the political and economic dimensions of a lasting peace. One lesson of World War I was that the nations that had started the war should be defeated and disarmed completely. Ambiguity had permitted German leaders to tell their people that they had not really been defeated in World War I but had been betrayed by their leadership, and that victory in a second round of fighting was plausible. A second lesson was that it was necessary to avoid the political and economic conditions that had led to totalitarianism in Germany, which in turn entailed a need for self-determination among the peoples of defeated nations, an imperative to prevent future economic depressions, and a need for some form of international governance. "American failure to join the League of Nations," Gaddis wrote, "had also contributed to the collapse of international order; therefore a third prerequisite for peace would be membership in a new collective security organization."[25] To achieve these goals, it would be necessary to maintain decent relations among the United States, the United Kingdom, and the Soviet Union once World War II was over.

According to Gaddis, "Roosevelt and his advisers clearly realized that their vision of the future would not materialize unless the members of the Grand Alliance, united now only by their common enemies, built relationships that could survive victory."[26] One might argue that the bomb poisoned the possibilities for enduring friendly relations, but one might equally argue that President Harry Truman used the bomb because he and his advisors had concluded that such prospects had already vanished.[27]

Ghosts of Depression Past and Future

Walter LaFeber defined the Cold War period as 1945–1996, but, like Gaddis, he looked back from 1945 to find its origins. The World War II alliance of the United States and the Soviet Union was a "shotgun marriage" preceded by a long history of conflict and animosity, much of it centered on trade. Late in the nineteenth century, LaFeber noted, Russia and the United States had "confronted each other on the plains of North China and Manchuria."[28] As the American economy expanded dramatically, Americans looked to Asia "as the great potential market for their magnificently productive farms and factories." Russians, however, after "annexing land in Asia," "tried to control it tightly by closing markets to foreign business people with whom they could not compete."[29] This control of competition, along with a distaste for Czarist repression sustained by horror stories carried to the United States by immigrants, had produced deep animosity in the United States toward Russia well before the 1917 revolution or the 1924 rise to power of Joseph Stalin. Russians, for their part, were not pleased by President Woodrow Wilson's refusal to open diplomatic relations after World War I, by his sending US troops in an attempt to overthrow Lenin, or by the creation in 1919 of the buffer states of Poland, Romania, Czechoslovakia, and Yugoslavia.[30]

The belief that the prosperity of the United States required an "open door" to Asia was reinforced by the Great Depression. As World War II came to a close, political leaders were mindful that the global economy had been had been pulled out of depression at least as much by the war as by the New Deal, and the urgency of international trade as a means to avoid a slide back into depression weighed heavily on allied minds. LaFeber saw trade as the central point of contention between the United States and the Soviet Union as World War II drew close. The US and its European allies, haunted by what LaFeber called "the Ghosts of Depression Past and Depression Future," were determined to keep global markets open. Secretary of State Dean Acheson put it this way: "We cannot expect domestic prosperity under our system without a constantly expanding trade with other nations."[31] Western leaders feared that without open markets the West would not only slide back into depression but would also slide into totalitarianism. Vice President Henry Wallace put it this way: "In the event of long continued unemployment, the only question will be as to whether the Prussian or Marxian doctrine will take us over first."

The idea that capitalism needed to expand indefinitely in search of markets was a central belief of Trotskyites and a major reason for the Soviet fear of "capitalist encirclement." Moreover, the specter of capitalist expansion ran headlong into the Soviet desideratum of a buffer zone of friendly socialist states in Eastern Europe. Russia had a long history of invasion by unfriendly neighbors, and had suffered devastating losses in World War II: more than 20 million fatalities, and more than 25 million left homeless.[32] Ideological clashes with the West aside, it was no surprise that the Russians were deeply concerned to end the war with safe and secure borders. LaFeber thus agrees with Gaddis that conflict was inevitable: "Roosevelt faced a choice: he could either fight for an open postwar world (at least to the Russian border) or agree with his ally's demands in Eastern Europe."[33] If he chose the first, Russian-American relations would collapse; if he chose the second, the world economy might collapse.

Roosevelt died before he had to make that choice, but his successor, Truman, decisively chose the first option. The atomic bomb figured heavily in his calculations, as Truman believed it had strengthened his hand and might enable him to wrest concessions from the Soviets. We know now that he misjudged. Stalin wasn't impressed by Truman's suggestion at Potsdam that the United States was in the possession of a uniquely destructive weapon; thanks to spying, he already knew it.[34] Short of actually using the bomb against the Soviets, it wasn't clear what advantage the bomb gave the United States, and Truman and his advisers "never figured out how to use the bomb to obtain concessions they wanted from the Soviets."[35] Meanwhile, the Soviets accelerated their own work on nuclear weapons. Whatever effect the bomb did or didn't have on the conclusion of World War II, its use clearly marked the beginning of what would become a long and costly arms race. It also dramatically altered relations between science and the modern nation-state—between nation and knowledge—in both the United States and the Soviet Union. And it changed what it meant to be a scientist in the new, security-driven nation-state.[36]

Nation and Knowledge

As the Cold War deepened, science and scientists were enlisted to support it in a variety of ways.[37] It is well documented that the Cold War provided the justification for massive increases in the US government's support, through existing and newly created federal agencies and through direct grants to researchers at colleges and universities across the country, for both basic and applied scientific research—some of it to be done at the newly established national laboratories.[38]

National security provided the justification for this huge increase in federal support for scientific research. The Office of Naval Research, created in 1946 from diverse wartime programs, explicitly authorized the Navy to plan, foster, and encourage "scientific research in recognition of its paramount importance [in] the preservation of national security." When the National Science Foundation was created four years later, in part on the ONR model of funding investigator-initiated projects, it was charged with fostering science to "to advance the national health, prosperity and welfare," but also " to secure the national defense." Science was also funded by new federal agencies, including the Atomic Energy Commission, the Advanced Research Projects Agency, and the expanded National Institutes of Health, whose extended research mandate included radiation sickness and nuclear medicine.[39]

All this makes it abundantly clear, that, whatever the role of science in the Cold War, the Cold War drove substantial changes in the scale and funding structure of American science. Yet, if the issue of the effect of science on the Cold War has long been argued, the reverse question (How did the Cold War affect science?) did not receive sustained academic scrutiny until relatively recently.

The Cold War's Effect on Science: Moving Past Miasma

Historians have long been interested in how cultural and political context affects the growth, development, and content of science. However, much of our work has suffered from what one might call a "miasma problem": it is easy to describe the culture surrounding a given science, much harder to demonstrate its causal effects. (No doubt there *were* miasmas in the nineteenth century, but that didn't prove that they caused the diseases that occurred in their midst.) One reason for this is the innate complexity of human experience: we rightly shy away from simplistic determinative accounts of complex historical developments. We recognize that the course of human events is long and winding. At the same time, we emphasize the necessity of placing the production of scientific knowledge in its full social, cultural, political, and even economic context, tending to be critical—sometimes harshly so—of histories that fail to do so. Our sociological colleagues go further, insisting that scientific knowledge and society are co-produced.[40] But if context is important, and certainly if knowledge is co-produced, then it behooves us not to simply use context as a kind of "background"—like the lakes and trees in Renaissance portraits—or even as a frame that highlights only some aspects of our picture, but rather to attempt to explain the *particular* ways it was important in any given situation. Put another way, what is the point of placing knowledge into its full historical context if that context doesn't help to explain how and why particular lines of inquiry were pursued, and other lines of inquiry were abandoned or left unpursued?[41] Our quarry is the development of scientific (and technical) knowledge, but is it even *possible* to demonstrate that a particular cultural setting played a determinative role in the content of knowledge produced in that setting?

At least one historian has tried. In 1971, Paul Forman put forward the controversial suggestion that the development of quantum mechanics—and specifically the assertion of acausality in quantum mechanics—was a direct result of the devastating defeat of Germany in World War I. In anger, frustration, and confusion about the inexplicable outcome, German intellectuals turned against determinism, rationality, and causality. Scientists, as intellectuals, weren't immune from this reaction, Forman argued, and they too began to doubt conventional rationality and to consider others forms of explanation. The Forman thesis—as it came to be known—was that this state of affairs led scientists "ardently to hope for, actively search for, and willingly embrace an acausal quantum mechanics."[42] Although there were, to be sure, peculiar quantum physical phenomena that required explanation, and which *could* be explained acausally (he wasn't suggesting that quantum mechanics was untrue), Forman proposed that scientists began to seek out accounts that were compatible with their cultural milieu, and they found it in acausality:

In the years after the end of the first world war, but before the development of an acausal quantum mechanics, under the influence of "currents of thought," large numbers of German physicists, for reasons only incidentally related to developments in their own discipline, distanced themselves from, or explicitly repudiated, causality in physics.[43]

As cultural conditions made the world seem increasingly inexplicable, physical phenomena were increasing viewed as inexplicable too. And as conventional notions of cause and effect lost their persuasive power in politics, they also began to lose their persuasive power in other domains—even in a domain that a previous generation of historians might have thought was immune to such considerations (and that many if not most scientists still think is impervious.) Previously, an explanation in physics was, by *definition*, causal; now, at least in the domain of quantum mechanics, it was acausal.

It was a striking reversal, and not all scientists found it congenial (Albert Einstein didn't), but Forman argues that many scientists embraced quantum mechanics not only happily but with a sense of *relief*. Speaking of the new theories in quantum mechanics, the mathematician Hermann Weyl, one of the founders of gauge theory and one of the first to apply group theory to quantum mechanics, wrote of the freedom to be found in quantum mechanics:

[T]he rigid pressure of natural causality relaxes, and there remains, without prejudice to the validity of the natural laws, room for autonomous decisions, causally absolutely independent of one another. ... The "decisions" are what is *actually real* in the world.[44]

Acausality was a startling break from historic tradition in physics, whose purpose, some might argue, was to give causal accounts of natural phenomena. Forman's argument that such a striking change—such an *abandonment* of historic goals and aspiration—requires explanation is not in that sense particularly radical: historians of science routinely accept that changes in intellectual commitments require accounts. What was radical at the time was that Forman found that account not in the improved appraisal of the phenomena of nature but in the cultural adaptation "of knowledge to the intellectual environment."[45]

The Forman thesis, which seemed to suggest that German scientists had capitulated to irrationality, offended most physicists and many historians. Yet it was and remains highly influential—a Google search of "Forman thesis" turns up 402,000 hits. It raised the question of why we bother to pay attention to the cultural context of science unless we believe that context has affected the content of science in a significant way.[46] Returning to our particular topic, we might therefore ask: Is it possible to distinguish what happened to science during the Cold War from what happened to it because of the Cold War?

In 1987, Forman took up the challenge again, this time addressing Cold War physics. His quarry was not so much any specific physical theory as the nature and character of physics as a discipline. Forman now suggested that US military funding had dramatically altered the nature of physics, causing its practitioners to shift from seeking a fundamental understanding of the laws of nature toward gadgeteering preoccupied with technical prowess.[47] Forman's starting point was something that scientists themselves had said and many historians had accepted as self-evident: "World War II was in many ways a watershed for American science and scientists. It changed the nature of what it means to do science and radically altered the relationship between science and government ... the military ... and industry."[48]

While scientists and historians accepted that World War II was a watershed, restructuring the relationship between science and government, they largely interpreted that change in quantitative and normative but not epistemic terms. It was obvious that the federal support for science had increased dramatically, and it was generally assumed that this was a good thing. Scientists needed money for science, so most scientists found it hard to see more money as problematic.[49] Historians of science in the 1960s and the 1970s generally admired and approved of science, so they tended to accept that appraisal. Left largely unanswered—indeed, largely unasked—was the question of how government patronage affected the content of scientific research and the character of the knowledge produced. For although it was widely supposed that the federal government was increasing its support for scientific research because of its value for national security, it was frequently (and paradoxically) asserted that federal support allowed scientists to pursue whatever their curiosity dictated.[50]

This paradox was scarcely noticed, much less examined. But if the federal government supported science *because of* its value for national security, wouldn't it stand to reason that it would privilege particular sciences (physics, electronics, computer science) that were obviously pertinent. Wouldn't it tend to neglect less pertinent sciences (ichthyology, botany)? And wouldn't it make sense that within individual sciences, such as physics, government patrons would tend to want to focus financial, logistical, and moral support into lines of inquiry deemed likely to produce valuable results? ("If oratorios could kill," the biochemist Erwin Chargaff quipped in 1978, "the Pentagon would long ago have supported musical research."[51]) Indeed, wouldn't it be a dereliction of duty if these agencies supported science without regard to national needs and priorities?[52]

Forman cited statistics on physics research in the United States at the height of the Cold War. From the end of World War II through the late 1950s, about 95–98 percent of the federal support for physics research came from either the Department of Defense or the Atomic Energy Commission. "The only significant support for academic physical research in the US were the Department of Defense and an Atomic Energy Commission whose mission was de facto predominantly military,"[53] Forman asserted. The growth of the National Science Foundation in the late 1950s and the 1960s changed the situation only modestly: the component of research support from the DOD and AEC dropped to around 90 percent.[54] "Thus, in the fifteen years following the war,"

Forman concluded, "the central fact of scientific life in physics was unprecedented growth based upon military funding."[55]

"What direction of the advance of science, and thus what kinds of science, result from military sponsorship?," Forman asked.[56] If he who pays the piper doesn't call the tune, then what is he paying for? If it was "a bit too crass" to assume the golden rule (that those with the gold rule), it was equally implausible that this huge transformation in the quantity and source of support for physics didn't alter the nature and the character of the physics done.

What kind of science *did* result? For Forman, the short answer was solid-state physics and quantum electronics, which expanded even more rapidly and more extensively than other areas of physics. Forman also suggested several mechanisms by which work in solid-state physics and quantum electronics was fostered. Most obviously, program managers in the Office of Naval Research, the Air Force Office of Scientific Research, and other agencies made choices about what projects would be funded and what projects would not. Less obviously, they encouraged and stimulated scientists to consider working in areas of military interest, in part through site visits to colleges and universities, in part by organizing workshops and conferences on particular themes, and in part through ongoing informal discussions. Scientists supported by those agencies were bound to consider what kinds of work and results would be likely to get continued support. "Whatever such program officers did beyond providing funds," Forman wrote, "must be reckoned as direction of research. The funding levels of their programs and the contentment of their table of researchers depended upon reconciliation of the interests of their military and their scientific constituencies, a reconciliation effected chiefly by envisaging and promoting military applications in and through basic scientific research. For the researcher himself, 'the mere need to defend what he is doing to a particular sponsor may be the factor which will trigger an important application.'"[57]

A scientist who valued the funding that he or she (although during the Cold War mostly he) was receiving would be sensitive to nuances of interest and applicability. Beyond the defensive motivation of accountability to patrons and the desire to be invited back to the table, there was also the positive motivation of the gratification that comes with knowing that one's work is valued and perhaps put to use.[58] Paul Edwards has used the term "mutual orientation" to describe the interactions and feedbacks by which scientists and military patrons found common ground. Describing Jay Forrester's work in Project Whirlwind, Edwards concludes that "the source of funding, the political climate, and their personal experiences oriented Forrester's group toward military applications, while the group's research eventually oriented the military toward new concepts of command and control."[59] Part of the job of the agencies was to stimulate scientists to work in areas of basic research that might prove useful to the military, if not immediately then perhaps in the long run; part of the work of scientific researchers was to find ways to connect their abilities to the needs and interests of their patrons. Harvey Brooks referred to this as "imaginative stimulation"; I have called it a "context of motivation."[60]

If work proved irrelevant to an agency's mission, program officers had the option of cutting it off, but the available evidence suggests that they seldom felt a need to do so. Some would take this as proof that the scientists *were* free to do what they wanted, but a more plausible explanation is that intelligent scientists would have been unlikely to propose lines of inquiry that were doomed to be rejected, and there were enough interactions between scientists and funders that any idea that didn't resonate would be unlikely to be developed sufficiently to reach the stage of overt rejection.[61]

For Forman, the net result was a science that was "effectively rotated ... towards techniques and applications." The construction of masers and atomic clocks and the improvement of microwave technologies and electronics constituted advances, to be sure, but in tools more than in conceptual understanding—one might even say in technology rather than in science, although Forman himself resists that characterization. The physics of the Cold War was an "instrumentalist physics of virtuoso manipulations and *tours de force* ... just such a physics as the military funding agencies would have wished."[62] This, then, is why Forman concluded that physicists "had lost control of their discipline." It was because the physics that physicists ended up with—the physics that they now found themselves doing—was focused *in areas that had not previously been viewed as priorities by physicists, but were priorities for their military patrons*.[63] Something had changed the priorities of physics and physicists, and that something, Forman argued, was the Cold War.

Again Forman's views proved controversial; the historian Dan Kevles, in particular, contested his claims.[64] Kevles agreed with Forman that physics had proved decisive in World War II; that American technological superiority in that war had been achieved primarily by civilian scientists working under the auspices of the federal government through the Office of Scientific Research and Development; that after the war there was broad agreement among leading scientists, politicians, and military officers that it would be important to maintain and foster the scientific-military alliance that had proved so valuable to the Allied victory; and that all this provided justification for a massive expansion of American physics and increase in federal financial support for physical science research. Above all, he agreed that there had been a "transformation of the relationship between science, especially civilian science, and the American state after World War II."[65]

What was at issue was the character of that transformation. Kevles strongly contested the suggestion that the federal government hadn't supported basic research. Indeed, he took it as a lesson learned during World War II that abstruse knowledge in pure science could prove important in unexpected ways, and that this provided a substantial part of the federal government's motivation to sustain basic science in the years to come. Kevles also took it as accepted by the historical actors that science and technological development were not either/or propositions, and that agencies and military patrons understood that advances in technology required advances in the underlying science that supported them. "Postwar national security required energetic federal programs of both pure and defense-related research," Kevles argued, suggesting that both were energetically supported. Finally, it was clear, although Kevles made this point only in passing, that the demand for large numbers of trained scientists and engineers was a major driving force for support of universities, where basic science continued to flourish.[66] "The government sponsored major programs of research in practical areas such as nuclear weapons and impractical ones such as high energy physics," and the net result was "a vital and balanced scientific enterprise."[67]

Although Kevles found large areas of agreement with Forman, he contested the claim that physicists had lost control of their intellectual agenda, had been "seduced" by the largesse of federal funding, or had fallen prey to the "self-delusion that they were engaged in basic research of intrinsic interest while in reality they were merely doing the military's bidding." The United States had always been a "practically oriented culture," Kevles noted, in which "the technological sciences had always tended to command more attention than the pure sciences," so it was hardly surprising that this remained the case as physics expanded under governmental largesse.[68] (Kevles didn't say, but it logically followed from his argument, that if physics changed between 1930 and the 1960s, one reason may have been that it changed from being dominantly a European activity to dominantly a North American one.)

Kevles emphasized that scientists served on many leading advisory boards and committees, including the crucial Science Advisory Committee created by President Truman and greatly strengthened by President Eisenhower. Some of them, by virtue of their positions on these boards and committees, were close to power and involved in decisions about the future and the direction of American science. Most of them were committed to the alliance of science and technology with the mission of national security and proactively advanced that agenda; they weren't pawns of the admirals and generals. The same was true of many rank-and-file scientists:

[F]or many of those physicists, national security was not a mere distraction. It was the life blood of their profession. ... One is hard pressed to imagine the great accelerator laboratories in the United States having come to exist and to flourish in the absence of the deep concern for national security that came to pervade the United States after World War II. Also, many physicists found abundant opportunities to do interesting physics by involving themselves in militarily-supported research of technological pertinence.[69]

In any case, Kevles concluded, it is counterfactual to argue on the basis of what scientists might have done in a different world. There is no essential definition of what constitutes physics. "Physics is what physicists do—or have done," Kevles concluded, not illogically but perhaps tautologically.

With the benefit of distance, it seems clear that Forman and Kevles *agreed* that the orientation of American physics during the Cold War became aligned with the national-security agenda. The agreed that during the Cold War knowledge was linked to the geopolitical ambitions of the American nation-state to an extent and a degree that it had not been before. Kevles allowed that physics was "restructured, its efforts diversified into intellectually promising areas made hot by the needs of national security."[70] Forman held that there was a "radical change in attitude toward science, toward national security, and toward the relationship between them on the part of both the military and the civilian leadership of the United States."[71] These claims seem entirely compatible.

Where Forman and Kevles disagreed was in the normative domain: They diverged on whether physicists were responsible for that alignment or victims of it, whether that alignment was a good thing, and whether scientists' self-image and self-appraisal was realistic or wishful. Kevles affirmatively characterized the re-organization of physics during the Cold War as a diversification that produced a "vital and balanced" scientific enterprise, which scientists themselves were largely responsible for directing. Forman concluded, less happily, that the ship was bigger, but narrower and tilted, and physicists were no longer steering it. Kevles saw the integration of physics into a national-security system as providing expanded opportunities for physicists to do physics; Forman agreed that physics was integrated, but saw that integration as a constriction and adjustment that altered the meaning of the word 'physics' in an unfortunate way. And perhaps the point on which they disagreed most strongly was scientists' self-perception of epistemic autonomy. Kevles believed that scientists were able to use their positions close to the center of executive power both to influence defense policy and "to represent the interests of the civilian-defense-science-enterprise." Forman believed that the "civilian-defense-science" was *precisely* what they represented, even while insisting, falsely, that they were representing unbounded "science."

Today few historians would consider the notion of unbounded science to be very useful; science, most of us would argue, is bounded, supported, sustained, and constrained by all the same social forces that bound, support, sustain, and constrain other human activities. Yet such a broad generalization only takes us so far, because we want to know how changes in human society change the activity we call science and the knowledge and insights that activity yields. We are also interested in how changes in scientific concepts and understandings change society. Whether or not scientific and technical knowledge was substantially responsible—through the agency of nuclear weaponry—for starting the Cold War, there is little doubt that science and technology enabled the arms race that became and sustained its center. Conversely, there is little doubt that science as we know it today was created in the Cold War, and that the Cold War expansion of science and technology continues to ramify through contemporary life.

Notes

1. P. M. S. Blackett, *Military and Political Consequences of Atomic Energy* (Whittlesey House, 1949), 120; Blackett, *Fear, War and the Bomb: Military and Political Consequences of Atomic Energy* (Whittlesey House), 139. This quotation has been repeated many times—see, e.g., Mary Jo Nye, *Blackett: Physics, War and Politics in the Twentieth Century* (Harvard University Press, 2004), 89.

2. For a recent discussion of the role of the atomic bomb in the early Cold War, see Michael Gordin, *Red Cloud at Dawn: Truman, Stalin, and the End of the Atomic Monopoly* (Farrar, Straus and Giroux, 2009).

3. Martin J. Sherwin, *A World Destroyed: Hiroshima and Its Legacies* (Stanford University Press, 2003); Richard Rhodes, *The Making of the Atomic Bomb* (Simon & Schuster, 1986); Rhodes, *Dark Sun: The Making of the Hydrogen Bomb* (Simon & Schuster, 1995); Gar Alperovitz, *The Decision to Use the Atomic Bomb* (Vintage Books, 1996); Gordin, *Red Cloud at Dawn*.

4. George Orwell, "You and the Atomic Bomb," *Tribune*, October 19, 1945.

5. See Rhodes, *Making of the Atomic Bomb*, esp. 592–593 and 599–603.

6. Blackett, *Military and Political Consequences*, 3.

7. See Blackett, *Military and Political Consequences*, 7–8: "The theme that atomic bombs are dangerous that humanity should be prepared to forego the advantages of atomic power in order to save itself from destruction by atomic bombs is being energetically propagated today in America." On the other hand, we could also argue that the same hysteria provided the justification for the massive increase of US federal government support for science, viz., "the rapid expansion of scientific knowledge ... may reasonably be said to be a major factor in national survival." Secretary of War Robert P. Patterson, quoted in Paul Forman, "Behind Quantum Electronics: National Security as Basis for Physical Research in the United States, 1940–1960," *Historical Studies in the Physical and Biological Sciences* 18 (1987): 149–229 (156, note 10).

8. Blackett, *Military and Political Consequences*, 141.

9. Ibid.

10. Famously by Herman Kahn in *Thinking about the Unthinkable* (Horizon, 1962) and later by various neo-conservatives such as Richard Perle (Herb York, personal communication.) Recently the argument was revived by several leading NATO generals—see Ian Traynor, "Pre-emptive Nuclear Strike a Key Option, NATO Told," *Guardian*, January 21, 2008. With breathtaking illogic, the generals asserted that "the first use of nuclear weapons must remain in the quiver of escalation as the ultimate instrument to prevent the use of weapons of mass destruction." See also Naomi Oreskes and Erik M. Conway, *Merchants of Doubt: How a Handful of Scientists Obscured the Truth on Issues from Tobacco Smoke to Global Warming* (Bloomsbury, 2010), chapter 2. The nuclear-

winter debate of the 1980s was fought against a backdrop of discussions of a "winnable" third world war.

11. On the Cold War beyond the US–USSR axis, see Odd Arne Westad, *The Global Cold War: Third World Interventions and the Making of Our Times* (Cambridge University Press, 2005).

12. Walter LaFeber, *America, Russia, and the Cold War, 1945–1996* (McGraw-Hill, 1997). It is difficult to say how much of the darkness of the Soviet period was a result of the Cold War and how much an extension of the long history of repressive government in Russia. Historians have noted the discrepancy between the image of the Soviet Union promulgated by Orwell and other Western Cold Warriors and the realities, which, while troublesome, were far from monolithically dark. Certainly, in the realm of science, a great deal of interesting and important work was supported. Besides the essays by Schmid and Siddiqi in this volume, see Alexei Kojevnikov, "Rituals of Stalinist Culture at Work: Science in Intraparty Democracy circa 1948," *Russian Review* 57, no. 1 (1998): 25–52; Alexej Yurchak, *Everything Was Forever Until It Was No More: The Last Soviet Generation* (Princeton University Press, 2006). Similarly, American McCarthyism and other Cold War anxieties had clear precedents in the nativist and anti-communist attitudes of earlier red scares, the Palmer Raids, the Smith Act, etc. American anti-communism didn't begin during the Cold War, although the Cold War certainly exacerbated it. See Paul Boyer, *By the Bomb's Early Light: American Thought and Culture at the Dawn of the Atomic Age* (Pantheon, 1985).

13. Of course, the Cold War altered many aspects of life, not only in the US and the USSR, but also in the European countries that offered the terrain where a third world war might be fought, in the Asian countries in which proxy wars were actually fought, and in developing nations viewed by the superpowers as prizes to be won in a global competition. But science played a role in the Cold War that music and dance did not—although scholars have shown how even music and dance weren't immune from Cold War considerations. See, e.g., Frances Saunders, *The Cultural Cold War: The CIA and the World of Arts and Letters* (New Press, 1999. Recently, scholars have begun to pay attention to sports as a playing field (pun intended) of Cold War competition. Stephen Wagg, David Andrews, and Robert Edelman, eds., *East Plays West: Sport and the Cold War* (Routledge, 2007).

14. The exact number of ballistic missiles depends on how you count (there are, for example, ICBMs, SLBMs, and intermediate-range missiles, and there is the problem of MIRVs), but SALT II limited strategic forces to 2,250 of all categories of delivery vehicles on both sides. See the introduction to Stephen I Schwarz, *Atomic Audit: The Costs and Consequences of U.S Nuclear Weapons since 1940* (Brookings Institution Press, 1998). On the history of attempts to control the arms race, see Committee on International Security and Arms Control, US National Academy of Sciences, *Nuclear Arms Control: Background and Issues* (National Academy Press, 1985).

15. President Dwight D. Eisenhower, farewell address, January 17, 1961 (http://www.ourdocuments.gov/doc.php?flash=true&doc=90).

16. "Dwight D. Eisenhower, From a speech before the American Society of Newspaper Editors, April 16, 1953" (http://www.quotationspage.com/quotes/Dwight_D._Eisenhower/).

17. LaFeber, *America, Russia*, 1.

18. Michael Gordin has recently emphasized that many "national" nuclear weapons programs were actually cooperative endeavors; for example, the Manhattan Project involved British and Canadian cooperation and many European émigré scientists. (See Gordin, *Red Cloud at Dawn*, esp. introduction and p. 121.) On the compunction felt by other nations to "be nuclear," see Gabrielle Hecht, *The Radiance of France: Nuclear Power and National Identity after World War II* (MIT Press, 1998); Hecht, *Being Nuclear: Africans and the Global Uranium Trade* (MIT Press, 2012).

19. Rhodes, *Making of the Atomic Bomb*; Sherwin, *A World Destroyed*; Gordin, *Red Cloud at Dawn*; Silvan S. Schweber, *In the Shadow of the Bomb: Bethe, Oppenheimer, and the Moral Responsibility of the Scientist* (Princeton University Press, 2000); Thomas P. Hughes, *American Genesis: A Century of Invention and Technological Enthusiasm, 1870–1970* (Viking, 1989).

20. Sherwin, *A World Destroyed*.

21. Boyer, *By the Bomb's Early Light*.

22. Gordin, *Red Cloud at Dawn*.

23. Westad, *The Global Cold War*.

24. John L. Gaddis, *The United States and the Origins of the Cold War, 1941–1947* (Columbia University Press, 1972). See also Gaddis, *The Cold War A New History*, new edition (Penguin, 2005).

25. Gaddis, *The United States and the Origins of the Cold War*, 2.

26. Ibid.

27. On the former, see Sherwin, *A World Destroyed*. On the latter, see Alperovitz, *Decision to Use the Atomic Bomb*.

28. LaFeber, *America, Russia*, 1.

29. Ibid., 1–2.

30. Here I follow LaFeber's *America, Russia*, one of the classic American sources on this issue. More recently, David Engerman has emphasized a contrasting point: that many American intellectuals romanticized the Russian modernizing model and closed their eyes to the (emerging) human costs. See David C. Engerman, *Modernization from the Other Shore: American Intellectuals and the Romance of Russian Development* (Harvard University Press, 2004). My own view is that these two perspectives are complementary rather than conflicting.

31. Quotations from LaFeber, *America, Russia*, 8.

32. LaFeber, *America, Russia*, 17. See also William I. Hitchcock, *The Bitter Road to Freedom: A New History of the Liberation of Europe* (Free Press, 2008).

33. LaFeber, *America, Russia*, 11.

34. Ibid., 23–25, Sherwin, *A World Destroyed*; Rhodes, *The Making of the Hydrogen Bomb*.

35. LaFeber, *America, Russia*, 24.

36. Michael Dennis, A Change of State: The Political Cultures of Technical Practice at the MIT Instrumentation Laboratory and the John Hopkins University Applied Physics Laboratory, 1930–1945, PhD dissertation, Johns Hopkins University, 1991.

37. Although science was enlisted in the Cold War, so were industry, labor, economics, Hollywood, and just about everything else. In that sense, the Cold War truly was a total war.

38. See Daniel J. Kevles, "Cold War and Hot Physics: Science, Security, and the American State, 1945–1956." *Historical Studies in the Physical and Biological Sciences* 20 (1990): 239–264; Paul Forman, "Behind Quantum Electronics: National Security as Basis for Physical Research in the United States, 1940–1960," *Historical Studies in the Physical and Biological Sciences* 18 (1987): 149–229; Hugh Gusterson, *Nuclear Rites: A Weapons Laboratory at the End of the Cold War* (University of California Press, 1996); Peter Galison, *Image and Logic: A Material Culture of Microphysics* (University of Chicago Press, 1997); Peter J. Westwick, *The National Labs: Science in an American System, 1947–1974* (Harvard University Press, 2003); Laurie Brown and Lillian Hoddeson, eds., *The Birth of Particle Physics* (Cambridge University Press, 1983); Robert Seidel, "A Home for Big Science: The Atomic Energy Commission's Laboratory System," *Historical Studies in Physical and Biological Sciences* 16 (1986): 135–175; Robert Seidel, "The Postwar Political Economy of High Energy Physics," in *Pions to Quarks: Particle Physics in the 1950s*, ed. Laurie Brown, Max Dresden, and Lillian Hoddeson (Cambridge University Press, 1989); Robert Seidel, "From Glow to Flow: A History of Military Laser Research and Development," *Historical Studies in the Physical and Biological Sciences* 18 (1987): 111–147.

39. No comprehensive history of the National Institutes of Health has been published, but some relevant material can be found in the following works: Buhm Soon Park, "The Development of the Intramural Research Program at the National Institutes of Health after World War II," *Perspectives in Biology and Medicine* 46, no. 3 (2003): 383–402; Park, "Disease Categories and Scientific Disciplines: Reorganizing the NIH Intramural Program, 1945–1960," in *Biomedicine in the Twentieth Century: Practices, Policies, and Politics*, ed. Caroline Hannaway (IOS Press, 2008). See also Advisory Committee on Human Radiation Experiments, *The Human Radiation Experiments: Final Report of the President's Advisory Committee* (Oxford University Press, 1996).

40. Sheila Jasanoff, ed., *States of Knowledge: The Co-Production of Science and Social Order* (Routledge, 2004). See also Steven Shapin and Simon Schaffer, *Leviathan and the Air-Pump: Hobbs, Boyle, and the Experimental Life* (Princeton University Press, 2011).

41. And, conversely, how and why that context didn't produce other knowledge, either deliberately or inadvertently. See Robert Proctor and Londa Schiebinger, eds., *Agnotology: The Making and Unmaking of Ignorance* (Stanford University Press, 2008).

42. Paul Forman, "Weimar Culture, Causality, and Quantum Theory: Adaptation by German Physicists and Mathematicians to a Hostile Environment," *Historical Studies in the Physical Sciences* 3 (1971): 1–115. Forman expanded on his original argument in "*Kausalität, Anschaulichkeit,* and *Individualität,* or How Cultural Values Prescribed the Character and Lessons Ascribed to Quantum Mechanics," in *Society and Knowledge*, ed. Nico Stehr and Volker Meja (Transaction Books, 1984).

43. Forman, "Weimar Culture," 268.

44. Cathryn Carson, Alexi Kojevnikov, and Helmuth Trischler, eds., *Weimar Culture and Quantum Mechanics: Selected Papers by Paul Forman and Contemporary Perspectives on the Forman Thesis* (Imperial College Press, 2011), 225.

45. Forman, "Weimar Culture," 63.

46. "Paul Forman," http://americanhistory.si.edu/profile/399. See also Carson, Kojevnikov, and Helmuth, *Quantum Mechanics and Weimar Culture*.

47. Forman, "Behind Quantum Electronics"; Paul Forman and José M. Sánchez-Ron, eds., *National Military Establishments and the Advancement of Science and Technology: Studies in 20th Century History* (Kluwer, 1996), 272–275 and note on 316. Forman didn't invent the word 'gadgeteering', but he highlights it as an actor's category and not necessarily an approbative one.

48. Forman, "Behind Quantum Electronics," 152, quoting the physicist Jerrold Zacharias.

49. There were notable exceptions. For an interesting negative appraisal of the role of military funding from within the physics community, see Vera Kistiakowsky, "Military Funding of University Research," *Annals of the American Academy of Political and Social Science* 502 (March 1989): 141–154. Biologists, too, expressed concerns about whether the model provided by physics was appropriate for their work; see Elena Aronova, Karen Baker, and Naomi Oreskes, "From the International Geophysical Year through the International Biological Program to LTER: Big Science and Big Data in Biology, 1957–present," *Historical Studies in the Natural Sciences* 40, no. 2 (2010): 183–224. Scientific ambivalence toward Big Science is taken up by Steven Shapin in *The Scientific Life: A Moral History of a Late Modern Vocation* (University of Chicago Press, 2010).

50. For a discussion of some of the contradictions inherent in this framework, see Naomi Oreskes, "Science, Technology, and Free Enterprise," *Centaurus* 52 (2010): 297–310, an essay inspired by John Krige's book *American Hegemony and the Postwar Reconstruction of Science in Europe* (MIT Press, 2008). See also John Krige, "NATO and the Strengthening of Western Science in the Post-Sputnik Era," *Minerva* 38, no. 1 (2000): 81–108.

51. Erwin Chargaff, *Heraclitean Fire: Sketches of a Life before Nature* (Rockefeller University Press, 1978).

52. The allegation that this was in fact happening would later become the justification for the Mansfield Amendment to limit military funding to research that directly supported military priorities.

53. Forman, "Behind Quantum Electronics," 194. On education, see John Rudolph, *Scientists in the Classroom: The Cold War Reconstruction of American Science Education* (Palgrave, 2002).

54. Forman, "Behind Quantum Electronics," 193–194.

55. Ibid., 198. On the growth of scientific manpower, see David Kaiser, "Cold War Requisitions, Scientific Manpower, and the Production of American physicists after World War II." *Historical Studies in the Physical Sciences* 33 (fall 2002): 131–159.

56. Forman, "Behind Quantum Electronics," 200.

57. Ibid., 209. The internal quotation is from remarks by Harvey Brooks on the occasion of the twentieth anniversary of the Office of Naval Research.

58. I am not suggesting this is unique to the realm of military patronage. Consider, for example, Robert Kohler's discussion of Rockefeller Foundation support for science in *Partners in Science: Foundations and Natural Scientists, 1900–1945* (University of Chicago Press, 1991). See also Kai-Henrik Barth, "The Politics of Seismology: Nuclear Testing, Arms Control, and the Transformation of a Discipline," *Social Studies of Science* 33, no. 5 (2003): 743–781.

59. Paul N. Edwards, *The Closed World: Computers and the Politics of Discourse in Cold War America* (MIT Press, 1996), chapter 3.

60. Forman, "Behind Quantum Electronics," 209–210, quoting from Harvey Brooks on twentieth anniversary of ONR; Naomi Oreskes, "A Context of Motivation: US Navy Oceanographic Research and the Discovery of Sea-Floor Hydrothermal Vents," *Social Studies of Science* 33, no. 5 (2003): 697–742.

61. An interesting example of this occurred in the 1960s when the Canadian marine biologist Frederick Aldrich proposed an investigation of the enigmatic giant squid using the submersible *Alvin*. The giant squid was one of the very subjects that had been used in the 1960s in the proposal to construct *Aluminaut*—an early submersible—in outlining its potential basic science applications, but the project was rejected on the grounds that it didn't fall under the mission profile under which *Alvin* was funded by the ONR. See Naomi Oreskes, *Science on a Mission: American Oceanography from the Cold War to Climate Change* (University of Chicago Press, forthcoming), chapter 6.

62. Forman, "Behind Quantum Electronics," 224.

63. See also Schweber's discussion of this point in *In the Shadow of the Bomb*.

64. Kevles, "Cold War and Hot Physics."

65. Ibid., 240.

66. Part of the problem arises from the very term "basic science." After all, what does it mean to refer to "basic research related to the development of guided missiles," or the "basic microwave research of the [MIT] Radiation Laboratory," when the Rad Lab's primary purpose was to advance the development of radar? See Kevles, "Cold War and Hot Physics," 242–244.

67. Kevles, "Cold War and Hot Physics," 241.

68. Ibid., 241.

69. Ibid., 263.

70. Ibid., 264.

71. Forman, "Behind Quantum Electronics," 152.

2 Atomic Tracings: Radioisotopes in Biology and Medicine

Angela N. H. Creager

There are three aspects to the use of atomic energy. Of these, the military aspect is familiar and has been discussed several times in these pages. ... The use of radioactive isotopes both directly and for research may well be the most important application of atomic energy in the long run. However, this second aspect of atomic energy does not involve large expenditures of funds or great concentration of technological effort. It will be felt in a multiplicity of small activities no one of which is very important in domestic or international politics. Distribution of radioactive isotopes and knowledge about them should and can play an important role in our atoms for peace program but it is only a part and the least controversial part of that program.[1]
Henry DeWolf Smyth, 1956

In June of 1946, the Manhattan Project announced that radioisotopes would soon be available for purchase to qualified civilian institutions, thanks to the government's decision to dedicate a reactor built for the bomb project in Oak Ridge, Tennessee for this purpose: "Production of tracer and therapeutic radioisotopes has been heralded as one of the great peacetime contributions of the uranium chain-reacting pile. This use of the pile will unquestionably be rich in scientific, medical, and technological applications."[2] That August, after President Truman signed the Atomic Energy Act, the Manhattan Engineer District began distributing radioisotopes. During the next ten years, the US Atomic Energy Commission (AEC), which inherited this operation upon its official establishment on January 1, 1947, sent out nearly 64,000 shipments of radioactive materials to more than 2,600 laboratories, clinics, and companies.[3]

In countless press releases and reports during the early postwar years, the AEC presented radioisotopes as civilian dividends of the military development of atomic energy.[4] To borrow a metaphor from Nicolas Rasmussen, this was the silver lining of the mushroom cloud: by supplying radioisotopes, the US government conveyed the message that atoms could be beneficial as well as harmful.[5] Though the quantity of radioactive materials associated with this program was minuscule in comparison with the amount of plutonium being produced for atomic weaponry (not to mention the amount of radioactive waste being generated), radioisotopes represented the AEC's civilian orientation. To put it more bluntly, the radioisotope program helped justify the fact that the production of nuclear weapons had been entrusted to a civilian agency in the first place.[6] In any event, the AEC had little else to hold out as evidence of the atom's peaceful benefits, since hopes for rapid development of a domestic nuclear power industry (with energy "too cheap to meter") soon faded.[7]

The US government's involvement in radioisotope supply intersected with popular hopes that atomic energy, having provided a decisive weapon in World War II, could be directed against cancer.[8] Scholars have already explored this aspect of the AEC's program, particularly its replacement of older radium sources for cancer therapy in hospitals with cobalt bombs, metaphorically echoing the Cold War.[9] However, this chapter will focus on a distinct—and arguably more pervasive—consequence of the AEC's distribution of radioisotopes to researchers in the life sciences and in medicine: the accelerating use of radioisotopes as tracers. These researchers used isotopic variants of common elements to tag compounds so as to follow their chemical transformations through biological processes within a cell, an organism, or an ecosystem. Scientists using tracers often represented changes over time—through biosynthesis or degradation of molecules, or the movement of elements through bodies or landscapes—as changes in space, through cycles or pathways.

Tracers were used in biochemistry before the atomic age. Scientists employed naturally occurring heavy radioelements in the 1920s and both artificial and stable radioisotopes in the 1930s. Yet these efforts remained small in scale before World War II, when the development of nuclear reactors made mass production of radioisotopes feasible. The Cold War context shaped the use of radioisotopes in two decisive ways.

The first has to do with scale. The AEC vastly increased the overall consumption of radioisotopes in the United States and allied nations by subsidizing the costs of production, providing technical training, and encouraging industrial participation. In this respect, new Cold War priorities dramatically reinforced and expanded certain pre-existing trajectories of research. Tracer research remained principally a benchtop activity, even as it was materially dependent on the massive infrastructure that had been developed to make nuclear weapons. Hans-Jörg Rheinberger has aptly described the dissemination of radioisotopes as "big science coming in small pieces."[10]

The second has to do with space. By 1951, Canada and Great Britain were selling radioisotopes in conjunction with their atomic-energy programs, and were exporting radioisotopes with fewer restrictions than the United States,[11] and the Soviet Union was supplying radioisotopes to institutions in satellite states as well as to its own institutions.[12] Consequently, by the 1950s the circulation of radioisotopes had become global, even as the networks of distribution were circumscribed by the geopolitics of the Cold War.

The explosion of biological and medical work with radiotracers after World War II had profound epistemological repercussions, particularly through facilitating a preoccupation with transformations at the molecular level. In biochemistry, the intensified use of isotopic tracers resulted in the mapping of hundreds of metabolic pathways, such as the Calvin-Benson cycle (which exploited newly discovered carbon-14). In both ecology and medicine, the availability of radioactive tracers ushered in new methods that mirrored the biochemical usage of tracers to study metabolism. Radioisotopes (especially phosphorus-32 and iodine-131) were used diagnostically to locate tumors and observe organ function. Ecologists, beginning with G. Evelyn Hutchinson, used radioisotopes such as phosphorus-32 to analyze the flow of materials and energy through ecosystems. Through the US government's program of radioisotope supply, which was embedded in the Cold War politics of atomic energy, radiotracers became a distinctive feature of postwar life science and medicine.

At another level, the AEC's radioisotope program illustrates that not all of the US government's atomic-energy activities were oriented toward military ends, or even (directly) to anti-communism. Viewed alongside the conspicuous growth of big science and the national-security state, the widespread use of government-produced radioisotopes illustrates how the Cold War shaped more quotidian aspects of research and clinical practice. The "civilian" development of atomic energy was just as deeply rooted in postwar politics as the stockpiling of nuclear weaponry. Radioisotopes served as symbols of the "peaceful atom" at the height of the arms race.[13] This is not to suggest that the civilian and military sides of atomic energy were entirely separable: they were two sides of the same coin, each implying the other. In fact, the US government's simultaneous commitment to the testing of nuclear weapons and to so-called peacetime applications of atomic energy spawned common research endeavors, especially as the AEC sought to manage environmental radioactivity and low-level human exposure without questioning their necessity. The AEC's supply of abundant, inexpensive radioisotopes had consequences both profound and trivial, from the selection of research problems and the ethics of human experimentation to routine assaying techniques and the disposal of laboratory waste.

Early Isotopic Tracers

George de Hevesy is credited with the first biological experiment using radioisotopes in 1922 (published in 1923) when he utilized lead-212 to follow the uptake of this element in plant tissues.[14] An array of similar studies followed this example (some conducted by Hevesy but many by other scientists), monitoring the incorporation and movement of heavy radioactive elements—such as bismuth, thorium, and polonium—into animal and plant tissues.[15] Humans were not exempt. In 1924, Herrmann Blumgart and co-workers at Harvard Medical School injected bismuth-214 into a clinical subject's arm and determined how long it took for the radioactivity to reach the other arm, detected using a Wilson cloud chamber.[16] In animals or in patients, researchers could examine where radioelements localized—in which tissues or organs—and measure how rapidly they were excreted.

There were practical incentives for studying the effects of radioactive materials used by industry. Toxicological studies of the distribution of radium in animals and humans dated back to the early twentieth century.[17] As the health hazards of radium became evident in the tragic suffering and deaths of watch dial painters in the 1920s, knowledge about the localization and biological effects of radioactive isotopes took on an urgent medical relevance.[18] Neither radium nor most of the heavy radioactive elements used in these early experiments were generally found in living organisms. Consequently, these studies didn't shed direct light on physiological processes. In order to use radioisotopes to study the dynamics of life, especially metabolism, scientists needed isotopes of lighter elements that were the main constituents of living matter. These first became available with the isolation of stable isotopes.

Harold Urey, a physical chemist, identified deuterium (a heavy isotope of hydrogen) in 1932 and used it to prepare "heavy water" (^2H_2O), prompting a spate of investigations of the effects of heavy water on various biological processes, such as the respiration of fish, the division of eggs, the growth of fungi, and the germination of plant seeds.[19] However, it was Urey's concentration of the naturally occurring isotopes oxygen-18, carbon-13, and nitrogen-15 that ushered in the thoroughgoing use of isotopes as metabolic tracers. By substituting these rare but stable isotopes for oxygen, carbon, or nitrogen atoms in biological molecules, one could track the fate of compounds marked with these heavy isotopic tags, even *in vivo*.[20]

Urey's search for naturally occurring isotopes coincided with the heightened interest in intermediary metabolism among biochemists.[21] Researchers studying metabolism investigated the myriad chemical transformations that occurred in the cell, both synthetic and degradative, opening up the organismal black box of nineteenth-century intake-output physiology.[22] Whereas nineteenth-century chemists had established the identity and the structure of many biological compounds (e.g., sugars, amino acids, fatty acids, dicarboxylic acids, and keto-acids), their successors focused their efforts on the chains of reactions, connecting those compounds *in vivo*, each step apparently controlled by a specific enzyme. Using laborious manometric techniques, Otto Meyerhof and W. Kiessling determined the reaction steps in anaerobic carbohydrate metabolism (glycolysis), soon termed the Embden-Meyerhof pathway. They published in 1936. The following year, Hans Krebs announced the steps of the citric acid cycle (now the eponymous Krebs cycle).[23] These became paradigmatic for biochemists, functioning as "exemplary achievements" guiding the field.[24]

Isotopes were even better suited than chemical micromanometers for detecting metabolites, which were produced in small quantities and quickly transformed. Harold Urey's colleagues at Columbia University were among the first to use stable isotopes to trace biochemical pathways. Rudolph Schoenheimer, David Rittenberg, and Mildred Cohn tagged biological molecules with deuterium, nitrogen-15, and carbon-13 in order to follow chemical transformations of fats and proteins within the cell, and found they were in a state of continuous flux.[25] Schoenheimer contended that vital compounds were being continuously broken down and regenerated from a metabolic pool, part of what he termed the "dynamic state of body constituents."[26] These techniques and findings, which did much to define biochemistry as the study of the molecular dynamics of life, exemplified tracer methodology in the life sciences.[27]

Artificial Radioisotopes as Tracers and for Therapy

Radioactive isotopes competed with stable isotopes as tracers. Soon after Frédéric and Irène Joliot-Curie discovered how to produce artificial radioisotopes, Ernest O. Lawrence and his group at the University of California's Radiation Laboratory (Rad Lab) in Berkeley bombarded table salt with deuterons in a cyclotron, generating sodium-24.[28] Lawrence hired two physicians—his brother John Lawrence and Joseph Hamilton—to explore this material's potential medical uses. Within two months of first producing radiosodium, Lawrence's group had made more than a millicurie, and they improved the efficiency of production further. In 1936 the cyclotron was able to generate 200 millicuries of radiosodium a day from rock salt worth less than a penny.[29] At a meeting of the American Physical Society in December of 1936, Paul Aebersold of the Rad Lab declared that "machines of science produce radiation equal to $5,000,000 worth of radium."[30] This helped justify the building of ever-larger accelerators. A 37-inch cyclotron became operational on August 18, 1937.[31]

The first clinical use of an artificial radioisotope occurred in 1936, when Joseph Hamilton and Robert Stone administered sodium-24 to two leukemia patients.[32] Although the patients didn't improve, neither did they appear to suffer ill effects. On the basis of the apparent safety (or, at least, nontoxicity) of radiosodium in these medical experiments, Hamilton launched a broader investigation of the rate of absorption of sodium in humans, feeding healthy subjects small amounts of sodium-24. His initial publication reported results from eight subjects, two of them women; most of the subjects had received radiosodium by mouth, the doses ranging from 80 to 200 microcuries.[33] A subject would put his or her left hand around a Geiger-Müller counter encased in lead, then use the right hand to drink the radioactive salt solution.

The appearance of radioactivity in the hand, detected by the counter a few minutes after ingestion, was used as an "indicator of absorption."[34] Ernest Lawrence set up live demonstrations of the absorption of sodium-24 in human volunteers in his public lectures, and touted its value of as a potential substitute for costly radium.[35] However, continuing experiments on the localization of radiosodium in healthy subjects

Figure 2.1
Joseph Hamilton (left) conducting one of the first isotope metabolism studies during the 1930s.
Credit: Lawrence Berkeley National Laboratory.

suggested this isotope was better suited for studying the vascular system and investigating the role of ions in water balance than for treating cancer.[36]

Phosphorus-32 also became available in 1936, and Berkeley scientists conducted similar tracer experiments with it. The previous year, Otto Chievitz and Georg von Hevesy had demonstrated that ingested radiophosphorus (obtained from a radon-beryllium source) concentrated in the bones and, to a lesser degree, the muscles of rats.[37] The Rad Lab's cyclotron generated phosphorus-32 of higher specific activity, and Lawrence soon made it available to collaborators in Berkeley and at the University of California's medical school in San Francisco, and to Hevesy and others in Europe.

In San Francisco, K. G. Scott and S. F. Cook fed phosphorus-32 to chicks to see if its selective localization (to bone) might make it useful for treating leukemia, lymphoma, and perhaps other blood-cell diseases.[38] Across the Bay, in Berkeley, John Lawrence studied the uptake of phosphorus-32 in inbred mice. He and Scott found that cancerous mice concentrated more radioactivity in their lymph glands and spleens than did healthy mice after both groups received small "tracer" doses.[39] This finding stoked hopes that radioisotopes would be selectively absorbed and localized in cancer patients and thus could be used to irradiate tumors. In fact, Lawrence was already experimenting with the therapeutic use of phosphorus-32 in patients with leukemia and polycythemia vera.[40]

In tandem with ongoing clinical experiments, phosphorus-32 was being used as a tracer to study metabolism. Israel L. Chaikoff of Berkeley's Department of Physiology collaborated with John Lawrence to study phospholipid turnover in the tumors of Lawrence's cancerous mice. The tumors showed different rates of phospholipid activity, but in each case turnover was at least as high as in normal tissue, and in some cases it was much higher.[41] Chaikoff and his colleagues also used the radiolabel to study phospholipid synthesis, tracking the percentage of tagged phosphorus that was recovered as phospholipid from the gastrointestinal tract, the liver, the kidneys, the brain, and the whole body of fasting rats.[42] The fact that so much radioactivity was absorbed and localized in adult animals indicated that there was rapid turnover of the tissues and molecules in the body. Along similar lines, and during the same time frame, the biochemist David Greenberg and his graduate student Waldo Cohn studied the absorption of phosphorus-32 through the digestive system and its assimilation in the organs of rats at the University of California at San Francisco.[43] Radioactive phosphorus brought into view "a dynamic system involving synthesis, transport, deposition, and breakdown of phospholipids in the tissues involved."[44]

Agricultural scientists were equally eager to use phosphorus-32 in physiological research. Daniel Arnon and his co-workers in Berkeley's Division of Truck Crops added phosphorus-32 (obtained from the Rad Lab in the form of sodium biphosphate) to unlabeled ammonium phosphate in a nutrient solution for tomato plants. The radiophosphorus was rapidly absorbed by seven-foot-tall tomato plants, accumulating most

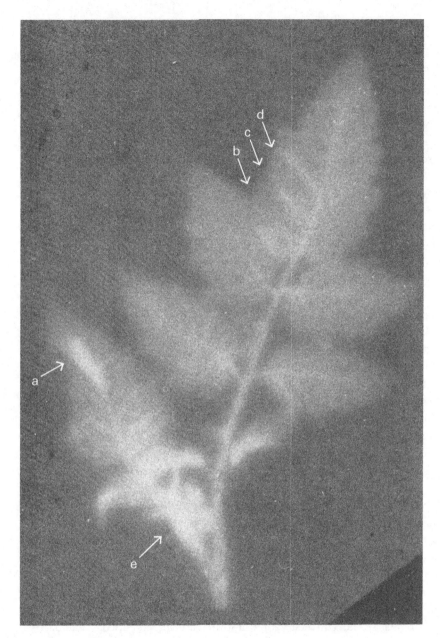

Figure 2.2
A contact radiograph of the young leaf of a tomato plant 36 hours after $^{32}PO_4$ was added to the nutrient solution. Notice the concentration of radioactivity (the lighter areas) in the parts of the plant that are growing. The light areas indicated by letters were caused by folds in the leaves (a–d) or by the bunching of several small leaflets (e). Image and caption from D. I. Arnon, P. R. Stout, and F. Sipos, "Radioactive Phosphorus as an Indicator of Phosphorus Absorption of Tomato Fruits at Various Stages of Development," *American Journal of Botany* 27 (1940): 791–798.

in the foliage and fruit in the upper portion of the plants—the region of most active growth. The smaller the tomato, the more radiophosphorus it took up.[45] As in the case of the mouse tumors, rapidly growing tissues concentrated more phosphorus-32 than slower-growing tissues.

Although these research uses yielded important biochemical knowledge, most of the requests Ernest and John Lawrence received for radiophosphorus came from physicians who wished to use it in therapy, and Berkeley became an important site for its clinical distribution. Besides phosphorus-32, the other radioisotope that physicians keenly sought in the late 1930s was radioiodine. Robley D. Evans, a physicist at the Massachusetts Institute of Technology, used a radium-beryllium neutron source (devised from discarded and donated medical radon needles and plaques) to make iodine-128. Evans' collaborators, Saul Hertz and Arthur Roberts of the Thyroid Clinic at Massachusetts General Hospital, performed the first biological tracer experiments with this isotope. In a 1938 paper, they reported the rapid, selective concentration of this isotope in the thyroids of 48 rabbits that had been injected with iodine-128.[46] Under conditions of thyroid stimulation, even more radioiodine was localized to the thyroid. The authors asserted that "the concentrating power of the hyperplastic and neoplastic thyroid for radioactive iodine may be of clinical or therapeutic significance," even though the half-life of this isotope—25 minutes—made the therapeutic prospect remote.[47]

Later in 1938, in Berkeley, J. J. Livingood and Glenn T. Seaborg announced the discovery of a longer-lived radioisotope of iodine, iodine-131.[48] Joseph Hamilton quickly put iodine isotopes to use in medical experiments, collaborating with Mayo Solley of the medical school in San Francisco to administer radioiodine orally to patients. Patients with overactive thyroids took up more than ten times as much radioiodine as healthy individuals did.[49] This finding laid the groundwork for the widespread use of iodine-131 in the treatment of hyperthyroidism.[50] One group reported that metastatic carcinoma of the thyroid accumulated radioiodine, but unfortunately only certain thyroid cancers selectively concentrated the isotope.[51] The most successful clinical applications of phosphorus-32 and iodine-131were for non-malignant diseases—polycythemia vera and hyperthyroidism, respectively.

One thus sees around the Berkeley cyclotron a variety of biological experiments alongside with attempts to develop therapies with radioisotopes. Human experiments were part of the patterns of use from the outset, and the same language of tracers was used to describe biochemical experiments in which radioisotopes were used to illuminate metabolic processes in animals and plants and nontherapeutic human experiments in which a small amount of a radioelement was administered to a subject to track its absorption and localization. The results of these experiments in turn fed into clinical trials, which generally used much larger doses of radioactivity in order to irradiate pathological (usually tumorous) tissue.

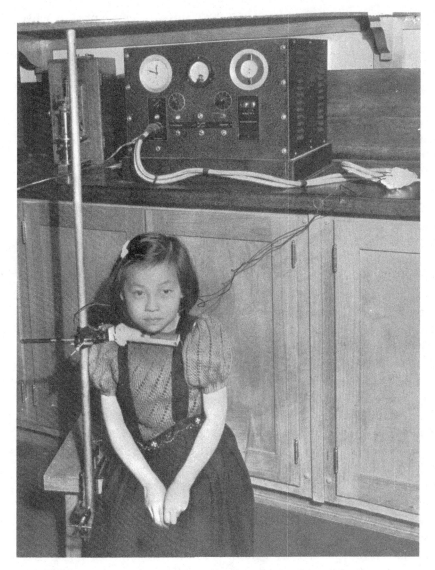

Figure 2.3
One of the first patients studied by Joseph Hamilton for thyroid uptake. After oral administration of radioiodine, the radioactivity of the thyroid was measured by placing a Geiger-counter tube over the gland. Credit: Lawrence Berkeley National Laboratory.

Cyclotrons at War

The mobilization of scientists in Lawrence's Rad Lab for work to aid the Manhattan Project, formalized by the US government in June of 1941, gave nuclear research and technology a new urgency.[52] The cyclotrons became important production sites for radioactive materials for laboratories throughout the country doing defense-related research.[53] The Manhattan Engineer District (MED) was established on August 13, 1942 in New York. That same month, Robert Stone left San Francisco to become head of the Health Division at the University of Chicago's Metallurgical Lab. On September 17, the Army appointed Colonel Leslie R. Groves to head this top-secret organization. In February of 1943, the Manhattan Engineer District contracted with the University of California to administer the Los Alamos Laboratory (Contract 36).[54] A few months later, Contract 48 between the University of California and the Army enlisted the Rad Lab as one of the central MED facilities. The production of plutonium from the Berkeley cyclotrons was critical for the early scientific work of the Manhattan Project.

By the time Contract 48 was signed, Joseph Hamilton was already conducting research on the metabolism and biological effects of plutonium and other fission products, through a contract with the Office of Scientific Research and Development. A component of the new Manhattan Project contract, 48A, assimilated this line of investigation.[55] Hamilton's group at Berkeley became part of a larger Plutonium Project aimed at establishing the occupational dangers for Manhattan Project employees of working with plutonium and the dozens of isotopes produced as uranium fission products.[56] Only one primary fission product—radioiodine—had been studied relatively well, and the amount of radioactivity to which Manhattan Project workers would be exposed was more than a million times that from industrial radium use worldwide.[57] Many fission products were radioactive isotopes of rare earths. For the most part, the metabolism of even the non-radioactive forms of these elements wasn't known.[58]

Hamilton's wartime study relied on use of the Berkeley cyclotrons to generate specific fission products by bombarding a uranium target. Because only small amounts were available, all his experiments of this sort were considered to be "tracer" studies. There was some overlap with his earlier research. Hamilton had already become interested in the possible use of radioactive strontium (also a fission product) for the clinical treatment of bone diseases.[59] But the work on fission products was larger-scale and more systematic, involving the testing of eighteen fission products each on twelve rats, exposed in groups of three at various times before sacrifice and analysis. By 1943, the pattern of accumulation (in various organs) of fourteen of the radioisotopes had been determined, as had their rates of elimination.[60] Several were found to localize in bones.[61]

As the Manhattan Project progressed, J. Robert Oppenheimer was increasingly concerned about the safety of plutonium—and the lack of knowledge about it. In

February of 1944, Hamilton was given 11 milligrams of plutonium for use in biological studies. His group rapidly ascertained that plutonium, like radium, was a bone-seeker and could be expected to cause bone cancer. Although its risk of absorption from ingestion was less than that of radium, when inhaled it persisted longer in the lungs.[62] In view of the urgency of plutonium's danger and the difficulty in extrapolating from rat to man, the Manhattan Project leadership decided to embark on research with human subjects. In January of 1945, Hamilton signaled his intention to begin "metabolic studies with [plutonium] using human subjects."[63] Just a few months later, the first injection of plutonium into a human subject under Contract 48A occurred at University Hospital in San Francisco. Albert Stevens, thought to be suffering from stomach cancer, was given 0.932 micrograms of a mixture of plutonium-238 and plutonium-239. He was later found to have a gastric ulcer rather than cancer. Stevens, designated CAL-1, was one of eighteen patients (including two more in San Francisco) who received injections of plutonium between April 1945 and July 1947 at several MED sites.[64]

These plutonium-injection experiments represent most vividly the ethical abuses associated with the US government's secret efforts during and after World War II to gather information about the dangers of fissionable material and radiation from research on humans.[65] Few of the patients injected with plutonium seem to have been informed of their exposure or their status as research subjects. How did leading researchers in the medical application of radioisotopes end up conducting these experiments for the military? In part, it was because these experiments built on civilian studies at Berkeley that had preceded them—the plutonium-injection research and similar studies were referred to in AEC reports as "human tracer experiments."[66] Clearly these wartime studies differed in crucial ways from those that went before—the selection of terminally ill patients as subjects signaled their potential danger. Yet the earlier pattern of human experimentation at Berkeley facilitated the subordination of research there to the emerging occupational health and safety requirements of the military. Even after the war, Joseph Hamilton was convinced that "under appropriate and suitable circumstances, it is highly desirable to conduct human studies with certain of the fission products and fissionable elements."[67] Yet even AEC officials understood that these experiments were ethically dubious and politically problematic.[68] These experiments illustrate the overlap between the AEC's civilian and military agendas in the management of atomic energy's dangers, an intersection not usually connected to tracer work with radioisotopes.[69]

Radiocarbon and the Proliferation of Metabolic Pathways

The non-medical investigations using radioactive tracers at Berkeley also launched important trajectories of postwar research. In the late 1930s in the Berkeley Rad Lab, the radiochemists Martin Kamen and Sam Ruben and the plant biochemist William Zev Hassid first used radioactive carbon—the isotope carbon-11—as a tracer to follow the fixation of carbon dioxide by both photosynthetic and heterotrophic organisms.[70] Earlier in the 1930s, Robin Hill at Cambridge University had demonstrated that the "light" and "dark" steps of photosynthesis could be studied separately.[71] Hassid, Kamen, and Ruben used radiocarbon to trace the "dark" steps—the conversion of carbon dioxide into carbohydrate. Using ^{11}C-labeled carbon dioxide, they found that the first product of photosynthesis in green plants was the transfer of carbon from CO_2 to a carboxyl group.[72] Experiments using this precious biological isotope required proximity to a cyclotron, as it had a half-life of only 21 minutes.

Carbon-14, first isolated by Kamen and Ruben in 1940, offered an excellent alternative—its half-life was 5,700 years. However, Kamen and Ruben's research collaboration was disrupted by the demands of World War II. Kamen became involved in the Rad Lab's work on uranium separation, and Ruben undertook work with poison gases, particularly phosphogene. An accident with phosphogene took Ruben's life in 1943.[73] Not until the end of the war was the first biological tracer experiment with carbon-14 published out of Berkeley, by Kamen and the plant biochemist Horace A. ("Nook") Barker.[74] Because of the ubiquity of carbon in living systems, the wide-ranging utility of carbon-14 in tracer experiments was recognized immediately. Aside from having a long half-life, carbon-14 could be diluted a billionfold and still be detected through its radioactive decay.[75] But when Kamen lost his security clearance as a result of suspicions about his association with leftist musicians and a meeting with a Russian consular official (in conjunction with arranging a shipment of phosphorus-32 from Berkeley to treat another official), the project again languished.[76]

Ernest Lawrence was keen to keep Berkeley at the forefront of radiotracer work on photosynthesis. Late in 1945, he persuaded the Berkeley chemist Melvin Calvin to undertake biological studies with the carbon-14 already available at the Rad Lab. Calvin recalled Lawrence telling him they should do something "useful" after their involvement in the Manhattan Project.[77] Calvin began the studies with a small vial of ^{14}C-labeled barium carbonate inherited from Ruben, and invited Ruben's former collaborator in photosynthesis research, Andrew Benson, to lead the effort.[78] They filed four papers on these earliest experiments in photosynthesis, done collaboratively with the Berkeley plant biochemists William Hassid and Horace Barker, as Manhattan District declassified reports.[79] As part of the work going on at the Berkeley Rad Lab, it was soon supported through an early AEC grant to Lawrence.[80] Calvin was also an early recipient of the AEC's carbon-14 (produced at Oak Ridge), which had a much higher specific activity than the material obtainable from the Berkeley cyclotrons.

Calvin's efforts in tracing carbon-14 through the photosynthetic pathway were immensely fruitful, leading to the elucidation of what came to be called the Calvin (or Calvin-Benson) photosynthetic cycle.

Calvin's group grew the single-celled green alga *Chlorella pyrenoidosa* in a culture suspension supplied with normal carbon dioxide in an apparatus called a "lollipop." ^{14}C-labeled carbon dioxide would be injected into the usual stream of carbon dioxide for a predetermined period of time, ranging from seconds to minutes. The algae would then be killed and their contents analyzed. Paper chromatography was the principal analytical tool—the algae juices would be separated in two dimensions, using two different eluting fluids. Different chemical compounds migrated in the two-dimensional space as discrete spots. Exposing the paper chromatogram to x-ray film enabled the research to pinpoint the radioactive compounds and to trace the appearance of radioactivity in new compounds with longer time periods after the exposure to labeled carbon dioxide.

This chromatographic method of analysis was much faster than traditional methods of organic chemistry, and it was visually impressive. Calvin and his co-workers

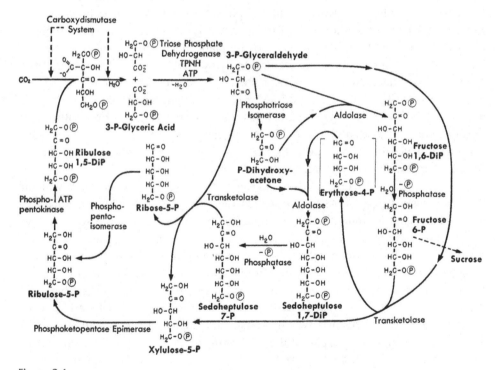

Figure 2.4
A schematic diagram of the photosynthetic carbon cycle ("Calvin-Benson cycle"). From Melvin Calvin, "Photosynthesis," in *Radiation Biology and Medicine: Selected Reviews in the Life Sciences*, ed. Walter D. Claus (Addison-Wesley, 1958). Reprinted by permission of Pearson Education Inc., Upper Saddle River, New Jersey.

showed that carbon dioxide was converted into phosphoglyceric acid, which then was converted, through several other biochemical steps, into fructose and other sugars. On August 26, 1948, the physicist Freeman Dyson attended a lecture by Calvin on these results. Afterward, he described in admiring terms the way the resulting pictures "show, in the most possible way, the progress of the delicate and transitory reactions through which the radio-carbon is assimilated." For Dyson, Calvin's research provided dramatic evidence of the advances atomic science had brought:

> The long-sighted people said, when nuclear energy first came on the scene, that the application to biological research would be more important than the application to power. But I doubt if anyone expected that things would actually get going as fast as they have. This blotting-paper-plus-radio-activity technique is completely revolutionary because it means that *any* substance can be fed to a cell and its transformations followed second by second in detail, even in quantities too small to be seen or weighted, and with substances too unstable to stand old-fashioned stewing and chemical extraction.[81]

The tone was exuberant, but Dyson's perspective wasn't hyperbolic. The photosynthetic pathway stood as an early example of how radioisotopes could unlock biochemical puzzles, and the AEC referred to it frequently.[82] Moreover, this work had taken place in one of the AEC's own laboratories, and had profited from Oak Ridge-produced carbon-14, so the AEC had a particular claim on the breakthrough.

But if the photosynthetic pathway gave early evidence of the promise of radiotracer work, a wealth of similar discoveries followed. The growing reliance on isotopic tracers resulted in "so elaborate a proliferation of metabolic pathways as to boggle the minds of students of biochemistry."[83] Carbon-14 was especially important to these metabolic studies because it could be used to tag almost any molecule of biological interest (nearly all of which contain carbon). The AEC regularly cited its sales of carbon-14 to illustrate the savings made possible by using a nuclear reactor to produce isotopes. In one report to Congress, the AEC estimated that, whereas the Oak Ridge pile could manufacture 200 millicuries of carbon-14 in a few weeks, at a cost of about $10,000, "it would take 1,000 cyclotrons to equal this output, and the operating cost would be well over a hundred million dollars."[84]

Nuclear Medicine

The early postwar publicity about the medical breakthroughs that radioisotopes would bring focused on their therapeutic uses, particularly in the treatment of cancer. The hope that isotopes would cure cancer was predicated on the notion that they would localize to specific tumors and deliver internal radiation. However, the expectation that radioisotopes would become so-called magic bullets to fight cancer didn't fully materialize, whereas radioisotopes did become important tools in the area of medical diagnostics.[85] The AEC's 1948 report to Congress contained the following passage:

Some malignant, abnormally growing tissues absorb certain elements in the body, such as phosphorus and iodine, faster than normal; others absorb certain elements more slowly than normal. Cancer specialists are taking advantage of this fact and using radioisotopes to help locate tumors. The University of California Medical School and the Cook County Hospital, Chicago, are using radiophosphorus to locate cancer in the breast; the radiations enable diagnosticians to distinguish between benign and malignant growths, since the latter take up phosphorus at a slightly greater rate.[86]

Radioisotopes could be used to identify cancer, even if they could not treat it.

As contemporary observers recognized, the use of radioisotopes in clinical diagnostics was essentially an application of tracer methodology to medicine, putting small amounts of isotopes to use in making physiological measurements or detecting abnormal tissue growth.[87] Tumors were not the only target. By 1955, radioisotopes were being used in a wide variety of diagnostic tests. In general, tracer applications of radioisotopes required much smaller amounts of radioactivity than the dosage required in therapeutic applications.[88] Indeed, at the biological level, it was the observation that very low-level amounts of radiation did *not* disturb fundamental living processes that legitimized the use of radioisotopic tracers as probes.[89]

One type of diagnostic technique involved dilution of the radioisotope in the body. For example, Joseph Hamilton's early work with sodium-24 laid the groundwork for diagnostic tests in which this radioelement was used to assess total exchangeable body sodium.[90] Under the same general rationale, chromium-51 or iron-59 was used to measure red cell mass, and ^{131}I-labeled serum albumin was used to measure plasma volume. Radioisotopes could also be used to measure the rate of flow in the circulatory system. Sodium-24 could be used to assess cardiac output or to diagnose peripheral vascular disorders, building on Hamilton's early study of its absorption and movement through the bloodstream to the extremities. Other diagnostic tests followed the metabolism of radiolabeled compounds in the human body. For example, patients with pernicious anemia didn't excrete as much vitamin B-12 in their urine as normal subjects. Consequently, reliable diagnosis could be achieved by administering cobalt-60-labeled vitamin B-12, then performing a precise urine test.[91]

Perhaps the best-known radioisotope-based diagnostics involved physiological localization, as exemplified by the use of iodine-131 to study thyroid physiology and dysfunction. The success of this procedure was established in the late 1930s, as has already been mentioned. In 1956, Paul Aebersold estimated that "over half a million thyroid studies have been done with iodine 131 since that time."[92] Similarly, the localization of phosphorus-32 in tumors had been noted by John Lawrence and other researchers during the 1930s, and diagnostic procedures using this isotope were developed for various forms of cancer. The challenge of using radioactivity to locate a tumor or to measure function in an internal organ, however, was one of detection: the tissues of the human body both absorbed and interfered with the radiation given off by isotopes.

Diagnostic administration of isotopes could be used in conjunction with surgery. When a patient's body was being opened up, radioactivity could be measured directly. In the case of surgery for brain tumors, researchers at Massachusetts General Hospital designed miniature Geiger counters, termed Selverstone-Robinson probes, to be placed directly into the brain.[93] The beta rays emitted from decaying phosphorus-32 don't penetrate the scalp or the skull, so this isotope, which localized well to tumors, was useful only once the brain had been opened up. The surgeon could then insert the probes (each only 1–3 millimeters long) at various depths to determine the exact location of the tumor, signaled by an increase in the counting rate from 5 to 36 times

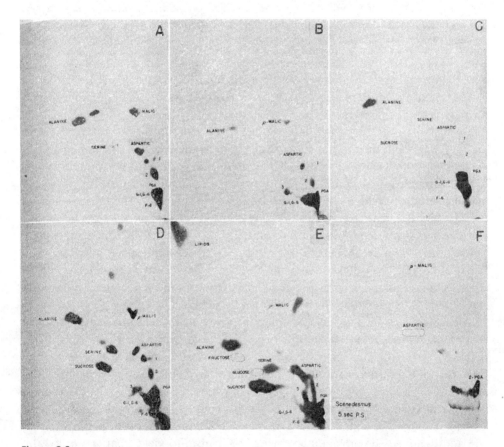

Figure 2.5
A two-dimensional chromatogram of extract from algae indicating uptake of radiocarbon during photosynthesis (30 seconds). From Melvin Calvin, "Photosynthesis," in *Radiation Biology and Medicine: Selected Reviews in the Life Sciences*, ed. Walter D. Claus (Addison-Wesley, 1958). Reprinted by permission of Pearson Education Inc., Upper Saddle River, New Jersey.

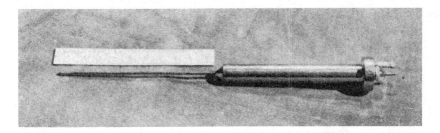

Figure 2.6
A Selverstone-Robinson probe counter. From William H. Sweet and Gordon L. Brownell, "The Use of Radioactive Isotopes in the Detection and Localization of Brain Tumors," in *Radioisotopes in Medicine*, ed. Gould Andrews, Marshall Brucer, and Elizabeth Anderson (Atomic Energy Commission, 1955).

over the background. Just a few years after the method was first published, William Sweet, one of its developers, asserted that "such counters are now used routinely at operations."[94]

Sweet and Gordon Brownell also worked to develop an isotope-based detection method that could be used on patients not undergoing surgery. They tested several positron-emitting isotopes, because, unlike phosphorus-32, when positrons are annihilated through interaction with electrons they emit radiation identical to x rays (gamma photons), which could be detected outside the skull. The most promising positron-emitting radioisotopes for this kind of diagnostic test proved to be arsenic-72 and arsenic-74, injected together intravenously at 20 microcuries per kilogram of body weight. (The amount of metallic arsenic involved was well below pharmacologic intoxication.) By 1953, 300 patients had undergone brain scans with radioarsenic. In 99 of 133 brain tumor patients tested, radioactivity concentrated detectably in intracranial lesions, whose locations were verified by surgery.[95]

The Geiger counter remained the main instrument for detecting radioactivity in medical diagnostics from the late 1920s through the early postwar years, when detection devices for the human body began to improve dramatically. In 1949, at the University of California at Los Angeles, Benedict Cassen designed a scintillation counter for the *in vivo* localization of radioiodine in patients.[96] Because it was 10–20 times more sensitive than Geiger counters, using this counter in diagnostic tests required less radioiodine to be administered. Cassen also developed a point-by-point counting grid and mounted the probe on a moving mechanism to yield what was called the rectilinear scanner, a device for visualizing organs that went into commercial production (by the Picker X-ray Company) in 1959. Hal Anger of the Donner Laboratory at Berkeley then built a device with ten scintillation detectors in a row that could scan the whole body.[97] The isotope technetium-99m was identified in 1960; by the 1970s

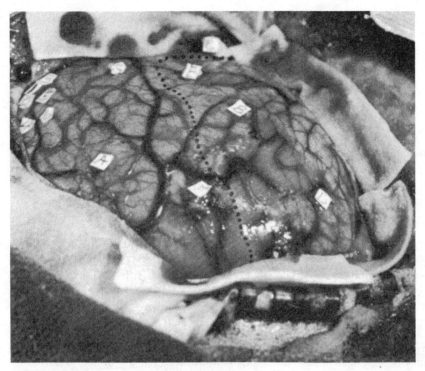

Figure 2.7
The operation field of Patient T.F. Labels indicate sites of insertion of probe counter. From William H. Sweet and Gordon L. Brownell, "The Use of Radioactive Isotopes in the Detection and Localization of Brain Tumors," *Radioisotopes in Medicine*, ed. Gould Andrews, Marshall Brucer, and Elizabeth Anderson (Atomic Energy Commission, 1955).

it was the most ubiquitous radioisotope in medical diagnostics, thanks to the energy of its emitted gamma ray and its short half-life.[98] Some retrospective histories of "nuclear medicine" skip over the late 1940s and the early 1950s and highlight technetium-99m and the invention of these body scanners.[99] Focusing on the earlier applications of radioisotopes in diagnostics highlights the commonality of the "tracer method" across biology and medicine. From the 1930s through the 1950s, many of the same radioisotopes and detection devices were being used for laboratory research and clinical diagnosis, actively promoted and supported by the AEC.

Tracing Ecosystems

The availability of radioisotopes also affected field sciences, particularly ecology. In adopting radiotracers as tools, ecologists sought to emulate how biochemists and physiologists used radioisotopes in the laboratory to elucidate pathways of metabolism. G. Evelyn Hutchinson pioneered the use of isotopes in fresh water as ecological tracers, releasing phosphorus-32 from the Yale cyclotron into a pond in order to investigate the cycling of phosphorus through the phytoplankton and inorganic matter.[100] First in 1941 and again in 1946, Hutchinson carried out this experiment with phosphorus-32 from the Yale cyclotron. After the war, he was among the first licensees to receive phosphorus-32 from Oak Ridge, with its unrivaled specific activity.[101] After undertaking a more thorough investigation with this AEC reactor-generated radiophosphorus in the summer of 1947, Hutchinson published the results in 1950.[102] The results confirmed his hypothesis that the overall metabolism of phosphorus in the lake was maintained at a steady state by the growth and death of algae.

Hutchinson's research on ponds and the stability of their aquatic communities built on an analogy Hutchinson had offered in 1940 in a review of *Bio-Ecology*, a book by Frederic Clements and Victor Shelford: "If, as is insisted, the community is an organism, it should be possible to study the metabolism of that organism."[103] In 1941, Hutchinson made good on this analogy, publishing an article titled "The Mechanisms of Intermediary Metabolism in Stratified Lakes."[104] The subsequent use of isotopes to trace the development and metabolism of aquatic communities rendered Arthur Tansley's 1935 notion of an ecosystem, including both biotic and abiotic components, in concrete terms.[105] But whereas Tansley had suggested "ecosystem" as a neutral category to rid ecology of the organismal term "biotic community," the use of isotopes to trace "metabolic pathways" in both physiology and ecology kept the organismal concept in play.

In addition, similar representational practices—mapping the patterns of circulation—manifested the epistemological links between metabolic biochemistry and biogeochemical studies of ecosystems, even as the ecological diagrams included nonliving components. Such uses of radioisotopes to trace the "metabolism" of aquatic systems informed Raymond Lindeman's notion of trophic-dynamics, which Eugene Odum and others made central to ecosystems ecology.[106] Between 1946 and 1953 Hutchinson (who had been Lindeman's advisor at Yale) attended the Macy Foundation conferences on cybernetics. In 1946 he wrote his influential paper "Circular Causal Systems in Ecology" for one of those meetings, emphasizing the self-regulating features of an ecosystem. The diagram of the global biogeochemical cycle of carbon presented in that paper shows a visual similarity to metabolic pathways at the time.

Many ecologists followed Hutchinson's example, using the intentional release of limited amounts of radioisotopes into the environment to trace the uptake and circulation of elements.[107] But, unlike biochemistry and nuclear medicine, ecology included many "experiments" which these scientists didn't set in motion. Radioactivity was entering the environment, often on a large scale, through the AEC's disposal of nuclear waste and atomic weapons tests, and ecologists began tracking the movement of these

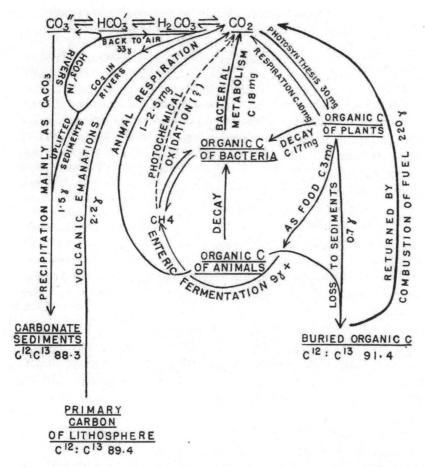

Figure 2.8
A schematic diagram of the global carbon cycle. From G. E. Hutchinson, "Circular Causal Systems in Ecology," *Annals of the New York Academy of Sciences* 50 (1948): 221–246. Reproduced by permission of John Wiley and Sons.

radioisotopes. In this respect, ecological tracing was more directly connected than other biomedical uses of radiotracers to the military development of atomic energy by the US government.

Some research along these lines began during the war. The Applied Fisheries Laboratory at the University of Washington in Seattle was established in 1943 under Professor Lauren Donaldson, with the support of the Manhattan Engineer District, to undertake laboratory studies on the biological effects of radioactivity on fish—specifically, to assess the possible impact of radioactive waste from the reactors at the

Hanford Reservation on the local salmon-fishing industry.[108] In 1946, Donaldson's group also participated, along with many other scientists, in an ecological and geological survey of the Bikini Atoll before and after the bomb tests as part of the Nuclear Testing Program in the Pacific. A follow-up survey at Bikini Atoll a year after the tests revealed the persistence of radioactivity in flora and fauna.[109]

Research on the biological effects of radioactive waste also took place on the Hanford Reservation itself.[110] In late 1944, Donaldson persuaded the MED leadership of the importance of beginning on-site research at the Columbia River. His graduate student Richard F. Foster was transferred to Hanford in June of 1945 to work at a new Aquatic Biological Laboratory operated by the DuPont Corporation with Donaldson as a consultant.[111] Initial laboratory studies by Foster, conducted in the second half of 1945, suggested that the addition of cooling effluent from the Hanford reactors, if it was diluted sufficiently (at least 1:50) by river water, would not threaten the salmon and trout populations.[112] Hanford scientists also studied the levels of radioactivity in the river water. After exiting the reactors, the effluent was held for 24 hours in retention basins in order to allow short-lived radioisotopes to decay. Most of these radioisotopes were not fission products from the reactor, but normal constituents of river water whose minerals became radioactive through activation (via neutron capture) when passing through the cooling vessels.[113] By the late 1940s, studies of radioactive contamination near Hanford yielded a disquieting result: several species, some aquatic and some terrestrial, accumulated radioactivity in high quantities, and several radioisotopes were concentrated as they moved up the food chain. Many terrestrial animals, including mammals, birds, reptiles, and insects, were exposed through the aerial release of radioactive waste gases.[114]

Hanford scientists emphasized in various publications that radioactive contamination in the Columbia River "never approached hazardous levels."[115] But Eugene Odum drew a more cautious conclusion: "[A]n isotope might be diluted to a relatively harmless level on release into the environment, yet become concentrated by organisms or a series of organisms to a point where it would be critical. In other words we could give 'nature' an apparently innocuous amount of radioactivity and have her give it back to us in a lethal package!"[116]

Another prominent program of ecological research was launched at Oak Ridge National Laboratory (ORNL). Ecology gained support at Oak Ridge because two health physicists there, Karl Morgan and Edward Struxness, considered their purview to include the study of radioactive contamination of the environment.[117] Stanley Auerbach arrived in 1954 to join their division, and by 1960 he had built up one of the largest ecology programs in the country. When he arrived, ORNL was pumping low-level radioactive waste into pits, from which it could seep into the surrounding soil. The expectation was that through binding of soil particles the radioisotopes would become immobilized.[118] Low-level radioactive wastes were also dumped into White

Oak Lake. In 1955 the lake was drained, and health physicists had an opportunity to study the fate of its radioactive contamination. In 1956, Auerbach began to investigate the movement of radioisotopes within the soil, plants, and fauna of the lake bed, and to examine the effects of contaminating radiation on the plants and animals.[119] The seepage of radioactive waste from the disposal pits was also studied. The discovery that radioisotopes were taken up by vegetation showed that radioactive waste couldn't be assumed to remain where it had been deposited.

Besides studying the environmental fate of radioactive waste, Auerbach followed the precedent of Hutchinson in using artificially produced radioisotopes, readily available at Oak Ridge, to investigate element cycling.[120] In May of 1962, a team of ORNL ecologists tagged an entire forest with cesium-137, intending to measure cesium

Figure 2.9
Oak Ridge National Laboratory ecologists inoculating a tulip poplar with cesium-137. The man at right is pouring radiocesium solution into a previously prepared trough, from which the isotope moves into the trunk through slits chiseled into the bark. The man second from left is adding water to facilitate the transfer of the isotope from the trough. The man at far left, in the foreground, is monitoring and timing the operation. From Onsite Ecological Research of the Division of Biology and Medicine at the Oak Ridge National Laboratory, compiled and edited by Stanley I. Auerbach and Vincent Schultz (AEC report TID-16890).

transfer between the components of the ecosystem. As Auerbach and two co-authors noted, tracer experiments had not previously been attempted in the United States "on a relatively large scale in field experiments."[121] Radiocesium was applied directly to the trunk of each tree in the forest for a total distribution of 467 millicuries. The results showed that cesium cycled out of the trees into the litter on the forest floor, but didn't rapidly re-enter the system through the roots.[122]

Such studies with isotopic tracers put ecosystems ecology on a quantitative footing.[123] Ultimately, computers were used to provide mathematical simulations of ecosystems, and the ORNL group led this effort. In particular, Jerry Olson, a research ecologist hired there in 1958, harnessed the computational resources available at a national laboratory to undertake mathematical modeling of nutrient cycling, focusing on the movement of radionuclides through ecosystems.[124] In fact, the 1962 radiocesium tagging of the Oak Ridge forest was done with the hope of inputting the data into Olson's computer model for mineral cycling.[125]

Eugene Odum carried out similar radiolabeling experiments on research tracts at the Savannah River nuclear site in Georgia. When the AEC decided to build a new plutonium-production plant on the Savannah River, the AEC's newly established Division of Biology and Medicine supported studying the plant's impact on the environment. Odum obtained a grant for the University of Georgia to participate in this work, and increased funding from the AEC in the next several years enabled him to establish a permanent laboratory there (the Savannah River Ecology Laboratory).[126] Work at Savannah River included field experiments with radioactive tracers. Odum wasn't constrained in the way that Auerbach was by the Oak Ridge focus on radioactive waste and the special concern with fission products such as cesium-137 and strontium-90.[127] Instead he could select an isotope to best measure the movement of material and energy between the various organisms in a terrestrial ecosystem.[128] Phosphorus, an element that was necessary for growth in all organisms, was a more desirable tracer than strontium or cesium.

Odum and Edward Kuenzler devised a method of laying out "hot quadrats" in which all individuals of a kind of plant in the designated area were labeled with phosphorus-32.[129] They radiolabeled three quadrats, each of a different plant species (heterotheca, rumex, sorghum) in the spring of 1957. By following the movement of radiophosphorus into higher trophic levels (animals), the researchers could isolate the food chain: Any animal that became radioactive had to belong to the food chain originating with the tagged plant species. In addition to showing how rapidly phosphorus (and thus energy) was transferred from plants (the primary producers) to grazing herbivores and ants (the primary consumers), and subsequently to small mammals such as mice, the experiment new shed light on the eating habits of snails, whose rapid acquisition of radioactivity indicated that grass was an important food source

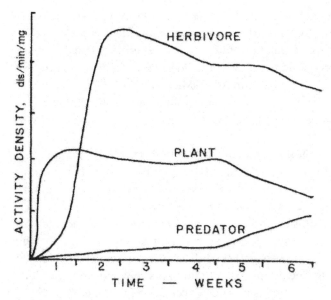

Figure 2.10
Typical activity density curves at three trophic levels resulting from the labeling of a single species of herbaceous plant at time 0 with phosphorus-32. The movement of the phosphorus-32 label from plants to carnivores is clearly visible. All curves are corrected for radioactive decay. From Eugene P. Odum, "Feedback between Radiation Ecology and General Ecology," *Health Physics* 11 (1965): 1257–1262.

for them.[130] The researchers' plots of radioactivity versus time in various species showed "the graphic separation of certain trophic and habitat groups."[131]

Continued investigations by Odum's group pointed to the promise of this radiolabeling method for determining "food web diversity for an entire community."[132] Odum also taught his methods to other ecologists interested in using radioisotopes.[133]

The effects of radiation and radioisotopes on ecology can be seen in the many symposia volumes published from the late 1950s through the early 1970s. The 1955 and 1958 International Conferences on the Peaceful Uses of Atomic Energy featured many papers by ecologists, most of them oriented toward the problems presented by radioactive waste from civilian nuclear power development. Concern about the hazards of radioactive fallout from tests of atomic weapons reinforced environmental investigation. A Symposium on Radioisotopes in the Biosphere held at the University of Minnesota on October 19–23, 1959 focused principally on the pathways of radioisotopes released into the atmosphere by atomic explosions. Similarly, the

International Symposium on Radioecological Concentration Processes, held in Stockholm in 1966, focused principally on the distribution and movement of radioisotopes in fallout.[134] But the importance of radiation studies to general ecology is also evident. The AEC supported three large-scale symposia on radioecology, the first held in Fort Collins, Colorado, in 1961, the second in Ann Arbor, Michigan, in 1967, and the third at Oak Ridge, Tennessee, in 1971.[135] Dozens of papers that featured the use of radioisotopes as tracers through aquatic or terrestrial ecosystems were presented, many of which represented AEC-sponsored work.

One of the legacies of radioecology was the conceptual transfer of the invisible danger from radioisotopes to other environmental contaminants. Rachel Carson's book *Silent Spring* posited a similarity between the hazards of radioactive fallout and the dangers of synthetic chemicals: "In this now universal contamination of the environment, chemicals are the sinister and little-recognized partners of radiation in changing the very nature of the world—the very nature of life."[136] The analogy between radioactivity and synthetic chemicals also informed how scientists approached the problem of understanding how industrial contaminants moved through ecosystems and entered food webs, both dispersing and concentrating in the environment. As it turned out, some insecticides were found to exhibit the same trait of bioconcentration as the compound moved up the food chain.[137] DDT became the exemplar of this phenomenon. In fact, George Woodwell at Brookhaven National Laboratory, having made his name studying how radiation affected forest ecosystems, subsequently demonstrated that DDT was concentrated up to 1.5 millionfold in an aquatic ecosystem on Long Island.[138] As Woodwell noted, the attention to one part per billion in the environment "was itself a revolution," and the realization that biotic studies required measurement in the "range of nanograms and picograms, nanocuries and picocuries" became a defining feature of the study of environmental contamination.[139] As two textbook authors noted in 1982, when it came to studying ecological processes involved in the spread of "smog, pesticides, and other chemical substances that may threaten the environment," radioisotopes served as a "model pollutant."[140]

Conclusion: The Material Culture of Cold War Science

In assessing the Cold War's significance for American biology, historians have focused particular attention on the influence, direct and indirect, of the military.[141] Susan Lindee has shown how the meaning of human mutation for the researchers on the Atomic Bomb Casualty Commission was shaped by the social realities of dealing with Hiroshima survivors.[142] Lily Kay has argued that the cracking of the genetic code in the 1960s bore signs of its casting a decade earlier as amenable to the techniques of code-breaking for military purposes.[143] Examples are far more numerous for physics and engineering—one thinks of Peter Galison's analysis of Norbert Wiener's cybernetics, and Paul Edwards' argument for the centrality of military command and control to the development of postwar computing, among other examples.[144] However, attention to the supply and the widespread uses of radioisotopes attests to the pervasive consequences of the Cold War for the material culture and practices of civilian biology and medicine. Both the symbolism and the outcomes of the AEC's radioisotope distribution challenged the prevailing image of atomic energy as a technology of war. Radioisotopes represented the atom's potential for humanitarianism and health, even in the midst of a nuclear arms race. At the same time, the public emphasis on medical benefits camouflaged the actual military utility of some radioisotope experiments, particularly studies of human exposure to plutonium, fission products, and radiation.

The tension between the benefits of radioisotopes for biological research and medical therapy and their possible military use dominated early debates about the exporting of radioisotopes, where the AEC's proposed policy collided with the politics of national security. Immediately after the passage of the Atomic Energy Act, there was substantial debate over whether the new agency should allow foreigners to purchase its radioisotopes. Republican critics of the civilian control of atomic energy, such as Senator Bourke Hickenlooper, were insistent that radioactive materials should not be shipped abroad, lest they speed the development of atomic weaponry elsewhere. The Atomic Energy Act of 1946 prohibited the export of "fissionable" materials, and even though the radioisotopes most useful to biology were not fissionable, congressional debates stalled the initiation of an export program until the fall of 1947.[145] When exports of radioisotopes commenced, President Truman's announcement justified them as enabling the "open, impartial, and truly international character of medical research [to] carry over into the realm of other problems of world concern."[146] President Eisenhower furthered the policy of using non-military nuclear resources in foreign diplomacy. This effort culminated in his Atoms for Peace initiative, which entailed bilateral agreements with developing nations to disseminate the other major dividend of atomic energy: nuclear power.[147] Eisenhower's "Atoms for Peace" speech, given on December 8, 1953, highlighted the AEC's foreign distribution of radioisotopes as a precedent.[148]

It was not just government propaganda that drove the widespread use of radioisotopes, in the United States or elsewhere. Researchers already wanted them. Radioactive elements could be detected with unprecedented sensitivity, and the detection methods (mostly Geiger counters and autoradiography in the 1940s) were more readily available than the mass spectrometers required for using stable isotopes. As compared with relying on chemical extraction and purification, the measurement of radiolabeled compounds could often be performed on an intact system, whether that was a cell, plant, animal, or landscape. One of the other great virtues of the isotopic tracer method was that it could be used in conjunction with many other laboratory techniques, including ultracentrifugation, electrophoresis, and chromatography. Radiolabeling allowed a researcher to pick out the molecule of interest—or its successive metabolic by-products—while using other instruments already at hand to separate biomolecules.

For these reasons, scientific interest in radioisotopic tracers both preceded and transcended the Manhattan Project. Yet the state's monopoly on nuclear technology, and the political pressure on the MED's successor, the AEC, to demonstrate the civilian benefits of the atom, meant that scientific research with radioisotopes developed much more rapidly than it would have otherwise, especially if the supply had continued to be tethered to cyclotrons. The US government also cultivated medical applications of radioisotopes, from installing cobalt-60 machines in hospitals to making radioelements used in cancer research, therapy, and diagnosis free from 1948 to 1952 and subsidizing them heavily thereafter.[149] The effect of the government's promotional efforts on research was rapid and dramatic. AEC-produced radioisotopes resulted in the publication of more than 10,000 papers during the first postwar decade, a majority of them on topics related to biomedical research.[150] In turn, the AEC touted the utility of radioisotopes as a demonstration of the peacetime benefits of atomic energy.

Other areas of research were directly created by the facilities that embodied the United States' commitment to atomic energy during the Manhattan Project and the Cold War. In the case of ecology, the landscapes around the AEC's atomic weapons plants and national laboratories became experimental test beds for ecological radiotracers, and problems of radioactive waste there were, in turn, understood in terms of ecosystems. More broadly, ecological studies of radioactive waste as well as the growing understanding of the dangers of low-level radiation exposure disclosed the environmental and occupational risks associated with using radioisotopes—information the AEC was initially reluctant to accept.[151] Radioactive tracers also made the problems of containing nuclear contamination newly visible; that, in turn, generated concerns about the disposal of radioactive waste from laboratories, clinics, and the government's large-scale atomic-energy and weapons facilities.[152]

Beginning in the 1950s, these ecological results and other developments complicated the many significations of "the peaceful atom." The debates about fallout changed the public perception of the relationship between atomic energy and cancer. Radioisotopes began to be understood as potential threats to health as well as medical bullets, as they had been viewed in 1947. Particular isotopes such as strontium-90, released through testing of atomic weapons and incorporated into the food chain through grazing livestock in the western United States, now symbolized the perils of atomic energy. Alongside the recognized utility of radioisotopes in biomedicine was an anxiety about radioactive contamination in the environment. The signing of the 1963 partial test ban treaty addressed the problem of radioactive fallout, even as the construction of nuclear power plants posed another source of environmental radioactivity. By the late 1970s, the anti-nuclear movement had effectively stalled the expansion of the nuclear power industry.[153]

During the same period, the building of civilian nuclear reactors, both in the United States and abroad, made the Oak Ridge reactor an increasingly obsolete machine for generating radioisotopes. Private and university reactors began producing radionuclides for the still-growing radiochemical and radiopharmaceutical market. In 1963, the AEC closed down the X-10 reactor at Oak Ridge National Laboratory, in acknowledgment that the demand for radioisotopes could be largely met outside the government infrastructure.[154] In 1974, the AEC was reorganized in order to segregate its promotional functions from its regulatory functions. The institutional and political vicissitudes besetting the AEC eclipsed the historic role of the US government in fostering the market for radioisotopes, even as these material traces of the Cold War remained central to biology and medicine.

Acknowledgments

This chapter includes material from my book *Life Atomic: A History of Radioisotopes in Science and Medicine* (University of Chicago Press, 2013); it is reprinted here with permission. Some passages were adapted from "Nuclear Energy in the Service of Biomedicine: The US Atomic Energy Commission's Radioisotope Program, 1946–1950," *Journal of the History of Biology* 39 (2006): 649–84 (© Springer 2006), with kind permission from Springer Science+Business Media B.V. I am grateful for comments and suggestions from John Krige, Naomi Oreskes, Michael Gordin, Susan Lindee, Dave Kaiser, Matthew Shindell, and Michael Barany, from the anonymous reviewers for the MIT Press, and from participants in the 2010 Francis Bacon Conference, "How the Cold War Transformed Science."

Notes

1. Henry DeWolf Smyth, "Nuclear Power and Foreign Policy," *Foreign Affairs* 35 (1956): 1–16, on 2–3.

2. "Availability of Radioactive Isotopes; Announcement from Headquarters, Manhattan Project, Washington, D.C." *Science* 103 (1946): 697–705, on 697.

3. Because many of these 63,990 shipments went to companies that prepared radiochemicals and radiopharmaceuticals for a secondary market, the ultimate number of radioisotope users was severalfold higher. US Atomic Energy Commission, *Eight-Year Isotope Summary*, volume 7 of *Selected Reference Material, United States Energy Program* (Government Printing Office, 1955), 2.

4. For example, in 1949 the AEC claimed that the "large-scale availability of isotopes is by far the most important constructive benefit which has yet been realized from the development of

atomic energy." US Atomic Energy Commission, *Isotopes: A Three-Year Summary of Distribution With Extensive Bibliography* (Government Printing Office, 1949), 1. See also Alfred Q. Maisel, "Medical Dividend," *Collier's* 119 (May 3, 1947): 14, 43–44.

5. Nicolas Rasmussen, "The Mid-Century Biophysics Bubble: Hiroshima and the Biological Revolution in America, Revisited," *History of Science* 35 (1997): 245–293. On the particular political meanings of radioisotopes during the Cold War, see John Krige, "The Politics of Phosphorus-32: A Cold War Fable Based on Fact," *Historical Studies in the Physical and Biological Sciences* 36, no. 1 (2005): 71–91; Krige, "Atoms for Peace, Scientific Internationalism, and Scientific Intelligence," *Osiris* 21 (2006): 161–181.

6. On the political struggle over civilian versus military control of atomic energy, see Richard G. Hewlett and Oscar E. Anderson Jr., *The New World: A History of the United States Atomic Energy Commission*, volume 1: *1939–1946* (University of California Press, 1989 [1962]).

7. See the statement issued by the General Advisory Committee to correct "unwarranted optimism" about an atomic power industry, published in the Atomic Energy Commission's *Fourth Semiannual Report to Congress* (Government Printing Office, 1948), 43–46. The oft-cited assertion that nuclear energy would be "too cheap to meter" is from an address given on September 16, 1954, in New York, by Lewis Strauss (then chairman of the Atomic Energy Commission) to the National Association of Science Writers. President Eisenhower had revived the idea of a domestic nuclear energy industry, and the prospect began to gain traction.

8. Stuart M. Feffer, "Atoms, Cancer, and Politics: Supporting Atomic Science at the University of Chicago, 1944–1950," *Historical Studies in the Physical and Biological Sciences* 22, no. 2 (1992): 233–261.

9. See Ellen Leopold, *Under the Radar: Cancer and the Cold War* (Rutgers University Press, 2009).

10. Hans-Jörg Rheinberger, "Physics and Chemistry of Life: Commentary," in *The Science-Industry Nexus: History, Policy, Implications*, ed. Karl Grandin, Nina Wormbs, and Sven Widmalm (Science History Publications, 2004), on 224.

11. Alison Kraft, "Between Medicine and Industry: Medical Physics and the Rise of the Radioisotope, 1945–65," *Contemporary British History* 20 (2006): 1–35; Néstor Herran, "Spreading Nucleonics: The Isotope School at the Atomic Energy Research Establishment, 1951–67," *British Journal for the History of Science* 39 (2006): 569–86; Herran, "Isotope Networks: Training, Sales and Publications, 1946–1965," *Dynamis* 29 (2009): 285–306.

12. Arnold Kramish, *Atomic Energy in the Soviet Union* (Stanford University Press, 1959), chapter 14.

13. Peter Galison and Bruce Hevly, *Big Science: The Growth of Large-Scale Research* (Stanford University Press, 1992); Stuart W. Leslie, *The Cold War and American Science: The Military-Industrial-Academic Complex at MIT and Stanford* (Columbia University Press, 1993); Jessica Wang, *American Science in an Age of Anxiety: Scientists, Anticommunism, and the Cold War* (University of North Carolina Press, 1999).

14. George de Hevesy, "The Absorption and Translocation of Lead by Plants: A Contribution to the Application of the Method of Radioactive Indicators in the Investigation of the Change of Substance in Plants," *Biochemical Journal* 17 (1923): 439–445; Hevesy, "Historical Progress of the Isotopic Methodology and Its Influences on the Biological Sciences," *Minerva Nucleare* 1 (1957): 189–200.

15. Engelbert Broda, *Radioactive Isotopes in Biochemistry* (Elsevier, 1960; originally published as *Radioaktive Isotope in der Biochemie* (Franz Deuticke, 1958), 1. For an overview of these experiments, see Robert Fink, ed., *Biological Studies with Polonium, Radium and Plutonium* (McGraw-Hill, 1950). The Austrian chemist Broda appears to have been a spy—see Andrew Brown, "The Viennese Connection: Engelbert Broda, Alan Nunn May and Atomic Espionage," *Intelligence and National Security* 24 (2009): 173–193.

16. See Paul Early, "Use of Diagnostic Radionuclides in Medicine," *Health Physics* 69, no. 5 (1995): 649–661, on 650; Herrmann L. Blumgart and Otto C. Yens, "Studies on the Velocity of Blood Flow. I. The Method Utilized," *Journal of Clinical Investigation* 4 (1927): 1–13; Herrmann L. Blumgart and Soma Weiss, "Studies on the Velocity of Blood Flow. II. The Velocity of Blood Flow in Normal Resting Individuals, and a Critique of the Method Used," *Journal of Clinical Investigation* 4 (1927): 15–31.

17. Broda, *Radioactive Isotopes in Biochemistry*, 1. For an example of such early work, see Harvey A. Seil, Charles H. Viol, and M. A. Gordon, "Elimination of Soluble Radium Salts Taken Intravenously and Per Os," *New York Medical Journal* 101 (1915): 896–898.

18. Barton C. Hacker, *The Dragon's Tail: Radiation Safety in the Manhattan Project, 1942–1946* (University of California Press, 1987), chapter 1; Claudia Clark, *Radium Girls: Women and Industrial Health Reform, 1910–1935* (University of North Carolina Press, 1997).

19. Robert E. Kohler, "Rudolph Schoenheimer, Isotopic Tracers, and Biochemistry in the 1930's," *Historical Studies in the Physical Sciences* 8 (1977): 257–298.

20. Heavy isotopes had applications beyond biochemistry. On their development and their use in geochemistry, see Matthew Shindell's chapter in this volume.

21. Frederic L. Holmes, "Manometers, Tissue Slices, and Intermediary Metabolism," in *The Right Tools for the Job: At Work in Twentieth-Century Life Sciences*, ed. Adele Clarke and Joan Fujimura (Princeton University Press, 1992), 151.

22. See Frederic L. Holmes, "The Intake-Output Method of Quantification in Physiology," *Historical Studies in the Physical and Biological Sciences* 17, no. 2 (1987): 235–270.

23. H. A. Krebs, "Cyclic Processes in Living Matter," *Enzymologia* 12 (1947): 88–100; Kärin Nickelsen and Gerd Graßhoff, "Concepts from the Bench: Hans Krebs, Kurt Henseleit and the Urea Cycle," in *Going Amiss in Experimental Research*, ed. Giora Hon, Jutta Schickore, and Friedrich Steinle (Springer-Verlag, 2009).

24. Frederic Lawrence Holmes, *Between Biology and Medicine: The Formation of Intermediary Metabolism* (Office for History of Science and Technology, University of California at Berkeley, 1992), 77.

25. Schoenheimer and Rittenberg used deuterated fatty acids to trace the metabolism of cholesterol and ^{15}N-labeled amino acids to follow protein synthesis. Rudolph Schoenheimer and David Rittenberg, "The Application of Isotopes to the Study of Intermediary Metabolism," *Science* 87 (1938): 221–226; Mildred Cohn, "Some Early Tracer Experiments with Stable Isotopes." *Protein Science* 4 (1995): 2444–2447.

26. Rudolph Schoenheimer, *The Dynamic State of Body Constituents* (Harvard University Press, 1942).

27. Harmke Kamminga and Mark Weatherall, "The Making of a Biochemist. I: Frederick Gowland Hopkins' Construction of Dynamic Biochemistry," *Medical History* 40 (1996): 269–292.

28. M. Stanley Livingston, "Early History of Particle Accelerators," *Advances in Electronics and Electron Physics* 50 (1980): 1–88, on 32; F. Joliot and I. Curie, "Artificial Production of a New Kind of Radio-Element," *Nature* 133 (1934): 201–202; Enrico Fermi, "Radioactivity Induced by Neutron Bombardment," *Nature* 133 (1934): 757. On the development of the Radiation Laboratory, see J. L. Heilbron and Robert W. Seidel, *Lawrence and His Laboratory: A History of the Lawrence Berkeley Laboratory* (University of California Press, 1989).

29. In fact, as Heilbron and Seidel note, the salt was actually free, donated by the Myles Salt Company of Louisiana (*Lawrence and His Laboratory*, 189).

30. Science Service release, December 23, 1936, as quoted in Heilbron and Seidel, *Lawrence and His Laboratory*, 190.

31. Herbert Childs to John Lawrence, 24 June 1966, John Hundale Lawrence papers, Bancroft Library, University of California, Berkeley, 87/86c, carton 10, folder 23.

32. Joseph G. Hamilton and Robert S. Stone, "The Intravenous and Intraduodenal Administration of Radio-Sodium," *Radiology* 28 (1937): 178–188. John Lawrence referred to this as a "stunt," because Hamilton had to know that it could have no therapeutic benefit. John H. Lawrence to Herbert Childs, 13 July 1966, John Hundale Lawrence Papers, Bancroft Library, University of California, Berkeley, 87/86c, carton 10, folder 23. See David S. Jones and Robert L. Martensen, "Human Radiation Experiments and the Formation of Medical Physics at the University of California, San Francisco and Berkeley, 1937–1962," in *Useful Bodies: Humans in the Service of Medical Science in the Twentieth Century*, ed. Jordan Goodman, Anthony McElligot, and Lara Marks (Johns Hopkins University Press, 2003).

33. One subject received 2,000 microcuries before Hamilton recognized that 5–10 percent of that amount of radioactivity gave results just as satisfactory. Joseph G. Hamilton, "The Rates of Absorption of Radio-Sodium in Normal Human Subjects," *Proceedings of the National Academy of Sciences* 23 (1937): 521–527.

34. Ibid., 523.

35. Heilbron and Seidel, *Lawrence and His Laboratory*, 190–191.

36. Ibid., 396.

37. O. Chiewitz and G. Hevesy, "Radioactive Indicators in the Study of Phosphorus Metabolism in Rats." *Nature* 136 (1935): 754–755. Chiewitz and Hevesy were in Copenhagen collaborating with Niels Bohr, who had been given a radium-beryllium source for his fiftieth birthday in 1935. "Within days," according to Marshall Brucer, "Hevesy was making P-32 for tracer studies in animals, but he could not make enough radioactivity to detect biological effects." *A Chronology of Nuclear Medicine 1600–1989* (Heritage Publications, 1990), 215. Hevesy's continuing work applying radioactive tracers (mostly in biomedical research) figured in his being awarded the Nobel Prize in 1944, twenty years after he was first nominated for his contributions to chemistry. See Gabor Pallo, "Scientific Recency: George de Hevesy's Nobel Prize," in *Historical Studies in the Nobel Archives: The Prizes in Science and Medicine*, ed. Elisabeth Crawford (Universal Academy Press, Tokyo, 2002).

38. K. G. Scott and S. F. Cook, "The Effect of Radioactive Phosphorus upon the Blood of Growing Chicks," *Proceedings of the National Academy of Sciences* 23 (1937): 265–272.

39. John H. Lawrence and K. G. Scott, "Comparative Metabolism of Phosphorus in Normal and Lymphomatous Animals," *Proceedings of the Society for Experimental Biology and Medicine* 40 (1939): 694–696.

40. John Lawrence first administered phosphorus-32 to a patient with chronic lymphatic leukemia in 1937, and he gave it to a woman suffering from polycythemia vera in 1938. Only the second clinical use seemed to be effective. John Lawrence to Edward B. Silberstein, 13 October 1978, Nuclear Medicine R&D Technical Documents, 434-92-66, ARO-2225, Lawrence Berkeley National Laboratory Archives and Records Office, 1 Cyclotron Rd. MS: 69R0102, Berkeley, California 94720, box 2, folder S; John H. Lawrence, "Early Experiences in Nuclear Medicine," *Journal of Nuclear Medicine* 20 (1979): 561–564. See also Jones and Martensen, "Human Radiation Experiments."

41. They used mice with several kinds of transplanted tumors: mammary carcinoma, lymphoma, lyphosarcoma, and sarcoma 180. H. B. Jones, I. L. Chaikoff, and John H. Lawrence, "Radioactive Phosphorus as an Indicator of Phospholipid Metabolism. VI. The Phospholipid Metabolism of Neoplastic Tissues (Mammary Carcinoma, Lymphoma, Lymphosarcoma, Sarcoma 180)," *Journal of Biological Chemistry* 128 (1939): 631–644.

42. I. Perlman, S. Ruben, and I. L. Chaikoff, "Radioactive Phosphorus as an Indicator of Phospholipid Metabolism," *Journal of Biological Chemistry* 122 (1937): 169–182.

43. Waldo E. Cohn and David M. Greenberg, "Studies in Mineral Metabolism with the Aid of Artificial Radioactive Isotopes. I. Absorption, Distribution, and Excretion of Phosphorus," *Journal of Biological Chemistry* 123 (1938): 185–198.

44. Leslie L. Bennett, "I. L. Chaikoff, Biochemical Physiologist, and His Students," *Perspectives in Biology and Medicine* 30 (1987): 362–383, on 367.

45. D. I. Arnon, P. R. Stout, and F. Sipos, "Radioactive Phosphorus as an Indicator of Phosphorus Absorption of Tomato Fruits at Various Stages of Development," *American Journal of Botany* 27 (1940): 791–798.

46. S. Hertz, A. Roberts, and Robley D. Evans, "Radioactive Iodine as an Indicator in the Study of Thyroid Physiology," *Proceedings of the Society of Experimental Biology and Medicine* 38 (1938): 510–513.

47. Ibid., 513.

48. J. J. Livingood and G. T. Seaborg, "Radioactive Isotopes of Iodine," *Physical Review* 54 (1938): 775–782.

49. Joseph G. Hamilton and Mayo H. Soley, "Studies in Iodine Metabolism by the Use of a New Radioactive Isotope of Iodine," *American Journal of Physiology* 127 (1939): 557–572; Joseph G. Hamilton and Mayo H. Soley, "Studies in Iodine Metabolism of the Thyroid Gland in Situ by the Use of Radio-Iodine in Normal Subjects and Patients with Various Types of Goiter." *American Journal of Physiology* 131 (1940): 135–143; Heilbron and Seidel, *Lawrence and His Laboratory*, 396–398.

50. The first two reports of effective treatment of hyperthyroidism by iodine-131 appeared side by side as abstracts for the 34th Annual Meeting of the American Society for Clinical Investigation, held in 1942: Joseph G. Hamilton and John H. Lawrence, "Recent Clinical Developments in the Therapeutic Application of Radio-Phosphorus and Radio-Iodine," *Journal of Clinical Investigation* 21 (1942): 624; Saul Hertz and A. Roberts, "Application of Radioactive Iodine in Therapy of Graves' Disease," *Journal of Clinical Investigation* 21 (1942): 624.

51. Joseph G. Hamilton, "The Use of Radioactive Tracers in Biology and Medicine," *Radiology* 39 (1942): 541–572, on 556; Paul C. Aebersold, "The Development of Nuclear Medicine," *American Journal of Roentgenology, Radium Therapy and Nuclear Medicine* 75 (1956): 1027–1039, on 1030. The original papers are Albert S. Keston, Robert P. Ball, V. Kneeland Frantz, and Walter W. Palmer, "Storage of Radioactive Iodine in a Metastasis from Thyroid Carcinoma," *Science* 95 (1942): 362–363; L. D. Marinelli, F. W. Foote, R. F. Hill, and A. F. Hocker, "Retention of Radioactive Iodine in Thyroid Carcinomas; Histopathologic and Radio-Autographic Studies," *American Journal of Roentgenology and Radium Therapy* 58 (1947): 17–32.

52. "History of the University of California Radiation Laboratory," Hardin B. Jones Papers, Bancroft Library, University of California, Berkeley, 79/112c, box 2, folder UCB—Lawrence Berkeley Lab, History, typescript p. 36; Martin D. Kamen, *Radiant Science, Dark Politics: A Memoir of the Nuclear Age* (University of California Press, 1985), 140–141.

53. "History of the University of California Radiation Laboratory," 35.

54. Advisory Committee, *The Human Radiation Experiments: Final Report of the President's Advisory Committee, Supplemental Volume 1* (Government Printing Office, 1995).

55. Ibid.

56. See Robert S. Stone, *Industrial Medicine on the Plutonium Project: Survey and Collected Papers* (McGraw-Hill, 1951).

57. The variety of isotopes was also an issue. According to Stannard, uranium fission produces more than 200 isotopes of 34 elements. About 60 radioisotopes are primary reaction products. J. Newell Stannard, *Radioactivity and Health: A History* (Pacific Northwest Laboratory, 1988), 299.

58. Hacker, *Dragon's Tail*, 43.

59. Hamilton, "Radioactive Tracers in Biology and Medicine," 566.

60. For a reproduction of a table from 1943 summarizing the early fission product studies, see Stannard, *Radioactivity and Health*, 305. Hamilton was also interested in the use of fission products in radiological warfare.

61. Joseph G. Hamilton, "A Report of the Past, Present, and Future Research Activities for Project 48-A-1" [c. 1948], Donald Cooksey Files Administrative (Director's Office), Accession Number 434-90-20, ARO-1537, Lawrence Berkeley National Laboratory Archives and Records Office, 1 Cyclotron Rd. MS: 69R0102, Berkeley, California 94720, box 4, folder 49 Medical Physics J. H. Lawrence's Group, General.

62. Hacker, *The Dragon's Tail*, 63.

63. Advisory Committee, *Suppl. Vol. 1*, 605.

64. See the table of these subjects in Stannard, *Radioactivity and Health*, 352. The most thorough account of these patients is Eileen Welsome, *The Plutonium Files: America's Secret Medical Experiments in the Cold War* (Random House, 1999). Similar government-sponsored experiments with polonium and uranium are described in Advisory Committee, *The Human Radiation Experiments: Final Report of the President's Advisory Committee* (Oxford University Press, 1996), chapter 5.

65. Advisory Committee, *Final Report*, chapter 5.

66. Stafford L. Warren, Report of the 23–24 January 1947 Meeting of the Interim Medical Committee, US Atomic Energy Commission, Department of Energy Opennet Acc NV0727195, 8.

67. Hamilton, "A Report of the Past, Present, and Future Research Activities for Project 48-A-1" [c. 1948], 9. See also Joseph G. Hamilton to Colonel E. B. Kelly, Subject: Summary of Research Program for Contract W-7405-eng-48-A, John Hundale Lawrence Papers, Bancroft Library, 87/86c, Film 2005, series 3, reel 5, 5:30, folder Correspondence H 1946; Jones and Martensen, "Human Radiation Experiments," 93–96.

68. Carroll L. Wilson to Stafford L. Warren, 30 April 1947, reproduced in Advisory Committee, *Supp. Vol. 1*, 71–72.

69. The topic of human experimentation has spawned a vast historical, ethical, and legal literature, which is well beyond the purview of this chapter. For some assessments of this historiography as it pertains to radiation studies, see Jonathan Moreno, *Undue Risk: Secret State Experiments*

on Humans (Freeman, 2000); Robert N. Proctor, "Human Experimental Abuse, in and out of Context," in *Science, History and Social Activism: A Tribute to Everett Mendelsohn*, ed. Garland E. Allen and Roy M. MacLeod (Kluwer, 2001); Gerald Kutcher, *Contested Medicine: Cancer Research and the Military* (University of Chicago Press, 2009).

70. S. Ruben and M. D. Kamen, "Radioactive Carbon in the Study of Respiration in Heterotrophic Systems," *Proceedings of the National Academy of Sciences* 26 (1940): 418–422. For a retrospective account, see Martin D. Kamen, "A Cupful of Luck, a Pinch of Sagacity," *Annual Review of Biochemistry* 55 (1986): 1–34.

71. See Doris T. Zallen, "Redrawing the Boundaries of Molecular Biology: The Case of Photosynthesis," *Journal of the History of Biology* 26 (1993): 65–87, on 71.

72. S. Ruben, W. Z. Hassid, and M. D. Kamen, "Radioactive Carbon in the Study of Photosynthesis," *Journal of the American Chemical Society* 61 (1939): 661–663.

73. Kamen, *Radiant Science, Dark Politics*, 165.

74. See Martin D. Kamen, "Early History of Carbon-14," *Science* 140 (1963): 584–590. The first publication of an experiment using carbon-14 as a biological tracer was H. A. Barker and M. D. Kamen, "Carbon Dioxide Utilization in the Synthesis of Acetic Acid by *Clostridium thermoaceticum*," *Proceedings of the National Academy of Sciences* 31 (1945): 219–225.

75. W. F. Libby, "The Radiocarbon Story," *Bulletin of the Atomic Scientists* 4, no. 9 (1948): 263–266.

76. See Kamen, *Radiant Science, Dark Politics*, 164, 168. For more on the problems of scientists with security clearances, see Wang, *American Science in an Age of Anxiety*; Naomi Oreskes and Ronald Rainger, "Science and Security Before the Atomic Bomb: The Loyalty Case of Harald U. Sverdrup," *Studies in History and Philosophy of Modern Physics* 31B (2000): 309–369.

77. Melvin Calvin, *Following the Trail of Light* (American Chemical Society, 1992), 51. According to Calvin, it wasn't being on the same faculty but, rather, working on uranium-plutonium fission product extraction for the Met Lab that brought him into close contact with Lawrence.

78. Ibid., 53; Glenn T. Seaborg and Andrew A. Benson, "Melvin Calvin, April 8, 1911–January 8, 1997," *Biographical Memoirs, National Academy of Sciences* 75 (1998): 3–21, on 9.

79. In the AEC bibliographies these are reported (undated) as A. Benson and M. Calvin, "Dark Reductions of Photosynthesis," MDDC 1027; S. Aronoff, A. Benson, W. Z. Hassid, and M. Calvin, "Distribution of C14 in Photosynthesizing Barley Seedlings," MDDC 965, also published in *Science* (see next note); S. Aronoff, H. A. Barker, and M. Calvin, "Distribution of Labeled Carbon in Sugar from Barley," MDDC 966; and S. Aronoff and M. Calvin, "Phosphorus Turnover and Photosynthesis," MDDC 1589.

80. It was AEC Contract W-7405-Eng-48. See S. Aronoff, A. Benson, W. Z. Hassid, and M. Calvin, "Distribution of C^{14} in Photosynthesizing Barley Seedlings," *Science* 105 (1947): 664–665, note 1.

81. Freeman Dyson, letter to family from Berkeley, August 26, 1948, Dyson personal papers.

82. For example, the AEC's fourth semiannual report to Congress featured isotopes, and listed photosynthesis as the first example of how tracers enable scientists "to follow in intimate detail nature's fundamental processes." US AEC, *Fourth Semiannual Report to Congress*, 5.

83. Joseph S. Fruton, *Molecules and Life: Historical Essays on the Interplay of Chemistry and Biology* (Wiley-Interscience, 1972), 446.

84. US AEC, *Fourth Semiannual Report to Congress*, 10.

85. John H. Lawrence and Cornelius A. Tobias, "Radioactive Isotopes and Nuclear Radiations in the Treatment of Cancer," *Cancer Research* 16 (1956): 185–193.

86. US AEC, *Fourth Semiannual Report to Congress*, 23.

87. Ibid.

88. On the thousandfold difference in radioactivity exposure between these kinds of applications, see Human Radiation Studies: Remembering the Early Years, Oral History of Biochemist Waldo E. Cohn, PhD, conducted January 18, 1995 through the Department of Energy by Thomas Fisher Jr. and Michael Yuffee, published at http://tis.eh.doe.gov.

89. See Martin D. Kamen, *Radioactive Tracers in Biology: An Introduction to Tracer Methodology* (Academic Press, 1951), 122. Biologists tended to use radioisotopes either as tracers or as sources of function-perturbing radiation, rarely as both.

90. Aebersold, "The Development of Nuclear Medicine," 1031.

91. Ibid., 1032.

92. Ibid., 1031.

93. Early, "Use of Diagnostic Radionuclides in Medicine," 651; B. Selverstone, A. K. Solomon, and W. H. Sweet, "Location of Brain Tumors by Means of Radioactive Phosphorus," *Journal of the American Medical Association* 140 (1949): 277–288; William H. Sweet, "The Use of Nuclear Disintegration in the Diagnosis and Treatment of Brain Tumor," *New England Journal of Medicine* 245 (1951): 875–878.

94. This assertion, made at a meeting in September 1953, was first published two years later. William H. Sweet and Gordon L. Brownell, "The Use of Radioactive Isotopes in the Detection and Localization of Brain Tumors," *Radioisotopes in Medicine*, ed. Gould A. Andrews, Marshall Brucer, and Elizabeth B. Anderson (Atomic Energy Commission, 1955), 211–218, on 211.

95. Sweet and Brownell, "The Use of Radioactive Isotopes in the Detection and Localization of Brain Tumors," p. 214. Sweet was subsequently involved with the AEC in uranium-injection experiments of patients with brain tumors; for details, see Advisory Committee, *Final Report*, chapter 5; Gilbert Whittemore and Miriam Boleyn-Fitzgerald, "Injecting Comatose Patients with Uranium: America's Overlapping Wars against Communism and Cancer in the 1950s," in *Useful Bodies: Humans in the Service of Medical Science in the Twentieth Century*, ed. Jordan Goodman, Anthony McElligot, and Lara Marks (Johns Hopkins University Press, 2003), 165–189.

96. Early, "Use of Diagnostic Radionuclides in Medicine," 652.

97. Ibid., 654.

98. The letter m at the end of this isotope designation means "metastable."

99. E.g., August Miale Jr., "Nuclear Medicine: Reflections in Time," *Journal of the Florida Medical Association* 82, no. 11 (1995): 749–750.

100. G. Evelyn Hutchinson and Vaughan T. Bowen, "A Direct Demonstration of the Phosphorus Cycle in a Small Lake," *Proceedings of the National Academy of Sciences* 33 (1947): 148–153. The initial radiophosphorus for this study was from the Yale cyclotron and was provided by the physicist Ernest F. Pollard.

101. Letter from Paul Aebersold, Isotopes Branch, to G. E. Hutchinson, 8 May 1947, AEC Records, National Archives Southeast Region—Atlanta, RG 326, MED CEW General Research Correspondence 1941–1948, Acc 67B0803, box 178, folder AEC 441.2 (R—Yale Univ.). A copy of Hutchinson's radioisotope license application, signed November 19, 1946, is in the AEC Records for the National Archives Southeast Region—Atlanta, RG 326, OROO Files for K-25, X-10, Y-25, Acc 671309, box 14, Certificates.

102. G. Evelyn Hutchinson and Vaughan T. Bowen, "Limnological Studies in Connecticut. IX. A Quantitative Radiochemical Study of the Phosphorus Cycle in Linsley Pond," *Ecology* 31 (1950): 194–203.

103. G. E. Hutchinson, "Bio-Ecology," *Ecology* 21 (1940): 267–268, on 268; Joel B. Hagen, *An Entangled Bank: The Origins of Ecosystem Ecology* (Rutgers University Press, 1992), chapter 4.

104. G. E. Hutchinson, "Limnological Studies in Connecticut. IV. The Mechanisms of Intermediary Metabolism in Stratified Lakes," *Ecological Monographs* 11 (1941): 21–60.

105. A. G. Tansley, "The Use and Abuse of Vegetational Concepts and Terms," *Ecology* 16 (1935): 284–307; Hagen, *Entangled Bank*, chapter 5.

106. R. L. Lindemann, "The Trophic-Dynamic Aspect of Ecology," *Ecology* 23 (1942): 399–418; E. P. Odum, *Fundamentals of Ecology* (Saunders, 1953).

107. For a retrospective assessment, see S. I. Auerbach, "Radionuclide Cycling: Current Status and Future Needs," *Health Physics* 11 (1965): 1355–1361.

108. F. W. Whicker and V. Schultz, "Introduction and Historical Perspective," in *Radioecology: Nuclear Energy and the Environment* (CRC Press, 1982), 4. For more on the origins and early research of the Applied Fisheries Laboratory, see Neal O. Hines, *Proving Ground: An Account of the Radiobiological Studies in the Pacific, 1946–1961* (University of Washington Press, 1962); Matthew W. Klingle, "Plying Atomic Waters: Lauren Donaldson and the 'Fern Lake Concept' of Fisheries Management," *Journal of the History of Biology* 31 (1998): 1–32.

109. See "What Science Learned at Bikini: Latest Report on the Results," *Life*, August 11, 1947: 74–87. The "Conclusions," authored by Stafford Warren, are titled "Tests Proved Irresistible

Spread of Radioactivity." On the value of these studies to oceanography, see Ronald Rainger, "'A Wonderful Oceanographic Tool': The Atomic Bomb, Radioactivity and the Development of American Oceanography," in *The Machine in Neptune's Garden: Historical Perspectives on Technology and the Marine Environment*, ed. Helen M. Rozwadowski and David K. van Keuren (Science History Publications, 2004).

110. Michele Stenehjem Gerber, *On the Home Front: The Cold War Legacy of the Hanford Nuclear Site* (University of Nebraska Press, [1992] 2002).

111. Hines, *Proving Ground*, 17.

112. Foster, "Some Effects of Pile Area Effluent Water on Young Chinook Salmon and Steelhead Trout," August 31, 1946, US AEC Report HW-7-4759, Hanford Engineer Works, Opennet Doc. No. NV0717097, 2; L. Donaldson, "Program of Fisheries Experiment for the Hanford Field Laboratory," July 1945, DUH-7287, OpenNet Acc. No. RL-1-336129. It should be noted that if the effluent wasn't diluted, it was highly toxic to the fish, but the scientists calculated that the dilution factor in the river was at least 1:100. Foster studied the effects on fish of placing them in water with a variety of dilutions with the effluent, from 1:3 to 1:1000.

113. Eugene P. Odum, *Fundamentals of Ecology*, second edition (Saunders, 1959), 467. As Foster observes, an aluminum jacket surrounding the fuel elements kept the cooling water from making direct contact with the uranium rods. R. F. Foster, "The History of Hanford and Its Contribution of Radionuclides to the Columbia River," in *The Columbia River Estuary and Adjacent Ocean Waters: Bioenvironmental Studies*, ed. A. T. Pruter and D. L. Alverson (University of Washington Press, 1972), 13.

114. These results were reported in classified documents, but were first published in the mid 1950s as part of the Atoms for Peace initiative. R. F. Foster and J. J. Davis, "The Accumulation of Radioactive Substances in Aquatic Forms," *Proceedings of the International Conference on the Peaceful Uses of Atomic Energy* 13 (1955): 364–367; W. C. Hanson and H. A. Kornberg, "Radioactivity in Terrestrial Animals Near an Atomic Energy Site," *Proceedings of the International Conference on the Peaceful Uses of Atomic Energy* 13 (1955): 385–388; J. J. Davis and R. F. Foster, "Bioaccumulation of Radioisotopes through Aquatic Food Chains," *Ecology* 39 (1958): 530–535; J. J. Davis, R. W. Perkins, R. F. Palmer, W. C. Hanson and J. F. Cline, "Radioactive Materials in Aquatic and Terrestrial Organisms Exposed to Reactor Effluent Water," in *Proceedings of the Second United Nations International Conference on the Peaceful Uses of Atomic Energy* 18 (1958): 421–428.

115. Davis and Foster, "Bioaccumulation of Radioisotopes through Aquatic Food Chains," 531. In a slightly earlier publication, Hanford scientists similarly stated "No effect from the small amounts of radioactivity present has been detected." Richard F. Foster and Royal E. Rostenbach, "Distribution of Radioisotopes in Columbia River," *Journal of the American Water Works Association* 46 (1954): 633–640, on 635. This same point comes through clearly in "Hanford Science Forum," a television broadcast (and sponsored by Hanford's contractor, General Electric), which featured an interview with Foster on the work of the Aquatic Biology Operations in a 1957 program. The interviewer introduced the venture as a special kind of "fishing" in the Columbia River. The telecast is available at http://www.archive.org/details/HanfordS1957.

116. Odum, *Fundamentals of Ecology*, second edition, 467.

117. Stephen Bocking, *Ecologists and Environmental Politics: A History of Contemporary Ecology* (Yale University Press, 1997), 65–68.

118. Ibid., 68.

119. Ibid., 69.

120. Leland Johnson and Daniel Schaffer, *Oak Ridge National Laboratory: The First Fifty Years* (University of Tennessee Press, 1994), 99–100.

121. S. I. Auerbach, J. S. Olson, and H. D. Waller, "Landscape Investigations Using Caesium-137," *Nature* 201 (1964): 761–764, on 761.

122. Stannard, *Radioactivity and Health*, 771.

123. Hagen, *Entangled Bank*, 112–115.

124. Chunglin Kwa, "Radiation Ecology, Systems Ecology and the Management of the Environment," in *Science and Nature: Essays in the History of the Environmental Sciences*, ed. Michael Shortland (Alden Press for British Society for the History of Science, 1993), 235–236; Bocking, *Ecologists and Environmental Politics*, 78–82; Jerry S. Olson, "Analog Computer Models for Movement of Nuclides through Ecosystems," in *Radioecology: Proceedings of the First National Symposium on Radioecology Held at Colorado State University, Fort Collins, Colorado, September 10–15, 1961* (Reinhold and American Institute of Biological Sciences, 1963) [hereafter *Radioecology*], 121–126. Olson used the National Laboratory Analog Computer Facility at Oak Ridge (Kwa, "Radiation Ecology," 243). The computer modeling he set in motion was further developed through the International Biological Program in the 1960s and the 1970s. See David C. Coleman, *Big Science: The Emergence of Ecosystem Science* (University of California Press, 2010), chapter 2.

125. Auerbach, Olson, and Waller, "Landscape Investigations."

126. The on-site laboratory was approved in 1960 and completed in 1961. Eugene P. Odum, "Early University of Georgia Research, 1952–1962," in *The Savannah River and Its Environs: Proceedings of a Symposium in Honor of Dr. Ruth Patrick for 35 Years of Study on the Savannah River* (E. I. du Pont de Nemours & Co. Savannah River Laboratory), 43–57; Kwa, "Radiation Ecology," 227–229; Eugene P. Odum, "Organic Production and Turnover in Old Field Succession," *Ecology* 41 (1960): 34–49.

127. Kwa, "Radiation Ecology," 230.

128. See Eugene P. Odum and Frank B. Golley, "Radioactive Tracers as an Aid to the Measurement of Energy Flow at the Population Level in Nature," in *Radioecology*, 403–410.

129. Eugene P. Odum and Edward J. Kuenzler, "Experimental Isolation of Food Chains in an Old-Field Ecosystem with the Use of Phosphorus-32," in *Radioecology*, 113–120.

130. Ibid., 118.

131. Ibid., 119.

132. Richard G. Wiegert and Eugene P. Odum, "Radionuclide Tracer Measurements of Food Web Diversity in Nature," in *Symposium on Radioecology: Proceedings of the Second National Symposium, Ann Arbor, Michigan, May 15–17, 1967*, ed. Daniel J. Nelson and Francis C. Evans (Clearinghouse for Federal Scientific and Technical Information, 1969), 710.

133. Stanley I. Auerbach, *A History of the Environmental Sciences Division of Oak Ridge National Laboratory* (Oak Ridge National Laboratory, 1993), 21.

134. *Radioecological Concentration Processes: Proceedings of an International Symposium held in Stockholm, 25–29 April, 1966* (Pergamon, 1967).

135. In addition to the symposium volumes from 1961 and 1967 cited above, see *Radionuclides in Ecosystems: Proceedings of the Third Symposium on Radioecology, May 10–12, 1971, Oak Ridge Tennessee* (National Technical Information Service, 1971).

136. Rachel Carson, *Silent Spring* (Houghton Mifflin, 1962), 6. See also Ralph H. Lutts, "Chemical Fallout: Rachel Carson's *Silent Spring*, Radioactive Fallout, and the Environmental Movement," *Environmental Review* 9 (1985): 210–225.

137. George M. Woodwell, "Toxic Substances and Ecological Cycles," *Scientific American* 216, no. 3 (1967): 24–31.

138. The concentration went from a dilution of 0.00005 ppm in water to 75.5 ppm in an immature ring-billed gull. G. M. Woodwell, C. F. Wurster, and P. A. Isaacson, "DDT Residues in an East Coast Estuary: A Case of Biological Concentration of a Persistent Insecticide," *Science* 156 (1967): 821–824. Odum reproduced some of the data in the report cited above in a figure on page 74 of the third edition of his textbook *Fundamentals of Ecology* (Saunders, 1971). On Woodwell's work at Brookhaven on forest ecosystems, see George M. Woodwell, "Effects of Ionizing Radiation on Terrestrial Ecosystems," *Science* 138 (1962): 572–577.

139. George M. Woodwell, "BRAVO Plus 25 Years," in *Environmental Sciences Laboratory Dedication: Daniel J. Nelson Auditorium, Feb. 26–27, 1979* (Oak Ridge National Laboratory, 1980), 62.

140. Whicker and Schultz, "Introduction and Historical Perspective," 2.

141. The major focus of discussions of biology in the USSR has been the effects of Lysenkoism. See David Joravsky, *The Lysenko Affair* (University of Chicago Press, 1970); Nikolai Krementsov, *Stalinist Science* (Princeton University Press, 1997). On how this ideological divide shaped genetics in the US, see Jan Sapp, *Beyond the Gene: Cytoplasmic Inheritance and the Struggle for Authority in Genetics* (Oxford University Press, 1987).

142. M. Susan Lindee, *Suffering Made Real: American Science and the Survivors at Hiroshima* (University of Chicago Press, 1994).

143. Lily E. Kay, *Who Wrote the Book of Life? A History of the Genetic Code* (Stanford University Press, 2000).

144. Peter Galison, "The Ontology of the Enemy: Norbert Weiner and the Cybernetic Vision," *Critical Inquiry* 21/1 (1994): 228–266; Paul N. Edwards, *The Closed World: Computers and the Poli-*

tics of Discourse in Cold War America (MIT Press, 1996). The role of defense-related funding of the physics and engineering has been a long-standing though contested theme in the historiography of Cold War science. For contending perspectives on this issue, see Paul Forman, "Behind Quantum Electronics: National Security as Basis for Physical Research in the United States, 1940–1960," *Historical Studies in the Physical and Biological Sciences* 18 (1987): 149–229; Daniel J. Kevles, "Cold War and Hot Physics: Science, Security, and the American State, 1945–1956," *Historical Studies in the Physical and Biological Sciences* 20 (1990): 239–264. For an assessment and another approach to Cold War science, see Naomi Oreskes, "A Context of Motivation: US Navy Oceanographic Research and the Discovery of Sea-Floor Hydrothermal Vents," *Social Studies of Science* 33 (2003): 697–642. I am merely sampling an extensive relevant historiography.

145. Angela N. H. Creager, "Tracing the Politics of Changing Postwar Research Practices: The Export of 'American' Radioisotopes to European Biologists," *Studies in History and Philosophy of the Biological and Biomedical Sciences* 33C (2002): 367–388; Krige, "Politics of Phosphorus-32"; Creager, "Radioisotopes as Political Instruments, 1946–1953," *Dynamis* 29 (2009): 219–239.

146. Telegram from President Truman to E. V. Cowdry, President of the Fourth International Cancer Research Congress, September 3, 1947, reprinted with US A.E.C. Press Release for September 4, 1947, "United States Atomic Energy Commission Announces First Shipment of Radioisotopes to a Foreign Country," AEC Records, National Archives, College Park, RG 326, E67A, box 47, folder 6 Foreign Distribution of Radioisotopes, volume 2. The first draft of this announcement was penned by Paul Aebersold, head of the Isotopes Branch in Oak Ridge. It was Aebersold who pointed out that this international conference would provide "an unexcelled opportunity for a highly favorable manner of public announcement." Paul C. Aebersold, Announcement of Foreign Distribution of Isotopes, August 20, 1947, AEC Records, National Archives, Atlanta, RG 326, MED CEW Gen Res Corr, Acc 67B0803, box 143, folder AEC 441.2 (R—Foreign), June 1–September 30, 1947.

147. See Krige, "Atoms for Peace." For a more extended analysis of the role of science and technology in US foreign policy during the early Cold War, see John Krige, *American Hegemony and the Postwar Reconstruction of Science in Europe* (MIT Press, 2006).

148. See *Atoms for Peace: Dwight D. Eisenhower's Address to the United Nations* (National Archives and Records Administration, 1990).

149. Press release for August 3, 1949, "AEC Distributes 8,363 Shipments of Radioactive and Stable Isotopes in Three Years," copy in AEC papers, National Archives, College Park, RG 326, E67A, box 45, folder 13, Distribution of Stable Isotopes Domestic. In 1952 the program was modified such that users paid 20 percent of production costs for radioisotopes used in the treatment, diagnosis, and study of cancer. US Atomic Energy Commission, *Twelfth Semiannual Report to Congress* (Government Printing Office, 1952), 32.

150. This number is based on published bibliographies: US AEC, *Isotopes: A Three-Year Summary*; idem., *Isotopes: A Five-Year Summary of Distribution with Bibliography* (Government Printing Office, 1951); US AEC, *Eight-Year Isotope Summary*.

151. For a good overview of the fallout debates, see J. Christopher Jolly, Thresholds of Uncertainty: Radiation and Responsibility in the Fallout Controversy, PhD dissertation, Oregon State University, 2003.

152. George T. Mazuzan and J. Samuel Walker, *Controlling the Atom: The Beginnings of Nuclear Regulation, 1946–1962* (University of California Press, 1985), chapter 12.

153. See Steven L. Del Sesto, *Science, Politics, and Controversy: Civilian Nuclear Power in the United States, 1946–1974* (Westview, 1979); J. Samuel Walker, *Containing the Atom: Nuclear Regulation in a Changing Environment, 1963–1971* (University of California Press, 1992).

154. Some radioisotopes continued to be produced in other Oak Ridge reactors. For a list of isotope-production reactors in the mid 1960s, see P. S. Baker, "Reactor-Produced Radionuclides," in *Radioactive Pharmaceuticals: Proceedings of a Symposium Held at the Oak Ridge Institute of Nuclear Studies as an Operating Unit of Oak Ridge Associated Universities, November 1–4, 1965*, ed. Gould A. Andrews, Ralph M. Kniseley, and Henry N. Wagner Jr. (US Atomic Energy Commission Division of Technical Information, 1966).

3 Self-Reliant Science: The Impact of the Cold War on Science in Socialist China

Sigrid Schmalzer

At first blush, Chinese science during the Cold War appears to reflect the same move toward "gadgeteering" that Paul Forman has documented in US physics.[1] After the communist revolution of 1949, many Chinese scientists who had previously pursued research in basic science began working instead on topics with immediate and direct potential applications. Entomologists shifted their focus from insect classification to insect control.[2] Physicists turned from research on theoretical questions to focus on developing China's weapons program.[3] When the political winds blew just right, influential scientists did manage to secure for basic science some level of state support, without which such research would have been impossible not only financially but also politically.[4] And some research areas had little hope for application but for ideological reasons nonetheless retained political favor even during the most anti-intellectual periods.[5] Overall, however, the move toward applied science in post-1949 China appears to be beyond dispute. Can we then say that the Cold War transformed science in China by causing a shift from basic to applied science? Only provisionally. More even than is generally the case in historical studies, the shift to applied science in China was profoundly overdetermined. Furthermore, "basic" and "applied" have a history that belies their deployment as naturalized categories. We know something of this history for the United States[6]; here I will discuss how it unfolded in socialist-era China, where what counted as "science" was even more subject to reinterpretation.

The relationship between basic and applied science emerged as an important concern in the revolutionary era as communist forces in the rural base areas struggled to develop necessary industrial and agricultural resources for use in the anti-Japanese and civil wars.[7] However, discourse on this relationship cannot be disentangled from myriad other concerns of the day. The decision to emphasize applied science was thoroughly intertwined with other, mutually reinforcing priorities, including the celebration of native techniques, mobilization of the masses, loyalty of scientists to the party-state, and achievement of self-reliance. By October 1, 1949, when Mao declared victory in Beijing, applied science carried the cachet of eschewing the ivory tower and securing China's liberation from foreign domination and feudal tradition by harnessing the knowledge of China's peasant masses. To capture this cluster of concerns, I would like to shift our focus away from the basic/applied dichotomy that informs our understanding of Cold War US science and instead employ "self-reliant science" as the overarching category most relevant to the case of China during the Cold War.[8]

The definition of science found in the materials explored here may not fit Western scholars' assumptions about distinctions between science and technology. Indeed, science vs. technology was not nearly as important a contradiction in Mao-era discourse as were the contradictions between foreign and native, theory and practice, and, by the late 1950s, expert and red. I use the word 'science' as an actors' category—that is, as it appears in the Chinese sources under investigation. As we will see, "science" in Mao-era China came to include activities far removed from understandings of the word dominant in capitalist countries. Even the collection and application of manure could count as "scientific farming," and horse breeding gained the noble appellation "scientific experiment." At the same time, we should not assume that the celebration of such practical activities as "science" arose from a purely utilitarian ideology. Rather, self-reliant science encompassed both an emphasis on practices of direct benefit to production and a decidedly non-utilitarian embrace of science as an agent of cultural revolution, i.e., a force capable of liberating society from oppressive old ways of thinking.

Returning to the problem of overdetermination, the dominance of "self-reliant science" and its component parts cannot be explained solely through reference to geopolitical patterns: a quick series of counterfactual tests clearly demonstrates the limits of a Cold War explanation. Even in the absence of conflict between the United States and the Soviet Union, a focus on application would have been of obvious practical importance for China as an impoverished "developing" country. Here China could readily be compared to any other country that faced immediate economic needs and had embraced a development ideology, whether socialist or capitalist.[9] Moreover, ideology—significant everywhere—played a far more explicit role in shaping science policy in socialist-era China than in the United States or even the Soviet Union, which was by the 1950s more technocratic than revolutionary.[10] Applied science, mass mobilization, and related priorities would—Cold War or no Cold War—have carried ideological significance in China. Mao's influential essay "On Practice" would still have provided the needed inspiration (and intimidation) for scientists to frame their scientific work in practical terms.[11] At the same time, and perhaps even in the absence of Maoist ideology, China's experience suffering more than 100 years of imperialist aggression—from the first Opium War through the War to Resist America and Aid Korea—would still have offered more than sufficient nationalist ideological incentive to celebrate the virtues of self-reliance through the development of native technical resources.

All these qualifications aside, Cold War geopolitics undoubtedly intensified such emphases. This chapter will thus examine the Cold War's effects on Chinese science within a web of related historical themes stretching back before the 1949 revolution and with attention to China's peculiar position during the Cold War.[12] Specifically, it will show that China's relative isolation during certain periods of the Cold War intensified the emphasis on self-reliance in science. Moreover, and despite the actual importance of transnational influences (as aptly recounted in Zuoyue Wang's contribution to the volume), the power of this representation fostered a belief in a uniquely socialist-Chinese approach to science.[13] With roots in the pre-1949 revolutionary period, this idea crystallized in 1958 and became even more sharply articulated through the international exchanges of the 1970s as foreign scientists eager to bring home exotic epistemologies participated in the promotion of Chinese uniqueness. In at least a few cases, such claims to uniqueness went beyond shaping the way people talked about science to change the actual character of scientific knowledge produced in Cold War China. Because the emphasis on self-reliance arose from directives of the Party Center, we find references sprinkled regularly through the discourses of all scientific fields. For this reason, I will offer examples from a number of areas explored in the secondary literature (including medicine, nuclear science, and bio-chemistry) in addition to a more thorough exploration of one area (agricultural science) that relates to my own current research.

Alternative Time Lines

The Cold War in China did not follow the pattern suggested by the "Cold War I and Cold War II" scheme advanced by Fred Halliday and embraced by historians of science such as Paul Edwards and Peter Westwick.[14] To make sense of China's experience, it is necessary to take at least two other time lines into account. The first follows China's changing position relative to the major Cold War powers: In the 1940s, the Chinese communists had uneasy relationships with both the United States and the Soviet Union; the 1949 revolution ushered in a period of "Soviet learning" that began falling apart in the late 1950s; after the Sino-Soviet split (c. 1960), and escalating with the Vietnam War, China maintained hostile relations with both major powers; beginning in 1971, China and the United States began cultivating a "friendship," culminating in normalization of relations in 1979; and in 1989, the first visit to Beijing by a Soviet head of state in thirty years was disrupted by the Tiananmen Square protests. The second time line tracks China's internal political changes—especially the Great Leap Forward (1958–1960), the post-Leap retreat of Mao and other radicals, the Cultural Revolution (1966–1976), and the 1978 rise of Deng Xiaoping, who developed a program of "modernization" (which had long been sought by other moderates) along with the new proposition of "market socialism," and whose 1989 crackdown on democratic protest signaled that Communist Party control in China would far outlast the celebrated "end" of the Cold War.

However, placing China in the greater international context of a Cold War chronology does present an important opportunity: it may help to break China scholars of the habit of seeing everything through an internal Chinese political framework. Most critically for our purposes here, China scholars are not accustomed to thinking about science as a part of the radical politics of Mao-era China. Rather, the standard historical narrative follows a pendulum-like alternation between "radical" periods (the Great Leap and most of the Cultural Revolution) during which political struggle stifled intellectual pursuits and economic development, making science virtually impossible, and "moderate" (or technocratic) periods during which steadier minds (especially those of Zhou Enlai, Liu Shaoqi, and Deng Xiaoping) prevailed and more liberal policies rekindled the hopes of beleaguered scientists.[15]

David Zweig, for example, depicts Maoist "radical policies" on agriculture to have been "fueled by an anti-modernization mentality that saw economic development as the antithesis of revolution."[16] Formerly sympathetic to Maoism, Zweig became disillusioned after the death of Mao and the fall of the "Gang of Four," and turned to modernization and rational choice theories to explain what went wrong.[17] Earlier analyses of Mao-era agricultural policy framed the history differently, and so found a great deal of continuity across radical and moderate periods. Writing in 1973, Benedict Stavis marked 1960–1962 as the watershed moment when China embarked on a "technological transformation of agriculture" that he found to be still going strong when he visited China in the early 1970s.[18] We now know much more about 1960s and 1970s China than Stavis was able to see; nonetheless, his conceptual frame helps to make sense of the history of agricultural science in socialist China. Indeed, the move to develop "scientific farming" (*kexue zhongtian*) began around 1961, during the heyday of the "moderate" technocrats, came into its own amid the intensifying radical politics of 1965, flourished throughout the Cultural Revolution, and remains relevant even today.[19] The "green revolution"—so much a part of the United States' engagement in the Cold War—thus progressed along much the same time line in China as elsewhere, and it did so in the very middle of China's continually unfolding "red revolution."

In fact, Maoist radicals were deeply committed to modernization and science; they just defined these goals differently. The Cold War thus presented at least three competing development paradigms, constructed in conscious comparison and contrast with one another. The first was the Leninist model of state-led economic development, based on a specific reading of Marxist philosophy of history and social development. The attractiveness of this model among Third World nations alarmed many academic and political leaders in the United States, inspiring Walt Rostow's tremendously influential "non-communist manifesto" *The Stages of Economic Growth* (1959). The parallels between Leninism and US modernization theory are clear.[20] Both were committed to modernization through technological development, and both depended on deterministic expectations that development would proceed through specific "stages." Though Mao considered himself a Leninist, his economic and political program—and the self-reliant "mass science" that went with it—departed in dramatic ways from modernization as pursued in the Soviet Union. Frustrated with the bureaucratic and technocratic structures of authority that formed in China during the period of Soviet learning, and with the rigid expectations of "stages" that slowed China's progress toward communism, Mao sought in the Great Leap Forward to abandon the determinism of staged growth and instead embrace a voluntarist faith in the power of the masses to channel their collective revolutionary will into rapid achievement of a truly communist economy.

My argument here is that acts of comparison and contrast similarly served as causal forces in transforming scientific practice. The Cold War created an expectation of ideological difference that was supposed to permeate even science. We see this clearly in several of the other contributions to this volume—for example, in Elena Aronova's treatment of Soviet philosophy of science and George Reisch's analysis of McCarthyism and the Intelligent Design movement in the United States. In China, a specific approach to science based on a cluster of related values—self-reliance, application, mass mobilization, nativism—emerged in a context of perceived isolation from the superpowers and then gained strength through repeated acts of contrast with American and Soviet examples. In the context of the Cold War, Maoist "self-reliant science" was meant to bolster domestic confidence in Chinese socialist science and also to offer an alternative model for Third World countries.

Revolutionary Roots

China's approach to science during the Cold War owed much to the experiences of the Chinese Communist Party during the 1940s as it struggled to mobilize people in the revolutionary base areas to fight two wars: the War of Resistance against Japan and the Civil War against Chiang Kai-shek's Nationalist Party.[21] With the emerging leaders of the Cold War either outright supporting Chiang Kai-shek (in the case of the United States) or at least committed to a policy of non-aggression with him (in the case of the Soviet Union), Chinese communists determined that the only sure course lay in the development of indigenous resources—material, methodological, and human—to meet pressing economic and military needs. In the revolutionary "cradle" of Yan'an, the commitment to self-reliance, applied science, native methods, and mass mobilization became intertwined in ways that were to last throughout the Mao era (1949–1976).[22]

In 1939, Chinese communists responded to an economic blockade by launching a movement for self-reliance in industry and defense.[23] Scientific knowledge had an obvious and important role to play in developing the means to produce such material necessities as matches, soap, candles, and explosives. Despite the inevitable orientation toward practical applications that this situation implied, for several years the Party maintained a commitment to basic scientific knowledge. This changed in mid 1942 with the major political upheaval of the Party Rectification Movement. As Mao was consolidating his power through criticism of "bourgeois" intellectuals and Party officials associated with the Soviet Union, the scientific leadership also underwent a profound shift.

Xu Teli was the head of the Natural Science Institute in Yan'an. His approach was rooted in a belief that teaching and research in basic science formed a necessary foundation for the development of revolutionary China's science and economy. The commitment to following the masses and learning from practical experience that came with Rectification doomed Xu's program. The chairman of the biology department at the Natural Science Institute, Le Tianyu, had embraced an approach far more consistent with what was newly in vogue. His success in establishing a factory for producing beet sugar entirely with local beets and handmade equipment had already made him something of a "local hero."[24] During the Rectification Campaign, Le took advantage of the political wind to argue for his own work as the model that the entire Natural Science Institute should follow. Le's criticisms focused on the institute's use of foreign textbooks, which was at odds with Mao's emphasis on self-reliance and learning through practice. In contrast, under Le's direction, the biology department required students to go among the peasants, learning from them how to manufacture dyes and medicines from local plants. This was applied science that mobilized the masses and made full use of local resources. Many faculty members and students rallied to defend Xu and basic science, but by early 1943 Le's approach to science had won the day, and the Natural Science Institute became a part of Yan'an University, which was fully controlled by the Party.[25]

Beyond agriculture and industry, the intertwined themes of self-reliance and nativism also profoundly influenced the field of medicine in the revolutionary base areas. Acupuncture in particular emerged as an indigenous practice that served the need for a self-sufficient medical system: requiring only needles and knowledge, acupuncture helped reduce reliance on medicines made scarce by the blockades. Developing China's native medical practices was not a rejection of "Western science"; in fact, Mao and others remained deeply committed to weeding out superstition, and in this sense the encouragement of native doctors—including so-called witch doctors—posed a potential problem. Thus, the approach established in this early period, which remained vitally important in later decades, centered on mobilizing local resources and adapting characteristically Chinese methods to achieve goals—modern science, public health, economic development—that were understood to be universal.[26]

The Sino-Soviet Split and the Second Wave of Self-Reliant Science

In 1949, the victory of the Chinese Communist Party in the Civil War appeared to change everything. Gone were the blockades that cut the communists off from major industrial centers. Communist schools joined a substantial existing educational network, including Qinghua University and Peking Union Medical College, whose deep connections to the United States were now officially severed.[27] Perhaps most importantly, the Soviet Union moved from being a reluctant sponsor to an "elder brother," and China's foreign policy moved in response from "emphasizing self-reliance" and "depending on our own organizational power" (as Mao famously said in 1945) to "leaning to one side" (i.e., toward the Soviet Union).[28]

During the period of Soviet learning (1949–1960), China received guidance from resident Soviet technical advisors in almost every field of the natural and social sciences. In biology, the Soviet Union immediately and insistently promoted Lysenkoism, which in China was called "Michurinism" after the man whose experiments had inspired Trofim Denisovich Lysenko.[29] This wasn't surprising: Lysenko had won his greatest battle in 1948, and in 1949 he was riding high on Stalin's support. Lysenko's chief Chinese proponent after the revolution was none other than Le Tianyu. Le's Yan'an-era ideas about science bore striking similarity to some of the more radical, peasant-based programs underway in the Soviet Union since the 1920s—the very approaches that had given Lysenko his start.[30] To what extent these precedents had influenced Le isn't clear, but in the early 1940s Le articulated his own peasant-based approach to science without highlighting Soviet examples. This is our third non-surprise: the Rectification Campaign was an important episode in Mao's struggle to chart a path away from Soviet leadership; reference to Soviet examples would hardly have served Le's purpose at that time. And, as we will see, the other two high points for Maoist "mass science," the Great Leap Forward and the Cultural Revolution, also were periods of rupture between Mao and the Soviet Union. Noticing this pattern, Laurence Schneider has concluded that "if Soviet Lysenkoism had not existed, the CCP would have invented something like it on its own."[31] I would add that it was important for Chinese radicals, no matter what the *actual* influence of foreign scientific models, to project an explicitly native, self-reliant form of mass science. (Here we see again the phenomenon of overdetermination.)

Despite the extraordinary level of Soviet assistance and the pervasive rhetoric about treating the Soviet Union as an "elder brother," Mao appears never to have fully lost his sense that ultimately China could rely on nobody but the Chinese people themselves. In 1955, bristling at the Soviet Union's unwillingness to share nuclear technology, Mao spoke of his commitment to developing nuclear energy "even if we have to do it on our own."[32] Sino-Soviet scientific collaboration continued until the final departure of the technical advisors in 1960, and the existence of 120 cooperative scientific agreements signed in 1957 and 1958 indicates that some people at least continued to see a future in the alliance.[33] But by 1958 Mao clearly had already launched China on a different path.

The Great Leap Forward (1958–1960) represented a bold departure from Soviet guidance. The rhetoric of self-reliance, application, nativism, and mass mobilization defined Mao's alternative vision. Although political agendas shifted substantially on several occasions,[34] this "Maoist" approach to science exerted enormous influence from the Great Leap Forward through the Cultural Revolution. And, importantly for our purposes here, after the Sino-Soviet split state policy and propaganda materials consistently identified this approach to science as evidence of China's commitment to upholding true revolutionary values, specifically in contrast not only with the "imperialist" United States but also with the "revisionist" Soviet Union. (On what Mao saw as "Soviet revisionism," see the chapters by Aronova, Schmidt, and Siddiqi.) Figures 3.1 and 3.2 chart the occurrence of relevant terms in *People's Daily*. As the major popular organ of the Chinese Communist Party, *People's Daily* offers a helpful indicator of the state's priorities—more specifically, what the state wanted the people as a whole to view as priorities.[35] This admittedly crude methodology nonetheless offers an indication of the relationship between self-reliance and the Cold War time line that would otherwise be difficult to capture. References to "self-reliance" (*zili gengsheng*) and

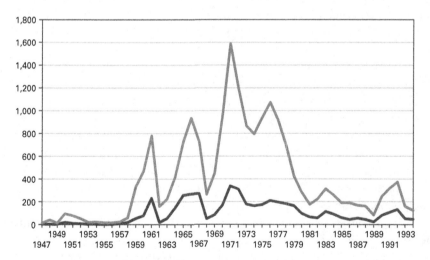

Figure 3.1
Incidence of Chinese terms for "self-reliance" and "self-reliance" + "science" in *People's Daily*. Lighter curve represents *zili gengsheng* ("self-reliance"); darker curve represents *zili gengsheng* + *kexue* ("self-reliance" + "science").

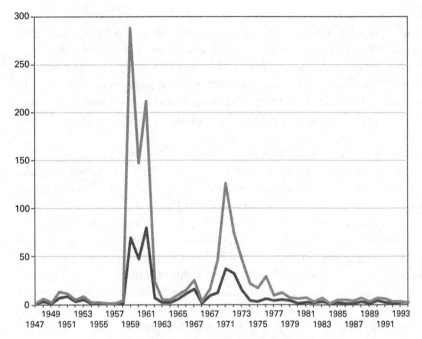

Figure 3.2
Incidence of Chinese terms for "native methods" and "native methods" + "science" in *People's Daily*. Lighter curve represents *tubanfa* ("native methods"); darker curve represents *tubanfa* + *kexue* ("native methods" + "science").

"native methods" (*tubanfa*) both skyrocketed from 1958 to 1960 with the Great Leap Forward and Sino-Soviet split, peaked again with the mid 1960s escalation toward the Cultural Revolution, and peaked yet again beginning in 1969 when Sino-Soviet relations took another turn for the worse.

I am not arguing that Mao-era scientific research was in fact utterly self-reliant. In view of China's extensive connections to transnational science (again, see Zuoyue Wang's chapter), it would be hard to claim that any area of Mao-era scientific research arose independently. Even acupuncture moved in new directions because of foreign influence.[36] Claims to "self-reliance" thus must be read critically. Even recent works by Chinese historians of science continue to display very clearly the nationalist stakes involved in debating the relative roles of foreign and Chinese in scientific achievements. Liu Jifeng, Liu Yanqiong, and Xie Haiyan, for example, devote an entire appendix in their book on Chinese nuclear science to the question of Soviet assistance. After outlining precisely what types of personnel, training, and material support the Soviet Union provided, they conclude that the Soviet Union acted as a kind of guide, pointing out the right direction, and so prevented the Chinese from wasting too much time on detours, but that it was the Chinese, through their own "gropings," who managed to resolve the crucial problems.[37] Yet self-reliance was not merely a rhetorical curtain obscuring dependency; the Chinese state really did face greater obstacles in pursuing scientific research because of its relatively isolated position during the Cold War. And the rhetoric that was crafted to turn that harsh reality into something ostensibly positive had tangible consequences for the approaches to science that the state supported.

The shift back to emphasizing self-reliance in science that occurred in the late 1950s took two somewhat different forms, one of which may be characterized as high-technology "big science" and one as low-technology "mass science." On one hand, the fetishizing of "bigness" that occurred during Great Leap era undergirded massive investment of resources into select projects, including the manufacturing of synthetic insulin and the development of nuclear weapons. In medicine and agriculture, on the other hand, the emphasis was on large-scale mobilization of "the masses" (and especially the peasant masses), employing "local" (*tu*) methods to surpass the achievements of the world leaders in both capitalist and communist spheres. The term *tu* referred potently to self-reliance on several levels: it connoted not only the immediate vicinity, but also "native" (as opposed to *yang*, which meant "foreign") and also "earthy" or "crude" (thus associated with peasants in contrast with elite intellectuals).

The decision to make the synthesis of insulin a priority came in 1958, and victory was pronounced in 1965. As described by Xiong and Wang, the project exemplified a specific Mao-era style of scientific research, which included a "military flavor," massive mobilization of human resources, influence of ideology, and heavy emphasis on planning and secrecy[38]—a kind of Cold War science with Chinese characteristics. Early in the project, the theme of self-reliance became prominent. Following the Party's lead, students in Beijing University's chemistry department criticized their professors' "Western slave mentality" and other faults.[39] When a team at Fudan University appeared to be on the verge of achieving synthesis of insulin's A and B amino acid chains, a high political official declared: "Some people say that what foreigners can't accomplish, Chinese people can never accomplish. Today we can say that Chinese people alone have accomplished something that foreigners have failed to do."[40]

In later years, China's success in manufacturing the world's first synthetic insulin became a shining example of self-reliance. A 1968 article in the *People's Daily* was titled "Self-reliance, Overtaking Advanced World Levels: Ousting the Chinese Khrushchev's 'Western Slave Philosophy.'"[41] (The "Chinese Khrushchev" was Liu Shaoqi, a former president and the most important early casualty of the Cultural Revolution, who had been targeted for his technocratic and internationalist political approach.) Another article, published in 1974, recounted the triumph of synthetic insulin even in the absence of the necessary raw materials of amino acids: "Researchers self-reliantly organized their own factory. With no foreign equipment available, they adopted native [*tu*] methods, and fighting bravely for several months were able to produce in the laboratory more than ten kinds of amino acids. With self-reliance and patriotic fervor, they ended in synthesizing the world's first entirely biologically active protein, crystalline bovine insulin."[42] Not emphasized in such articles was the continued significance of transnational connections. As Wang discusses in his chapter, many of the scientists involved had returned to China after receiving their degrees in the West.[43]

The year 1958 also brought a profound shift in nuclear science. Already prepared to "do it on our own," Mao began talking more directly about developing a uniquely Chinese approach to the problem. In mid 1958, Mao approved eight guidelines for developing nuclear weapons; the fourth guideline specifically warned against "imitating other countries" in the effort to "catch up to world levels." The central concept, in a nutshell, was that the goal was assumed to be universal, but the methods used to get there would be Chinese. In a speech to military officials, Mao criticized the Soviet approach: "At present, the things worked out by the Soviet military advisers (such as operational plans and thinking) are all of an offensive nature, based on victory; no provision is made for the defensive and for defeat." Pointing to the strength of China's indigenous military strategies, Mao argued that "we don't have to learn from the Soviet Union."[44] The link between self-reliance and applied science was as tight then as it had been in 1942. In 1960 Chinese physicists working in the Soviet Union met to discuss Mao's call to embrace self-reliance. They wrote a joint letter to the State Council approving of the plan and pledging to "change their professions to meet China's need even at the cost of giving up research on basic theory in which they had been engaged."[45]

Like the synthesis of insulin, the development of nuclear weapons involved massive deployment of technical and human resources. Moreover, nuclear physics was also relatively insulated from political upheaval. The natural sciences in general suffered less than the humanities and social sciences, but nuclear physics—along with weapons research more generally—was especially privileged in this way. The combination of great state investment and shielding from political campaigns resulted in China's own version of "big science," which produced the "two bombs, one satellite" program, including Qian Xuesen's famous "Silkworm" missile.[46]

Even in this biggest of big sciences, the local, the crude, and the masses played important roles. One big hurdle that Chinese nuclear science had to overcome was a lack of uranium. Just as the state organized peasants in the Great Leap to create "backyard furnaces" in an attempt to surpass the British in steel production, it also asked them to collect and prepare uranium. A *People's Daily* article provides insight into what uranium mining probably looked like. In developing smelting facilities, Sichuan Province "sought out local methods [*tubanfa*] that fully relied on local folk technological power and were crude and simple, and so economically organized production."

Local mudstone was used to make the furnaces, and the technicians were all local people.⁴⁷ Despite considerable waste of resources and pollution of local environments, scientific personnel remember the significant contribution such activities made in providing the nuclear program with uranium.⁴⁸ The spirit of self-reliance reportedly inspired technicians at the nuclear testing site, too—they sought to make the base self-sufficient by manufacturing monitoring instruments on site with available materials.⁴⁹

In other branches of physical sciences too, native and crude methods gained ground. An article published in the *People's Daily* in October of 1958 began with the common Great Leap expression "uniting native and foreign" (*tu yang jiehe*), then argued that the "native" could replace and even create the "foreign," as was the case in Beijing University's physics department, where young professors and students used native smelting methods and native materials to manufacture an electrostatic particle accelerator. "If foreign experts [*yang zhuanjia*, meaning experts in 'foreign' types of knowledge] take frequent breaths of 'native' [*tu*] air, this will help break the fetters of dogmatism and prevent the ossification of thought."⁵⁰

The local and crude side of self-reliance found far greater expression in other fields. During the Great Leap Forward, the renewed emphasis on self-reliance created unprecedented interest in Chinese herbal medicine and acupuncture, and even in the kind of "home-grown remedies" that Mao had once associated with "witch doctors" and superstition. This was also the period when local people trained in primary care—known in the Cultural Revolution as "barefoot doctors"—became an important component of the health-care system.⁵¹ Here again we find the tight interweaving of self-reliance, indigenous knowledge, "crude" (*tu*) methods, and mobilization of local peasants that emerged from the revolutionary experience and became the hallmark of "mass science" during the Cold War.

Just as the term *tu* acted multivalently to forge conceptual links among native, local, and crude, the nationalist rhetorical power of "self-reliance" simultaneously worked to encourage local economic independence. That is, the state used the slogan "self-reliance" to urge not just making it without foreign help, but also making it without assistance from the central state or other regions. In 1970, when the Red Flag Canal required maintenance, *People's Daily* reported that local revolutionary cadres struggled with representatives of the "traitor" Liu Shaoqi and class enemies, saying: "Are we moving forward or abandoning it? Are we self-reliant or depending on the nation-state? Are we using 'native' ['*tu*'] construction methods by making do with local materials, or are we using 'foreign' ['*yang*'] methods of reinforced concrete? Are we mobilizing the masses, or blindly believing in a few 'experts'?"⁵² Thus the isolating context of the Cold War lent a patriotic energy to the notion of self-reliance, which in turn served the Chinese state's domestic needs, both by spurring scientific innovation and production and by dissuading people from expecting too much from the central government.

"Self-Reliance and Arduous Struggle": "In Agriculture, Learn from Dazhai"

Agriculture offered perhaps the richest field for the growth of self-reliant mass science. Increasing dependence on agro-chemicals (*nongyao*) had caused demand to outstrip supply. A number of different approaches, involving not only scientists in research institutes but also rural political leaders and grassroots experiment teams, addressed this mounting problem. Scientists at universities and institutes developed biological control regimens to reduce the need for chemical insecticides.[53] Experiments replacing "foreign chemicals" with "native chemicals" further uncovered a wide assortment of locally available materials useful in combating insect pests.[54] At agricultural conferences, local leaders trumpeted the success of "poor and lower-middle peasants" in demonstrating through scientific experiment the effectiveness of green-fertilizing crops such as Chinese milk vetch (*ziyunying*) in resolving fertilizer problems and achieving self-reliance.[55] Labor-intensive observation of insect activity, often known as "insect pest forecasting," helped peasants time the application of chemicals for optimal efficiency and thus husbanding of this precious resource. Here the knowledge of "old peasants" sometimes proved invaluable (and was almost always said to be invaluable, whatever its actual worth).[56] The manual elimination of insect pests and their eggs was another technique consistent with a program of mass mobilization and self-reliance, as was the establishment of "local-method" (*tufa*—and here the meaning of "crude" is clearly indicated) factories for producing microbial agents to combat insect pests.[57] And throughout the 1960s and the 1970s, rural experiment teams used simple, locally available resources to produce certain agro-chemicals, the most common being the plant hormone gibberellin (called 920 in Chinese) and a microbial fertilizer known as 5406.[58]

But in agriculture, perhaps more clearly than in any other field, the call to be self-reliant meant not just socialist Chinese independence from untrustworthy world leaders, but still more the need for locales to make do without assistance from the central state. And so requests for funding the "mass movement for scientific experiment" highlighted plans to "organize the development and production of simple local [*tujian*] instruments" such as light-traps for monitoring insects and soil analysis instruments.[59] In the other direction, memos announcing the distribution of funds for scientific experiment often included encouragements to realize the slogans of "self reliance, arduous struggle" and "practice thrift, using less to do more," or exhortations to be "self reliant, with the spirit of diligence and thrift, practicing meticulous planning and careful accounting, and being conscientious in management and use [of funds]."[60]

The slogan "self-reliance, arduous struggle," often associated with the "Yan'an spirit" and immortalized as a chapter title in the "little red book" of Mao quotations, gained its greatest currency with the policy "In agriculture, learn from Dazhai" (*nongye xue Dazhai*). Dazhai was a production brigade in the northern province of Shanxi celebrated for its success in building terraces to reclaim mountainous land for agriculture. After 1967, Dazhai was the most important agricultural model in China until its leftist foundations were repudiated early in the Deng era. Countless local political and scientific leaders visited Dazhai to learn about mobilizing the masses and organizing agricultural production. Often they returned to their locales to impose Dazhai-like terracing programs or to institute the policy of "taking grain as the key link"—the growing of grain instead of other crops so as to achieve local self-sufficiency. And often these projects, poorly suited to local conditions, wreaked havoc on local environments and economies.[61] The bitter irony of this case was that Dazhai's remarkable success owed not just to local ingenuity and hard work, but also to generous state subsidies designed to create a glowing example for the rest of China and the world. However, it would be a mistake to emphasize the disastrous effects of the "Learn from Dazhai" movement without also noting the ways in which calls for self-reliance often helped in *resisting* inappropriate imposition of external models. Propaganda materials frequently highlighted the need for attention to the environmental conditions of specific places. Experiment teams at the village level were often called upon to test seeds from other places for local suitability or even to produce new hybrid strains tailored to local conditions. Here was the epitome of self-reliance: local people breeding local plants using locally available resources.[62]

Creative use of limited resources was a repeated theme in propaganda designed to inspire the development of "scientific farming." Scientists, young peasants, and urban youth "sent down" to the countryside had to make do under crude working conditions. In a story published in 1966, a "sent-down" youth hybridized two existing strains of rice, "Nation's Wealth" and "Atomic #2" (note the nuclear connection) in 1958. He followed directions found in an agricultural textbook, but because he did not have access to a thermometer or a watch he used his fingers to test the temperature and the school bell to measure the time.[63] A report delivered at a 1965 Beijing-area conference on rural scientific experiment groups named "self-reliance and arduous struggle" a "fundamental policy" in "mass scientific experiment activities" and called upon all participants to "conscientiously implement this policy and study and develop the spirit of Dazhai." Leaders should provide some necessary support, but otherwise they should "encourage group members to take initiative" in devising ways of producing "crude and simple" (*yinlou jiujian*) equipment and "replacing the foreign with the local" (*yi tu dai yang*). In some places, the scientific experiment groups were using test tubes as levels, cooking pots as kilns, and ceramic plates as Petri dishes.[64]

The specific terminology used to discuss agricultural science was indicative of the revolutionary refiguring of scientific practice in Cold War China. According to Mao in 1963, "scientific experiment" was one of the "three great revolutionary movements" that would protect Chinese socialism from bureaucracy, revisionism, and dogmatism—a clear reference to China's split from the Soviet Union.[65] But "scientific experiment" did not necessarily mean scientists in ivory towers with lab coats, or even research in the pursuit of new generalizable knowledge. Despite significant interruptions from political campaigns, the kind of professional research international scientific communities would recognize continued in research centers around the country, and the knowledge those centers produced served agriculture in important ways. However, the notion of "scientific experiment" encompassed a much broader variety of activities. For example, the production of plant hormones and microbial fertilizers counted as "scientific experiment." Such activities required basic laboratory skills, but they were not "experiments" in the conventional sense. The goal was not to produce new scientific knowledge, though the participants certainly acquired new knowledge in the process of production, and the challenge of using only crude, locally available resources created a degree of unpredictability and thus an aura of "experiment." Some projects, such as the testing of new seed varieties and the production of new hybrids, were more clearly experimental. Others (soil improvement through application of manure, weather prediction, pest observation and control, troubleshooting malfunctioning machinery, animal husbandry) were perhaps less so. Claiming these often mundane practices as "scientific experiment" was itself a revolutionary act that brought science down from the ivory tower and into the realm of rural laborers.[66]

Science was also revolutionary when pursued in defiance of traditional prejudices and class enemies. In 1971, a group of ten women of the minority Zhuang nationality established the March Eighth Agricultural Science Group (named after International Women's Day). Their average age was 19. According to an article in an agricultural science journal, the young women plowed and fertilized, braving inclement weather and the sexism of class enemies to get sand from a river, fertile mud from caves, and manure from the noisome manure pit. Through such "scientific farming," they transformed the hardened clay fields into fertile land again.[67] Women who worked with livestock risked sexually charged verbal abuse. A report from a 1965 conference on rural youth in scientific experiment reported that some people scolded young women engaged in livestock breeding, saying "You spend all day mating donkeys and horses."[68] Practicing scientific agriculture was said to be a way of overthrowing sexism and conservative thinking.

The vast majority of available materials documenting agricultural scientific work during the Mao era are state documents and propaganda; they are useful for understanding how the state sought to portray agricultural science. However, there is also evidence to show how fluently people spoke the state's language on science. The published diary of Shen Dianzhong, who was among the approximately 14 million urban youth "sent down" to engage in agricultural work during the Cultural Revolution, contains detailed descriptions of his participation in "scientific experiments" involving gibberellin and microbial fertilizer. On June 13, 1972, after more than a year of emotional hardship coping with the difficulties of the work and the frequent failure of the experiments, Shen wrote an extensive summary of his experiences. His second itemized point (after an initial reference to using Mao Zedong Thought and uniting theory with practice) read "920 [gibberellin] work brings into play the proletarian revolutionary spirit of using local methods, starting from scratch, self-reliance, hard work, not fearing failure, and overcoming hardships." He continued: "Local methods and starting from scratch: you just have to look at the conditions, facilities, equipment, materials (*cailiao*), raw materials (*yuanliao*), operations, etc. over the course of one year [of experiments], and you will understand this point." As for "self-reliance and hard work," he noted that almost all the activities were accomplished during midday siesta or in the evening, which went to show that "a revolutionary spirit infused all the work."[69]

Sino-US Rapprochement and the Production of Socialist-Chinese Uniqueness in Science

A world removed from Shen Dianzhong's rural laboratory, a major geopolitical shift was underway. In 1969, tensions between China and the Soviet Union came to a head, and Mao began to seek renewed relations with the United States. Rapprochement meant not just strategic partnerships but also opening doors to cultural and scientific exchange, not just with the United States and other Western countries but also through the United Nations. China's admission to the UN in 1971 allowed for participation in international scientific collaboration to a degree that its unique position in the Cold War had previously made impossible.

One might expect that the dramatic change in China's global position would have resulted in an equally dramatic decrease in the emphasis on "self-reliance" in socialist Chinese science. Significantly, that was far from the case. In figure 3.1 we see that incidence of the term "self-reliance" shoots up in *People's Daily* articles in 1969 and remains high through 1977 before plummeting in 1979, after Deng Xiaoping took the reins. Figure 3.3 shows a propaganda poster from 1975 articulating messages virtually indistinguishable from the discourse on self-reliance and scientific experiment of ten years earlier. The reports of dozens of American visitors (delegations of scientists, journalists, activists, and others lucky enough to secure invitations) during what we might think of as the courtship period of the 1970s are filled with references to China's consistent emphasis on "self-reliance."[70] Figure 3.4, an impressive example of the Chinese art of paper-cutting purchased by an American visitor in 1978, represents Maoist perspectives on rural production and scientific experiment. The palm trees suggest a southern locale, but the emphasis on terracing clearly indicates the movement to "learn from Dazhai" in order to achieve self-reliance in agriculture.

Self-reliance thus remained a badge of honor for Chinese science; moreover, it was promoted as the basis for a uniquely socialist-Chinese style of science from which

Self-Reliant Science

Figure 3.3
The sixth of a set of 1975 posters designed to be displayed in common areas of communes around China to inspire scientific experimentation. The title translates to "Self-Reliance; Scientific Research through Hard Work and Frugality." The upper text celebrates the policy of "self-reliance and arduous struggle" and praises the county of Huarong for "persistently drawing on local resources, using local methods, and improvising equipment, such that they met the needs of agricultural scientific research and drove forward mass-based scientific farming activities." The explanation for the left picture reads: "In spring 1971, in order to popularize cultivating seedlings in greenhouses, Huarong County established a 'model' greenhouse, but because it was too expensive to build, they couldn't popularize it. Xinjian Brigade in Xinhe Commune substituted mud bricks and wood for red bricks and reinforced concrete, membrane to replace glass, and reeds for seedling trays, thus spending little more than 10 yuan. This kind of 'native [*tu*] greenhouse' was warmly welcomed by the masses and very quickly became popularized throughout the county." The explanation for the right picture reads: "At each level of the agricultural science organization, the masses are mobilized to select methods that are crude and simple, substituting the native for the foreign, and in this way resolve the equipment needs of scientific experiment. They use [old-fashioned] balance scales to replace [scientific] scales, clay bowls for seedling containers, and warming on the stove in place of incubators. These are educated youth from Jinggang Commune using clay bowls to conduct scientific experiment." Source: Xinhua tongxun she, ed., *Dagao kexue zhongtian, jiasu nongye fazhan: jieshao Hunan Huarong xian siji nongye kexue shiyan wang* [*Greatly Undertake Scientific Farming, Accelerate Agricultural Development: Introducing Hunan Province, Huarong County's Four-Level Agricultural Scientific Experiment Network*] (Renmin meishu chubanshe, 1975).

Figure 3.4
A paper-cut (18 × 32 inches) depicting the transformation of agriculture in socialist China. Most of the activities represented are related to the construction of terracing to reclaim mountainous land for agriculture, but in the lower left corner we see two people engaged in scientific experimentation, one using a microscope and the other pouring a liquid through a funnel into a flask. (See detail at right.) In the full image, note the weathervane above the experiment station—weather prediction and reporting were sometimes the responsibility of local scientific experiment groups. Collected by Britta Fischer on a 1978 tour of China organized by the US-China People's Friendship Association. Grateful acknowledgment to Britta Fischer. In author's possession.

other countries could learn. This was an extension into the détente era of China's desire to present a "third way" to the world—an alternative (not only for the Third World but also for Sweden and other potential European allies) to the options offered by the two superpowers.[71] A perfect example is China's most ambitious technology-transfer project: the TAZARA Railway, which linked Tanzania and Zambia, bypassing apartheid South Africa. The project, initiated in 1967 and carried out between 1970 and 1975, exported not only China's scientific know-how but also its philosophy of self-reliance.[72] Similarly, in Liberia, Sierra Leone, and Gambia, agricultural assistance from China emphasized this theme. The president of Sierra Leone returned from a visit to China inspired by the rhetoric on self-reliance, while Chinese experts in West Africa supervised the production of locally made rice threshers, demonstrated composting and use of animal manure for fertilizer, and raised chickens and pigs to feed themselves, all the while calling attention to these activities as examples of self-reliance. (As in China, West African political leaders recognized the usefulness of a philosophy that not only stoked anti-imperialist sentiment but also encouraged locales not to depend on aid from the central government.[73]) Maoist approaches to science also were influential in Mozambique, where the revolutionary leader Samora Machel celebrated the wisdom of peasants and mechanics and decried the "arrogance" of experts who kept themselves apart from the masses, making themselves into a "privileged class." Their intelligence, Machel asserted, became "sterile, like those seeds locked in the drawer."[74]

The Chinese state deeply valued the propaganda opportunities afforded by technological assistance to Third World countries. Visits from foreign delegations offered similar possibilities. In an internal serial publication titled *Reference Materials for Propaganda Directed at Foreigners* (*Dui wai xuanchuan cankao ziliao*), state officials tracked the published accounts of foreign visitors and commented on the degree to which they reflected the messages about Chinese socialism that the Chinese state intended to convey. For example, in 1973 the journal published a translation of a Japanese scientist's report on his recent visit. The editor's note explained: "The author examines rural changes in China with respect to politics. Although he is writing about agricultural science, he is able to conduct an analysis of our country's planning policies, and moreover is able to form a contrast with Soviet revisionism, in order to enlighten his audience."[75]

If Chinese political and scientific elites were excited to present the scientific achievements that Chinese socialism had fostered, foreign visitors were, for their own diverse reasons, often equally excited to bring such examples home. The passage from the Japanese report that inspired the Chinese propagandists' appreciative note compared Chinese and Soviet manufacture of herbicide. An herbicide factory at a commune the Japanese scientist visited had an annual output of 1,300 tons and was still under expansion. According to the Japanese scientist,

This situation, compared with the 2,000 tons of herbicide the Soviet Union purchased from Japan over the past several years, can offer such a deep awareness! The Soviet Union should be a very advanced socialist country, but in fact imports this kind of pesticide from foreign countries; on the other hand, in the so-called industrially backward China, peasants themselves are able to produce it. When I visited China in 1966, I saw the slogan "Class struggle, struggle for production, scientific experiment." China calls these the three great revolutionary movements. But at that time I did not understand why scientific experiment was called a revolution or what use peasants and workers could make of it. Now I've discovered the crux of the issue. This agrochemical factory is a concrete reflection of China's pursuit of new-style scientific experiment through reliance on the masses.[76]

At the same time, foreign scientists had to reconcile their enthusiasm for the exotic with the uncomfortably obvious ways in which science in Cultural Revolution-era China departed from some of their own scientific assumptions and values. One of the most significant of these involved the relative importance of basic research, technical application, and popularization. As Naomi Oreskes has argued, building on John Krige's work, the emphasis placed on basic science by American scientists emerged not only from a belief in the necessity of basic research before technological development but also from a commitment to fostering a form of science "resonant with the American way of life."[77] Indeed, as early as the first decade of the twentieth century, efforts by the United States to promote scientific development in China had emphasized laboratory research and had presumed a clear connection between the ideal of research science and positive social transformation.[78] Traveling to China in 1974, the American Plant Studies Delegation noted that some of the work they witnessed, though "termed experimental," was "actually demonstrational: for instance, plantings of improved seeds next to other varieties in order to show peasants the advantages of the new over the old."[79] A 1975 delegation of agricultural scientists from the United Nations Food and Agriculture Organization, committed to "leaving our mental luggage behind" in order to "learn from China," approached the issue from another angle: "The Chinese put it quite succinctly: 'In China, all agriculture is extension.'"[80] Defending Chinese agricultural science from the charge that research was too often neglected, a Chinese-American entomologist writing for a UN publication explained that in China scholarly publication tended to follow applications in the field and extension to farmers, whereas in the United States scholarly publication came first.[81] In another article, this one in *Science*, the entomologist suggested that "the image of Chinese entomology as ignoring basic research may be an oversimplification," and that a more accurate assessment would acknowledge "the priority the Chinese give to putting scientific results into operation."[82]

For many foreign visitors, China's experience appeared to offer something valuable that the West lacked. Many visitors with leftist or left-leaning politics specifically sought inspiration in China's socialist approach to medicine.[83] Others were drawn to

Chinese medicine for different reasons. Western interest in "Traditional Chinese Medicine" (or TCM, a term itself obviously created for foreign consumption) emerged along with the growth of the New Age movement. For many Westerners, TCM represented China's long tradition of "holistic" philosophy and thus offered a powerful antidote to the overspecialized and reductionist medicine that had become "mainstream" in the West. But this was never how the Chinese state framed the role of Chinese medicine. Rather, the state selected acupuncture anesthesia as the exemplar of what China could uniquely *contribute to modern science*.[84] Based in indigenous knowledge, but rendered scientific, acupuncture anesthesia offered an effective and economical means of serving the people's medical needs. It was a perfect example of China's self-reliance: replacing scarce and costly "foreign medicines" with widely available materials embedded in an indigenous practice that was as useful to surgeons in operating rooms as to peasant paramedicals in the fields.[85]

The 1970s also brought an increase in environmentalism in the West. Insect scientists were anxious about the consequences of ever-increasing use of chemical pesticides, and many scientists, especially in the US, were angry about the power chemical corporations had in setting research agendas. Socialist China appeared to offer hope of a different way. In the absence of corporate capitalism, and making a virtue of the necessity of extreme thrift, Chinese insect scientists had succeeded in working with peasants to develop an "integrated" system of pest control that minimized use of toxic chemicals. "Clearly," the entomologists on the 1975 US Insect Control Delegation reported, "the Chinese have progressed beyond levels attained in the United States both in widespread enthusiasm for integrated control and, in many respects, in the application of the ecological principles fundamental to its development."[86] One British delegate reportedly told his Chinese hosts: "In Western countries people talk a lot about integrated control but do very little of it. You do so much work; you are our model."[87] The official report of the Swedish delegation similarly posited the relative backwardness of biological control in Sweden and suggested that knowledge be sought in China, where biological methods and integrated pest control were more developed.[88] China's bag of insect-controlling tricks included light traps, parasitic wasps, mobilization of peasants for insect forecasting and manual elimination, and, most popular of all, insect-eating ducks. Foreign delegations were treated to special demonstrations of this last method—and to roast duck in the cafeteria—at a commune outside of Guangzhou, where the US-trained entomologist Pu Zhelong had organized a number of biological control projects. So charming were these feathered representatives of Chinese ingenuity that the magazine *Environment* ran an article by an American insect-control delegate under the title "China Unleashes Its Ducks."[89] Thus did foreign scientists participate in the construction of a uniquely socialist-Chinese vision of scientific practice.

Conclusion and Epilogue

What connects the humble bug-eating ducks of Guangzhou with Qian Xuesen's imposing "Silkworm" missile? Self-reliance. Though Cold War politics was not the only factor, it unquestionably contributed to the significance of self-reliance for science in Mao-era China. Not only did the Cold War result in isolating China at certain historical moments; it also produced an assumption of ideological difference and thus an expectation that science in socialist China would offer a distinct alternative to existing models. From 1958 to 1971, Mao's decision to part ways with both superpowers entailed a commitment to finding a Chinese path for Chinese science. After 1971, with the resumption of international scientific exchange accompanying Sino-US rapprochement, China no longer truly needed to go it alone; now Chinese political and scientific leaders sought to demonstrate what China had to contribute to international science.[90] During the 1970s, both foreigners and Chinese people contributed to the notion of a uniquely socialist-Chinese approach to science, though the two sides did not always share a common understanding of what this meant.

The rhetoric of self-reliance in socialist Chinese science was intense and pervasive enough to mask the surprisingly transnational character of much scientific work in Mao-era China. And rhetoric is important. To what extent the rhetoric actually represented significant epistemological differences and research results is more difficult to judge. In broad terms, we could hazard that Cold War pressures contributed to an experience, shared by most scientific fields in Mao-era China, of increased emphasis on application over basic research. Moreover, in some cases (notably the synthesis of insulin and the development of nuclear technologies) such pressures also helped to produce a kind of "big science" approach comparable to that pursued by the Cold War superpowers.

A more fine-grained analysis yields a more complex picture. For example, despite the very different priorities Chinese proponents of Traditional Chinese Medicine held, the "holistic" approach that some Westerners derived from TCM certainly offered a profoundly "alternative" epistemology. However, for these Westerners the difference was less about Cold War ideologies than about their perceptions of "Western materialism" and "Eastern spiritualism." The most we can say is that Chinese commitments to self-reliance (which were strengthened by Cold War realities) drove Chinese medical practitioners and policy makers to promote TCM, and that this promotion helped fuel Western interest in TCM as an "alternative medicine." On the Chinese side, a Marxist commitment to seeing science and progress as universal, in combination with nationalist pride and the need for self-reliance, produced a desire to demonstrate the usefulness of Chinese practices such as acupuncture to modern medical science. Interest in this approach emerged during a period of geopolitical isolation in the revolutionary base area of Yan'an, increased during a second period of isolation from the superpowers (1958–1971), then took on new meaning after China's reconciliation with the United States, its admission to the UN, and its emergence as a participant in a larger international science community.

In agriculture, the emphasis on self-reliance and the related concern for mass-based, practical approaches encouraged the development of some technologies that might not otherwise have emerged. Pest management based on close monitoring of insect populations and labor-intensive agricultural and biological control mechanisms is one example, and Western participants in 1970s scientific exchange recognized it as such. In some cases, agricultural technologies developed during the Mao era have continued to be of scientific interest not only in China but also in other parts of the world. This is true, for example, of microbial fertilizers such as 5406, which played a prominent role in Cultural Revolution-era rural-based youth experiment projects and which served the purposes of self-reliant science because they could be manufactured locally and so reduce the need for imported chemical fertilizer. The fertilizer 5406 is now used by scientists at the International Nature Farming Research Center in Japan.[91] However, when present-day scientists turn to China for inspiration in agriculture, they are far less likely to highlight China's socialist experience and more likely to revive the visions of F. H. King, the American soil scientist whose 1911 book *Farmers of Forty Centuries, Or Permanent Agriculture in China, Korea, and Japan* extolled the ancient wisdom of Chinese farming practices and inspired the budding organic farming movement.[92]

The publication of the 1987 book *Learning from China? Development and Environment in Third World Countries* represented a turning point in international perspectives on socialist Chinese science. *Learning from China?* originated at a 1983 conference in West Berlin that brought scientists and scholars from different countries together to speak on subjects ranging from biogas technology to development policy. By then, people around the world had begun to lose interest in socialist China as a model; that helps explain why the conference's organizers felt obliged to put a question mark at the end of the title. China had changed. The very real negative aspects of the Mao era— especially the political persecutions of many millions of people, including most of China's top scientists—had become harder to ignore in the post-Mao era, when the Chinese state was, for its own political reasons, increasingly calling attention to them. And if it was more difficult to draw unambiguous lessons from China's socialist past, it was also increasingly clear that China's new road differed little from that of any other developing country with its sights set on industrialization along typical Western lines. "At a time when China is busy emulating Taiwan and South Korea," Vaclav Smil wrote in his review of the volume, "what is one to learn from China's experience? Since the late 1970s many critical and courageous Chinese scientists and economists have documented the enormity of pre-1978 environmental degradation and economic mismanagement. They have been the driving force behind the current reforms and the spirit of learning from abroad." Smil went on to characterize biogas as "a large-scale failure" and biological pest control as "vastly exaggerated efforts while pesticide poisonings are common and traditional farming methods are disappearing fast."[93]

Though I would argue that Smil dismissed the agricultural innovations of the Mao era too readily and too absolutely, he was undoubtedly right that by the 1980s the time for China to serve as a socialist model for other countries had passed. And with Deng Xiaoping's ascendance in 1978, self-reliance had ceased to serve as an important inspiration for science within China—though it has been used in new ways to excuse the central state from responsibility for local economies.

A study of science in China during the final decade of the Cold War would look very different from the history of the Mao era discussed here. Interested readers could do no better than to consult Susan Greenhalgh's book *Just One Child*, a fascinating study of the role of missile scientists in crafting the population science and policy of the Deng era. Whereas Mao had called for scientists to rely on China's masses, Deng called on scientists to control the numbers of those masses, now agreed to be entirely too massive, using theories and technologies with the clearest of connections to Cold War science.[94]

Acknowledgments

I thank Naomi Oreskes, Zuoyue Wang, John Krige, other participants in the Francis Bacon Conference, and an anonymous reviewer for their helpful suggestions, and Charlotte Goor for her invaluable assistance. This research has been funded in part by a Franklin Research Grant from the American Philosophical Society and a Faculty Research Grant from the University of Massachusetts. All Chinese names, except for overseas Chinese, are rendered as pinyin, with family names first and given names second.

Notes

1. Paul Forman, "Behind Quantum Electronics: National Security as Basis for Physical Research in the United States, 1940–1960," *Historical Studies in Physical and Biological Sciences* 18, no. 1 (1987): 149–229.

2. Tien-Hsi Cheng, "Insect Control in Mainland China," *Science* 140, no. 3564 (1963): 269.

3. See note 45 below.

4. Zuoyue Wang, "Physics in China in the Context of the Cold War," in *Physics and Politics*, ed. Helmuth Trischler and Mark Walker (Franz Steiner Verlag, 2010), 262.

5. Paleoanthropology is one example of such a field. Sigrid Schmalzer, *The People's Peking Man: Popular Science and Human Identity in Twentieth-Century China* (University of Chicago Press, 2008).

6. See, for example, Ronald Kline, "Construing 'Technology' as 'Applied Science': Public Rhetoric of Scientists and Engineers in the United States, 1880–1945," *Isis* 86, no. 2 (1995): 194–221.

7. It was also a major concern of the Guomindang. For a thoughtful analysis, see James Reardon-Anderson, *The Study of Change: Chemistry in China, 1840–1949* (Cambridge University Press, 1991).

8. Although the concept of self-reliant science was found widely in Cold War-era China, the use of this term as an overarching category to characterize the Maoist approach to science is my intervention. The concept of "self-reliance" itself is long overdue for an extensive transnational historical analysis. It is well known as a cornerstone of North Korean leader Kim Il-sung's political philosophy; although the broader concept is typically rendered *juche* in Korean, the Chinese term for self-reliance, *zili gengsheng*, also appears frequently in North Korean documents directly translated as *charyok kaengseng*. Scholars have typically traced the North Korean concept of self-reliance to the influence of Chinese Communists in the Yan'an area (see below), and we also know that many groups around the world (including the Black Panthers in the United States) were inspired by Mao's writings on self-reliance. However, evidence suggests that the term has a more complex history: in the 1930s, it was adopted also by the Chinese Nationalists (Guomindang) and by the Saito government in Japan. I am indebted to Bruce Cumings for sharing his leads on this subject. See also Gordon Mark Berger, *Parties Out of Power in Japan, 1931–1941* (Princeton University Press, 1977), 69.

9. Note, however, the interesting counterexample represented by post-1945 Europe, which faced significant economic needs and was committed to economic development, but which, under the influence of United States hegemony, became oriented toward basic rather than applied science. See John Krige, *American Hegemony and the Postwar Reconstruction of Science in Europe* (MIT Press, 2008).

10. On the greater importance of ideology in Mao's China compared with the Soviet Union of the late 1940s and later, see Chen Jian, *Mao's China and the Cold War* (University of North Carolina Press, 2001), 4. On the emphasis of ideology on science in socialist China, see Michael Gordin, Walter Grunden, Mark Walker, and Zuoyue Wang, "'Ideologically Correct' Science: French Revolution, Soviet Union, National Socialism, WWII Japan, McCarthyism, and People's Republic of China," in *Science and Ideology: A Comparative History*, ed. Mark Walker (Routledge, 2002). On technocracy in the Soviet Union, see Joel Andreas, *Rise of the Red Engineers: The Origins of China's New Class* (Stanford University Press, 2009), 266; Kendall Bailes, *Technology and Society under Lenin and Stalin: Origins of the Soviet Technical Intelligentsia, 1917–1941* (Princeton University Press, 1978).

11. "On Practice" is usually said to have been written at Yan'an in 1937, but it was first published in 1950. For an overview of the debates surrounding the origins of "On Practice," see Joshua Fogel, *Ai Ssu-ch'i's Contribution to the Development of Chinese Marxism* (Council on East Asian Studies, Harvard University, 1987).

12. In Odd Arne Westad's alternative way of analyzing the Cold War, the "Third World" collectively forms a third force, disrupting the notion of a bipolar conflict. See Westad, *The Global*

Cold War: Third World Interventions and the Making of Our Times (Cambridge University Press, 2005).

13. My own research further supports Zuoyue Wang's conclusions. Sigrid Schmalzer, "Insect Control in Socialist China and Corporate US: The Act of Comparison, The Tendency to Forget, and the Construction of Difference in 1970s US-Chinese Scientific Exchange," *Isis* 104, no. 2 (2013): 303–329.

14. Fred Halliday, *The Making of the Second Cold War* (Verso, 1983); Paul Edwards, *The Closed World: Computers and the Politics of Discourse in Cold War America* (MIT Press, 1996); Peter Westwick, "The International History of the Strategic Defense Initiative: American Influence and Economic Competition in the Late Cold War," *Centaurus* 52, no. 4 (2010): 338–351.

15. This interpretation follows the "two-line" analysis advanced by Mao-era radicals themselves. The post-Mao anti-radical historiography reverses the signs and depoliticizes the rhetoric. My understanding of Liu Shaoqi, Deng Xiaoping, and others as "technocrats" follows Andreas, *Rise of the Red Engineers*.

16. David Zweig, *Agrarian Radicalism in China, 1968–1981* (Harvard University Press, 1989), 192. Zweig sees peasants in Popkin's terms as economically rational actors, which he seems to equate with an interest in economic development and modernization, whereas the radicals were committed to ideology over material concerns and sought to impose these values on the peasants.

17. Ibid., ix, 3, 190–192.

18. Benedict Stavis, *Making Green Revolution: The Politics of Agricultural Development in China* (Rural Development Committee, Cornell University, 1974).

19. The first reference to "scientific farming" (*kexue zhongtian*) in *People's Daily* is from July 22, 1961. In 1965 there were eleven references.

20. Nils Gilman, "Modernization Theory, the Highest Stage of American Intellectual History," in *Staging Growth: Modernization, Development, and the Global Cold War*, ed. David C. Engerman et al. (University of Massachusetts Press, 2003), 48–49.

21. Wu Heng, *Kang Ri zhanzheng shiqi jiefangqu kexue jishu fazhan shi ziliao* [Historical materials on scientific and technological development in the liberated areas during the War of Resistance], multiple volumes (Zhongguo xue shu chu ban she, 1983–).

22. Stuart Schram cautions that the links between Yan'an and later Maoist perspectives on economic development represent "existential continuity" but "no intellectual continuity in terms of detailed policy formulations, and certainly no unbroken chain of development in Mao's own thinking." I do not seek to argue this point beyond the level of "existential continuity"; it is sufficient, I think, to recognize the lasting impact of both the experience of Yan'an and the heroic narrative woven around the "Yan'an Way." Stuart Schram, *The Thought of Mao Tse-tung* (Cambridge University Press, 1989), 93; Mark Selden, *The Yenan Way in Revolutionary China* (Harvard University Press), 1971.

23. Reardon-Anderson, *The Study of Change*, 323.

24. Laurence Schneider, *Biology and Revolution in Twentieth-Century China* (Rowman & Littlefield, 2005), 105. Note that Le Tianyu's family name has previously been incorrectly Romanized as "Luo" (in pinyin) and "Lo" (in Wade-Giles) by a number of scholars, including me.

25. Excellent discussions of this episode can be found in Reardon-Anderson, *The Study of Change* (352–359) and in Schneider, *Biology and Revolution* (104–108).

26. Kim Taylor, *Chinese Medicine in Early Communist China 1945–1963: A Medicine of Revolution* (Routledge Curzon, 2005), 24.

27. These events effectively relieved People's University of its responsibilities for science in the new era. Andreas, *Rise of the Red Engineers*; Mary Brown Bullock, *An American Transplant: The Rockefeller Foundation and Peking Union Medical College* (University of California Press, 1980).

28. Mao Zedong, "Kangri zhanzheng shengli hou de shiju he women de fangzhen" [The Situation and Our Policy After the Victory in the War of Resistance against Japan], August 13, 1945.

29. On the early emphasis on Lysenkoism in the PRC, see Schneider, *Biology and Revolution*, 120–134.

30. For an utterly unsympathetic account, see David Joravsky, *The Lysenko Affair* (Harvard University Press, 1970), 54ff.

31. Schneider, *Biology and Revolution*, 177.

32. Wang, "Physics in China," 259.

33. Ibid., 263.

34. Richard P. Suttmeier, *Research and Revolution: Science Policy and Societal Change in China* (Lexington Books, 1974).

35. The availability of a full-text-searchable database of the entire run of the paper makes possible analysis that is far more difficult and less reliable using other sources (e.g., scientific journals).

36. The most striking example is the emergence during the Great Leap Forward of "ear acupuncture," which did not exist in Chinese traditions and which owed directly and explicitly to French innovation. Elisabeth Hsu, "Innovations in Acumoxa: Acupuncture Analgesia, Scalp and Ear Acupuncture in the People's Republic of China," *Social Science and Medicine* 42, no. 3 (1996): 421–430.

37. Jifeng Liu, Yanqiong Liu, and Haiyan Xie, *Liang dan yi xing gong cheng yu da ke xue* [The project of "two bombs, one satellite": A model of big science] (Shandong jiaoyu chubanshe, 2004), 195.

38. Xiong Weimin and Wang Kedi, *Hecheng yi ge danbaizhi: Jiejing niuyi daosu de rengong quan hecheng* [Synthesize a protein: The story of total synthesis of crystalline insulin project in China] (Shandong jiaoyu chubanshe, 2005), 56–66.

39. Ibid., 31.

40. Ibid., *Hecheng*, 37.

41. "Zili gengsheng, ganchao shijie xianjin shuiping" [Self-reliance, overtaking advanced world levels], *Renmin ribao* [*People's Daily*], August 18, 1968.

42. "Duli zizhu, gaoge mengjin" [Independent self-governance, triumphant advance], *Renmin ribao* [*People's Daily*], October 17, 1974.

43. Xiong and Wang, *Hecheng*, 17, 21.

44. John Wilson Lewis and Xue Litai, *China Builds the Bomb* (Stanford University Press, 1988), 70–71.

45. Ibid., 123. The quoted material appeared in a *People's Daily* article from 1984. How closely it represented the original statement as expressed in the 1960 letter is not clear. Gu Mainan, "Shiwan fen zhi yi: Ji Zhongguo kexueyuan xin ren fuyuanzhang Zhou Guangzhao" [1 in 100,000: On the new vice-president of the Chinese Academy of Sciences, Zhou Guangzhao] *Renmin ribao* [*People's Daily*], July 23, 1984.

46. Liu, Liu, and Xie, *Liang dan yi xing*; Iris Chang, *Thread of the Silkworm* (Basic Books, 1996). The "two bombs" referred to the two distinct technological projects of nuclear bombs and missiles.

47. "Quanmin jian tulu" [The entire population builds local furnaces], *Renmin ribao* [*People's Daily*], August 30, 1958.

48. Lewis and Xue, *China Builds*, 87–88.

49. Ibid., 205.

50. "Yi tu dai yang, yi tu sheng yang" [Using the local to replace the foreign, using the local to create the foreign], *Renmin ribao* [*People's Daily*], October 17, 1958. *Yang zhuanjia* could refer broadly to people whose knowledge came from foreign sources (including foreign publications, hence by extension the professional research establishment), but could also more specifically refer to intellectuals who had returned from abroad or to Soviet advisors.

51. Taylor, *Chinese Medicine*, 116–18.

52. "Yi ke hongxin, liang zhi shou" [A red heart and two hands], *Renmin ribao* [*People's Daily*], September 7, 1970.

53. See Schmalzer, "Insect Control."

54. See, for example, "Tu nongyao fangzhi daowen xiaoguo hao" [Native agricultural chemicals show good results in controlling rice paddy diseases], *Nongye kexue tongxun* [*Agricultural Science Bulletin*] 1958, no. 9: 503. Note that producing native plant-based pesticides was of interest already in the Republican era. Yun-pei Sun and Ming-tao Jen conducted research in this area at the University of Minnesota, and Ting Wu (Mrs. T. Shen) applied for a fellowship from the China Foundation (funded by the Boxer Indemnity) to conduct similar research (I am unsure whether she was successful). Letter from H. H. Shepard to the China Foundation for the Promotion of

Education and Culture, dated 22 December 1939, UMN archives. Collection 938: Division of Entomology and Economic Zoology. Box 10. Folder: "China—Miscellaneous, 1937–1945."

55. Chen Qinde, "Pinxia zhongnong shi nongye kexue shiyan de jianbing" [The poor and lower-middle peasants are the vanguard of agricultural scientific experiment], Guangdong sheng pinxia zhongnong daibiao he nongye xianjin danwei daibiao huiyi fayan zhi shiyi [eleventh speech at Guangdong Provincial Conference for Representatives of the Poor and Lower-Middle Peasants and Advanced Work-Units], 1965, Guangdong Provincial Archives, 235-1-365-034~036.

56. See, for example, Heilongjiang sheng Binxian Xinlisi dui keyan xiaozu, "Bai ying dadou wang de xuanyu" [The selection of "white-breast soybean king" (variety of soybean)], *Nongye keji tongxun* [Agricultural science and technology bulletin] 1973, no. 12, 4.

57. Guangdong sheng kejizu, "Zazhong youshi liyong he shengwu fangzhi liangxiang huizhan jinzhan qingkuang" [Progress in the two campaigns of utilizing superior hybrids and biological control], May 28, 1972, Guangdong Provincial Archives, 306-A0.02–41–85; "Sihui xian Dasha gongshe shuidao bingchonghai zonghe fangzhi qingkuang huibao" [Report on the integrated control of rice-paddy diseases and insect pests in Dasha Commune, Sihui County], Guangdong Provincial Archives, 306-A0.06–12–42~47.

58. See, among many other examples, *Nongcun zhishi qingnian kexue shiyan jingyan xuanbian* (Beijing renmin chubanshe, 1974), 31, 35, 66.

59. Guangdong kejizu, "Guanyu shenqing kexue shiyan qunzhong yundong jingfei de baogao" [Report on applying for funds for the mass movement for scientific experiment], October 15, 1974, Guangdong Provincial Archives, 306-A0.05–22–16.

60. Sheng nonglinshui zhanxian weiyuanhui, "Xiada 1970 nian kexue shiyan buzhu fei" [1970 allocation of supplementary funds for scientific experiment], July 10, 1970, Guangdong Provincial Archives, 277–2-22–1~1; Zhongkeyuan, Caizhengbu, "Zeng bo qunzhongxing kexue shiyan huodong buzhu jingfen de han" [Letter regarding the additional appropriation of supplemental funds for mass-scientific experiment activities], October 31, 1975, Guangdong Provincial Archives, 306-A0.05–22–91.

61. See, for example, Jacob Eyferth, *Eating Rice from Bamboo Roots: The Social History of a Community of Handicraft Papermakers in Rural Sichuan, 1920–2000* (Harvard University Press, 2009).

62. *Nongcun zhishi qingnian*, 37–39. For another example in which local needs trumped the call to prioritize grain production, see Peter Ho, "Mao's War Against Nature? The Environmental Impact of the Grain-first Campaign in China," *China Journal* 50 (July 2003): 37–59.

63. *Kexue zhongtian de nianqing ren* [Youth involved in scientific farming] (Zhongguo qingnian chuban she, 1966), 33.

64. "Gaoju Mao Zedong sixiang hongqi gengjia guangfan shenru de kaizhan nongcun qunzhongxing kexue shiyan yundong (cao)" [Hold high the red flag of Mao Zedong Thought in order to increase, broaden, and deepen the development of the rural mass-scientific experiment

movement (draft)]," November 15, 1965, Documents on the Beijing Municipal Rural Scientific Experiment Group Positive Elements Conference, Beijing Municipal Archives, 002-022-00031, 1-15.

65. Mao Tse-tung, *Quotations from Chairman Mao*, second edition (Foreign Languages Press, 1966), 40.

66. The broad and politically potent significance of "experiment" was related to its relevance within Chinese communist political philosophy, with origins in the revolutionary period. See Sebastian Heilmann, "From Local Experiments to National Policy: The Origins of China's Distinctive Policy Process," *China Journal* 59 (2008): 1-30.

67. "Zhuangzu guniang xue Dazhai: Kexue zhongtian duo gaochan" [Zhuangzu girls study Dazhai: Scientific farming reaps bumper harvests], *Guangxi nongye kexue* [Guangxi agricultural science] 1975, no. 7, 32-35.

68. Gongqingtuan zhongyang qingnong bu, ed., *Wei geming gao nongye kexue shiyan* [Agricultural scientific experiment for the revolution] (Zhongguo qingnian chubanshe, 1966), 8.

69. Shen Dianzhong, *Sixiang chenfu lu* [The ebb and flow of my thoughts] (Shenyang: Liaoning renmin chubanshe, 1998), 286-89. Shen Dianzhong later became the director of the Institute of Sociology at the Liaoning (Province) Academy of Social Sciences.

70. Sigrid Schmalzer, "Speaking about China, Learning from China: Amateur China Experts in 1970s America," *Journal of American-East Asian Relations* 16, no. 4 (2009): 313-352.

71. Perry Johansson, "Mao and the Swedish United Front against USA," in *The Cold War in Asia: The Battle for Hearts and Minds*, ed. Zheng Yangwen, Hong Liu, and Michael Szonyi (Brill, 2010). On Chinese efforts to enroll Mexico, see, in the same volume, Matthew Rothwell, "Transpacific Solidarities: A Mexican Case Study on the Diffusion of Maoism in Latin America."

72. Jamie Monson, *Africa's Freedom Railway: How a Chinese Development Project Changed Lives and Livelihoods in Tanzania* (Indiana University Press, 2009).

73. Deborah Bräutigam, *Chinese Aid and African Development: Exporting Green Revolution* (St. Martin's Press, 1998), 1-2 and 176-179.

74. Michael Mahoney, "Estado Novo, Homem Novo (New State, New Man): Colonial and Anticolonial Development Ideologies in Mozambique, 1930-1977," in Engerman, *Staging Growth*, 191.

75. "Wenhua da geming hou de Zhongguo: Nongye kexue jianwen" [China since the Cultural Revolution: The view from agricultural science], *Duiwai xuanchuan cankao ziliao* [Reference materials on external propaganda] 13 (June 25, 1973): 1-4. I thank Michael Schoenhals for sharing this extraordinary source.

76. "Wenhua da geming hou," 1-2. The Japanese scientist was Tamura Saburō.

77. Naomi Oreskes, "Science, Technology, and Ideology," *Centaurus* 52 (fall 2010): 297-310.

78. Peter Buck, *American Science and Modern China, 1876–1936* (Cambridge University Press, 1980).

79. American Plant Studies Delegation, *Plant Studies in the People's Republic of China: A Trip Report of the American Plant Studies Delegation* (National Academy of Sciences, 1975), 164.

80. Food and Agriculture Organization of the United Nations, *Learning from China: A Report on Agriculture and the Chinese People's Communes* (Food and Agriculture Organization of the United Nations, 1977), 55.

81. H. C. Chiang, "Pest Management in the People's Republic of China: Monitoring and Forecasting Insect Populations in Rice, Wheat, Cotton, and Maize," *FAO Plant Protection Bulletin* 25, no. 1 (1977), 3.

82. H. C. Chiang, "Pest Control in the People's Republic of China," *Science* 192, no. 4240 (1976): 676.

83. See, for example: Science for the People, *China: Science Walks on Two Legs* (Avon, 1974); Victor Sidel, *Serve the People: Observations on Medicine in the People's Republic of China* (Beacon, 1973); Arthur Galston, *Daily Life in the People's Republic of China* (Crowell, 1973).

84. Taylor, *Chinese Medicine*, 141.

85. Cultural Revolution-era claims about the effectiveness of acupuncture anesthesia have since been widely challenged.

86. American Insect Control Delegation, *Insect Control in the People's Republic of China: A Trip Report of the American Insect Control Delegation, Submitted to the Committee on Scholarly Communication with the People's Republic of China* (National Academy of Sciences, 1977), 142.

87. Mai Baoxiang, "Pu Zhelong jiaoshou zai Dasha de rizi" [Professor Pu Zhelong's days in Dasha], posted to Sun Yat-sen University website Zhongda chunqiu [Annals of Sun Yat-sen University], http://baike.sysu.edu.cn. This was apparently Professor M. J. Way.

88. Per Brinck, ed., *Insect Pest Management in China* (Ingenjörsvetenskapsakademien, 1979), 2. On the generally very positive impressions foreign visitors had of Chinese science during the late Cultural Revolution, see Boel Berner, *China's Science through Visitors' Eyes* (Research Policy Program, 1975).

89. Robert L. Metcalf, "China Unleashes Its Ducks," *Environment* 18, no. 9 (1976): 14–17. Some of the material in this paragraph is taken from Schmalzer, "Insect Control."

90. Schneider, *Biology and Revolution*; Sigrid Schmalzer, *The People's Peking Man*.

91. Zhengao Li and Huayong Zhang, "Application of Microbial Fertilizers in Sustainable Agriculture," in *Nature Farming and Microbial Applications*, ed. Hui-lian Xu, James F. Parr, Hiroshi Umemura (Food Products Press, 2000), co-published simultaneously as *Journal of Crop Production* 3, no. 1 (2000).

92. In addition to advocates of "nature farming" and followers of permaculture (the term derives from King's description of "permanent agriculture" in East Asia), proponents of "integrated farming" (or "agro-ecological farming") point to examples of traditional farming techniques in China, the most famous example being the mulberry-silkworm-fish pond systems of the Pearl River Delta. See, for example, Li Wenhua, ed. *Agro-Ecological Farming Systems in China* (UNESCO, 2001); George L. Chan, "Aquaculture, Ecological Engineering: Lessons from China," *Ambio* 22, no. 7 (1993): 491–494.

93. Vaclav Smil, Review of *Learning from China? Development and Environment in Third World Countries*, ed. Bernhard Glaeser, *Journal of Asian Studies* 47, no. 3 (1988): 595.

94. Susan Greenhalgh, *Just One Child: Science and Policy in Deng's China* (University of California Press, 2008).

4 From the End of the World to the Age of the Earth: The Cold War Development of Isotope Geochemistry at the University of Chicago and Caltech

Matthew Shindell

In the late 1960s, the isotope geochemist Clair "Pat" Patterson—already famous among earth scientists for determining a precise age for the planet—reported the disturbing results of his study of lead concentrations in the ice of northern Greenland.[1] Using ice cores that contained uninterrupted sequences of annually deposited snows reaching back several centuries, Patterson and his colleagues had determined a record of environmental lead that showed the increase of lead pollution from the Industrial Revolution to the 1960s. The lead in the snow told the story of the large-scale contamination of Earth's atmosphere and surface—contamination that had been proceeding at an increasingly rapid pace since the 1940s. Furthermore, Patterson reported, much of this dramatic rise in lead was due to the addition of tetraethyl lead to gasoline by the petroleum industry in the 1920s.[2]

Patterson's ice-core data helped to reopen a controversy about leaded gasoline that had long been closed. Lead poisoning was a recognized occupational hazard of many industries in the early twentieth century, and many of the health risks of tetra-ethyl lead were already known in the United States at the time of its introduction into gasoline. Some advocates of industrial medicine, such as Alice Hamilton of Harvard University, had seen the effects of lead poisoning on refinery workers and emphasized how easily the mucus membranes of the respiratory tract absorbed lead. Hamilton tried to raise alarm about the possible broader health risks of tetraethyl lead in the environment, especially for developing children, by extending the known effects of lead on workers to the possibility of contamination at gas stations, in garages, and on city streets.[3]

Despite Hamilton's efforts, the petroleum and auto industries successfully convinced the government and the public that the only risks posed by the additive were to those who worked with the material in its concentrated form in the refineries. If it was simply a workplace concern, then no substantial regulations or safety measures were needed outside of factories and refineries. Forty years later, Patterson's work demonstrated that the lead contained in gasoline, once released, reached not only human mucous membranes, but nearly every part of the world—making its way onto and into food and water sources and the air. Evidence and congressional testimony by Patterson resulted in regulatory measures that eventually phased out the use of leaded gasoline in the United States.

Patterson's discovery of lead contamination in the Greenland ice cores changed the way we live and do business as few scientific discoveries have done. Another similar example from the twentieth century is the discovery by Frank Sherwood "Sherry" Rowland that chlorofluorocarbons (CFCs) released into the atmosphere cause ozone depletion.[4] Both Patterson and Rowland utilized physical methods that hadn't been available to Alice Hamilton.

It is no coincidence that Patterson and Rowland received their PhDs from the University of Chicago only a year apart. As junior members of Chicago's postwar Institute for Nuclear Studies (later renamed in honor of Enrico Fermi), working under the Manhattan Project alumni Harold Urey, Harrison Brown, and Willard Libby, they participated in the development of the foundational methods of geochemistry and atmospheric chemistry. Nor is it a coincidence that the US Atomic Energy Commission, one of the primary patrons of Cold War earth science, supported both chemists' careers. Patterson would later claim that his discovery of lead contamination began as an unanticipated by-product of his attempts to determine the age of the Earth using lead-isotope ratios, basic research that he claimed the AEC supported unwittingly.[5] However, historical examination makes it clear that basic research and the internal logic of scientific development alone don't account for the rise of the methods or the scientific communities that made Patterson's (or Rowland's) work possible. Patterson's story, understood in the context of the postwar rise of isotope geochemistry in the United States, is a Cold War story.

The rise of military and government funding for the disciplines that would become known as the earth sciences in the United States after World War II brought research in these disciplines levels of support that were second only to those received by physics. This support created what Ronald Doel refers to as a "new intellectual map" for the earth sciences, "a new set of challenges, guided by military and national-security needs, which elevated the fortunes of certain fields of the physical environmental sciences and decreased opportunities in others."[6] Both Doel and Naomi Oreskes agree that one important development that resulted from the military push was the increasing domination of deductive physical methods, particularly those of geophysics. Oreskes has described this push as a move from the field to the laboratory, involving an adoption of "the concomitant values of exactitude and control that laboratory work suggests."[7] Doel and Oreskes further assert that the rise of physical laboratory methods in the earth sciences during this period had little to do with their historical successes at settling controversies within the geosciences, but rather came as "the result of an abstract epistemological belief in the primacy of physics and chemistry, coupled with strong institutional backing for geophysics premised on its concrete applicability to perceived national-security needs."[8]

Geochemistry, often treated by its sponsors as a branch of geophysics, was supported generously by military, governmental, and industrial patronage during the Cold War. But whereas geophysics can be described as one of the two competing traditions within American geology before the war, isotope geochemistry arose almost entirely after the war. After World War II, isotope geochemistry developed rapidly from the purview of a handful of physicists and physical chemists into a transformative force for university geology departments throughout the United States. This development was due in part to the increased availability of mass spectrometers, which made precise isotopic measurements possible. Mass spectrometers had existed before the war, but only a few laboratories had had sufficient expertise to build and maintain them. Wartime development of mass spectrometers by US industries under military contract and postwar tinkering by interested engineers, scientists, and oil companies effectively black-boxed the technology (although a trained technician was still required to operate it), increased its precision dramatically, and put it within reach of any university department willing to pay for the still relatively expensive instrument.

But the success of geochemistry is not just the story of the migration of new instruments into an existing geosciences community. It is also the story of a set of outsiders—atomic chemists and physicists—who turned their attention to the geosciences, forged new bonds with geological and geophysical departments and institutions, and helped to train a new generation of hybrid scientists to use isotopes and the techniques of mass spectrometry to define new lines of inquiry. Many of the atomic scientists who turned to the geosciences after the war were turning away from the technical work they had contributed to the Manhattan Project. But although geochemistry may in some ways have represented an escape to pure or basic science from weapons work, or an avoidance of future conscription in the production of weapons of mass destruction, these scientists and their students were nonetheless enlisted in the construction of Cold War science.

More often than not, government or military contracts paid for the mass spectrometers and the salaries of those who used them. The questions to which these technologies were applied tended to be related to the concerns of the contracting agency. At least on paper, geochemistry research programs often evolved around questions central to such activities as the search for and understanding of nuclear fuel sources, the use of isotopes as tracers for explosions, and, when the Navy paid for research, the characteristics of the sea floor and ocean circulation.[9] As in other areas of Cold War contract research, geochemists found ways to fit their own questions within the scope of their research contracts. However, even when the patrons of geochemistry got something other than what their contracts specified, they seem to have felt they got their money's worth—at least to the extent that they remained willing to renew funding for these same projects year after year.

What follows is not a chronological history of the development of geochemistry. Rather, I present two episodes from the early history of postwar geochemistry that illustrate the forces that helped to establish geochemistry in the early Cold War. These episodes focus on the development of geochemistry at two institutions: the University of Chicago's Institute for Nuclear Studies and the California Institute of Technology's Geology Division.

The first episode I examine is Harold C. Urey's entry into the field of isotope geochemistry. I highlight the aspects of this move that illustrate the causal links between the Cold War and Urey's new paleotemperature research program at Chicago. Urey's move to the University of Chicago's new Institute for Nuclear Studies after the war brought him into one of the nation's first Cold War institutions. I suggest that the University of Chicago was ahead of the national curve in its promotion of the new alliance between science, industry, and the government in order to support large-scale individual research programs. The postwar research budgets at the Institute for Nuclear Studies reflected the scale of research funding that became available during the Cold War, made possible by the postwar model of government-funded research.

Examining the various channels through which Urey funded his research program, and how he took advantage of the varying interests of industrial, military, and governmental patrons, I also suggest that these interests weren't stable throughout the 1950s. While interested oil companies and the Office of Naval Research (ONR) supported the initial development of Urey's new techniques and research program, it is doubtful that his program would have survived had the Atomic Energy Commission not emerged as a patron. These funding agencies clearly had different interests in Urey's work, and their interests by no means promised permanent funding. Urey's success was due to his ability to align these interests in his new project and in the larger purpose of the Institute, and ultimately to his ability to wield his clout as an atomic insider—an expert on the deuterium and heavy water so important to the AEC—to frame his research program as an extension of this expertise.

Though not all universities would adopt the Institute for Nuclear Studies' model of scientific research, the funding structure and the new hybrid sciences initiated in the Institute would later aid in the transformation of university departments around the country. The second episode I examine deals with the importation of geochemistry to the California Institute of Technology (Caltech) from Chicago. The arrival of Caltech's new president, Lee A. DuBridge, ended an era of reliance upon private money. DuBridge insisted that Caltech's researchers actively raise their own funds. In so doing, he ushered in a new era of contract research at Caltech. Caltech's Geology Division responded to this challenge by turning to geochemistry and the AEC. Recruiting Harrison Brown from Chicago to initiate a research program in geochemistry had an immediate effect on the Geology Division. As DuBridge and Brown's recruiters had hoped he would, Brown brought with him a sizable contract from the AEC. Under the pretext of solving the AEC's uranium-supply problems, Brown supported a broad geochemistry research program, including the lead-isotope work that led to Patterson's determination of the age of the Earth and his realization of the extent of industrial lead pollution.

Brown's contract not only supported his own research program; it also redirected the focus of existing research programs, raised the Geology Division's prestige within the university, and permanently changed the division's institutional culture. These changes at Caltech (along with earlier trends in Chicago's Geology Department) are illustrative of changes within the entire discipline of geology as its practitioners and institutions, encouraged by their patrons, moved toward laboratory methods, began favoring the geophysical and geochemical over the traditionally geological, and joined the emerging earth sciences.

Shaping the University of Chicago's Institute for Nuclear Studies for Postwar Research

In the early postwar years, the University of Chicago became a hotbed of activity in the emerging field of isotope geochemistry. In the respective laboratories of Harold Urey, Harrison Brown, and Willard Libby, the atomic scientists and their research teams devised the oxygen thermometer, uranium-lead dating, and carbon-dating techniques that would become methodological mainstays of earth science and planetary science during the Cold War. These three men, along with Enrico Fermi, Edward Teller, Joseph Mayer, Maria Goeppert Mayer, and a handful of other atomic scientists, were drawn to Chicago after the war by the university's new Institute for Nuclear Studies. Most of the founding members of the Institute had distinguished themselves through wartime service to the Manhattan Project. To no small degree, the early success of the INS was due to these scientists' wartime achievements, and much of the structure and activity of the INS were inspired by wartime work.

The University of Chicago had been a good place to be a physicist before the war, a fact reflected in the university's roster of eminent physicists. The wartime uranium-related contracts were due in no small part to the Nobel laureate Arthur Holly Compton, whose efforts to establish uranium work at Chicago drew upon the reputations of fellow University of Chicago physicists such as Arthur Dempster, Samuel Allison, and William Zachariasen. Compton became the architect of the wartime model of research at Chicago. He styled himself as a "bridge-builder between three diverse and separate sorts of people, each inclined to be rather suspicious of the others: to wit, government and the military, business and engineering, and a hastily assembled array of academic physicists and chemists."[10] And it had been his decision, a little more than a month after he had been put in charge of the uranium project, to move many of the scientists who had been working on achieving a chain reaction at Princeton and Columbia to the University of Chicago campus under the umbrella of the Metallurgical Laboratory (called, informally, the Met Lab).[11]

As early as 1943, Compton imagined that the Met Lab might have a postwar life as a leading institution in the maintenance of scientific and technical leadership in nuclear research. He persuaded the University of Chicago's president, Robert M. Hutchins, to pull the university out of military research as soon as possible, but to retain management of Argonne Laboratory and build a complimentary academic program. Under Compton's guidance, the Met Lab prepared for its postwar life.[12] For their own part, the atomic scientists were evidently happy to make their home at the University of Chicago. Even before being approached about the INS, many of them felt that something like it would be possible after the war, and felt that Chicago was well suited for it.

The atomic scientists felt that Hutchins' university—already baptized by fire into the world of atomic research—"understood the trend of the times" and "would not confine the activities of basic research to the meager laboratories and still more inadequate funds available before the war."[13] The wartime experience had transformed the University of Chicago. During the war, as a result of research and training contracts from the government's Office of Scientific Research and Development (OSRD) and the military, the university's annual budget had swelled to $32,290,945 at its height in the year 1944–45, approximately three times its prewar level.[14] About $3 million of this funding was overhead for the university's role in the Manhattan Project, for which the university not only coordinated the initial uranium studies but also oversaw the industrial plants at Oak Ridge and Hanford.[15] That much money couldn't help but alter the expectations of the faculty and administration of the university, who before the war had been used to working with modest research budgets.

Compton—who didn't remain at Chicago to see the postwar Institute for Nuclear Studies—had nonetheless inaugurated there a *modus operandi* very similar to what Rebecca Lowen has described as the "Cold War University"—administrators, scientists, and an array of patrons that included the government and the private sector together defined research goals, and the university benefited from the overhead that the new contract system generated.[16] Samuel Allison, the INS's first director, made it clear to new recruits that the purpose of the INS was to maintain the scale of wartime research without the wartime atmosphere and its "emphasis on technical details, haste, and military applications."[17] Despite the INS's three main foci—nuclear physics, radiochemistry, and isotope separation—the lab wasn't organized around specific research programs. Instead, each member of the institute was promised the independence to develop whatever research program they desired.

Paying for the $12 million Institute for Nuclear Studies, its luminaries, and its large technical staff would require one final—and crucial—continuation of the wartime research model: the research contract.[18] The earliest major funders of the INS included, of course, the Office of Naval Research. As Roger Geiger and Jacob Hamblin have described, the Navy was particularly eager to catch up in developments in nuclear science, and found ways of appealing to the idealism toward research that prevailed at Chicago after the war.[19] The INS also tapped into the atomic enthusiasm of industry. The university offered companies the opportunity to buy "memberships" in the INS at prices beginning at $20,000 per year. The companies didn't enter into these memberships without expectations. "For their money," the *New York Times* reported, "they share in facilities they couldn't buy for the same sums individually. They have a share in what is described as the world's largest nuclear studies program of a privately supported university."[20]

From Isotope Separation to Isotope Geochemistry

Harold Urey came to the University of Chicago from Columbia University, where as wartime director of the Manhattan Project's Substitute Alloy Materials Laboratory (SAM Lab) he had been in charge of developing plans for the separation of uranium isotopes via thermal diffusion. Before the war Urey was one of the nation's foremost experts on methods of isotope separation. After receiving the 1934 Nobel Prize in chemistry for the discovery of deuterium (a heavy isotope of hydrogen with mass number of 2), he had devoted much of his career to developing and refining separation techniques. When Urey first entered negotiations with Chicago for his postwar position in the INS, the research plan he presented to Chicago was for an isotope-separation program that required roughly $68,000 per year for salaries, an equal amount for general laboratory apparatus, and $100,000 for the construction of new mass spectrometers and other specialized instruments.[21]

By the time the war ended, however, Urey had been so traumatized by his experience as director of the SAM Lab that he could no longer muster any enthusiasm for the prospect of continuing isotope-separation work.[22] In part, Urey's trauma stemmed from his having been forced from an active scientific role in the uranium project to a managerial one. The stress of management was compounded by his constant head butting with his superiors, namely James B. Conant and General Leslie R. Groves, over issues of planning and secrecy. Urey later told the historian John Heilbron: "I was most unhappy during the war. I had bosses in Washington who didn't like me, and I had people working for me who didn't like me. Imagine a more miserable situation— where you can't resign, but nobody wants you around!"[23] His wartime experience had pushed him "very close to a nervous breakdown," and his health had deteriorated to such a degree that it became a matter of concern for Groves' personal physician.[24] All this was followed by the extreme demoralization that came with the use of the atomic bomb on Hiroshima and Nagasaki.

After the war, Urey wanted little to do with the isotope-separation work that had won him his fame. His fellow physical chemist Hans Suess remembered that, whereas most scientists were able and eager to return to their prewar research programs, Urey was "anxious to get away as far as possible, in time as well as in space, from everything connected with weaponry and means of destruction," including his prewar work on isotope separation.[25] According to Joseph Mayer, the trauma of war work stuck with Urey for some time, even after he took up residence at the INS, and caused him to "drift, looking for new fields to conquer."[26]

Aimlessness and angst weren't characteristic of Urey, who before the war approached his scientific projects with great enthusiasm and what his colleagues described as childlike curiosity.[27] In his Nobel address, more than a decade earlier, Urey had reported excitedly on the thermodynamic properties of isotopes and speculated upon the possible methods of separating the isotopes based on these differences. Once put into practice at Columbia with a string of graduate students, lab assistants, and grants from private foundations such as the Carnegie Institution, this work came to define the research program in Urey's lab up through World War II. This prewar work was of interest mainly to a relatively small number of physicists and chemists who were interested in the structure and behavior of the elements and their isotopes, and to an even smaller number of biologists who were interested in using these isotopes as experimental tracers. During the war, however, these efforts were accelerated, and Urey's workforce grew larger and more difficult to manage as he took on the directorship of the SAM Lab.

If Urey's war trauma was the primary reason he was aimless in the immediate postwar years, his activity on behalf of the control of atomic weapons was a close second. Urey's involvement with the various organizations of the scientists' movement consumed him in the first few years after the war. "I've dropped everything to try to carry the message of the bomb's power to the people," he told *The New Yorker*.[28] This was no small commitment. Urey had been a popular public speaker before the war, but his postwar pace, combined with the urgency of the atomic problem, was difficult for him to handle. After two years of working for world governance of atomic weapons, Urey wrote to his scientific hero, Albert Einstein, co-founder of the Emergency Committee of Atomic Scientists, that his doctors had ordered him to avoid outside activities: "I find that I am able to carry my university work and that is about all. Otherwise I become very tired, unable to sleep, and generally quite unable to take care of any of my work."[29]

These years were not entirely aimless. At the end of 1946, while still in search of a new line of active scientific work, Urey prepared and delivered that year's Liversidge Lecture before the Chemical Society of the Royal Institution in London. The Liversidge Lecture was one of Urey's last outstanding prewar commitments. In it, Urey chose to update the earlier isotope exchange equilibria that he and Lotti J. Greiff had calculated and published in the 1930s. This time Urey employed a more sophisticated method that had been developed for the SAM Lab by Jacob Bigeleisen and Maria Goeppert Mayer. In the 1930s, Urey and Greiff had shown that relatively large differences in the physical and chemical properties of isotopic compounds could be detected—differences that were then exploited in the various separation techniques developed in the intervening years.[30] Revisiting the thermodynamic properties of isotopes, now with his postwar aversion to separation, Urey instead emphasized another way that these chemical differences could be exploited.

Urey turned his attention to the geological abundances of the isotopes of carbon and oxygen. He noted that certain processes in nature tended to result in isotope enrichment. Aquatic carbonate-precipitating organisms, which used oxygen in their metabolic processes, tended to concentrate oxygen-18 (the more common of oxygen's two heavy isotopes) preferentially. The shells of these organisms often contained up to 4 percent more of the isotope than their surrounding waters. This enrichment was sensitive to temperature, Urey's tables suggested—a temperature change of 25°C resulted in a change in the oxygen-18/oxygen-16 ratio of 1.004 relative to the water. "These calculations suggest investigations of particular interest to geology," Urey commented.[31] He further speculated that, with the mass spectrometers that had been developed during the war by the University of Minnesota physicist Alfred O. C. Nier (work that Urey had overseen), a researcher could determine the oxygen-isotope ratio of carbonate rock samples to within an error of ±0.001, and perhaps discover the temperature at which the rock was deposited with a certainty of within 6°C or less. Urey admitted that there was still a great deal of experimental investigation left to perform before the method could be put to use, but he felt confident that oxygen-isotope abundances were well suited for the determination of historic temperature changes. He concluded his 1947 paper on this subject by stating that the same small differences in the thermodynamic properties of isotopes and their compounds that "make possible the concentration and separation of the isotopes of some of the elements [in the laboratory]" might "have important applications as a means of determining the temperatures at which geological formations were laid down."[32]

Although Stephen Brush, in his account of the postwar rise of geochemistry and cosmochemistry, mentions Urey's ability to use his prestige to attract researchers to the new fields that he pioneered, in the beginning of his work on paleotemperature Urey seems to have had difficulty finding younger scientists to work in his new research program.[33] The first postdoctoral fellow Urey managed to attract was Samuel Epstein, a young Polish-born Canadian chemist with mass-spectrometer experience. Epstein had studied in Canada under Urey's former research assistant Harry Thode, and it was Thode who convinced Epstein to work on Urey's new research project. As Epstein later remembered it, even though Urey had already publicized his speculation about the possibility of using isotopes in carbonate rocks to determine paleotemperature, people weren't lining up to work with him on the problem. Of course, one thing that drew Epstein to Urey's lab was the promise of funding, and as the amounts of available funding grew, so too did the line of graduate students and postdocs eager to learn and practice the new geochemistry. For the time being, however, the research team was small: Epstein, Urey, Charles McKinney (an electrical engineer), and John McCrea (a graduate student). It was McKinney's job to work with Nier's designs and produce a working mass spectrometer. By 1948, Urey had his first such instrument.

Once research got underway, Epstein witnessed Urey's "comeback in the scientific academic world": "He never walked up a set of stairs one step at a time, always two steps at a time. … I clearly remember him coming into the laboratory dressed meticulously in a white shirt and coming home with a shirt stained with oil because he couldn't resist the temptation of changing a dirty oil pump or some other work that was usually left to the younger set."[34] Now feeling at home in the INS, and excited again by what promised to be a fruitful research program, Urey was able to leave behind the traumas of war work.

Moving into geological territory meant that Urey had to develop a new network of scientific contacts and collaborators. In addition to Epstein, McKinney, and McCrea, Urey also drew upon colleagues in Chicago's Department of Geology. Before the war, Chicago's geologists already had tended to be more lab-oriented than field-oriented. The Chicago geologists considered their geophysical program to be one of the strongest in the country, housing one of the only working high-temperature petrology labs outside of the Geophysical Laboratory at the Carnegie Institution of Washington. The Chicago Department of Geology's close ties with the Carnegie Institution's Geophysical Laboratory were embodied in its first postwar chairman, Norman Levi Bowen. A practitioner of thermodynamic geochemistry before the war, Bowen had left Chicago from 1942 to 1944 for war work at the Geophysical Laboratory, and after two postwar years as chair of the Chicago Department of Geology returned to Washington. His chairmanship was brief, but it brought the Department of Geology an unprecedented increase in funding—at Bowen's request, the University of Chicago increased the department's typical expense and equipment budget of $1,500 per year to $45,000 for the first three postwar years.[35] This influx of money enabled the department to invest in new equipment and allowed for the conversion of some existing facilities into state-of-the-art laboratories for analytical chemistry.

After Bowen's departure, Walter H. Newhouse became chairman of Chicago's Geology Department. Newhouse made it his mission to modernize the department and to eliminate traditionalism within it. Starting in 1946, the department, now feeling itself to be in competition with its counterparts in physics and chemistry, adopted Newhouse's attitude that "anyone on the staff who was not opening up brand-new fields was a piece of dead wood."[36] To facilitate change, the department hired new faculty members, including the geochemists Julian Goldsmith, Hans Ramberg, and Kalervo Rankama. These men, particularly Goldsmith, would work closely with Urey, Libby, and Brown to bridge the gap between the INS and the Chicago geologists, and all would assist in proposing a joint curriculum in geochemistry for students who wished to become geochemists.[37] As early as 1947 the department was receiving ONR contracts to do geophysical research. The Navy even put some "Paperclip Specialists" (German scientists who had worked under the Nazi regime) under the care and supervision of Chicago's geology faculty.[38]

In 1947 Urey secured the cooperation of the German-born paleoecologist Heinz Lowenstam, who had left Germany before the war and was working for the Illinois State Geological Survey. In 1948 the Department of Geology hired Lowenstam specifically to work with Urey on his paleotemperature studies, and Lowenstam's salary was paid through Urey's research contracts.[39] Beginning in 1948, Urey asked scientists at the Scripps Institution of Oceanography in La Jolla, and at other marine laboratories, for shells and information about the waters in which they had been deposited. Epstein and Lowenstam worked to develop methods for preparing uncontaminated samples of carbon dioxide gas from calcium carbonate shells, then worked to establish a temperature scale for oxygen isotopes. The first published results of this work appeared in 1951, shortly before Epstein and Lowenstam left Urey and Chicago to join Harrison Brown at Caltech.[40]

Urey's lab also benefited from the growth and expansion of the University of Chicago's Department of Geology. In 1950 Cesare Emiliani completed his PhD work in the department and went to work with Urey's paleotemperature group in the INS. Emiliani extracted foraminifera shells from long deep-sea cores. Using those shells, the group studied temperature variations in the Pleistocene and estimated the length and severity of the ice ages. The acquisition of the deep-sea cores was evidence of Urey's diverse and expanding scientific network and of his connection to the emerging earth-science network. In 1950, Urey's lab began collaborating with Columbia University's newly established Lamont Geological Observatory, a "quintessential Cold War institution" that Columbia had established in order to take advantage of military support for geophysics research.[41] There, with substantial support from the Office of Naval Research, Maurice Ewing had developed a method for piston coring seafloor sediment. The Navy gave Ewing access to broad swaths of the deep ocean, and Ewing's research program was shaped by the Navy's priorities. Throughout the 1950s Ewing and his colleague David Ericson sent core samples to Urey's lab, where Emiliani and the lab's technician, Toshiko Mayeda, prepared and analyzed the samples in the mass spectrometer. But, as an examination of Urey's research funding makes clear, Urey's connection to Cold War military contract research went far beyond his connection to the Lamont Geological Observatory.

Funding the New Program

In its early years, Urey's new research program benefited from the close alliance between the Institute for Nuclear Studies and industry. As Ronald Doel points out, the petroleum industry was a major supporter of geophysical and geochemical research during the Cold War.[42] Both Shell and Standard Oil had bought memberships in the INS, for which they were promised the right of first refusal on any patents or practical applications developed there.[43] During a tour of the INS in 1947, a Shell representative met with Urey and heard about his research plans. The Shell rep came away from the meeting impressed. Writing to the chairman of the American Petroleum Institute's Advisory Committee on Fundamental Research on Occurrence and Recovery of Petroleum, he characterized Urey's research as "of considerable interest, since, if successful, it will help measure one more of the many unknown variables of importance to the origin of oil."[44] Furthermore, he noted, Urey's program might find a place within the API's ongoing Project 43, a broad investigation of the transformation of organic matter into petroleum that included a research team at MIT investigating the effects of radioactivity on the transformation of marine organic materials into petroleum hydrocarbons.[45]

Urey's research program didn't become part of Project 43, but the American Petroleum Institute nonetheless had the impression that it might contribute to the understanding of the processes that produced oil. That impression was attributable to Urey. In his initial courting of API funding, Urey had offered this speculation: "It may be that oil deposits occur in places where the temperature at which they were deposited was unique in some way, and if this should be the case then it might furnish one additional tool for geological exploration for oil."[46] It is also possible that the oil companies were interested in developments in mass spectrometry generally, as the method had been introduced within the petroleum industry in the early 1940s and had proved highly useful as an accurate way of analyzing hydrocarbon mixtures.[47] Urey requested $12,000 for the construction and maintenance of his instruments, but the API was only willing to grant him $5,000 for 1948–49.[48] That amount fell well short of what Urey estimated it would cost just to build his first mass spectrometers, much less do anything with them.

In the summer of 1947, Urey requested funding from the Geological Society of America's Penrose Bequest, playing up the possibility that his work would replace existing qualitative methods of determining paleotemperature—namely paleoecological studies of the fossil organisms found within geological samples—with more quantitative methods.[49] The Geological Society granted Urey $17,900 for salaries to support one chemist, one physicist, and three technicians.[50] That amount, even when combined with the API funding listed above, still didn't approach the $50,000 to $100,000 that Urey estimated he would need in order to build all the necessary instruments and establish the new methodology.

The traditional sponsors of geological and geophysical research—the petroleum industry and the Geological Society of America—provided some support for Urey's work in these early years, but they were either unable or unwilling to provide the amount of money Urey needed in order to launch his new research program in earnest. Eventually they withdrew their support. As the American Petroleum Institute explained to Urey, there were "several other more desirable projects which are basically fundamental in nature, but are still closer to our immediate problems" than was Urey's.[51] Military patrons, however, were both willing and able to make the investment.

In 1949 the amount of funding Urey had at his disposal increased dramatically as he began a new contract with the Office of Naval Research. Urey had participated as a scientific observer in the Navy's Operation Crossroads atomic bomb test at Bikini Atoll in 1946. There he had met Roger Revelle, future director of the ONR's Geophysics Branch. He no doubt also became acquainted with the Navy's attitude that "almost all fields of oceanographic research had potential Navy applications."[52] In 1948 Urey made his first contract proposal to the ONR, asking for about $105,000 for an "investigation of natural abundances of stable isotopes with the primary objective of measuring paleo-temperatures." The proposal was vague about the practical applications of paleoclimate research to the Navy's mission, but Urey did manage to frame the more general aspects of isotope abundance measurements as having the potential to contribute to the Navy's existing mapping program and to develop natural tracer techniques that could be employed in the ocean.[53] The Navy agreed to provide roughly $30,000 per year for four years—a much larger sum than Urey's industrial or private sponsors had yet provided.[54]

The Office of Naval Research was, in some ways, an ideal funding agency for the early years of Urey's research program. The Navy preferentially funded research into the development of new methods and techniques. From the Navy's point of view, Urey's work might help them to better understand the ocean's basic geochemical features and assist them in the development and maintenance of Naval technologies. However, once Urey's methods had been established, the ONR informed him that they were no longer willing to fund his research.[55]

The withdrawal of ONR funding put pressure on Urey to find a new funding agency to take its place. He was able to find two funding agencies that together were able to raise his funding level to still greater heights. In 1953–54, Urey received $55,956 from the AEC and $21,400 from the National Science Foundation (which had established an Earth Science Program in 1953).[56] With more than $75,000 in contract funding, 1953–54 was a banner year for Urey's research program. In his remaining years at

Chicago, before he left for the University of California at San Diego in 1958, Urey would keep his external funding at or slightly above this new level.

Although Urey had decided to leave isotope-separation work behind, much of his clout with the Atomic Energy Commission and the military was attributable to his expertise in the field of isotope separation and his past position as the head of Columbia's SAM Lab. For this reason, it was not only difficult but also impolitic for Urey to completely close the door on isotope separation. In fact, Urey had been involved in the formation of the AEC and had been working under contract with it since November of 1950, first as a consultant on a Heavy Water Production Processes Survey for the AEC's Division of Research.[57] Remaining connected to the AEC's concerns about heavy water and isotope separation—and flexing his expertise in this area at the AEC's behest—helped Urey to maintain the prestige he had earned from his wartime service. It also allowed him to keep abreast of the AEC's concerns (and even at times to define these concerns), and made it easier for him to frame his new projects in language that would garner the AEC's approval. This relationship, which encouraged fundamental research connected to the AEC's concerns, was symbiotic. While Urey received support for non-separation-related research, not only was the AEC satisfied that his new interests were close enough to the AEC's interests that his work merited funding; it was also able to enlist him in the work of advising and planning the AEC's activities. Unclassified projects such as Urey's also gave the AEC examples of AEC-supported research that could be discussed and promoted before Congress and the public.[58]

One example of this symbiosis at work is Urey's reluctant agreement to chair the Committee on Isotope Separation for the AEC's Division of Research in early 1951.[59] In a letter to Kenneth Pitzer, the division's director, Urey wrote: "Long ago I developed a subconscious reaction to all separation jobs. It is, first, that any separation project is an enormous amount of hard and uninteresting work, and second, that it is very likely that all new schemes for separating isotopes will not work."[60] Nonetheless, accepting the position allowed Urey to exert some influence on the direction of isotope work in the United States and put him in constant contact with Pitzer. The Committee on Isotope Separation had a high priority within the AEC's Division of Research. It was responsible for reviewing the literature on isotope separation, the techniques used in it, the atomic energy program's immediate and long-range needs for the separation of isotopes, and the work in progress on isotope separation within the AEC, and it was charged with recommending to the Division of Research what steps should be taken for the investigation and development of specific isotope-separation techniques.[61]

In addition to keeping Urey and his fellow members of the Committee on Isotope Separation connected to the Division of Research, the work also kept them connected to classified materials and places of atomic research. As a contractor and a consultant, Urey maintained the security clearance that had been granted to him during the

Manhattan Project. The AEC installed facilities in the offices of committee members for the storage of classified documents (if they didn't already have such facilities) and initiated clearance procedures for secretaries and technical assistants. The members also received a classified bibliography of sources held in classified libraries at the National Laboratories and the associated universities.

The first meeting of the Committee on Isotope Separation took place at Oak Ridge in February of 1951, and during their stay at Oak Ridge the committee's members were given a full tour of the facilities. Later meetings took place in the New York Operations Office of the AEC and at the University of Chicago. Under the auspices of the AEC's Division of Research, members of the committee also toured the DuPont laboratories in Wilmington, Delaware, where they discussed problems related to heavy water. They met with scientists and technicians at General Electric, Yale, and Brookhaven National Laboratory, and they visited the Washington headquarters of the AEC to learn about the raw-materials situation.[62]

With his knowledge of the inner workings of the Atomic Energy Commission, Urey was able to construct proposals for isotope geochemical work that were directly related to the AEC's concerns, enlisted Urey's prestige as the discoverer of heavy hydrogen and heavy water, and also satisfied his own research goals. In 1949 Harrison Brown floated a "Proposed Program for the Accumulation of Quantitative Data Concerning: the Chemical Composition of Meteorites and the Earth's Crust; the Relative Abundances of Elements in the Solar System; the Ages of the Elements and Planets," and hoped that the AEC would at least fund those parts of the program that were performed at its Argonne facility. The AEC demurred. Urey's first proposal was far more politically savvy in both name and form. Urey's proposal for "Research on the Natural Abundance of Deuterium and Other Isotopes in Nature" outlined an intentionally broad research program that included work to be done on meteorites, igneous rocks, and fossils, with the stated aim of discovering how the abundance of hydrogen isotopes had changed over time.[63] In addition to addressing the AEC's concerns about deuterium and heavy water and their abundances in nature, Urey's proposal also emphasized the scientific attention that his initial work on paleotemperature was receiving, thus tapping into the AEC's desire for visible scientific rewards from unclassified and non-military projects.

Caltech, Lee DuBridge, and the Government as a Customer for Research

Funding from the Atomic Energy Commission helped to sustain and expand Urey's research program and those of other members of the Institute for Nuclear Science, but the work didn't remain in Chicago. Younger members of the INS brought the new techniques and the promise of money to new institutions. The techniques, too, moved independently once reliable mass spectrometers became commercially available and geology departments began competing for newly available funding from the AEC and from the National Science Foundation. In many cases, the new money and the new science contributed to the restructuring of entire university departments. In the case of Caltech's Division of Geology, AEC funding helped to raise a relatively small department in a relatively small institution to national recognition, and by the end of the 1950s had positioned the division to be one of the leading participants in the National Aeronautics and Space Administration's space science program.

In the late 1940s, Caltech's Division of Geology was far from an unknown entity. With a $25,000 grant from the Carnegie Corporation, and under the direction of the physicist Robert A. Millikan (president of Caltech in all but title), the university had in 1925 created the Division of Geology as part of its efforts to expand its scientific mission while building close and mutually supportive ties with local wealth and industry.[64] Two dominant forces in the division were its founding chairman, John Buwalda, and his successor, Chester Stock. Buwalda, a traditional "hard rock" geologist, had come to Caltech from the University of California in 1926, recruited by Millikan himself. While in Berkeley recruiting Buwalda, Millikan had met and been impressed with Stock, a paleontologist who specialized in the mammalian fossils of the Western United States and who made his name in the excavation of the La Brea tar pits.[65] Buwalda and Stock were expected to establish the Division of Geology's research and teaching program with funding from Caltech and from private sources. The two chairmen developed an impressive undergraduate program that emphasized some fundamental training in the physical sciences, followed by specialized training in mineralogy, petrology, paleontology, and geophysics. In contrast with the University of Chicago's department, Buwalda's program required a student to take part in two summer field camps and a year-long course in field methods before graduation.

The Division of Geology lived up to Millikan's expectations and was well regarded by local government and industry. Most of its graduates went on to careers in the petroleum industry, state and federal geological surveys, and mining.[66] The division also enjoyed international prestige for its Seismological Laboratory. Supported by the Carnegie Institution, the Seismological Laboratory employed Harry Wood, Beno Gutenberg, Charles Richter, and Hugo Benioff, world-renowned pioneers in geophysical research and earthquake studies.[67] Although the division produced only a few PhDs, in 1951 its faculty was considered one of the strongest schools of geophysics in the United States.[68]

But the early postwar years brought many changes to Caltech, changes that challenged Buwalda and Stock's program and the funding model upon which it was built. Chief among these changes was the appointment of Lee DuBridge to Caltech's presidency in 1947. DuBridge, a physicist, came to Caltech after directing the MIT Radiation Laboratory's wartime research program. W. Patrick McCray described DuBridge as "part of the interlocking system of boards and committees that shaped postwar science."[69] A close ally of Vannevar Bush, DuBridge was a firm believer in Bush's model of American scientific growth as laid out in the highly influential treatise *Science, the Endless Frontier*.[70] DuBridge approached the postwar growth and expansion of Caltech's scientific mission as his own small scientific frontier in the West.[71] DuBridge agreed with Bush that American scientific ascendancy was possible with the support of the federal government. Moreover, he believed that the federal government had a responsibility to support the production of American science, noting that in the past the United States had been primarily a "consumer" of science from abroad.[72] DuBridge looked for opportunities to get the government to support expansion of Caltech's existing programs.

DuBridge insisted that changes would have to be made within the Division of Geology if it was to continue building its international reputation. DuBridge's attitude toward the Division of Geology can be discerned from his retrospective assessment of Chester Stock's paleontology research program. He brought to this assessment the stereotypical prejudices of a physicist; in Stock's program he saw a "tremendous collection of fossils," but also "a one-man show, practically, with a couple of assistants."[73] That Stock was a well-regarded expert in his field and a member of the National Academy of Sciences meant little to DuBridge if Stock wasn't running a laboratory filled with graduate students and research assistants, using the most modern equipment, and building a research program with government funds. DuBridge also felt that the division's geophysics program wasn't as strong as its proponents claimed. Although the seismologists had built a fine laboratory, the more traditional field approach to geology represented by Buwalda and his hires was preventing the division from "initiating some new and more modern activities."[74] DuBridge later claimed, quite bluntly, that he was responding to "feelings around the campus and outside that the [Caltech] geologists were still back in the nineteenth century, analyzing rocks."[75] Of course, there was great merit to the work that Stock and his less geophysically inclined contemporaries were doing, but their lines of inquiry were to wither as sponsors pushed individuals and institutions away from the field.

Caltech's new president perceived that the days of supporting a world-class, competitive research program entirely with funds from the Institute and from private foundations were coming to an end. In its earlier years, Caltech's research program had relied upon the generosity of private foundations and the wealthy businessmen of Pasadena and Los Angeles, often facilitated by Millikan and Caltech's close affiliation with the National Research Council. However, those private funding sources had fallen short of Caltech's scientific ambitions during the Great Depression and World War II.[76] The war had initiated a new period in Caltech's history, as it became a wartime institution and a recipient of government contracts worth more than $80 million (much of which went to the Jet Propulsion Laboratory).[77] After the war, DuBridge faced the challenge of ending the disruption of wartime mobilization and reestablishing peacetime research while at the same time maintaining the high level of funding to which the Institute for Nuclear Science had by then become accustomed.

With his experience as the wartime director of the MIT Radiation Lab and his intimate knowledge of the Atomic Energy Commission, DuBridge initiated at Caltech a new era of contract research. Though he didn't approach contract research without caution, DuBridge nonetheless used the contract system as a tool for expanding and reshaping Caltech's research mission. By 1951 he was willing to make this claim: "Government contracts have been by far the most important single factor in the post-war improvement and expansion of science in American universities."[78] While congressional debates held up the establishment of Vannevar Bush's proposed central government funding agency, the National Science Foundation, DuBridge estimated that military and government sponsors such as the Office of Naval Research, and the Atomic Energy Commission were already investing about $50 million per year in university research. Though a modest sum in comparison with the billions of dollars spent annually on military research, that still put the NSF's proposed 1950 budget of $3.5 million to shame. Caltech was already heavily invested in contract research. Government contracts provided nearly a third of Caltech's budget.[79] Without the contract money, and in particular the government's inclusion of overhead expenses (customarily calculated at 40 percent of the amount budgeted for salaries), Caltech wouldn't be able to cover the many bureaucratic costs of administering large-scale research, let alone conduct the research itself. DuBridge warned that if the Institute for Nuclear Science had to suddenly withdraw from contract research and rely solely upon more traditional funding sources, it would be immediately bankrupt.

DuBridge did worry about the implications of contract research, however. Government contracts might lead to overemphasis of some areas of research at the expense of others, and DuBridge insisted that it was the responsibility of the Institute to maintain a well-rounded research program. If government funds were used in some areas, Caltech should seek out other sources of funds for neglected areas (a lofty ideal that doesn't seem to have been realized at most universities). Another protection from government control of the research agenda was the diversity of the funding agencies themselves. DuBridge insisted that the various government agencies were "all independent of each other and to some extent in competition with each other for the good will of scientists."[80] Employing a capitalist sensibility, he suggested that Caltech and its faculty be sure not to get too involved with any single agency, and should instead encourage diversity and market competition: "A 'contract to purchase research services' offers less possibility of government control than would a direct educational 'subsidy.' In other words the government may be welcomed as a 'customer' but not as a 'stockholder.'"[81]

From Old Bones to New Machines and Utopian Visions

By 1950, forces within the Division of Geology were willing to consider changes to their research program and to consider new ways of courting the government's custom. At the end of 1950, Chester Stock unexpectedly died from a cerebral hemorrhage. The remaining members of the division held a series of internal discussions about new directions and determined that the division had "a great need for developing the field of geochemistry."[82] Although they described this "great need" intellectually, as an "opportunity of closer association with our sister science of chemistry" similar to the association between their geophysics program and physics, it is clear that something more was desired.[83] What the division really needed was money and growth. The division's next director, Robert P. Sharp, may have sincerely believed that building a geochemistry program at Caltech "[did] not mean an eclipse of the field of geology, but rather a broadening of its horizons and a strengthening of its abilities to cope with problems in the earth sciences."[84] However, Sharp also clearly expected geochemistry to bring more than simply a new approach to supplement the work Caltech's geologists were already undertaking. Simply invoking the term "earth science" implied a complete reorganization of Caltech's activities.

It was Robert Sharp who spearheaded efforts to find a suitable geochemist. He wanted to see the successes of the Chicago group replicated at Caltech.[85] After consulting with Linus Pauling (the chairman of the Division of Chemistry, and a confidant of Lee DuBridge), the geologists settled on the nuclear chemist Harrison Brown and invited him to visit them in Pasadena.[86] DuBridge supported the choice of Brown enthusiastically and, once Brown's visit to Caltech had concluded, traveled to Chicago to assist in Brown's recruitment. Convinced that Caltech was serious about establishing a first-rate geochemistry program, Brown accepted. Chester Stock's research materials, including his collection of more than 50,000 fossil specimens, were moved out of the Mudd Laboratory to make room for Brown's new geochemical labs.

Harold Urey's junior by nearly 25 years, Brown was nonetheless developing a reputation every bit as distinguished as Urey's in the new field of geochemistry. Before the war, like Urey, Brown had worked primarily on isotope separation. During his graduate days at Johns Hopkins, Brown had worked on the isotopic separation of uranium—an element that later, of course, became crucial to the Manhattan Project. At the request of Glenn Seaborg, Brown came to the University of Chicago during the war to work on Seaborg's plutonium project. After the war, Brown turned from isotope separation to the investigation of elemental abundances, employing the new techniques of neutron activation and mass spectrometry.[87]

Brown shared Urey's unease over atomic warfare and worked alongside his senior colleagues in the Chicago-centered efforts to control atomic weapons. Though he didn't yet have a public reputation equal to Urey's, Brown became one of the INS's most outspoken atomic scientists. In 1946 he published a monograph (*Must Destruction Be Our Destiny?*) in which he argued that control of the world's atomic weapons by the United Nations was the only way to prevent the destruction of mankind.[88] Brown's turn toward meteorites and elemental abundances at the INS probably reflected an aversion to weapons-related work similar to Urey's.

Brown was most interested in geochronology and the distribution of trace elements in nature. The equipment he requested from the Geology Division included two mass spectrometers suitable for uranium-lead and potassium-argon age determination, carbon-dating equipment, Geiger counters, and various other laboratory supplies. This initial equipment budget amounted to $56,000. In addition, Brown estimated that he would need at least five employees, including a silicate analyst, a physicist/mass spectroscopist, an analytical chemist, and a chemist familiar with isotopic age determination techniques. The annual personnel budget amounted to $20,500.[89] These numbers at first seemed daunting to Ian Campbell, the division's acting chairman, who wrote to Brown "We cannot wave a wand and say 'Let there be a geochemical program as outlined by Harrison Brown,' and hope to have it overnight!"[90] Still, Campbell knew that Brown's outline for a geochemical program would carry a great deal of weight with Lee DuBridge. Furthermore, Brown knew from his experience at Chicago that the government could be persuaded to support this work. DuBridge and Sharp knew that Brown "spent a lot of time in Washington," that he "had his fingers on the pulse of a lot of activities there," and that he "had a very keen sense of where new opportunities were likely to be forthcoming and what to do about them."[91] They were banking on Brown's ability as an atomic insider to tap into these new opportunities.

Brown's experience at Chicago had taught him that the government would support a general research program if it were appropriately framed as being in the national interest. As has already been noted, in 1949 he had unsuccessfully requested a far smaller amount from the Atomic Energy Commission to study the ages and distributions of elements in meteorites and in the Earth's crust and the abundances of elements in the solar system.[92] Perhaps it was this rejection, or perhaps it was DuBridge's advice, that led Brown to accentuate his uranium expertise and frame his new research proposal as a study of "critical" materials in nature—uranium in particular—and economic processes for their isolation. With help from DuBridge, Brown drafted and submitted a research proposal to the AEC that went beyond what he asked of Ian Campbell. Instead of the $76,500 he had requested from Campbell, Brown's proposal to the AEC outlined a two-year research program that would cost more than $350,000. The new equipment budget allowed for a third mass spectrometer, and the number of scientific personnel he wanted had increased to nine.[93]

By "critical" materials, Brown meant uranium and other valuable elements typically associated with it. Brown's proposal predicted long-term heavy demand for uranium in America's energy and weapons programs, and contrasted this to the low average surface concentration of the element (which Brown estimated to be no more than four grams per ton of rock). It was essential, Brown argued, that the AEC invest in the development of methods for economically and efficiently refining uranium from materials that held the element in low concentrations, thus establishing its long-range supply. The economical means that Brown proposed, and what he claimed his project would develop, was a method of processing low-grade ores for multiple elements simultaneously.

The logic was simple. Mining low-grade ores for uranium alone would be prohibitively expensive, and would involve disposing of great amounts of waste materials. However, if other valuable metals could be isolated from the ore during processing, the proportional cost of the uranium would be decreased. And although Brown's proposal began by addressing the AEC's main concern, it went on to address all the metals on the "critical" list. Brown argued that supplies of high-grade ores of many important metals were dwindling. He suggested integrated refining operations that would target all the valuable materials found in low concentrations on the surface of the Earth.

Devising such involved mining operations would require a comprehensive research program concerned with the general geochemistry of uranium and other "critical" elements. Brown's proposal included using isotope analysis to determine the natural abundances and concentrations of uranium and other elements in various types of rock, and using radiometric age determinations of rocks to study geochemical processes as a function of time. The program would be highly inclusive, as Brown's research team would produce a quantitative picture of the geochemical cycle of uranium, examining the developmental paths taken by uranium in the solidification of molten rocks, the weathering of igneous rocks, shales, sandstones, and limestones, and the formation of evaporates, hydrolyzates, and bioliths. Brown also planned to incorporate biogeochemical studies into his program, determining quantitatively the extent to which plant forms such as algae enrich uranium and other trace elements.

Not only was Brown's proposed research program extensive; his vision of its application was utopian. The general geochemistry program Brown described was to be completed in the first year of funding, followed in the second year by application of the knowledge gained to the development of the refining process. Brown had an idea of the shape his ideal process would take. He provided the AEC with a hypothetical example that involved isolating uranium from limestone and also isolating calcium, magnesium, iron, aluminum oxides, manganese, strontium, barium, copper, lead, and large amounts of carbon dioxide. The process he imagined would use sea water and air (both harmless and "essentially in infinite supply") as the main reagents, and would be powered by clean nuclear energy. The refinery would be attached to an algae farm that would use all of the refinery's carbon dioxide and nitrate wastes to produce food and useful organic by-products.[94] It would be a perfect system, wasting nothing and producing no pollution. It also perfectly illustrated Bush's model of the value of pure research to applied projects.

From Utopia to the Age of the Earth and the Birth of a "Lead Man"

In hindsight, outside of the context of Cold War anxieties about uranium supplies and optimistic visions of the possible applications of nuclear technologies, the purpose of Brown's research program seems unrealistic to say the least. Nonetheless, it made sufficient sense to the Atomic Energy Commission at the time that the AEC committed not only to the initial two years of funding Brown requested, but also to numerous extensions throughout the 1950s.[95] This gave the new geochemistry program the distinction of being supported by one of Caltech's largest unclassified research contracts.[96]

The boost in the Geology Division's funding gained the geochemists considerable prestige at Caltech. The experience of the division reflected that of the geosciences as a whole during the Cold War as universities' geology departments throughout the United States set aside traditional field work to become laboratory geology programs. So internalized did the values of laboratory research become within the geosciences community in the next few decades that the community soon went from recognizing that geophysics and geochemistry were fields that the military and the government funded to arguing that these were the fields that their patrons *should* fund at the expense of field work and mapping.[97]

The injection of new money, personnel, and equipment transformed Caltech's Geology Division. During the 1950s, when a faculty position opened up, Robert Sharp tended to give priority to hires that would further help to establish geochemistry at Caltech. More of the younger set from Chicago who found that there was no way to move up the chain at the Institute for Nuclear Studies (crowded as it was with senior statesmen of atomic science) came to Caltech. In the coming years, the faculty of the Geology Division grew from about a dozen researchers to nearly thirty. Sharp estimated that nearly half of the staff, including himself, shifted their own research programs toward geochemistry, or employed the geochemical equipment and expertise now available to them.[98]

In addition to supporting individual research projects, the government and military contracts allowed the Geology Division to purchase expensive equipment that other geology departments couldn't afford and to support a technical staff and an administrative staff. "Before there were government contracts," Robert Sharp explained, "the money for such support had to come out of the [division's] budget. Now much of the technical and administrative help comes out of grant budgets. We now have a large group of secretaries and aids who keep the wheels grinding in our Geology Division. Without them, we'd come to a halt tomorrow."[99]

These changes didn't come without resistance. Older faculty members complained to Sharp that he was "gutting geology to build geochemistry."[100] Gerald J. Wasserburg, who came from Urey's Chicago laboratory to Caltech in 1955, remembers that the attitude among alumni and older members of the Caltech faculty was that "geochemistry was certainly not real geology!"[101] But although the critics may have seen this Cold War transformation as the end of the golden era that had begun under Millikan and Buwalda, the new era brought more and more prestige and resources to the division. By the end of the 1950s, the investment in geochemistry had positioned Caltech's Geology Division as a major player and contract winner in the NASA-funded space science that accompanied the emerging "space race."

Contracts had become essential to the operation of the Geology Division. When in 1957 it came time to replace Beno Gutenberg as director of the Seismology Laboratory, rather than passing the position to Hugo Benioff or Charles Richter the division recruited Maurice Ewing's protégé Frank Press. Press was chosen because he "understood the modern world of government contracts" and was expected to use this understanding to modernize the lab, increase its staff, and bring it "to the forefront in the modern world."[102] What this meant in practice was that Press was able to connect the Seismology Laboratory's expertise to the detection of atomic weapons tests.

An integrated refining process for "critical" materials such as Harrison Brown had envisioned never emerged, though Brown spent much of the remainder of his career involved in political projects concerned with resource development. The initial research, however, proceeded with gusto, though not always with Brown's participation. It isn't likely that the Atomic Energy Commission—advised as it was by Brown's peers (Harold Urey reviewed at least one of Brown's renewal requests, and Willard Libby was appointed to the AEC in 1954)—failed to recognize early on that Brown's proposed research program was unlikely to produce a useful refining process. It is far more likely that the AEC didn't see Brown's potential failure to produce a refining process as a major setback. The AEC got its money's worth from Brown, in the development of new technological methods for the study of isotopes in nature and in the increased understanding of the geochemistry of uranium and other associated metals. Thus, the AEC had no reason to worry when the publications Brown listed in his progress reports were more the type that one might expect from basic research proposals, and didn't actually demonstrate significant progress toward his contract's stated goals.

The AEC funds, along with money from the Guggenheim Foundation, allowed Brown to bring with him from Chicago two members of Urey's research team: Charles McKinney, the engineer who had built Urey's mass spectrometers, and Sam Epstein, the postdoc who had helped Urey to develop the oxygen thermometer. The paleoecologist Heinz Lowenstam, who had worked with Urey and Epstein, also received an appointment in the division. McKinney built the new mass spectrometers that would soon make the division a world-renowned center for isotope geochemistry. Meanwhile, Epstein and Lowenstam continued the oxygen thermometer work initiated in Chicago. Robert Clayton and Lee Silver, two up and coming researchers, also joined the roster. But perhaps the most famous work to be done by the research team was the determination of the age of the Earth by Brown's former graduate student and postdoc Clair Patterson.

Patterson gained expertise in the use of mass spectrometers while working in the Manhattan Project's electromagnetic separation plant at Oak Ridge. After the war, he decided to follow the atomic scientists to the University of Chicago and pursue a PhD. There he met Brown, who put him to work on projects related to Brown's interest in meteorites and the age of the solar system. The first project involved using mass spectrometers to measure the small amounts of uranium and lead isotopes in zircon crystals embedded in rocks in order to determine their geologic age. Brown put Patterson and another graduate student, George Tilton, on the zircon work. The two students split the work, and Patterson applied his talents in mass spectrometry to measuring the isotopic compositions of lead in the samples. "I was the lead man," Patterson later recalled, "and Tilton was the uranium man."[103]

Brown believed that the lead in iron meteorites was primordial lead—lead that had been preserved unchanged within the meteorites from the time of the solar system's formation. The forming planets had accumulated both lead and uranium, but the iron meteorites contained no uranium. Whereas terrestrial rocks contained two types of lead (primordial lead and the lead created by the radioactive decay of uranium and thorium), the meteorites contained only the original lead of the solar system. Brown convinced Patterson that once he had perfected his lead techniques with the zircon samples, he would be able to apply his methods to an iron meteorite. The isotopic composition of the meteorite would yield the original isotopic composition of primordial lead at the time of the Earth's formation. By comparing that composition against the present isotopic composition of terrestrial rocks, Patterson would be able to determine the age of the Earth. The concept was so simple, Brown assured Patterson, that the work would be "duck soup."[104]

But the work turned out not to be simple at all. In principle, the methods should have worked. However, Patterson was working with microgram samples of lead, and was attempting to adapt spectrometer techniques that required milligrams. Moreover, he was discovering that all his samples were contaminated with lead from industrial processes. After many failures, Patterson had to put the meteorite work aside while in Chicago, turning instead to geochronometric work on granites.

After receiving a PhD in 1951, Patterson decided to make another attempt at the meteorite research. He wrote a proposal to the AEC requesting funds to support a postdoc at Chicago for the work. The AEC turned him down: "They said they weren't interested in measuring the age of the Earth."[105] But Brown was able to fit Patterson's meteorite work under the umbrella of his new AEC contract. At Caltech, Brown had new labs built for Patterson. Patterson was able to work in a cleaner space, and in the time since his initial meteorite work he had developed new contamination control techniques. During his granite work, Patterson had measured the isotopic composition of contaminant lead and had developed a technique for separating radiogenic lead from contaminant lead in the mass-spectrometer readout. In 1953, Patterson received his first meteorite sample. He extracted the lead from the meteorite in his new lab in Pasadena, then traveled back to Chicago to run the lead through one of Mark Inghram's mass spectrometers at Argonne National Laboratory. Shortly afterward, Patterson announced the age of the Earth: 4.55 billion years.[106]

Conclusion

Clair Patterson preferred to describe his own research program as completely basic, not guided by the desires of sponsors, and not concerned with practical applications. He told an interviewer that his motivation had been purely "Science, science, science!"[107] He described Brown's AEC contract proposals as useful "fibs" that allowed him to do the research that he deemed worthwhile:

[Brown] went through all these calculations, and he told the Atomic Energy Commission how there was enough uranium in ordinary igneous rock that if you ground that rock up and then leached it with hydrochloric acid you would get enough uranium to use in an atomic generator that would be equivalent in energy to 10,000 tons of coal. It would pay for the energy not only of grinding up the rock, which required energy, but you would have left over huge amounts of extra energy. In other words, 10,000 tons of coal would equal the amount of energy of the uranium in one ton of granite. ... They bought that! And it was that kind of sales pitch he used. ... I would say [in my proposals], "Well, I want to know how this chunk of North America evolved and then got thrown around and came over here, and how this other chunk came up later. And we want to know when this chunk came up and when that chunk came up, and how they were related to each other. What was their ancestry?" And the Atomic Energy Commission would say to me, "To hell with you, Patterson! We don't care about that stuff at all." ... And I never got funded. But Harrison would get them funded for me.[108]

However, as was noted above, it is naive to believe that the AEC was oblivious to what it was actually buying. It is also unrealistic to divide, as Patterson did, Brown's fund-raising efforts from the work done in his laboratory. Work in the social studies of science has demonstrated that such divisions nearly always break down upon investigation.[109]

Patterson's demonstrated ability to follow an isotope through time and space, and to produce a global picture of how that isotope moved through nature, was directly related to his sponsors' Cold War concerns. Furthermore, the prestige of Patterson's science and his quantitative data were bestowed not by the methods themselves, but by the work done by the AEC and other patrons to promote geochemistry and "earth science" within academia. It is clear that Caltech didn't invest in isotope geochemistry because of its proven usefulness; most of the dramatic discoveries that vindicated the new science—including Patterson's determination of the age of the Earth—were yet to come at the time the Geology Division recruited Brown. What was obvious at the time was not the new field's proven success, but its proven ability to attract large sums of contract money. Furthermore, this money was made available to projects not on the basis of whether they addressed the most timely or important scientific questions, but on the basis of whether they addressed the most pressing national-security concerns. And although we might take Patterson at his word that the AEC money didn't substantially distort his scientific agenda, we must also recognize that the money did place him within a privileged group of scientists whose work was promoted at the expense of other work less related to these concerns (or to other national-security and geopolitical concerns). One need only consider the fate of paleontology at Caltech to realize that not all branches of geology benefited from the support of government or military patrons during the Cold War. When Chester Stock's La Brea specimens were sold in 1957 to the Los Angeles County Museum, the money was used to improve the geochemical facilities that had displaced them.[110]

Harold Urey, Harrison Brown, and other atomic scientists may have turned to geochemistry because it offered an escape from weapons-related work and an opportunity to ask fundamental questions about the natural world. Their activities truly did move from fearing the end of the world to determining the age of the Earth. However, the work these scientists did and the schools that were founded around them were made possible because their research programs addressed Cold War military and political concerns. And though Stephen Brush may claim that it was the prestige of the atomic scientists that convinced younger scientists to take the new fields of geochemistry and cosmochemistry seriously, the mass movement into these fields was clearly more a product of institutional changes in response to funding sources.

The University of Chicago's Geology Department was transformed not only by its encounter with Urey, but also by its financial relationship with the Office of Naval Research. In 1966, Caltech's Geology Division—significantly affected by its encounter with Cold War geochemistry, seismology, and eventually space exploration—began planning the construction of a new 65,000-square-foot Geophysics and Planetary Science Laboratory.[111] Similar changes took place at universities around the United States as the Cold War transformed geology, now dominated by geophysical and geochemical approaches, into a component discipline of the new earth and planetary sciences.

Notes

1. Clair C. Patterson, "Lead Contamination of the Atmosphere and the Earth's Surface," *Proceedings of the American Philosophical Society* 114, no. 1 (1970): 9; M. Murozumi, Tsaihwa J. Chow, and Clair C. Patterson, "Chemical Concentrations of Pollutant Lead Aerosols, Terrestrial Dusts and Sea Salts in Greenland and Antarctic Snow Strata," *Geochimica et Cosmochimica Acta* 33, no. 10 (1969): 1247–1294.

2. The other major contributing factor was the increase in lead smelting activity for industry. See National Research Council Environmental Studies Board Commission on Natural Resources, *Lead in the Human Environment: A Report Prepared by the Committee on Lead in the Human Environment* (National Academy of Sciences, 1980), 289.

3. Alice Hamilton, "Nineteen Years in the Poisonous Trades," *Harper's Monthly Magazine* 159 (October 1929): 3–14; Alice Hamilton, Paul Reznikoff, and Grace M. Burnham, "Tetra-Ethyl Lead," *Journal of the American Medical Association* 84, no. 20 (1925): 1481–1486.

4. Mario J. Molina and F. Sherwood Rowland, "Stratospheric Sink for Chlorofluoromethanes: Chlorine Atom-catalysed Destruction of Ozone," *Nature* 249, no. 5460 (1974): 810–812.

5. Clair C. Patterson, "Duck Soup and Lead," *Engineering and Science* 60, no. 1 (1997): 20.

6. Ronald E. Doel, "Constituting the Postwar Earth Sciences: The Military's Influence on the Environmental Sciences in the USA After 1945," *Social Studies of Science* 33, no. 5 (2003): 635–666, at 636.

7. Naomi Oreskes, *The Rejection of Continental Drift: Theory and Method in American Earth Science* (Oxford University Press, 1999), 289.

8. Naomi Oreskes and Ronald E. Doel, "Physics and Chemistry of the Earth," in *The Cambridge History of Science*, vol. 5: *The Modern Physical and Mathematical Sciences*, ed. Mary Jo Nye (Cambridge University Press, 2003), 538.

9. Naomi Oreskes, *Science on a Mission: American Oceanography in the Cold War and Beyond* (University of Chicago Press, forthcoming). See also Oreskes, chapter 11 in this volume.

10. William H. McNeill, *Hutchins' University: A Memoir of the University of Chicago, 1929–1950* (University of Chicago Press, 1991), 105.

11. Jack M. Holl, *Argonne National Laboratory, 1946–1996* (University of Illinois Press, 1997), 8.

12. Holl, *Argonne National Laboratory, 1946–1996*, 33.

13. Samuel K. Allison, "Thoughts on 10th Anniversary of First Chain Reaction," 7 November 1952, Samuel King Allison Papers, Box 25, Folder 7, Special Collections Research Center, University of Chicago Library.

14. Robert M. Hutchins, "The State of the University: A Report by Robert M. Hutchins," 25 September 1945, Robert M. Hutchins Papers, Box 26, Folder 6, Special Collections Research Center, University of Chicago Library.

15. McNeill, *Hutchins' University*, 99.

16. Rebecca S. Lowen, *Creating the Cold War University: The Transformation of Stanford* (University of California Press, 1997).

17. Samuel K. Allison to Don M. Yost, 4 December 1945, Harold C. Urey Papers, Box 3, Folder 3, University of California, San Diego Special Collections Library, La Jolla.

18. The $12 million figure comes from Thomas E. Mullaney, "Atomic Research Shows Progress," *New York Times*, June 26, 1949.

19. Roger L. Geiger, "Milking the Sacred Cow: Research and the Quest for Useful Knowledge in the American University Since 1920," *Science, Technology, & Human Values* 13, no. 3–4 (1988): 332–348; Jacob D. Hamblin, *Oceanographers and the Cold War: Disciples of Marine Science* (University of Washington Press, 2005).

20. "Industry, Science Team in Research," *New York Times*, May 6, 1951.

21. Harold C. Urey to Walter Bartky, 6 August 1945, Harold C. Urey Papers, Box 12, Folder 29.

22. 'Traumatized' is the word Urey's colleagues chose to use when describing his postwar state of mind. See Karl P. Cohen et al., "Harold Clayton Urey. 29 April 1893–5 January 1981," *Biographical Memoirs of Fellows of the Royal Society* 29, November 1983: 622–659.

23. Harold C. Urey, Interview with Dr. Harold Urey, Session 2, interview by John L. Heilbron, March 24, 1964, Niels Bohr Library & Archives, American Institute of Physics, College Park, Maryland.

24. Ibid.

25. Hans E. Suess, "Harold C. Urey," copy, December 1981, Samuel Epstein Papers, Box 10, Folder 12, Caltech Archives, Pasadena.

26. Joseph Mayer, "[Biography of Harold C. Urey]," 1970, Harold C. Urey Papers, Box 1, Folder 11.

27. James R. Arnold, "Harold C. Urey Chair in Chemistry, Inaugural Address," 29 April 1983, Stanley L. Miller Papers, Box 139, Folder 9, University of California, San Diego Special Collections Library, La Jolla.

28. "The Talk of the Town, Notes and Comment," *The New Yorker*, December 15, 1945, 23.

29. Harold C. Urey to Albert Einstein, 15 April 1947, Emergency Committee of Atomic Scientists, Records, Box 2, Folder 13, Special Collections Research Center, University of Chicago Library.

30. Harold C. Urey and Lotti J. Greiff, "Isotopic Exchange Equilibria," *Journal of the American Chemical Society* 57, no. 2 (1935): 321–327.

31. Harold C. Urey, "The Thermodynamic Properties of Isotopic Substances," *Journal of the Chemical Society* (1947): 562–581.

32. Ibid., at 581.

33. Stephen G. Brush, *Fruitful Encounters: The Origin of the Solar System and of the Moon from Chamberlin to Apollo* (Cambridge University Press, 1996), 144.

34. Samuel Epstein, untitled, 28 January 1981, Samuel Epstein Papers, Box 10, Folder 12.

35. Daniel Fisher, *The Seventy Years of the Department of Geology, University of Chicago, 1892–1961* (University of Chicago, 1963), 52; Julian R. Goldsmith, "Some Chicago Georecollections," *Annual Review of Earth and Planetary Sciences* 19 (1991): 1–18, at 4.

36. Fisher, *The Seventy Years of the Department of Geology*, 58.

37. Harrison Brown, "Formal Establishment of Training in Geochemistry," 4 April 1950, Harold C. Urey Papers, Box 15, Folder 37.

38. Goldsmith, "Some Chicago Georecollections," 9.

39. Fisher, *The Seventy Years of the Department of Geology*, 61.

40. Samuel Epstein, Ralph Buchsbaum, Heinz A. Lowenstam, and Harold C. Urey, "Carbonate-Water Isotopic Temperature Scale," *Bulletin of the Geological Society of America* 62 (April 1951): 417–426; Harold C. Urey, Heinz A. Lowenstam, Samuel Epstein, and Charles R. McKinney, "Measurement of Paleotemperatures and Temperatures of the Upper Cretaceous of England, Denmark, and the Southeastern United States," *Bulletin of the Geological Society of America* 62 (April 1951): 399–416.

41. Ronald E. Doel, Tanya J. Levin, and Mason K. Marker, "Extending Modern Cartography to the Ocean Depths: Military Patronage, Cold War Priorities, and the Heezen–Tharp Mapping Project, 1952–1959," *Journal of Historical Geography* 32, no. 3 (2006): 605–626, at 608. See also Oreskes, *Science on a Mission*, chapters 5 and 7.

42. Ronald E. Doel, "The Earth Sciences and Geophysics," in *Companion to Science in the Twentieth Century*, ed. John Krige and Dominique Pestre (Routledge, 2003), 398.

43. Mullaney, "Atomic Research Shows Progress."

44. H. Gershinowitz to E. G. Gaylord, 8 April 1947, Harold C. Urey Papers, Box 6, Folder 1.

45. The history of Project 43 research at MIT is described in chapter 20 of Robert Rakes Shrock, *Geology at MIT 1865–1965: A History of the First Hundred Years of Geology at Massachusetts Institute of Technology*, volume 2: *Department Operations and Projects* (MIT Press, 1982).

46. Harold C. Urey to T. V. Moore, 25 June 1947, Harold C. Urey Papers, Box 86, Folder 36.

47. Carsten Reinhardt, "The Chemistry of an Instrument: Mass Spectrometry and Structural Organic Chemistry," in *From Classical to Modern Chemistry: The Instrumental Revolution*, ed. Peter J. T. Morris (Royal Society of Chemistry, 2002), 231.

48. C. A. Young to Harold C. Urey, 9 March 1948, Harold C. Urey Papers, Box 6, Folder 1.

49. Harold C. Urey, "Penrose Application," 17 July 1947, Harold C. Urey Papers, Box 37, Folder 10.

50. Henry Aldrich to Harold C. Urey, 22 September 1948, Harold C. Urey Papers, Box 37, Folder 10.

51. American Petroleum Institute to Harold C. Urey, 7 January 1952, Harold C. Urey Papers, Box 6, Folder 1.

52. Ronald Rainger, "Science at the Crossroads: The Navy, Bikini Atoll, and American Oceanography in the 1940s," *Historical Studies in the Physical and Biological Sciences* 30, no. 2 (2000): 349–371, at 366. As Rainger points out, this meant that "studies of ocean bottoms, surface layers, coastlines, and almost any topic other than biological oceanography were, at once, both intellectually meaningful to oceanographers and operationally useful to the Navy."

53. Harold C. Urey, "Proposal for Task Order Under Contract No. N6 Ori - 20," 24 November 1948, Harold C. Urey Papers, Box 94, Folder 19.

54. Ibid.

55. Office of Naval Research to Harold C. Urey, 22 August 1952, Harold C. Urey Papers, Box 95, Folder 1.

56. Harold C. Urey, "Proposal for Extension of Research on the Natural Abundance of Deuterium and Other Isotopes in Nature," February 1953, Harold C. Urey Papers, Box 92, Folder 9; National Science Foundation to Harold C. Urey, 20 July 1953, Harold C. Urey Papers, Box 68, Folder 1.

57. George M. Cableman to Roy B. Snapp, "Implementing Action - Dr. Harold C. Urey," n.d., Atomic Energy Commission, Records, Box 8, Folder 095 (12-11-46), National Archives and Records Administration, Washington.

58. See Angela Creager's chapter in this volume for a more detailed discussion of the AEC's attitude toward unclassified contracts.

59. Kenneth S. Pitzer to Harold C. Urey, 9 February 1951, Harold C. Urey Papers, Box 92, Folder 8.

60. Harold C. Urey to Kenneth S. Pitzer, 1 January 1951, Harold C. Urey Papers, Box 92, Folder 8.

61. Kenneth S. Pitzer to Harold C. Urey, 9 February 1951, Harold C. Urey Papers, Box 92, Folder 8.

62. Joseph Platt to Harold C. Urey, 23 February 1951, Harold C. Urey Papers, Box 92, Folder 8.

63. Harrison Brown, "A Proposed Program for the Accumulation of Quantitative Data Concerning: the Chemical Composition of Meteorites and the Earth's Crust; the Relative Abundances of Elements in the Solar System; the Ages of the Elements and Planets," 1949, Harold C. Urey Papers, Box 15, Folder 37; Harold Urey, "Atomic Energy Commission Contract No. AT(11-1)-101 To Investigate the Natural Abundance of Deuterium and Other Isotopes in Nature," February 1953, Harold C. Urey Papers, Box 92, Folder 9.

64. Judith R. Goodstein, *Millikan's School: A History of the California Institute of Technology* (Norton, 2006), chapter 7; Robert H. Kargon, "Temple to Science: Cooperative Research and the Birth of the California Institute of Technology," *Historical Studies in the Physical Sciences* 8 (January 1977): 3–31.

65. Richard J. Proctor, J. David Rogers, and Allen W. Hatheway, "Saluting Some of the Outstanding Pioneers of Engineering Geology in Southern California," in *History of the Association's First 50 Years* (Association of Environmental and Engineering Geologists, 2007), 92; George Gaylord Simpson, "Chester Stock, 1892–1950," *Biographical Memoirs of the National Academies of Science* 27 (1952): 340–341.

66. This information about the training program and student composition of the division comes from a 1947 report to the president of Caltech written by Chester Stock. Excerpts from the division's annual reports were compiled in Robert P. Sharp, "Excerpts from Division Annual Reports," 1966, Clair C. Patterson Papers, Box 22, Folder 2, Caltech Archives.

67. The Seismology Laboratory predated the Division of Geology by five years, having been founded by Wood with support from the Carnegie Institution in 1921. On the influence of Gutenberg and his colleagues at the Seismological Laboratory on American geology, see Oreskes, *The Rejection of Continental Drift*, chapter 8. Oreskes discusses Gutenberg's move from Göttingen to Pasadena in 1930 and his importing of European ideas of continental drift.

68. See Ian Campbell, in Sharp, "Excerpts from Division Annual Reports," 3.

69. W. Patrick McCray, "Project Vista, Caltech, and the Dilemmas of Lee DuBridge," *Historical Studies in the Physical and Biological Sciences* 34, no. 2 (2004): 339–370, at 343. As McCray notes, DuBridge was a member of the National Science Board and the Atomic Energy Commission's General Advisory Committee, as well as science advisor to Presidents Truman and Nixon.

70. Vannevar Bush, *Science, the Endless Frontier* (Government Printing Office, 1945). DuBridge made his agreement with Bush explicit in "Some Observations on the Government Contract Situation at Caltech," February 1952, Lee A. DuBridge Papers, Box 235, Folder 1, Caltech Archives.

71. This much was signaled in DuBridge's book *Frontiers of Knowledge: Seventy-five Years at the California Institute of Technology* (Newcomen Society in North America, 1967).

72. DuBridge, "Some Observations on the Government Contract Situation at Caltech," 3. DuBridge became even more insistent on this point when the threat of defunding reared its head—see Lee A. DuBridge, "Science and a Better America," *Bulletin of the Atomic Scientists* 16, no. 8 (1960): 340; DuBridge, "The Government Role in Science Education," *Bulletin of the Atomic Scientists* 22, no. 5 (1966): 16–20; DuBridge, "The Future of University Research," *Bulletin of the Atomic Scientists* 25, no. 1 (1969): 39.

73. Lee A. DuBridge, interview by Judith R. Goodstein, part I, February 19, 1981, Oral History Project, Caltech Archives.

74. Ibid.

75. Ibid.

76. Kargon, "Temple to Science," 31.

77. McCray, "Project Vista, Caltech, and the Dilemmas of Lee DuBridge," 343.

78. DuBridge, "Some Observations on the Government Contract Situation at Caltech," 5.

79. Ibid., 17.

80. Ibid., 14.

81. Ibid., 17.

82. Sharp, "Excerpts from Division Annual Reports," 5.

83. Ibid., 4.

84. Ibid., 7.

85. Gerald J. Wasserburg, "Isotopic Adventures—Geological, Planetological, and Cosmic," *Annual Review of Earth and Planetary Sciences* 31 (2003): 1–74, at 11.

86. Robert P. Sharp, "Robert Sharp's History of the Geology Division," n.d., Division of Geology and Planetary Science Papers, Box A10, Folder 10, Caltech Archives.

87. John P. Holdren, "Introduction," in *Earth and the Human Future: Essays in Honor of Harrison Brown*, ed. Kirk R. Smith, Fereidun Fesharaki, and John P. Holdren (Westview, 1986), 3–5.

88. Harrison Brown, *Must Destruction Be Our Destiny?* (Simon and Schuster, 1946).

89. Harrison Brown to Ian Campbell, 1 June 1951, Lee A. DuBridge Papers, Box 9, Folder 1.

90. Ian Campbell to Harrison Brown, 7 June 1951, Lee A. DuBridge Papers, Box 9, Folder 1.

91. Sharp, "History of the Geology Division," 5.

92. Brown, "A Proposed Program for the Accumulation of Quantitative Data."

93. Harrison Brown, "Proposal for Research Project, Study of the Fundamental Geochemistry of Critical Materials and the Development of Economic Processes for Their Isolation, to the United States Atomic Energy Commission," September 1951, Caltech Geology Division Papers, Box 14.

94. Ibid.

95. Harrison Brown, "Proposal for a One Year Extension of Contract AT(11-1)-208: Study of the Fundamental Geochemistry of Critical Materials and the Development of Economic Processes for Their Isolation," October 15, 1957, Harold C. Urey Papers, Box 92, Folder 4.

96. "Government Sponsored Research at C.I.T.," 1951, Lee A. DuBridge Papers, Box 235, Folder 1.

97. Oreskes, *The Rejection of Continental Drift*, 290–291.

98. Sharp, "History of the Geology Division," 28.

99. Ibid.

100. Ibid., 6.

101. Wasserburg, "Isotopic Adventures," 12.

102. Sharp, "History of the Geology Division," 17.

103. Clair C. Patterson, interview by Shirley K. Cohen, March 5, 1995, Oral History Project, California Institute of Technology Archives, 15.

104. Ibid., 16.

105. Ibid., 20.

106. Clair C. Patterson, "Age of Meteorites and the Earth," *Geochimica et Cosmochimica Acta* 10, no. 4 (1956): 230–237.

107. Patterson, "Duck Soup and Lead," 30.

108. Clair C. Patterson, interview by Shirley K. Cohen, 24–25.

109. For a classic discussion of this false distinction, see Bruno Latour, *Science in Action: How to Follow Scientists and Engineers Through Society* (Harvard University Press, 1987). Oreskes has demonstrated that such a distinction is especially unrealistic when military patronage is involved, even if basic science discoveries are accomplished. See Naomi Oreskes, "A Context of Motivation: US Navy Oceanographic Research and the Discovery of Sea-Floor Hydrothermal Vents," *Social Studies of Science* 33, no. 5 (1, 2003): 697–742.

110. Finding Aid, Chester Stock Papers, Caltech Archives.

111. Sharp, "Excerpts from Division Annual Reports," 22.

5 Changing the Mission: From the Cold War to Climate Change

Naomi Oreskes

In 1996, the historian Paul Forman argued that military patronage in physics had fostered a science of technical mastery and gadgeteering in physics; in 1965, the oceanographer William von Arx had come to the same conclusion about oceanography.[1] Military patronage was problematic, von Arx argued, because it fostered a culture of technological bravado at the expense of conceptual understanding. This could be remedied, however, by changing the focus of oceanographic research. The particular change von Arx wanted was from warfare to weather (and climate). "This refreshing change of 'mission' in ocean research," he wrote optimistically, "would draw a different sort of people into marine science. There would be more thought-centered effort and less thing-centered preoccupation as with deep submersibles … and other elements of technological derring-do which 'big science' tends to encourage."[2]

Foreshadowing conclusions that would soon become commonplace, von Arx noted that human effects on the natural environment were increasingly evident and worthy of investigation. "Man," he wrote, "is altering the radiation balance [of the atmosphere] by his vigorous consumption of fossils fuels."[3] This was worthy of serious scientific attention.

Although von Arx's complaint foreshadowed Forman's, his views weren't typical. The majority of oceanographers active at that time had mostly good things to say about their Cold War military patrons. Many have since described the Cold War as a "golden age" of oceanography, and it is hard for a historian to disagree strongly with that view. Many significant discoveries were made and advances in conceptual understanding emerged that would not have been possible without the financial and logistical support of the US Navy and the intellectual motivation provided by Cold War geopolitical demands.[4] Moreover, the purpose of patronage, military or otherwise, is to adjust the focus of attention and influence the direction of work, so there are bound to be those who dislike or disagree with that adjustment. The question for the historian of scientific knowledge is this: In what specific manner has a patron adjusted priorities and focused attention, and what epistemic consequences, if any, did those adjustments and change in focus have?

Both von Arx and Forman alleged that during the Cold War there was a loss of conceptual understanding in favor of technological prowess. This chapter examines that claim by exploring what happened to oceanographers at the end of the Cold War, when they belatedly took up von Arx's recommendation in the form of a project called Acoustic Tomography of Ocean Climate (ATOC). Explicitly conceptualized as an attempt to turn swords into plowshares, ATOC addressed the question of whether the oceans were warming in response to increased atmospheric concentrations of carbon dioxide from the burning of fossil fuels. It ran aground, however, when environmentalists and biologists suggested that the proposed investigations might harm marine mammals, and when members of the lay public interpreted the project as a cover story for further secret military-scientific projects. The scientific and public opposition to the Acoustic Tomography of Ocean Climate project suggests that military patronage had both epistemic consequences and also social and political ones. And some of these consequences remained salient even after the Cold War was over.

Anthropogenic Climate Change as a Scientific Opportunity

William von Arx was neither idiosyncratic nor clairvoyant in calling attention to anthropogenic climate change. It had been known since the nineteenth century that CO_2 was a greenhouse gas—highly transparent to visible light, fairly opaque to infrared—and that its presence in the atmosphere made Earth a comfortably warm planet. Among physicists, oceanographers, geologists, and geophysicists it had also become broadly accepted that changing concentrations of atmospheric CO_2 could affect the climate by altering Earth's radiative balance. That led Charles David Keeling, a geochemist at the Scripps Institution of Oceanography in San Diego, to begin systematic measurement of atmospheric carbon dioxide in 1958.[5]

Keeling was a junior colleague of the oceanographer Roger Revelle and the geochemist Hans Suess, who had emphasized the historically unprecedented character of mid-twentieth-century human activities. Humans had become geological agents, they argued, returning to the biosphere in just a few centuries organic carbon that had accumulated in rocks over the course of hundreds of millions of years. In hindsight, a 1957 paper by Revelle and Suess is often cited as an early warning of the dangers of global warming, but in fact, consistent with the Cold War spirit of making scientific virtue out of political necessity, Revelle and Suess were primarily making the point that global warming presented a scientific *opportunity*:

[H]uman beings are now carrying out a large scale geophysical experiment of a kind that could not have happened in the past nor be reproduced in the future. Within a few centuries we are returning to the atmosphere and oceans the concentrated organic carbon stored in sedimentary rocks over hundreds of millions of years. This experiment, if adequately documented, may yield a far-reaching insight into the processes determining weather and climate.[6]

As the human population and its use of resources continued to increase, the rising atmospheric concentration of CO_2 probably would be large enough to produce detectable climatic effects; Revelle and Suess suggested that scientists try to document those effects:

> In contemplating the probably large increase in CO_2 production by fossil fuel combustion in coming decades, we conclude that a total increase of 20-40 in atmospheric CO_2 can be anticipated. This should certainly be adequate to allow a determination of the effects, if any, of changes in atmospheric carbon dioxide on weather and climate throughout the Earth.[7]

How large these effects would be would depend on the fraction of CO_2 accumulated in the atmosphere relative to the fraction taken up by the biosphere and absorbed by the oceans. Within a few years, Keeling's data showed that about half of the released CO_2 was "missing" and presumed to have been absorbed into the oceans or taken up by plants. The remainder stayed in the atmosphere, where its concentration was on the rise.

In the mid 1960s, most Earth scientists—particularly geologists focusing on geological rather than human time scales—believed that the planet was heading naturally toward an ice age. If they considered human impacts (and most did not), perhaps they expected accelerated cooling caused by sulfate aerosols and other particulate emissions; with coal the dominant source of fossil-fuel energy, these effects looked to be larger than any possible warming effect. The geophysicist Gordon MacDonald later wrote: "In 1969, it seemed plausible that our activities could either lead to a disastrous ice age or to an equally disastrous melting of the polar ice caps."[8]

Things changed, however, as more scientists began to learn of Keeling's measurements, which showed that absolute values of atmospheric CO_2 were steadily climbing.[9] By the 1970s, a number of scientists were building numerical simulation models to predict when a detectable climate signal might occur. In 1978, *Oceanus*, the official journal of the Woods Hole Oceanographic Institution in Massachusetts, dedicated a special issue to "Oceans and Climate"; in the introduction, Robert M. White, chairman of the National Research Council's Climate Research Board, wrote:

> We now understand that industrial wastes, such as carbon dioxide released during the burning of fossil fuels, can have consequences for climate that pose a considerable threat to future society. The Geophysics Research Board of the National Research Council in its recent report, "Energy and Climate," foresees the possibility of a quadrupling of the CO_2 content of the atmosphere in the next two centuries with a possible increase of 6 degrees Celsius in global surface temperatures. ... [E]xperiences of the past decade have demonstrated the consequences of even modest fluctuations in climatic conditions [and] lent a new urgency to the study of climate. ... The scientific problems are formidable, the technological problems, unprecedented, and the potential economic and social impacts, ominous."[10]

In 1979, the World Meteorological Organization held the first World Climate Conference, issuing an "appeal to nations" to "foresee and prevent potential man-made changes in climate that might be adverse to the well-being of humanity"[11] and focusing particularly on "the burning of fossil fuels, deforestation, and changes in land use [that] have increased the amount of carbon dioxide in the atmosphere."[12] When would such changes occur? Few scientists were prepared to say; most who ventured a guess imagined not before 2000. When the 1980s turned out to be the warmest decade on record (thus far), and the midwestern states experienced major heat waves and droughts, some scientists concluded that a detectable signal *had* occurred and that the costs of climate change were beginning to be felt.[13]

Still, heat waves were nothing new; could one say that this was something other than natural variability?[14] "Mathematical models of the world's climate indicate that the answer is probably yes," Roger Revelle concluded in 1982, "but an unambiguous climate signal has not yet been detected."[15] To find that unambiguous climate signal, oceanographers proposed an ambitious project called Acoustic Tomography of Ocean Climate.

Taking the Ocean's Temperature

Scientists today tell us that anthropogenic global warming is "unequivocal," but it has taken decades to reach this conclusion.[16] One reason is that there is no thermometer that permits direct measurement of Earth's temperature; scientists calculate the temperature from diverse historical records and geological proxies, and these calculations involve numerous inferences and assumptions.[17] But what if you *could* measure Earth's average temperature, more or less directly? This question was posed in the early 1980s by a group of oceanographers at the Scripps Institution of Oceanography and the Woods Hole Oceanographic Institution. Their answer was that you could, using the speed of sound in the oceans.

The speed of sound in water is temperature dependent, so if the oceans are warming then the speed of sound in them should be increasing. Since the oceans cover about 70 percent of the Earth's surface, an average ocean temperature would be pretty close to an average global temperature, and it would be more reliable than an atmospheric average because the oceans are less temporally and spatially variable. Moreover, sound can travel very long distances in the ocean so its speed over those distances is a measure of the average temperature of the water along the way. Release sound from a high-intensity source and record the travel time to a receiver, and in effect you measure the average temperature of the water mass through which the sound has traveled. A long-range transmission, say from Honolulu, Hawaii to Half Moon Bay, in California, could provide an integrated assessment of the thermal conditions of the water between those two points. So acoustic velocity is a thermometer.[18]

Ocean acoustic tomography was developed in the 1970s by Walter Munk and Carl Wunsch.[19] The technique is acoustic because it relies on sound waves; it is tomographic because it creates an image using vertical slices through the water column, and the measurements from the numerous pathways, or slices, are integrated to create a picture of the ocean through which the sound has traveled.[20] The technique relies on low-frequency sound waves (which travel efficaciously over long distances) by integrating information from sound waves that have traveled over various possible "ray paths." It works over long distances because the ocean sound channel permits the propagation of low-frequency sound with minimal attenuation.[21]

If scientists could take measurements around the world, collect the ray paths, and analyze the travel times, then they would come close to measuring the average ocean temperature at that moment. If they did it repeatedly for a decade, then they would have an independent assessment of whether the ocean--and thus the planet—was warming.

The acoustic thermometer depended on a precise understanding of the relationship between sound velocity and ocean conditions, but that relationship was very well understood. As early as World War I, scientists in Germany, Russia, and the United States worked on underwater sound transmission, and the topic became a major focus of the US National Defense Research Committee during World War II, inspired by the exigencies of submarine warfare.[22] With the development, during the Cold War, of SOSUS (SOund SUrveillance System—the secret US underwater acoustic system that tracked Soviet submarines) and of submarine-launched ballistic missiles, these programs flourished; some scientists at the Scripps Institution of Oceanography jested that it should be renamed the Scripps Institution of Underwater Listening and Location.[23] At both Scripps and Woods Hole, Navy funding flowed in for studies of underwater sound.[24] By the 1980s, Office of Naval Research support for acoustic research at Woods Hole amounted to $1.5 million per year, the largest single project at the Institution.[25] ATOC built on this history of Navy largesse and focus. Because the SOSUS network provided the equipment needed to detect the sound transmissions, the scientists would be relying on a well-tested, well-maintained technology with global reach.[26]

William von Arx had thought that shifting from warfare to weather would enable oceanographers to escape the Navy yoke, but things didn't exactly work out that way. On the contrary, the ATOC project was designed to exploit existing Navy technology and facilities, and the scientists involved turned to their existing Navy patrons for funding. The Office of Naval Research would support the use of data processing and recording equipment at the University of Michigan's Cooley

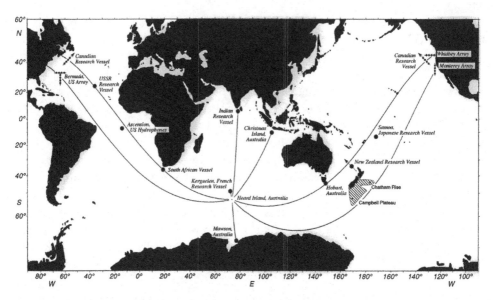

Figure 5.1
A map showing ray paths from Heard Island across the world ocean to oceanographic research institutions on five continents. Source: Walter Munk and A. M. G. Forbes, "Global Ocean Warming: An Acoustic Measure?" *Journal of Physical Oceanography* 19 (1989): 1765–1777.

Electronics Laboratory and at the Navy's communications facility at Centerville Beach, California. The proof of concept would occur at Heard Island in the southern Indian Ocean, from which there were unimpeded ray paths that could reach research stations in Brazil, India, South Africa, Australia, and the east and west coasts of the United States.[27]

Meanwhile, the oceanographer John Spiesberger and his colleagues at Woods Hole had preliminary results from the northeast Pacific demonstrating the concept's feasibility. They summarized these results in 1991 in the *Journal of Geophysical Research* and in a Woods Hole internal report, concluding in the latter that the results demonstrated that it was "possible to accumulate long records of arrival times that might be used to … measur[e] oceanic thermal changes associated with global warming brought on by increases in greenhouse gases."[28]

A Risk to Marine Mammals?

The Heard Island Feasibility Test was scheduled for January of 1991. "The issues in [the Heard Island Feasibility Test]," Walter Munk explained a few years later, "were: can signals generated by currently available acoustic sources be detected at ranges of order 10 Mm [mega-meters, i.e., thousands of kilometers], can coded signals be 'matched filtered' to measure travel time to better than 0.1 [second], and can this be done without harm to local marine life?"[29] This, however, was a bit of revisionist history, because in the early ATOC proposals there was no discussion of potential harm to marine life. Tomography had been conceptualized and developed by physical oceanographers and engineers to whom the prospect of interfering with marine life evidently hadn't occurred or hadn't seemed strong enough to pursue. Perhaps this was because they were using familiar technology whose safe operation they took for granted. Perhaps, because no biologists were involved in developing the project, the issue never came up. Or perhaps because, as J. Robert Oppenheimer said about the hydrogen bomb, it was technically so sweet that it just drew you in.

However, in the United States the Marine Mammal Protection Act of 1972 and the Endangered Species Act of 1973 prohibit the "taking" of any federally listed endangered or threatened species without authorization, and "taking" is defined very broadly as any activity that tends to harass, harm, pursue, shoot, wound, kill, trap, capture, collect, or to attempt to do any such thing.[30] Quite a few species of marine mammals, including some species of endangered or threatened whales, hear low-frequency sound, so the National Marine Fisheries Service concluded that the ATOC sounds might "take" whales, and that authorization was required. Yet the project was ready to go, and to reschedule it wouldn't be easy. So the participants set sail from Fremantle, Western Australia, hoping that the required permits would arrive in time, while colleagues, hoping to expedite the process, got in touch with contacts in Washington.[31]

Eight days before the scheduled start of transmissions, the permits were approved, but with the stipulation that a monitoring team look for adverse affects on marine life in the area. Four marine-mammal observers (one of them a bio-acoustician) were to be aboard a dedicated survey vessel; three additional observers traveling with the sound source would "monitor the effects of the transmission on marine mammals close to the source," with comparative observations made before and after each transmission. If marine mammals were "sighted or heard … within 5 km of the source," or "in the event of injury or mortality of one animal," the transmissions were to be delayed or suspended. The results of the biological observations were to be submitted to the National Marine Fisheries Service within 90 days of completion of the experiment.[32] With little time to spare, the Scripps oceanographers recruited Ann Bowles, a junior biologist from the nearby Hubbs Sea World in San Diego, to serve as bioacoustician and to supervise the monitoring program.[33] (Bowles would later earn a PhD at Scripps for a study of vocal recognition in emperor penguins, but at the time she had only a BA in linguistics.[34]) She found that sperm whales' sonar clicks were noticeably absent during the transmission periods, but she and her team concluded that the whales' silences weren't "associated with long-term effects."[35] With that, the scientists made plans to launch the Acoustic Tomography of Ocean Climate project, an ambitious worldwide program, running over a decade, to definitively detect global warming.

From Heard Island to ATOC

The Acoustic Tomography of Ocean Climate project would be as expensive as it was ambitious, but this ambition was made realizable by the creation of a new federal program (advocated by Senator Albert Gore Jr.), the Strategic Environmental Research and Development Program (SERDP), which had been established by Congress in 1990 as a program within the Department of Defense's Advanced Research Projects Agency. The goal of SERDP was to "harness some of the resources of the defense establishment ... to confront the massive environmental problems facing our nation and the world today"—that is, to turn Cold War swords into scientific plowshares.[36] By this time, the idea of ATOC had been around for a decade, but its implementation was made possible by the release of military hardware for civilian purposes at the end of the Cold War. In February of 1993, the Scripps Institution was awarded a $35 million contract to run the ATOC project.

Heard Island's acoustic access to the ocean was impressive—its signals were detected at sites around the globe—but this was outweighed by the logistical difficulties of working in such a remote location. The realized project would be moved closer to home, exploiting existing Navy facilities in Hawaii and California and relying primarily on bottom-mounted horizontal hydrophone arrays maintained by the Navy as part of SOSUS.[37] During the proposed initial two-year project period, a 20-minute signal would be released every four hours, up to six times per day, seven days a week. To communicate more clearly the intent to detect to climate change, the acronym ATOC was now said to stand for Acoustic *Thermometry* of Ocean Climate.

To manage a large project with such complex logistics, the Scripps Institution contracted with Science Applications, Inc. (SAIC), a private science and engineering firm with extensive military contracts, who helped to prepare the necessary permit applications. This time scientists knew in advance that they would need permits, particularly because the California source would be located on Sur Ridge, within the boundaries of the Monterey Bay National Marine Sanctuary, a federally protected marine area, and linked by a 22-mile cable to the Point Sur Naval Facility at the southern end of the Monterey Bay, in an area of California famous for its spectacular coastline and abundant marine life and a human population highly attuned to environmental issues. Before a full-scale program could go forward, biological concerns would have to be addressed.

One major question was whether the oceanographers would have to obtain a formal legal release from the laws prohibiting activities that might adversely affect threatened or endangered marine species. According to a Fish and Wildlife Service official who reviewed the proposal, ATOC might harass marine mammals, since by law the term 'harass' would include any "intentional or negligent act or omission that creates the likelihood of injury to wildlife by annoying it to such an extent as to significantly disrupt normal behavioral patterns which includes breeding, feeding, or sheltering."[38] Since the acoustic transmission would penetrate a large portion of the Pacific, the potential for "harassment" was considerable. Ann Bowles disagreed, arguing to the National Marine Fisheries Service's Office of Protected Resources that small cetaceans and monk seals have poor hearing in the relevant frequency range, and that humpback whales, which do hear well in that range, swam through the region only for a few months each year. "Within the zone of influence, humpback whales may alter swim direction and exhibit subtle changes in behavior until they habituate," she speculated, but this was too minor to constitute a "take."[39] "We do not anticipate any deleterious effects to the hearing, migration, communication or reproduction of marine mammals," Bowles explained, "although some species may avoid the transmission site during the first few months of exposure."[40] Representatives of the National Marine Fisheries Service rejected this interpretation, concluding that behavioral changes did represent a "take" and that the ATOC team would have to apply for a "small take exemption." Public hearings would be required, then a 60-day comment period, then a 120-day lag between comments and agency response, and finally additional time for the Office of Management and Budget to approve the legal exemption.[41] Bowles and her colleagues were shocked at the prospect of a long delay for what seemed to them to be purely bureaucratic reasons, and they tried to find another approach. Exceptions to the "taking" rule were permitted for a small number of specific reasons falling into two main categories. The first category excepted commercial activities; ATOC clearly was not such an activity.[42] The second category excepted bona fide scientific activities that supported the overall conservation goals of the Marine Mammal Protection Act, including survival and recovery efforts such as captive breeding.[43] That wasn't the purpose of ATOC, either, although the idea of pretending that it was had occurred to ATOC scientists. As Ann Bowles explained to David Hyde, a physical oceanographer who had been hired to serve as ATOC project manager: "You can't get a scientific research permit because ... NMFS already knows you are doing the work for other reasons."[44]

Bowles whined that none of the required steps—public hearings, public comments, agency responses—were "really necessary to protect the marine mammals; in fact they foster resistance among agencies and commercial operations to approach NMFS at all. Other sources of noise, such as shipping, most tomographic experiments, and private vessels are completely unregulated. Therefore, as usual, the scientific community is getting picked on." Bowles noted that the Marine Mammal Protection Act was up for re-authorization, and suggested that the Department of Defense "put some lobbyist to work to try to get the regulation changed."[45]

Bowles' comments revealed a number of presumptions. One was that the ocean was already filled with noise, so what additional difference would ATOC make? To single out ATOC when so many other sources of noise were allowed seemed arbitrary. She and her colleagues had a point, insofar as noise in the ocean was poorly studied, but that did not mean that it did no harm. Her comments also suggest that she believed that scientific research—perhaps because its aims were altruistic rather than mercenary—should be exempted from legal restrictions. As the historian Etienne Benson has recently noted, this view was common among scientists of a certain generation and bent; for example, Carl Hubbs—for whom Hubbs Sea World was named—had opposed an early version of the Marine Mammal Protection Act because it lacked exceptions for scientific and educational work.[46]

Bowles seemed to suggest that scientists should try to change laws that didn't suit them—a view that, however justified, would not, in the end, serve her or her colleagues well. After all, if scientists tried to change laws to suit them—even at the expense of the research subjects they claimed to care about—weren't they just another interest group? (Many citizens would later come to just that conclusion.) Whatever the case, once the National Marine Fisheries Service determined that ATOC would require an exemption, Bowles focused her attention on getting it. If the scientists succeeded, then at least they would have "no problems with stupid and useless monitoring requirements above and beyond what we have already agreed to, if we can just get through the regulatory paperwork."[47]

Guidelines set by the National Marine Fisheries Service required the scientists to calculate the numbers of animals likely to be "taken" on the basis of the numbers of marine mammals and other species that inhabited the waters, that swam at depths where the transmissions could be heard, and that heard in the relevant frequency range. Making such estimates wasn't trivial, particularly because data on marine mammal numbers and habits were scant, and because the ATOC signal was *designed* to penetrate the whole ocean, so on some interpretations, ATOC might end up "taking" just about all the whales. "It looks," Bowles admitted, "like we're going to get a permit to take an astronomical number of whales."[48] Bowles thought this wouldn't be a problem ("apparently no one raised an eyebrow about the 386,000 marine mammals we were supposed to disturb at Heard Island"), but Bowles did realize that an early draft application made it seem as if ATOC was a done deal—something that might prove problematic. "I'd like to suggest a couple of minor changes that may save us some trouble later," she wrote. "[Y]ou should emphasize that you are developing this Acoustic Observatory rather than treating it as a *fait accompli*. This covers your backside and emphasizes the fact that this project is really research."[49]

Was ATOC really research? More to the point, was the *biological* side of the program research? This became a point of discussion among the ATOC scientists, particularly Christopher W. Clark, a biologist-engineer at Cornell specializing in the acoustic response of marine mammals, and Daniel P. Costa, a professor of Ecology and Evolutionary Biology at the University of California, Santa Cruz. Clark had worked with the Navy on "Whales 93," an initiative that had used the Navy's Integrated Undersea Surveillance System to locate and track whales on ocean basin scales; Costa was now working with the Office of Naval Research on the announcement of a new program in low-frequency sound, and was in the process of developing a proposal for studying the effects of sound on elephant seals.[50]

Both Clark and Costa were disturbed by the ATOC research plan, which did indeed look like a fait accompli. They also drew a distinction between monitoring and research. Bowles had formulated the work as a monitoring program; Clark and Costa argued that this was a bad approach, both scientifically and politically. Scientifically, it didn't (in their view) constitute research, for it wouldn't answer any basic scientific questions. If effects were observed, there wouldn't be any way to explain why; if effects were not observed, the scientists would be stuck with the logical problem of equating the absence of evidence with evidence of absence. Politically, the ATOC scientists *did* appear to be covering their derrières. Clark and Costa agreed with Bowles that the project needed to be "really research," and they didn't think that as it stood it was.

What ATOC needed, Christopher Clark argued, was a well-structured scientific research program capable of determining whether systematic behavioral changes were occurring, and whether those changes were comparable with other oceanographic or meteorological variables. The studies, Clark suggested, should analyze species-specific vocal rates and repertoires, locate and tracking individual whales, characterize whales' migration tracks and corridors, and evaluate species-specific spatial and temporal distribution in enough detail to produce scientifically meaningful results. Solid data on these matters would both enable scientists to determine whether or not ATOC had changed marine mammals' behavior and contribute to the basic understanding of those mammals. This would also increase the likelihood that both biologists and agency officials would see the program as beneficial. Costa agreed. "The goal [should be to get] solid data, so that 2 years from now you can go to the table and say, "look here are data that show. ..."[51] After all, Clark concluded, whales were "*not* going to go away."[52]

The ATOC scientists began to design a $2.9 million program for studying marine mammals—which they now called their Marine Mammal Research Program (MMRP)—to accompany the first phase of ATOC.[53] The revised permit application emphasized that full-scale acoustic tomography would not proceed until the MMRP had resolved the question of impacts. In September of 1993, the scientists submitted a draft application to the National Marine Fisheries Service for a Scientific Research Permit with a small "take" exemption.[54] By the Scripps Institution's own estimate, the ATOC's "take" could be up to 670,000 animals per year, encompassing ten different species of whales (including blue, fin, sei, gray, right, sperm, minke, and humpback), eight species of dolphins and porpoises, and various seals, sea lions, otters, and turtles, some of threatened or endangered species.[55]

"Whale Lovers Went Wild"

Public hearings on ATOC were announced in the *Federal Register* of February 3, 1994 and scheduled for March 22. As word of the project spread, opposition grew among marine biologists, conservationists, and, especially, whale aficionados. As one conservationist put it, "whale lovers went wild."[56] Led by Hal Whitehead (a biologist at Dalhousie University in Nova Scotia) and Linda (Lindy) Weilgart (a postdoctoral fellow in Cornell University's Bioacoustics Research Program), opponents of the project took to the Internet, drawing on a listserv of persons interested in marine mammals that had more than 1,500 subscribers (marmam@uvvm.uvic.ca).[57] Postings warned of potentially severe damage to marine mammals and suggested that the rushed nature of the original permitting process was a deliberate attempt to avoid public scrutiny.[58]

Because it seemed likely that the hearing process would be highly contentious, Scripps Director Edward Frieman sent a letter defending the project to a long list of senators and representatives, urging them to express support for ATOC to the National Marine Fisheries Service (NMFS). Frieman supplied a sample letter and the NMFS's fax number. He sought to steer the conversation away from whales and back to global warming, emphasizing that ATOC could provide concrete data that would enable researchers to assess whether climate models, which suggested that warming was already underway, were correct. "The current projections of global warming," Frieman wrote, "are largely based on computer modeling [and] there are no measurements of ocean temperature which can be used to assess the modeling predictions. ATOC's ability to measure annual change in ocean temperature … will fill in a critical missing piece in the global warming puzzle."[59]

John R. Potter, a scientist at Scripps, blamed the public outcry on an article in the *Los Angeles Times*, published on March 22 to coincide with the first public hearings, in which Lindy Weilgart had asserted that ATOC could make whales deaf. But the risk of deafening whales had already been placed into public conversation in a feature story in the *San Diego Union-Tribune,* where the possibility had been raised not by hysterical environmentalists or sentimental whale lovers, but by a Navy veterinarian named Sam Ridgway. Ridgway, the chief veterinarian in the Navy's Marine Mammal Research Program, had worked for more than 30 years at the Naval Control and Ocean Surveillance Center at Point Loma.[60] The *Union-Tribune* article explained that the Navy had, for some time, been training dolphins, whales, and sea lions to one day "help fight a war." The Point Loma facility housed dozens of marine mammals, more than fifty of which were considered "surplus—some of them retired from active duty stretching back to patrolling Vietnam's Cam Ranh Bay." Three beluga whales, captured in 1977, had been used in experiments related to submarine warfare beneath the polar ice caps. Dolphins had been used in sonar research. Whales had had wires attached to their heads to record their brain waves. Ridgway's long experience made him uniquely suited to evaluate the impact of the ATOC transmissions, and he felt that it *was* possible for whales' hearing to become overloaded by the ATOC sounds. He concluded: "Continued exposure to this degree of sound could result in some degree of deafness."[61]

Walter Munk and David Hyde argued otherwise. Munk insisted there was "a great deal of scientific literature that would suggest that the sound levels we're generating do not do any damage." Hyde was quoted as flatly asserting that the ATOC transmissions "cannot cause long-term hearing damage."[62] The next day, the *Los Angeles Times* ran a front-page story under the headline "Undersea Noise Test Could Risk Making Whales Deaf." The debate was cast as a clash of the Titans: between scientists promising to solve the problem of global warming and marine biologists wanting to save the whales. The whales were represented by Lindy Weilgart, who insisted that ATOC's sounds might cause deafness in nearby whales, "leaving them unable to navigate or find food." She noted that the ATOC broadcasts at 195 decibels were "10 million times as loud as the 120-decibel levels that were known to disturb some whales." "We are invading an ocean habitat that so far has been untouched by man," she continued. "It's an experiment of tremendous implications and we are doing it without a clue of what it would do."[63] She concluded with a line that was widely quoted: "A deaf whale is a dead whale."[64] ATOC scientists would later cite these comments as evidence that the public had become inflamed on the basis of a misunderstanding. Weilgart's comments about the logarithmic decibel scale would have been true had they referred to sound transmission in air, but transmission through water is different—195 decibels in water doesn't have the same effect on an eardrum as 195 decibels in air.[65] Anyone who had had anything to do with Navy's extensive undersea programs—or knew anything about the history of whaling or fishing or telegraphy—knew that to say that the deep ocean habitat was "untouched by man" was just plain wrong.[66] On the other hand, by ATOC scientists' own account, the ATOC signal *was* equivalent to 110 dB in air—a level of noise comparable to that produced by a rock band.[67] While the signal might not be deafening, even to a human it would be very, very loud.

The *Los Angeles Times* article was syndicated in local papers across California, in the *Orlando Sentinel*, in the *Detroit News*, in the *Denver Post*, and in other papers, often under headlines even more inflammatory than the original one. Several referred to the acoustic source as a "boombox," while the *Portland Press Herald* presented a risk as a fact: "Sound-Blast Proposal Imperils Sea Creatures: The High-Decibel Experiments, Part of Global-Warming Research, Would Harass and Kill Whales and Dolphins." Meanwhile, scientists on both sides of the issue prepared for the hearings.

NMFS Hearings, March 1994

Initial hearings were held at the headquarters of the National Marine Fisheries Service in Silver Spring, Maryland. In his testimony, Walter Munk said that the ocean was not only an important reservoir of global heat and carbon dioxide but also a "reservoir of ignorance." ATOC could diminish that ignorance, and help to ensure that policy decisions were made on the basis of complete information. Robert M. White, president of the National Academy of Engineering, vice chairman of the National Research Council, and former head of the National Center for Atmospheric Research who had warned of the dangers of global warming in 1978, wrote a seven-page letter in support of the project, emphasizing that information on global ocean temperatures was needed to determine "whether climate warming is unequivocally occurring."[68] Although protecting marine mammals was important, he argued, the need for information about global climate change was urgent, and land-based and satellite measurements weren't likely to be as conclusive as ATOC.[69]

Christopher Clark testified that biological effects probably would be small and he assured the assembled group that any harm to marine mammals would be to him "a particularly acute concern."[70] But few biologists stood with him. Besides Hal Whitehead and Linda Weilgart, others testifying against ATOC included Robbins Barstow, a past president of the Cetacean Society, who criticized the oceanographers for resisting public scrutiny, asserting that both the public and other scientists—particularly marine biologists—had the right to "question and debate the merits of this request and its implications for marine mammals and ocean ecology."[71]

Newspaper editors seemed to agree. The day after the hearings, the *Los Angeles Times* article was picked up by more West Coast papers, including the *Seattle Times* and the *Oregonian*, and by the Associated Press; "A deaf whale is a dead whale" was spreading around the country. The *Los Angeles Times* published a follow-up article describing how activists were mobilizing to stop the ATOC project. A marketing director in Los Angeles was quoted: "This is a nightmare. I've been calling everyone I know. I've been calling senators and the governor. It would be criminal to do this."[72] This article, too, was widely syndicated. In the *San Jose Mercury News*, the quotation from the Los Angeles marketing director was printed under the headline in large boldface type. ATOC was no longer a scientific project being evaluated by scientists on scientific terms.[73] It was now a public affair, even a cause célèbre.

A spokesman for the National Marine Fisheries Service described the public response as "unprecedented." Letters, faxes, phone calls, and email poured into the NMFS's offices, not only from ordinary citizens but also from members of Congress. On March 23, a group of congressional representatives of Pacific Rim constituencies—Patsy Mink of Hawaii and George Miller, Ron Dellums, and Sam Farr of California—wrote to Secretary of Commerce Ronald Brown requesting an extension of the public comment period on the permit applications. (The NMFS is part of the National Oceanic and Atmospheric Administration, which is part of the Department of Commerce.)[74] Senator Barbara Boxer asked for the public hearings to be held in California; her request was seconded by George Miller, the Democratic chairman of the House Natural Resources Committee, by Gerry Studds, chairman of the House Merchant Marine and Fisheries Committee, and by California's senior Senator, Dianne Feinstein.[75] Both Boxer and Feinstein also wrote to the NMFS, Feinstein asking whether the experiment could be done elsewhere and with less impact on marine mammals.[76]

The Scripps Institution's communications office blamed the situation on the news media, particularly the widely syndicated *Los Angeles Times* article, and adverse coverage was certainly continuing. Editorials against the project appeared across the country. The *San Francisco Examiner* took a particularly critical position: "Imagine what it would be like if aliens from space decided—in the name of science—to target the Earth's inhabitants from their orbiting ships with megadecibel blasts of noise that could frighten or deafen many of the people below. Substitute humans for aliens, and you pretty much have the scenario for an experiment proposed by the Scripps Institution of Oceanography."[77]

The *Ventura Star* called the project "frightful," the *Seattle Post-Intelligencer* called it "goofy." Alluding to claims that many marine mammals were deaf in the 70-hertz frequency range, the *Los Angeles Times* concluded that it was the Scripps scientists who were deaf. The *San Francisco Chronicle* concluded that they were both deaf and dumb: "Whales and dolphins, which are known to have a high degree of intelligence, must be wondering just how lethally dumb their terrestrial mammalian cousins can get."[78] An op-ed writer in the *Santa Barbara News-Press* wrote: "These people are all supposed to have college degrees, aren't they? The only rational explanation for this scheme is that the Scrippsites have already run this experiment on themselves, scrambling their brains beyond recognition."[79]

On March 31, the Advisory Council of the Monterey Bay National Marine Sanctuary called for a delay until more information was gathered.[80] On April 5, the *New York Times* quoted Sylvia Earle, a distinguished marine biologist and a former chief scientist of NOAA, as saying "If you further damage the patient, the Earth, while you try to take its temperature, then maybe the method is flawed."[81]

In a letter sent to more than 68 members of Congress, Ed Frieman tried to counter the "deeply disturb[ing]" media coverage by outlining the various steps that had been taken to minimize impacts, and to detect any impacts that did occur.[82] He also wrote to the E.W. Scripps Associates—prominent individuals who supported the institution morally and financially, including the medical researcher Jonas Salk, the former newscaster Walter Cronkite, the philanthropists David Packard and Cecil Green, and the actor Ted Danson—to reassure them that the ATOC scientists had not had their brains scrambled. ATOC was "the most significant effort to date to determine if greenhouse gases are indeed causing a heat transfer to the oceans as part of global warming," Frieman explained. "Until global warming is better understood, governments will not be able to take effective steps to counteract its negative impact."[83]

Damage Control

The Scripps Institution now launched an organized public-relations effort. A communications officer named Cindy Clark prepared a package of materials to be sent to her counterparts at Woods Hole and other oceanographic institutions. Blaming the media for creating "public hysteria," she explained that "ATOC scientists were unable to convince this audience [at the NMFS hearings] that the project would do no real harm to the marine ecosystem" and she asked that the other institutions' communications officers "help to serve as the point of contact for your local media and government offices.[84] She included model letters to officials and suggestions for specific actions, such as calling local science writers.

Clark also asked for feedback on an ATOC "fact sheet" in the form of a set of questions and answers, released over the signature of Walter Munk.[85] One portion of the fact sheet read as follows.

Q: How are ATOC's acoustic sources designed to minimize impact on marine mammals?
A: The sources will radiate about 200 watts of acoustic energy, much less than many sonars, communications, and geophysical research sound sources which have been in use for many years. The ATOC signal is about the same level as radiated by an individual large ship traveling at 20 knots speed. ... The ATOC source transmits a very low frequency sound, spread from 60 to 90 Hz, which sounds like a distant rumbling to the human ear. Its energy is in the frequency band below the range most animals hear. ...
Q: Will the ATOC underwater sounds deafen whales, dolphins, seals or sea lions?
A: Definitely not. No physiological damage will occur to marine life as a result of ATOC sounds, even if they dive deep. Ships pass by animals hundreds of times a day without their sounds harming them.

This "fact" sheet was circulated among Woods Hole scientists, one of whom sent it to Chris Clark (no relation to Cindy).[86] In response to the question of how ATOC's acoustic sources were "designed" to minimize impacts, Clark wrote: "I think it is unwise and slightly untrue to say that ... the source characteristics were designed to minimize impact. ... The reasons were based on [the] oceanographic experiment's needs. ..." Next to the question whether ATOC signal would deafen mammals, Clark wrote "This is scientifically not a true response," annotating the fact sheet in italics:

No physiological damage will occur to marine life as a result of ATOC sounds, even if they dive deep. Ships pass by animals hundreds of times a day without their sounds harming them *(we don't know this)*. ... Scientific data *(what data?)*. ...

He suggested the following rewording:

Although there are no specific scientific experimental data relating low-frequency underwater sound levels with auditory damage, we believe that there is little chance that animals will suffer physiological damage as a result of ATOC sounds. The animals (i.e., blue, finback, humpback whales, etc.) believed to have the most sensitive hearing in the frequency ranges of the ATOC sounds are not known to dive deeply enough to come with a range (~500–600 feet) of the loudspeaker that might cause temporary loss of hearing. The animals (i.e. toothed whales, sea lions, turtles, seals) that are known to dive to great depths probably have poor hearing in the frequency range of the ATOC sound. However, scientific knowledge on this subject of the effect of loud, low-frequency sounds and marine mammals is extremely limited, and it is for this reason that we are supporting marine mammal research.[87]

Chris Clark's response was forwarded to Woods Hole's director, Robert Gagosian, who concluded that Woods Hole should not sign the press release. "If [our] people don't agree with the SIO answers, then we don't want SIO as our spokesperson. ... I don't want us to be associated with what we consider incorrect answers."[88]

Meanwhile, the ATOC scientists had assembled a Scientific Advisory Board and, with that board's guidance, had agreed to prepare Environmental Impact Statements for the California and Hawaii sites, and to delay operation of the ATOC system until after the Marine Mammal Research Program had submitted its results.[89] The National Marine Fisheries Service now announced that no permits would be issued until after the Environmental Impact Statements had been submitted and assessed, and that additional hearings would be held in Hawaii and in California. These hearings that went much the same way as the previous ones—which is to say, not well for the advocates of ATOC.[90]

The conflict reached a new level as a consortium of environmental groups—the Natural Resources Defense Council, the Sierra Club Legal Defense Fund, the Environmental Defense Fund, the Earth Island Institute, the American Oceans Campaign, the League for Coastal Protection, and the Humane Society of the United States—filed suit in federal court to stop the project. The plaintiffs accused the researchers of violating the National Environmental Policy Act, the Marine Mammal Protection Act, and the Endangered Species Act.[91] Scientists who had cast themselves as environmental heroes found themselves cast by their opponents as environmental villains.

The Draft Environmental Impact Statements

In late 1994 the Draft Environmental Impact Statements for the Point Sur and Kauai sites were released. Extensive and detailed, they responded to many issues that had been raised, but in the end they remained committed to the project and stood by the claim that there would be "no significant impacts."[92]

What *was* new was the proposed biological research program, which had been revised in several ways to try to produce bona fide scientific research, as well as to address the double bind that the only way to determine whether ATOC would do harm was to undertake the very activities that were alleged to cause that harm.[93] It did this by dividing the research into three phases. In the first phase, lasting from April to October of 1995, biologists would collect base-line data, with no ATOC transmissions. In the second phase, ATOC transmissions would begin, but the source would remain under the biologists' control and the transmission cycle would be greatly reduced from what had been proposed in the original plan: only one 20-minute transmission every four days, allowing any affected animals at least three days of respite. After a month, preliminary results would be compared with the base-line data, with a commitment to modify or abort the project if the results provided evidence of disturbance or harm. In the absence of such evidence, the third phase would be a two-year ATOC experiment, in 1996 and 1997, more or less as originally planned. Biologists would continue to remain involved, and the experiment could be halted at any time if unacceptable impacts were detected.[94] ATOC had now been made dependent on the Marine Mammal Research Program, rather than the other way around.

ATOC's opponents appreciated the changes and the fact that the new report included helpful bibliographies and summaries of existing knowledge about the effects of low-frequency sound, including a "crash course in marine bioacoustics."[95] Still, questions remained.

Field Supervisor Craig Faanes of the US Fish and Wildlife Service noted that the proposed six-month base line might "not be of sufficient duration to determine whether a species is negatively affected," particularly if it didn't include a mating season. What "methods and criteria" would be used to determine whether a species had been affected? How would the scientists determine effects on organisms that "are difficult to observe, not present during the time of year the data are gathered, or those for which little information is known on their behavior patterns prior to the project?"[96] Above all, how could the scientists be sure that no *observed* effects equaled no effects? Although there were provisions for the study to be shut down if "an unacceptably significant disruption of the behavioral patterns of a marine animal" was observed, who would decide what constituted unacceptably significant disruption?[97]

The Canadian biologist Paul K. Anderson noted that the problem of optimistic bias remained: "Although the objectives are formulated in classic null-hypothesis format ... the introductory paragraphs suggest a philosophical bias. The stated objective is to "validate" the assumptions that "reactions from marine mammals are unlikely at ATOC received levels < 120 dB at distances of > 20 km."[98] Just as ATOC scientists had trouble abandoning the claim that scientific evidence suggested that ATOC would do no harm, they also had trouble abandoning the presumption that the Marine Mammal Research Program would *demonstrate* that ATOC would do no harm. Indeed, when the members of the MMRP's advisory board had first been appointed, their charge had been described as follows : "[A] key program objective will be to *demonstrate* that the planned global network can be operated without any adverse effects on marine mammal populations."[99] That was in 1993; in the Environmental Impact Statement, the same optimistic bias was still evident.

The biggest question facing the National Marine Fisheries Service, however, was not how ATOC should proceed, but whether it should proceed at all. Although it might seem unrealistic to expect scientists to evaluate the option of not doing their science, the law in fact required it. Regulations developed by the US Council on Environmental Quality stipulate that considering alternatives and presenting them in comparable ways is the key to "sharply defining the issues and providing a clear basis for choice among options by the decisionmaker and the public."[100] The law required the "development of a range of reasonable alternatives that ... satisfy the basic project purpose (collecting data on global warming) but have fewer potential adverse effects on marine biological resources."[101] The Environmental Impact Statement did not do this. It evaluated alternatives in terms of geographic sites, duty cycles, and other technical details and components of the proposed ATOC framework, but it didn't seriously evaluate options to collect data on climate change in other ways, nor the option of doing nothing at all.

It is hardly surprising that the ATOC scientists didn't argue the case against their project with vigor (and it may well be unrealistic to expect people to argue the case for alternatives with the same vigor as the case for the thing they wish to do). But it is notable that at this point in the debate they began to soften their claims about what ATOC would achieve. Walter Munk, for example, told Lindy Weilgart: "[M]y views of what we should focus on have been modified over the last two years. A stand-alone detection and mapping of the greenhouse-induced changes over and above the ambient changes will take a long time. ... I now think our emphasis should be to test, and help improve, current climate models.[102] Munk made a similar qualification when he wrote to the California Coastal Commission in June. Rather than suggesting that ATOC would detect, prove, or even confirm global warming, he now argued that the point was to test climate models:

ATOC is intended to observe the ocean on the large space scales that characterize climate—3,000 to 10,000 kilometers—so that modelers will be able to:

• test their models against the changes seen by ATOC over a few years
• and, if, and when, the models prove adequate at hindcasting, use those same models to make climate predictions.

By testing and improving climate models now, ATOC can make progress toward greenhouse predictions later.[103]

The original claim had been that ATOC could detect global warming and would do so within a few years; the new claim was that ATOC could be used to benchmark the models used to predict climate change. This modified position was certainly more intellectually defensible than the claim that ATOC would conclusively prove the reality of global warming; perhaps Munk and his colleagues sensed that their opponents were doubting not only the wisdom of the project but also the assertive claims being made for it. Or perhaps the opposition had caused them to think more deeply about what the project could realistically be expected to achieve. Whatever the reason, this more modest epistemic position, although intellectually honest, backfired. Opponents of ATOC now asked: If ATOC couldn't prove global warming, or at least provide very strong evidence of it, then why risk harm to whales? Why do it at all?

California Coastal Commission Hearings: Absence of Evidence or Evidence of Absence?

In January of 1995, the California Coastal Commission scheduled public hearings in Santa Cruz and invited written comments. The hearings would be based on the environmental impact study for the Point Sur site. Cards, letters, and faxes flooded in from scientists, conservation organizations, and ordinary citizens; the vast majority were negative.[104] Virtually all of the comments submitted by environmental groups were thoughtful and detailed. Virtually all shared the concern about global warming. Yet not one environmental group supported ATOC. ATOC was being presented as an "environmental project," but environmental groups all viewed it as dismissive of one of the central concerns of modern environmentalism--protection of threatened and endangered species--and nearly all of them noted that the ATOC scientists were making a classical logical fallacy: using the absence of evidence as evidence of absence. This point was central to a detailed critique by the Washington-based Center for Marine Conservation.[105] The draft Environmental Impact Statement (EIS) repeatedly "defaults to a conclusion of no expected significant impact," they noted, when the very reason why the Marine Mammal Research Program was needed was the dearth of scientific basis for predicting the impact. Noting the EIS's frequent use of the adjectives 'nonexistent', 'negligible', and 'minimal' to describe potential impacts, the Center for Marine Conservation concluded that the "tendency to dismiss uncertainty exacerbates rather than alleviates questions regarding impacts." And although the Environmental Impact Statement claimed that the ATOC program was designed to minimize adverse effects, elsewhere it revealed that the choice of the source location was based on economics and pragmatics, such as a "minimum cable run to shore" and "close logistical support." The authors' advocacy of the project had undermined their capacity to produce an objective report, resulting in a document whose "ambiguities, inaccuracies, and treatment of uncertainties has intensified rather than quelled concern."[106]

A similar argument was presented by Rodney Fujita on behalf of the Environmental Defense Fund (EDF). Fujita, a PhD marine ecologist, had worked with the Intergovernmental Panel on Climate Change on the effect of elevated sea-surface temperatures on coral reefs, so he was acutely aware of the risks that global warming posed to marine life, as was the organization for whom he worked. But Fujita argued that ATOC scientists had not been forthcoming about the central fact that "there exists virtually no evidence bearing on the question of how marine organisms might respond to the ATOC sound source." "The key to good policy making on this issue," Fujita asserted, "is to freely acknowledge the great uncertainties surrounding the potential impacts of ATOC and work to reduce them, rather than attempting to paint a rosy picture that shows that the impacts are likely to be insignificant. ... [The] EIS consistently makes the error of concluding that if no evidence for a significant impact exists, the impact must be non-existent."[107]

Various commentators noted that whales weren't the only marine life that could be affected. The Pacific Fishery Management Council pointed out that extensive scientific literature documented the effects of sound on fish. They, too, found the EIS dismissive in suggesting that injury to fish was insignificant because any injured fish would simply be more easily caught by their predators and all fish get eaten sooner or later—as if that were not a disruption to the ecosystem.[108] The Center for Marine Conservation noticed that, although the EIS claimed that "no information exists on noise impacts to salmon," there was in fact "abundant evidence that salmonids hear and behaviorally respond to low-frequency sounds." Indeed, "repetitive low-frequency sounds" were being used to direct the paths of juvenile salmon in the Sacramento–San Joaquin Delta; that this could hardly be efficacious if salmon were deaf!

Because advocates justified the potential costs to marine life on the grounds of the benefits of the data that ATOC would produce, a central question for opponents of ATOC was how definitive the results would actually be. The Kauai Friends of the Earth noted that even by the scientists' own reckoning ATOC was no sure thing. Scientists had provided detailed discussions of uncertainties about spatial resolution and the analysis of ray paths and about the interface between ATOC measurements and global circulation models, satellite data, and sea-surface temperatures. This was normal scientific practice (any grant proposal is expected to include discussion of uncertainties and sources of error), but the Kauai Friends of the Earth raised what might have been considered the most urgent question about the project: whether measuring basin-scale deep ocean temperature would resolve the question of human versus natural drivers of climate change. They quoted Walter Munk's statement, in the *Journal of the Acoustical Society of America*, that "it is important to emphasize that acoustic thermometry addresses the issue of measuring climate change (ambient or otherwise) in the oceans; it does not tell us anything about the underlying causes."[109]

Munk's statement was admirably forthright and honest: acoustic thermometry could tell you if the ocean was warming, but not why. The evidence for drivers had to come from other sources. The scientists were asking for a potential sacrifice of marine life for the sake of answering key questions about climate change, but if ATOC couldn't actually answer those questions—particularly the *politically* crucial one of whether the discernible warming was natural or anthropogenic—then what was the justification for the sacrifice of marine life?[110] If the claim was simply that ATOC could *help* to answer significant questions, then the justification was significantly weakened from what had originally been claimed.

The central presumption of ATOC's supporters—only occasionally stated outright but implicit in the entire project—was not simply that ATOC would provide the "unambiguous signal" for which Roger Revelle had longed, but that *this would make a difference to public debate*. The presumption was that political action was impeded by the lack of clear science, and that better science would lead to better policy. Robert White had asserted that ATOC would reveal "whether climate warming is unequivocally occurring;"[111] Edward Frieman had argued that until warming was better understood "governments would not be able to take effective steps to counteract its negative impact."[112]

But was this presumption correct? Was a lack of scientific understanding preventing governments from taking action? Would crisp science lead governments to crisp action? ATOC's critics wondered.[113] The more sophisticated among them emphasized that even if the evidence were persuasive to *scientists*, this didn't mean that it would be persuasive *politically*. Rodney Fujita made this point most acutely, arguing that, although the ATOC scientists accused their opponents of irrationality, they had fallen into an irrationality of their own, or at least a position that was counterfactual: the presumption that knowing the scientific facts would lead to political action. Scientists, he argued, needed to "come to grips with the limitations of science." Fujita urged the National Marine Fisheries Service to "recognize that uncertainty about the impacts of ATOC will always remain, because the habits of marine mammals, the complexity of the marine environment, and the difficulty of doing controlled experiments that isolate cause and effects relationships in the ocean will often prevent the drawing of strong inferences."[114] But even if those scientific uncertainties could be resolved, it didn't follow that this would move governments to action. From a political perspective, one had to conclude that the "potential for sweeping changes in global warming policies resulting from the ATOC data is low."[115] Fujita explained:

None of us should be overly optimistic that data generated by ATOC, no matter how accurate or precise, will result in a dramatic improvement in climate change *policies*. ATOC could reduce key uncertainties about ocean heat uptake, [and] while a reduction in this uncertainty, better climate models, and a more definitive indication that global warming is occurring—all potential benefits of ATOC—would definitely be helpful, they are probably not the most important factors limiting progress toward taking action to prevent global climate change. Vast economic and political interests continue to resist significant changes in the current patterns of fossil fuel use and deforestation that are driving climate change, and they are not expected to disappear in the foreseeable future.[116]

That was in 1995. In hindsight, it certainly seems that Fujita was right.

The comments from ordinary citizens overlapped with those of the environmental groups but had several additional elements: offense at the tone of the scientists' assertions (which were viewed as arrogant and dismissive), skepticism as to whether ATOC was even needed, and, more interestingly, doubt that ATOC was what scientists said it was.[117]

Accusations of arrogance and hubris peppered the public comments, sometimes as the main complaint, sometimes as an extra source of irritation added to other concerns. Numerous citizens noted the irony of oceanographers' asking the public to respect and welcome their expertise while they disrespected the expertise of their biological colleagues. Others objected to the hubris of scientists' thinking they were above the law, as evidenced by their failure to apply for permits for Heard Island, their attempt to claim exemption from the Marine Mammal Protection Act, and their having at one point denied the legal authority of the California Coastal Commission.[118] Derek Cole, a retired radar and sonar engineer, wondered how scientists could be so sure that animals would not be harmed. "I have yet to encounter a scientist that can communicate with a whale," he wrote, "yet we purport to know what they hear and how they interpret it."[119] Cole wasn't the only one to think it inappropriate to place a high-intensity acoustic source inside a marine sanctuary. In the Environmental Impact Statement, it was argued that the sanctuary was a particularly *good* place to study the potential adverse effects, because there were a large number of marine mammals there. Most respondents considered that logic perverse; some thought it smacked of killing a patient to cure a disease.[120]

The cultural status of whales as exemplars of animal intelligence, loyalty, and even musical ability added insult to the injury that many citizens felt. Defending ATOC on the basis that it would add only a small increment to the background hum of existing noise pollution from tankers, ships, seismic exploration, and other human activity was broadly rejected as akin to justifying more air or water pollution on the grounds that the air and water were already polluted. If people hadn't previously known the extent of ocean noise pollution, they did now, and it proved that "there is an immediate need for noise reductions to make the oceans quieter."[121]

Various citizens argued that, whether or not ATOC harmed animals, it simply wasn't needed. Many cited the 1992 UN Framework Convention on Climate Change, which committed its signatories (including the United States) to preventing "dangerous anthropogenic interference" in the climate system.[122] Although many citizens were not tracking these developments, most environmentalists were; those who were paying attention to ATOC were more likely than the average American to know that scientific evidence of anthropogenic climate change was already pretty strong. Some argued that NASA satellite data could be used to evaluate global warming; others argued that by the time ATOC provided a "definitive" result, warming would be well underway and difficult, if not impossible, to reverse. Many respondents argued that ATOC's $35 million budget would be better spent on developing solar, wind, or tidal power. One characteristic letter read: "We already have plenty of studies and data showing that there is global warming. That is a given. We do not need another experiment that proposes to tell us what we already know."[123]

Whatever benefits oceanographers claimed for ATOC, for some members of the public they were moot because those citizens didn't believe that ATOC's purpose was to detect global warming. For the scientists, the swords-into-plowshares aspect of the project was something to boast about, but to many people it was grounds for suspicion. It was a characteristic feature of Cold War science to blur the boundary between "basic" and "applied" research, and some citizens wondered if that was being done with ATOC.[124] Bill Dietrich, a reporter for the *Seattle Times*, noted that "rather than taking this as an example of post Cold-war conversion, critics already unhappy with Navy experiments with captive dolphins regarded it with suspicion."[125] Stanley Flatté, a professor of physics at UC Santa Cruz, put it more bluntly: "Folks thought it was some kind of secret Navy project."[126]

Various respondents wondered why, if ATOC was environmental research, it was being funded by the Department of Defense rather than the Department of Interior or the Environmental Protection Agency. "What other projects have been funded by the Advanced Research Projects Agency?" one citizen asked. "Do they have a track record in environmental science they would care to share with us?"[127] Others wondered why portions of ATOC were classified, expressing suspicion that the public wasn't being told the whole truth, or even any part of the truth. Many citizens believed that ATOC was a secret military project, even a weapons system. Sarah Miquiabas of Kapaa, Hawaii began a handwritten letter objecting to the installation of an ATOC source near Kauai with this sentence: "I am writing to you about the underwater bomber."[128] (Presumably she had confused "boomer" with "bomber.") Others expressed the same idea in more sophisticated fashion. David N. Seielstad of Princeville, Hawaii wrote at length about the many reasons why the ATOC program seemed suspicious:

From the beginning the ATOC proposal has had the aroma of a military research project. It is funded by DOD monies. It is administered by the US Navy. The originators of the project seem to be going to great lengths to disguise and conceal the true nature and the purpose of the project. In the proposal (p. 62), provision is made to "manage classified aspects of the project. ..." The Johns Hopkins University Applied Physics Laboratory (a major Navy research and development contractor) is to use its clearance and store [any classified] data.[129]

Seielstad was right about the associations (the Applied Physics Laboratory *was* a major Navy contractor), and various aspects of ATOC were classified and had not been well explained (among them its links to the Pacific Missile Range Facility, first denied and then affirmed).[130] Noting that when scientists are "less than candid" the public naturally gets suspicious, Seielstad concluded: "What is ATOC really? It is being promoted as a study of global warming. ... Who could be opposed to that? If something is cloaked in the aura of environmental research I guess we are all expected to stand up and applaud it as good science. [But] ATOC is ... only masquerading as environmental research."[131]

Similar sentiments were expressed throughout the public comments. One commenter asked the organizers to "stop insulting the intelligence of the human race" with their "global warming greenwash."[132] Another insisted that "if global warming was the true priority, then the use of tax dollars would be more wisely spent in the areas of clean energy and ... efficiency" and asserted that "the 'classified' nature of the ATOC implies that this has nothing to do with global warming, rather it is a military operation intended to improve submarine detection."[133] Another suggested "you should be honest with the American public about the true nature of these experiments. If the purpose is to learn more about global warming, why the classified designation?" and asked "Please *respond*."[134] Another wrote "if the Navy wants to sell us defense research cloaked as environmental concern, they should have gone to the CIA or NSA and kept their mouth shut."[135] And another expressed amazement "that the public is viewed as being so stupid that we would believe that the Navy is suddenly concerned about global warming."[136]

Perhaps the best evidence of the widespread belief that ATOC was a military project is that the few expressions of support for it were mostly based on the corollary that ATOC was necessary to defend the United States. In a comment on the Kauai Environmental Impact Statement, Dave Clewett wrote: "I am not opposed to this research project, and I encourage you to press on with it. I believe in our military, and the importance of being defensively prepared. I do not agree with the efforts of Greenpeace and others of the liberal left to cripple our ability to defend our country. I am a conservative American, so if they are against your project, I am for it." Ronald Peet and Sandra Castro similarly wrote: "We are adamantly FOR the ATOC project. Do *not* let the Santa Cruz Marxists stop this important work."[137]

Resolution and Results

Amid all the claims and counterclaims—the voices of oceanographers and climate scientists mostly on one side, the voices of biologists, conservation associations, and citizens mostly on the other—one eloquent individual attempted to find a *via media*: Sylvia Earle, scientist, engineer, deep-sea diver, and grande dame of American marine science, whose diverse accomplishments ranged from setting the woman's world record for deep diving to serving as NOAA's Chief Scientist from 1990 to 1992. Smart, beautiful, and articulate, Earle was known for her unwavering dedication to protecting the oceans as the "blue heart" of our planet.

Earle didn't accept the argument that ATOC would do no harm, suggesting instead that it seemed "obvious that the proposed research will, in fact, have some impact on the behavior of marine organisms." Like many others, she had "deep concerns about adding additional stresses to ocean ecosystems already modified by recent human activities ranging from overfishing to various kinds of pollution including high levels of noise pollution." However, she was also convinced that the "greatest threat to the health of the oceans and to the planet as a whole is lack of knowledge and the profound mistakes in judgment that result from ignorance." Therefore, she asserted, it was "important to try to resolve the problems associated with ATOC, if possible, and find ways to fill the enormous gaps in understanding the nature of the ocean and the effects of human activity on marine life."[138]

Earle's comments seem to have moved the California Coastal Commission, because in making their final decision they quoted from them at length, finally concluding that "given the potential scientific and environmental benefits from the research proposed, and since the only way to determine the project's impacts is to proceed in the short term and study its impacts, the authorization of a two-year initial ATOC project is warranted." The approval was conditional on the "combination of the monitoring and protective measures incorporated into the project, the up-front commencement of the MMRP, and the relatively short (two-year) duration of the project prior to seeking any further permanent authorization."[139] The Scripps scientists were required to inform the commission of any significant modifications to the project and of any developing evidence "documenting adverse effects on marine resources," and to request explicit permission for any extension beyond the approved two-year period. Effects on fish would be included in the monitoring and the analysis, and the Scripps scientists would develop a clear set of "termination criteria" for curtailing the transmissions.[140] Finally, the scientists would move the sound source out of the Monterey Bay Sanctuary to the Pioneer Seamount, 48 nautical miles west of the Pillar Point Air Force Tracking Station (which was near Half Moon Bay) and twice as far from land as the original site.[141]

These terms also provided the basis of an out-of court settlement with the parties that had sued the ATOC consortium.[142] The settlement added the additional stipulations that the Marine Mammal Research Program would continue for the entire 18–24 months of the initial research period rather than just 6 months, that the sound source would be controlled by MMRP biologists and not by physical oceanographers, that there would be no claim that the MMRP would be able to prove or disprove long-term impacts on marine mammals, and that any proposal for a long-term ATOC program after the initial research period be subject to full environmental review (including the preparation of a new Environmental Impact Statement).[143] With these stipulations in place, the National Marine Fisheries Service issued permits for the initial phase of the project to begin. This phase, now referred to as ATOC-MMRP, would study impacts on marine life, and would not attempt to measure ocean warming.

The ATOC-MMRP project began officially in October of 1995, and work was underway early in 1996. A mid-1996 interim review revealed no significant problems, and the project continued for two years, after which the results were reviewed by a committee of the National Research Council. The committee took as its charge to review both the specific results of the MMRP and any independent advances in understanding the impacts of low-frequency sound since 1995. The conclusions of the 100-page report, *Marine Mammals and Low Frequency Sound*, were summarized in the first paragraph of its Executive Summary:

Some of the MMRP observations, such as movements of humpback whales in near-coastal areas off Kauai and the abundance of some whale species near the Pioneer Seamount source off California, showed no statistically significant effects of ATOC transmission. For these observations, the Committee could not distinguish among true lack of effect, and insufficient observations, small sample sizes, and incorrect statistical treatment of data. ... Some statistically significant differences between control and exposure were found for other species, including (1) an increase in average distance of humpback whales from the California source and (2) increased dive duration for humpback whales off Hawaii. The MMRP found no obvious catastrophic short-term effects as a result of transmissions from either source, such as mass strandings or mass desertions of source areas.[144]

The Executive Summary took the understated tone characteristic of NRC reports, but the report itself was noticeably critical, highlighting the problem that ATOC's critics had long pointed out (and Chris Clark had tried to address): that the project's design wasn't adequate to answer the biological questions. MMRP was essentially a "retrofit" onto a program designed for other reasons, and the result was predictable—indeed, had been predicted. The project was too short-term to elucidate long-term effects, and its design wasn't based on maximizing the relevance to marine mammals or on minimizing the impacts, but rather on the project's primary goal of detecting global warming. "As a consequence," the NRC concluded, "the results of the MMRP do not conclusively demonstrate that the ATOC signal *either* has an effect *or* has no effect on marine mammals in the short or the long-term."[145] After five years, months of hearings, hundreds of comments, and millions of dollars spent, the question of whether ATOC would harm marine life remained unanswered—and oceanographers hadn't demonstrated whether the ocean was warming, either.[146]

It was beyond the committee's charge to recommend whether ATOC should continue, but they concluded that the Scripps Institution had been, at best, premature in its assertions that ATOC would do no harm. Even now, there weren't adequate grounds for making such judgments, and developing those grounds would require "a more sustained and integrated approach than has been the case in previous research."[147] In an implicit rebuke of the manner in which the MMRP had been created, if not of the entire idea of scientific research under the aegis of mission-driven agencies, the committee argued that the necessary biological research "should be sponsored by the agencies that fund basic and applied biological research," not by the Navy. If mission-oriented agencies within the Department of Defense were to fund research, they should "ensure that the research they sponsor will not only contribute to their immediate missions but also answer basic scientific questions," and "all of these projects should receive strict peer review and be evaluated on the quality of the science proposed."[148]

This point was reiterated in the report's Findings and Recommendations section, in a conclusion that could well be applied to the whole of the Navy's oceanographic research program, not just to marine-mammal research. It was a conclusion with which William von Arx and Paul Forman would have heartily concurred:

Most marine mammal studies are funded from mission-oriented sources. At this time the greatest source of funding for marine mammal research is ONR. However, by its nature, ONR-funded research tends to be focused on questions of practical importance to the Navy, and is not necessarily responsive to the broad interests of scientists seeking to learn more about the basic biology of marine mammals. Scientist-driven fundamental research could significantly improve our understanding of hearing and the effects of low-frequency sound on marine mammals, as well as our overall understanding of the acoustic behavior of these animals.[149]

On the basis of the NRC findings, the National Marine Fisheries Service declined to extend the permits for ATOC. The project ended in August of 2000 when a winch operator named Ron Hardy died after being struck in the head by a piece of equipment while trying to remove the 12,000-pound transmitter from the sea floor.[150] Two years later, a federal judge halted a project (funded by the National Science Foundation) that was using intense blasts of compressed air to study the structure of the sea floor in the Gulf of California after two beaked whales were found dead on the nearby Mexican coast with evidence of hearing damage. The ruling by US District Magistrate James Larson was based in part on evidence that the Navy considered sounds above 180 dB to be "potentially harmful to marine mammals."[151]

Discussion

ATOC scientists repeated the "no evidence" claim for years, but in October of 1994 Ann Bowles and colleagues had published results that plainly refuted it. In an article in a special volume of the *Journal of the Acoustical Society of America*, Bowles et al. had reported that sperm-whale "clicks, clangs and a few codas" were detected during 24 percent of the base-line period but that no such sounds were detected during the transmissions, and that several individual marine mammals had been observed changing course to avoid the direction from which the transmissions were coming. "The results," they wrote, "suggest that ... whales could have altered their distribution in the immediate vicinity of the HIFT transmissions." Bowles et al. emphasized that these results were very limited: indeed, they tried their best to explain them away, noting that the whales "returned or were replaced by new individuals quickly when transmissions stopped." But if the full-scale ATOC program were implemented, transmissions would be continual for a decade. How would this affect them? Bowles et al. speculated that "in the long run animals might have habituated well to the transmissions," but the operative word here was 'might'. The available evidence suggested that marine mammals *had* been affected: whales went silent and changed their courses. In the words of Bowles et al.: "Changes in behavior of pilot whales and sperm whales provided *unequivocal evidence of behavioral effects of the transmissions.*"[152] Yet as ATOC went forward, Bowles and her colleagues buried these findings, insisting that marine mammals would be unaffected by ATOC.

Why did the oceanographers make assertions that at best were unsupported and at worst were refuted by their own data? Most of them referred back to their long history of Navy-sponsored work, suggesting that none of those earlier projects had done harm. But they didn't really *know* that. None of those earlier projects had been subject to the same degree of scrutiny, either by colleagues in other disciplines or the public. Most had not been subject to public scrutiny at all. This, perhaps, is the best explanation for why the oceanographers behaved the way they did.

Throughout the Cold War, scientists had been accountable to their Navy patrons, and had been judged by experts in their own field. Studies of underwater acoustics were peer-reviewed by other experts in acoustics, but not by cetacean biologists, not by the National Marine Fisheries Service, and certainly not by the public. If a project was classified, there was still less external scrutiny. Even members of Congress often lacked information about classified scientific work; President Franklin Roosevelt was famously secretive about the Manhattan Project, and after Roosevelt's death his successor, Harry Truman, had to be told about it.

As a patron of science, the Navy wanted projects to produce reliable knowledge and accepted that this entailed giving scientists some intellectual latitude.[153] The Navy also accepted that it was in its interest to permit publication of scientific results from the projects it funded to the extent that publication was compatible with military interests. But the Navy didn't *encourage* scientists to reach out to colleagues in other fields, much less to speak to the public, write popular accounts, or explain the meaning and significance of their work in broader venues and terms. To a historian of twentieth-century science, this stands out as a significant shift. Before World War II,

American Earth scientists routinely wrote popular books and articles, even on seemingly arcane topics. William Bowie, chief of the Geodesy Division of the US Coast and Geodetic Survey, wrote pieces for the *New York Times* and for *Popular Science*, and even had a radio program on geodesy—hardly the most accessible of scientific topics.[154] Others scientists gave public lectures, particularly if their work involved expeditions to exotic or dangerous places. Sometimes they did this for money—fees for lectures and newspaper articles could be important sources of financial support, particularly for research that had an expeditionary component. Before World War II, funds for scientific research were scant; payments for popular accounts and public lectures were one means of supporting scientific work, and such accounts and lectures might draw the further interest of private patrons.[155] Some scientists, including William Bowie, were civil servants who thought that the citizens who paid their salaries deserved to know what their monies were being spent on.

After World War II, this changed. As funding for scientific research increased, outward communication from the scientific community decreased. While it is challenging to read historical silences, one cannot help but notice this one: In the postwar years, Earth scientists simply didn't engage the general public as they previously had. Before the war, scientists found support for research hard to come by and thought that reaching out to the public might pay off. During the Cold War, reaching out was tricky, and a steady stream of money from the Navy made it unnecessary. Scientists who had worked with the Navy for decades were used to proceeding without public scrutiny, and the idea of the public as their ultimate patron was rarely if ever raised.[156] Indeed, one rarely if ever sees expressed the idea that the true patrons of American science during the Cold War were the American people.

When the Cold War ended, things changed again, and ATOC brought that shift into sharp relief. Walter Munk acknowledged that he had been quite unprepared for the public outcry, and that he "had never experienced such press interest or scrutiny" in anything else he had ever done. For four decades oceanographers had worked with the US Navy on all varieties of acoustic matters, and these issues had never come up—at least not in this form. Scientists working with the US military hadn't concerned themselves with environmental impacts. It was not so much that they didn't care about marine life as that they had never *had* to care. Protecting the environment wasn't part of the mission.

What oceanographers failed to grasp when they attempted to change their mission was that they would have to change their strategies and tactics as well. Scientists supported by taxpayers might reasonably have expected to explain the significance, the impacts, and the risks of their work, but these men had no habit and no experience of doing so. When asked to explain themselves, their responses essentially amounted to "Trust us, we're experts" and "Trust us, our intentions are good."

But the public didn't trust them and didn't accept that their intentions were good. If the ATOC scientists weren't deaf and dumb to others' concerns, they evidently didn't hear well in alternative frequency ranges, and the reason lies at least in part in four decades of disuse. Scientists working with the Navy had rarely confronted the implicit values behind their research program. The question of the *values* of their research—beyond the immediate *value*—was rarely discussed, and almost never overtly, even within the tight circles of Navy oceanography, much less in public. Because of military secrecy, some of it *couldn't* be discussed. Oceanographers ridiculed Lindy Weilgart for her naive view that the deep ocean was untouched by man, but many Americans probably shared her naiveté. How were they to know that the Navy had, for decades, been secretly laying cables, hydrophones, and instrumentation of all sorts beneath the waves? Who knew what ocean acoustic surveillance was? For decades, most Americans knew nothing of the work that oceanographers had done, ostensibly on their behalf. When they began to learn about it, they didn't necessarily feel grateful, especially when they discovered that it potentially threatened things that they loved. And if Weilgart was naive about the physical operating conditions of the deep ocean, so, it seems, were oceanographers naive about the social, political, and cultural operating conditions of American life at the end of the Cold War. They seemed unaware that their personal and professional histories cast them in a certain light, as men with certain affinities and affiliations.

As far as the available historical evidence shows, ATOC was not a weapons program disguised as basic research. But it *was* a military project in the sense of relying on hardware, facilities, and funding supplied by the Department of Defense. That affected how others viewed it, and it also affected how they viewed the men who were its advocates. If environmentalists, biologists, and lovers of whales didn't trust them, it is not hard to see why. Oceanographers who for decades had been studying the ocean as a theater of warfare simply weren't credible when they presented themselves as guardians of the ocean as an abode of life. Naomi Rose, a biologist with the Humane Society, put it this way: "The oceanographers asked: 'Why would you even think we would hurt the environment?' and we responded, 'Why would we think you wouldn't?'"[157]

Forty years of military patronage were not just epistemically consequential; they were socially and culturally consequential as well. Among other things, they produced a scientific community accustomed to various forms of internal accountability but unaccustomed to public scrutiny. At the end of the Cold War, when they faced broader scrutiny, they found themselves lacking both crucial skills and sensibilities and the ability to develop those skills and sensibilities quickly. The net result was both a political and an epistemic failure. Politically, oceanographers failed to garner the support they needed for the ATOC project. Epistemically, the lack of that support left them unable to answer the scientific questions they wished to answer. Forty years of freely flowing military funding, restricted interactions with other communities and their concerns, and little or no external accountability had produced researchers who were not necessarily gadgeteers, but whose horizons were constricted on several social, cultural, and intellectual dimensions. ATOC might have offered a gratifying conclusion to the era of Cold War oceanography: techniques, knowledge, and technology developed in pursuit of military power would be turned toward peaceful purposes. It held the promise of a conclusive answer to a scientifically challenging and socially important question. But it crashed on the shoals of Cold War legacies of secrecy and hubris.

To return to the question of the consequences of military patronage, and what the military did and did not support: *Of course* the Navy focused on matters of practical importance to the Navy. How could it be otherwise, unless the Navy neglected *its* political, social, and legal mandate? At times that mandate aligned with issues of scientific import, creating robust and vital domains of knowledge; at other times it did not, leaving significant domains of ignorance. The impact of underwater sound on marine life was one of those domains of ignorance. The Navy had spent many millions studying the propagation of underwater sound, and had studied animals to better understand that propagation, but the basic biological science that might have enabled scientists to predict ATOC's impact had never been done.[158] And so the science that would have been needed to determine whether ATOC could proceed without harm to marine life simply did not exist. ATOC was both a product and a victim of the Cold War.

Acknowledgments

I am extremely grateful to the numerous scientists and environmentalists who spoke to me honestly and openly about what was, for virtually all involved, a difficult and disturbing chapter. I am especially grateful to the late Edward Frieman, who provided feedback on an earlier version of this work, to Robert Gagosian for permission to access relevant WHOI archival materials, and to the extraordinary staff of the archives of the Scripps and the Woods Hole Oceanographic Institutions. This work was suported in part by the National Science Foundation (grant SBE 01-96022).

Notes

1. Paul Forman, "Into Quantum Electronics: The Maser as 'Gadget' of Cold-War America," in *National Military Establishments and the Advancement of Science and Technology*, ed. Paul Forman and José M. Sánchez-Ron (Kluwer, 1996). See also Forman, "Behind Quantum Electronics: National Security as Basis for Physical Research in the United States, 1940–1960," *Historical Studies in the Physical and Biological Sciences* 18 (1987): 149–229.

2. "A Science in Bondage," February 1965, Papers of William Stelling von Arx, 1942–1977, Manuscript Collection 24 (MC24), Woods Hole Oceanographic Institution Archives (hereafter WHOIA), Box 1, folder 14. Paul Forman made the same argument about twentieth-century physics in "Into Quantum Electronics" and "Behind Quantum Electronics."

3. "A Science in Bondage."

4. Naomi Oreskes, *Science on a Mission: American Oceanography from the Cold War to Climate Change* (University of Chicago Press, forthcoming). On the "golden age" of oceanography, see "Oceanography: The Making of Science," Oregon State University Archives, Corvallis. See also Oreskes, "A Context of Motivation: US. Navy Oceanographic Research and the Discovery of Sea-Floor Hydrothermal Vents," *Social Studies of Science* 33, no. 5 (2003): 697–742.

5. G. N. Plass, "The Carbon Dioxide Theory of Climate Change," *Tellus* 8, no. 2 (1956): 140–154; Plass, "The Influence of the 15u Carbon-Dioxide Band on the Atmospheric Infra-Red Cooling Rate," *Quarterly Journal of the Royal Meteorological Society* 82, no. 353 (1956): 310–24; J. R. Fleming, *Historical Perspectives on Climate Change* (Oxford University Press, 1998); Spencer Weart, *The Discovery of Global Warming* (Harvard University Press, 2004); Charles D. Keeling, "Rewards and Penalties of Monitoring the Earth," *Annual Review of Energy and the Environment* 23 (1998): 25–82.

6. Roger Revelle and H. E. Suess, "Carbon Dioxide Exchange Between Atmosphere and Ocean and the Question of an Increase of Atmospheric CO_2 During the Past Decades," *Tellus* 9, no. 1 (1957): 18–27, on 19–20.

7. Ibid., 26.

8. Gordon MacDonald, "Climatic Consequences of Increased Carbon Dioxide in the Atmosphere," in *Power Generation and Environmental Change*, ed. David Berkowitz and Arthur Squires (MIT Press, 1971). See also MacDonald, "Pollution, Weather and Climate" in *Environment: Resources, Pollution, and Society*, ed. William W. Murdoch (Sinauer, 1971); MacDonald et al., *The Long-Term Impacts of Increasing Atmospheric Carbon Dioxide Levels* (SRI International, 1979); MacDonald, "Scientific Basis for the Greenhouse Effect," *Journal of Policy Analysis and Management* 7, no. 3 (1988): 425–444.

9. MacDonald, "Climatic Consequences of Increased Carbon Dioxide."

10. Robert M. White "Oceans and Climate: An Introduction," *Oceanus* 21 (1978): 2–3.

11. *Declaration of the Climate Conference* (World Meteorological Organization, 1979), 1.

12. Ibid., 2.

13. MacDonald, "Scientific Basis for the Greenhouse Effect." See also John S. Perry, "Energy and Climate: Today's Problem, Not Tomorrow's," *Climatic Change* 3, no. 3 (1981): 223–225; J. Hansen et al., "Global Climate Changes as Forecast by Goddard Institute for Space Studies Three-Dimensional Model," *Journal of Geophysical Research* 93 (1988): 9341–9364.

14. Paul N. Edwards, *A Vast Machine: Computer Models, Climate Data, and the Politics of Global Warming* (MIT Press, 2010). See also John Knauss to Richard Hallgren, 28 July 1971, Box 3, folder

5–7: Correspondence June–Sept 1971 (2 of 4), Papers of Henry Melson Stommel, 1946–1996, Manuscript Collection 06 (MC6), WHOIA; Roger Revelle, "Carbon Dioxide and World Climate," *Scientific American* 247, no. 2 (1982): 33–41.

15. Revelle, "Carbon Dioxide and World Climate," 33.

16. S. Solomon et al., "Summary for Policy Makers," in *Climate Change 2007: The Physical Science Basis* (Cambridge University Press, 2007).

17. J. T. Houghton et al., *Climate Change 1995: The Science of Climate Change* (Cambridge University Press, 1995; J. T. Houghton et al. *Climate Change 2001: The Scientific Basis* (Cambridge University Press, 2001); Weart, *The Discovery of Global Warming*; Edwards, *A Vast Machine*; Naomi Oreskes, "Behind the Ivory Tower: The Scientific Consensus on Climate Change," *Science* 306, no. 5702 (2004): 1686.

18. On the history of the thermometer, see Hasok Chang, *Inventing Temperature: Measurement and Scientific Progress* (Oxford University Press, 2004).

19. Walter H. Munk and Carl Wunsch, "Ocean Acoustic Tomography," *Deep Sea Research* 26A (1979): 123–161; Munk and Wunsch, "Ocean Acoustic Tomography: A Scheme for Large Scale Monitoring," *Deep Sea Research* 26A (1979): 439–464.

20. Munk and Wunsch, "Ocean Acoustic Tomography"; Munk and Wunsch, "Ocean Acoustic Tomography: A Scheme for Large Scale Monitoring." See also John Spiesberger et al., "Global Acoustic Mapping of Ocean Temperatures (GAMOT)," *Journal of the Acoustical Society of America* 94, no. 3 (1993): 1803; R. C. Spindel and P. F. Worcester, "Ocean Acoustic Tomography," *Scientific American*, 263, no. 4 (1990): 94–99; Walter H. Munk, P. F. Worcester, and Carl Wunsch, *Ocean Acoustic Tomography* (Cambridge University Press, 1995).

21. There is a long history of studying variation in the speed of sound over long distances in the ocean. See, for example, J. Lamar Worzel and Maurice Ewing, *Propagation of Sound: Explosion Sounds in Shallow Water* (Geological Society of America, 1948); Ewing and Worzel, "Long Range Sound Transmission," Interim Report No. 1, Contract Nobs-2083, August 25, 1945, declassified March 12, 1946; Ewing and Worzel, "Long Range Sound Transmission," *Geological Society of America* 27 (1948); Maurice Ewing, C. Iselin, and J. L. Worzel, "Sound Transmission in Sea Water," *Woods Hole Oceanographic Institution Report*, February 1941; R. H. Johnson, "Synthesis of Point Data and Path Data in Estimating SOFAR Speed," *Journal of Geophysical Research* 74 (1969): 4559–4570; V. A. Del Grosso, "New Equation for the Speed of Sound in Natural Waters (with Comparisons to Other Equations)," *Journal of the Acoustical Society of America* 56, no. 4 (1974): 1084–1091; G. R. Hamilton, "Time Variation of Sound Speeds over Long Paths in the Ocean," in *International Workshop on Low-Frequency Propagation and Noise, Woods Hole, Massachusetts* (Department of the Navy, 1977). For a discussion of comparable Russian work, see Oleg Godin and David R. Palmer, *History of Russian Underwater Acoustics* (World Scientific, 2008). There is some ambiguity as to how the idea of using acoustic tomography to detect climate change first developed. Like many scientific ideas, it seems to have been discussed by quite a few people before being taken up and advanced by a subset of them. In 1979, Munk and Wunsch promoted the idea of using acoustic tomography for large scale monitoring in "Ocean Acoustic Tomogra-

phy" and in "Ocean Acoustic Tomography: A Scheme for Large Scale Monitoring," and later they explicitly proposed to monitor ocean temperature related to anthropogenic climate change—see Walter Munk and A. M. G. Forbes, "Global Ocean Warming: An Acoustic Measure?" *Journal of Physical Oceanography* 19 (1989): 1765–1777; Munk, "Global Ocean Warming: Detection by Long Path Acoustic Travel Times," *Oceanography* 2 (1989): 40–41. In 1983, John Spiesberger, T. G. Birdsall, and Kurt Metzger proposed the development of an acoustic thermometer in a proposal to the ONR—see J. L. Spiesberger et al., "Measurements of Gulf Stream Meandering and Evidence of Seasonal Thermocline Development Using Long-Range Acoustic Transmissions," *Journal of Physical Oceanography* 13 (1983): 1836–1846. Also J. Spiesberger, email communication to Naomi Oreskes, February 15, 2012; Spiesberger et al., "Stability and Identification of Ocean Acoustic Multipaths," *Journal of the Acoustical Society of America* 67 (1980): 2011–2017; Spiesberger and Metzger, "Basin Scale Tomography: A New Tool for Studying Weather and Climate," *Journal of Geophysical Research* 96, no. C3 (1991): 4869–4889; Spiesberger and Metzger, "A Basin-Scale (3000 km) Tomographic Section of Temperature and Sound Speed in the Northeast Pacific," *Transactions of the American Geophysical Union* 71, January 7, 1990; Spiesberger and Metzger, "A New Algorithm for Sound Speed in Seawater," *Journal of the Acoustical Society of America* 89 (1991): 2677–2688; Spiesberger and Metzger, "New Estimates of Sound-Speed in Water," *Journal of the Acoustical Society of America* 89 (1991): 1697–1700; Spiesberger et al., "Listening for Climatic Temperature Change in the Northeast Pacific: 1983–1989," *Journal of the Acoustical Society of America* 92 (1992): 384–396.

22. John Cloud, "Imagining the World in a Barrel," *Social Studies of Science* 31, no. 2 (2001): 231–252. See also Lichte, 191, cited in Spiesberger et al., "Measurements of Gulf Stream Meandering"; Bureau of Ships, Navy Department, *Prediction of Sound Ranges from Bathythermograph Observations: Rules for Preparing Sonar Messages* (National Defense Research Committee, 1944); Ewing and Worzel, "Long Range Sound Transmission," *Interim Report No. 1*; Ewing and Worzel, "Long Range Sound Transmission," *Geological Society of America*; Carl Eckart, ed., "The Refraction of Sound," in *Principles and Applications of Underwater Sound*, Originally Issued as Summary Technical Report of Division 6, Volume 7, National Defense Research Committee (reprinted by Department of the Navy); Lyman Spitzer Jr., ed., *Physics of Sound in the Sea, National Defense Research Committee Summary Technical Reports*, Originally Issued as Division 6, Volume 8 (The Research Analysis Group, Committee on Undersea Warfare, National Research Council, 1969).

23. Robert A. Frosch, "Underwater Sound: Deep-Ocean Propagation," *Science* 146, no. 3646 (1964): 889–904; R. J. Urick, *Sound Propagation in the Sea* (Government Printing Office, 1979); Fred Noel Spiess, *Seeking Signals in the Sea: Recollection of the Marine Physical Laboratory* (Marine Physical Laboratory of the Scripps Institution of Oceanography, 1997).

24. Hersey, in *International Workshop on Low-Frequency Propagation and Noise*, 4.

25. "Congressional Testimony, House Armed Services Committee," 28 March 1985, Box 4, folder 12, Paper of Charles Davis Hollister Papers, 1967–1998, Manuscript Collection 31 (MC31) Unprocessed papers, WHOIA.

26. J. L Spiesberger, T. G. Birdsall, and K. Metzger, "Acoustic Thermometer Proposal," Submitted to the Office of Naval Research, May 3, 1983, Woods Hole Oceanographic Institution, Box 3,

Folder 18: Correspondence, Stommel papers, MC6, WHOIA. This was later referenced as ONR Research Contract N00014-82-C-0019, 1983; see Spiesberger and Metzger, "Basin-scale Tomography: A New Tool for Studying Weather and Climate"; Spiesberger, Spindel and Metzger, "Stability and Identification of Ocean Acoustic Multipaths"; D. Behringer et al., "A Demonstration of Ocean Acoustic Tomography," *Nature* 299 (1982): 121–125.

27. Munk and Forbes, "Global Ocean Warming: An Acoustic Measure?"; Munk, "Global Ocean Warming: Detection by Long Path Acoustic Travel Times."

28. Summary of "Acoustic Measurements of Climatic Variability in the Northeast Pacific," 4 October 1991, Box 4, folder: 26–27, Correspondence 1991 (1 of 2), Stommel papers, MC6, WHOIA.

29. Walter Munk, R. C. Spindel, A. Baggeroer, and T. G. Birdsall, "The Heard Island Feasibility Test," *Journal of the Acoustical Society of America* 96 (1994): 2330–2342.

30. On the history of the Marine Mammal Protection Act, see Gregg Mittman, *Reel Nature: America's Romance with Wildlife on Film* (Harvard University Press, 1999); Mark Barrow, *Nature's Ghosts: Confronting Extinction from the Age of Jefferson to the Age of Ecology* (University of Chicago Press, 2009); Etienne Benson, "Endangered Science: The Regulation of Research by the United States Marine Mammal Protection and Endangered Species Acts," *Historical Studies in the Natural Sciences* 42, no. 1 (2012): 30–61.

31. J. R. Potter, "ATOC: Sound Policy or Enviro-Vandalism? Aspects of a Modern Media-Fueled Policy Issues," *Journal of Environment and Development* 3 (1994): 47–62. See also Acoustic Thermometry of Ocean Climate (ATOC) AC32 12: Applications for permit, January 1991–August 1 1993, Scripps Institution of Oceanography Archives (hereafter SIOA).

32. William W. Fox Jr. to Walter H. Munk, December –7 (not legible), 1990, ATOC AC32 12: Applications for permit, January 1991–August 1, 1993, SIOA.

33. Hubbs Sea World Research Institute, "Anne E. Bowles, PhD" (http://www.hswri.org/index.php?pr=Ann_E_Bowles_PhD). Bowles got her PhD in 1994 from SIO, according to her home page.

34. A Google Scholar search reveals three publications by Anne E. Bowles in the years up to and including 1990: F. T. Awbrey and Anne E. Bowles, *The Effects of Aircraft Noise and Sonic Booms on Raptors: A Preliminary Model and a Synthesis of the Literature on Disturbance*, Noise and Sonic Boom Impact Technology, Technical Operating Report No. 12 (US Air Force Air Force Systems Command, 1990); A. E. Bowles, W. G. Young, and E. D. Asper, "Ontogeny of Stereotyped Calling of a Killer Whale Calf During Her First Year," *Rit Fiskideildar* 11 (1988): 251–275; A. E. Bowles, F. T. Awbrey, and Robert Kull, "A Model for the Effects of Aircraft Overflight Noise on the Reproductive Success of Raptorial Birds," in Inter-noise 90, Proceedings of the International Conference on Noise Control Engineering, Göteborg, Sweden, 1990.

35. Munk et al., "The Heard Island Feasibility Test." The results were published in a special issue of the *Journal of the Acoustical Society* of *America* in 1994. Eighteen papers were published; one dealt with the biological results.

36. Sam Nunn, Senate Floor Speech, June 28, 1990 (see "The Environment and National Security: Remarks as Prepared for Delivery by Sherri Wasserman Goodman, Deputy Under Secretary of Defense, National Defense University August 8, 1996" at http://www.loyola.edu); Potter, "ATOC: Sound Policy or Enviro-Vandalism?"; ATOC Records 1990–1998, ATOC AC32, SIOA.

37. On the use of the SOSUS system as the "bottom receiver," as well as the use of deliberately ambiguous language, see appendix A in Munk et al., *Ocean Acoustic Tomography*.

38. Craig Faanes to Marilyn Cox, 20 July 1994, ATOC AC32 2: California EIS 1994, SIOA.

39. Ibid.

40. Ann Bowles to Nancy Foster, National Marine Fisheries Service, 6 March 1992, ATOC, AC32 12: Marine Mammal Correspondence Pre-September 1993, SIOA.

41. Ann Bowles to David Hyde, 26 May 1993, ATOC AC32 12: Marine Mammal Correspondence Pre-September 1993, SIOA.

42. Swartz and Hofman, 1991, "Marine Mammal and Habitat Monitoring: Requirements; Principles; Needs; and Approaches," in ATOC AC32 12: Correspondence Sept–Oct 1993, SIOA, esp. p. 5. The Marine Mammal Protection Act allows for accidental taking of small numbers of nonthreatened species and populations during fishing and other commercial activities. In 1986 this was amended to permit the taking of depleted species and stocks as well, if it were shown to be inconsequential, which in turn required site-specific monitoring programs. For example, in 1991 the Fish and Wildlife Service had agreed to an unintentional take of small numbers of walruses and polar bears incidental to oil and gas exploration in the Chukchi Sea, near Alaska, so long as steps were taken to minimize losses, particularly in areas of traditional subsistence hunting, no walruses were taken during the spring migration period, and all of this was monitored and subject to peer review.

43. NOAA Fisheries, "Marine Mammal Protection Act (MMPA) of 1972" (http://www.nmfs.noaa.gov/pr/laws/mmpa/). See also discussion in Swartz and Hofman, "Marine Mammal and Habitat Monitoring: Requirements; Principles; Needs; and Approaches," ATOC AC32 12: Correspondence Sept–Oct 1993, SIOA. For example, Adam S. Frankel of the University of Hawaii at Manoa was given authorization for up to 1,000 "takes," by harassment, of humpback whales, to study their response to low-frequency sound. The approval had come with strong caveats to "exercise extreme caution in approaching mother/calf pairs," and a requirement for a detailed report of all underwater approaches to whales, and the number and sex of the whales involved. Michael F. Tillman to Adam S. Frankel, February 1, 1993, ATOC AC32 12: Application for permits, January 1991–August 1, 1993, SIOA.

44. Ann Bowles to David Hyde, 26 May 1993, ATOC AC32 12: Marine Mammal Correspondence Pre-September 1993, SIOA.

45. Ibid.

46. Etienne Benson, "Endangered Science: The Regulation of Research by the US Marine Mammal Protection and Endangered Species Acts," *Historical Studies in the Natural Sciences* 42, no. 1 (2012): 30–61.

47. Ibid. The claim that industry routinely worked without permits couldn't have been entirely correct; elsewhere Chris Clark noted that he frequently reviewed permits from industry. See Chris Clark to David Hyde, 27 July 1993, ATOC AC32 12 Marine Mammal Correspondence Pre-September 1993, SIOA.

48. Ann Bowles to David Hyde, 26 May 1993, ATOC AC32 12: Marine Mammal Correspondence Pre-September 1993, SIOA. In fairness to Bowles, other conservation biologists also objected to the bureaucratic paperwork required by the Marine Mammal Protection Act, see Benson, "Endangered Science: The Regulation of Research by the US Marine Mammal Protection and Endangered Species Acts."

49. Ibid.

50. Daniel Costa to David Hyde, email message, August 13, 1993, ATOC AC32 12: Marine Mammal Correspondence Pre-September 1993, SIOA. The use of IUSS in Whales 93 is also discussed in Marine Acoustics, Inc., Proposal MAI 208P, 28 September 1993. Faxed copy to Cornell University Ornithology Lab, in ATOC AC32 12: Correspondence Sept–Oct 1993, SIOA.

51. Daniel Costa to David Hyde, email message, August 13, 1993, ATOC AC32 12: Marine Mammal Correspondence Pre-September 1993, SIOA.

52. Christopher Clark, ATOC Task 7.0, "Acoustic Monitoring and Experimental Studies on the Effects of ATOC transmissions on marine mammals," ATOC AC32 13: Original Viewgraph Presentations, 7 June 1993, ARPA Review, SIOA. (In the original, 'not' was underscored.)

53. Various documents, ATOC AC32 12: Marine Mammal Correspondence Pre-September 1993, SIOA. This also included a study of humpback whales by Adam S. Frankel of the University of Hawaii at Manoa. See ATOC AC32 12: Application for permits, January 1991–August 1 1993, SIOA; "Draft Study Plan, Acoustic Thermometry of Ocean Climate (ATOC)," ATOC AC32 12: Marine Mammal Correspondence Pre-September 1993, SIOA. See also Christopher Clark to David Hyde, 6 July 1993, ATOC AC32 12: Marine Mammal Correspondence Pre-September 1993, SIOA.

54. David Hyde to Nancy Foster, "Marine Mammal Permit, Sept 14, 1993, Current version submitted to NMFS," ATOC AC32 12: NMFS, 9 September 1993, SIOA. See also Forbes California Coastal Commission, 29 November, 1994, ATOC AC32 2: California Coastal Commissions, Nov–Dec 1994, SIOA, which details the permit history. The permit signatories were Hyde, Bowles, Clark, and two others; William Kuperman (a professor of physical oceanography at Scripps and an expert in noise and signal processing) and Sue Moore (a master's-degree-level scientist employed by SAIC who had previously worked on small take exemptions for the Navy). The application was submitted with a cover letter signed by Hyde. Anne Bowles to David Hyde, 26 May 1993, ATOC AC32 12: Marine Mammal Correspondence Pre-September 1993, SIOA.

55. Various materials including ATOC NMFS permit, chart, copy in Box 2 Folder California Coastal Commission Correspondence Nov–Dec 1994, ATOC AC32, SIOA; Potter, "ATOC: Sound Policy or Enviro-vandalism?"; Louis M. Herman, "Hawaiian Humpback Whales and ATOC: A Conflict of Interests," *Journal of Environment and Development* 3, no. 2 (1994): 63–76.

56. Naomi Rose, telephone interview with author, 2001.

57. A search of marmam@uvvm.uvic.ca on January 16, 2001 turned up 1,937 messages under the heading "ATOC."

58. Potter, "ATOC: Sound Policy or Enviro-vandalism?" ATOC AC32 12: MARMAM, Feb 1994–July 1995, SIOA.

59. Edward Frieman, "Draft Letter to The Honorable <Field>, 18 March 1994," SIO Office of the Director (Frieman) Records, 1988–2002 (hereafter ODR) 3: Director's Office, Correspondence, March 1994, SIOA.

60. On Ridgway's career, see "Dr. Sam Ridgeway—2009 Morris Award Winner" at http://www.marinemammalscience.org.

61. Neal Matthews, "Sound Effects: Navy Whales Making Sure Noises Won't Hurt Their 'Civilian' Brothers," *San Diego Union-Tribune*, March 16, 1994.

62. Ibid.

63. Richard Paddock, "Undersea Noise Test Could Risk Making Whales Deaf," *Los Angeles Times*, March 22, 1994.

64. Ibid.

65. Potter, "ATOC: Sound Policy or Enviro-vandalism?" 54.

66. Helen M. Rozwadowski, *Fathoming the Ocean: The Discovery and Exploration of the Deep Sea* (Harvard University Press, 2005); Graham D. Burnett, *The Sounding of the Whale: Science and Cetaceans in the Twentieth Century* (University of Chicago Press, 2011); Helen M. Rozwadowski, and David K. van Keuren, eds., *The Machine in Neptune's Garden: Historical Perspectives in Technology and the Marine Environment* (Science History Publications, 2004).

67. See, for example, discussion by Lindy Weilgart of ATOC responses to MARMAM email messages, ATOC AC32 12: MARMAM, February 1994–July 1995, April 7, 1994, SIOA.

68. Statement by Robert M. White, 22 March 1994, ATOC AC32 11: California Coastal Commission Public Hearing, June 15, 1995, Agenda, SIOA; See also *Ocean Science News*, March 22 1994, in ATOC AC32 2: News clippings, March–April 1994, SIOA.

69. Statement by Robert M. White, 22 March 1994, ATOC AC32 11: California Coastal Commission Public Hearing, June 15, 1995, Agenda, SIOA.

70. Quoted in *Ocean Science News*, March 22 1994, in ATOC AC32 2: News clippings, March–April 1994, SIOA.

71. *Ocean Science News*, March 22 1994, ATOC AC32 2: News clippings, March–April 1994, SIOA.

72. Richard C. Paddock, "Critics Mobilize in Effort to Block Ocean Noise Tests," *Los Angeles Times*, March 23, 1994, in ATOC AC32 2: News clippings, March–April 1994, SIOA.

73. "Permits are Delayed for Underwater Noise," *Pleasanton Valley Times*, March 26 1994, ATOC AC32 2: News clippings, March–April 1994, SIO.

74. Patsy Mink et al. to Ron Brown, Copy of Faxed letter, 24 March 1994, Woods Hole Oceanographic Institution Office of the Director (hereafter WHOI ODR) 136: 2: ATOC, WHOIA.

75. Barbara Boxer to Ron Brown, Copy of Faxed letter, 23 March 1994, WHOI ODR 136: 2 ATOC, WHOIA; See also ATOC AC32 2: News clippings, March–April 1994, SIOA.

76. Paddock, "Block Ocean Noise Tests," in ATOC AC32 2: News clippings, March–April 1994, SIOA.

77. "Don't' Deafen Whales," editorial, *San Francisco Examiner*, March 24, 1994, in ATOC AC32 2: News clippings, March–April 1994, SIOA.

78. Various editorials, ATOC AC32 2: News clippings, March–April 1994, SIOA.

79. John Lankford, "Save a Few Harpoons for Scripps Eggheads," *Santa Barbara News Press*, March 30, 1994, in ATOC AC32 2: News clippings, March–April 1994, SIOA.

80. "Sanctuary Panel Urges Delay in Sonar Project," *Monterey Herald*, March 31, 1994, ATOC AC32 2: News clippings, March–April 1994, SIOA.

81. "2 Environmental Camps," ATOC AC32 2: News clippings, March–April 1994, SIOA.

82. Edward Frieman to ---, 28 March 1994, ODR, Manuscript Collection 77 3: Director's Office, Correspondence, March 1994, SIOA.

83. Edward Frieman, "Dear SIO Associate," 29 March 1994, ODR MC77 3: Director's Office, Correspondence, March 1994, SIOA.

84. Cindy Clark, "Memo and attachments sent to S. Lauzon at WHOI," SIO Communications Office, 2 April 1994, WHOI ODR 136: 2 ATOC, WHOIA.

85. Ibid.

86. Source: http://www.birds.cornell.edu/brp/ResWhale.html (now inactive). This refers to Christopher Clark, who later became director of the Marine Mammal Research Program. Professor Clark's current Web address is http://vivo.cornell.edu/display/individual5549.

87. Christopher Clark to Peter Tyack, Fax Correspondence, WHOI ODR 136: 2 ATOC, WHOIA.

88. Robert Gagosian to Shelley Lauzon, 8 May 1994, WHOI ODR 136: 2 ATOC, WHOIA.

89. The advisory board was headed by W. John Richardson, Executive vice president of LGL Limited, a private environmental consulting group based in Ontario, and a member of the US National Research Council's committee on low-frequency sound and marine mammals. Among the other members were William Ellison (an expert in underwater acoustics at Marine Acoustics, Inc.), Jeff Laake (a population biologist at the National Marine Mammal Laboratory in Seattle who had developed techniques to estimate population densities), Jeannette Thomas (a biologist at Western Illinois University with expertise in cetacean and pinniped hearing), and Judy Zeh (a professor of Statistics at University of Washington).

90. "Acoustic Thermometry of Ocean Climate Program Update," press release, 13 May 1994, ATOC AC32 6: Publications Folder 1993–2002, SIOA. See also "Scoping Process Summary, 15 June 1995, California Coastal Commission, Revised Staff report and Recommendation," ATOC AC32 2: California Coastal Commissions April=Sept 1994 [sic]; additional copy in ATOC AC32 2: California Coastal Commission June–November 1995, SIOA. Scripps scientists were reluctant to undertake the enormous effort involved in preparing an EIS, but legal counsel advised that failure to do so would increase exposure to litigation, and in the long run probably would cost more time and money. See various documents in ATOC AC32 12: Kauai DEIS correspondence, May–June 1994, SIOA.

91. Various documents, ATOC AC32 12: Kauai DEIS correspondence, May–June 1994, SIOA; Naomi Rose, telephone interview with author, 2001. There were additional legal issues involving Hawaii's claim to territorial integrity between the islands, and the Hawaii Environmental Policy Act, which occupied a great deal of time of the legal counsel hired by the University of California. These details are omitted here.

92. Draft Environmental Impact Statement, and various correspondences about it, ATOC AC32 2: Records, California EIS, 1994, SIOA.

93. "Statement of Work, Cornell University," 28 July 1994, ATOC AC32 12: MMRP Research Protocols, April 1994–November 1995, SIOA.

94. "Final Draft Environmental Impact Report," August 16, 1994, ATOC AC32 2: California EIS 1994, SIOA.

95. Paul Anderson, "An update on the Acoustic Thermometry of Ocean Climate (ATOC) project," email message, January 30, 1995, ATOC AC32 12: MARMAM February 1994–July 1995, SIOA.

96. Craig Faanes to Marilyn Cox, 14 July 1994, ATOC AC32 2: California EIS (1994), SIOA.

97. Paul Anderson, email message, June 16, 1995, ATOC AC32 12: MARMAM February 1994–July 1995, SIOA.

98. Ibid.

99. Walter Munk and David Hyde to John Richardson, 17 November 1993, ATOC AC32 12: Correspondence Nov–Dec 1993, SIOA.

100. David R. Tomsovic to Marilyn Cox, 15 June 1994, ATOC AC32 2: California EIS (1994), SIOA, emphasis added.

101. Ibid.

102. Paul Anderson, email message, June 16, 1995, ATOC AC32 12: MARMAM February 1994–July 1995, SIOA. This point was raised in several public comments after the January 1995 California Coastal Commission public hearings, at which Walter Munk spoke, emphasizing the importance of ATOC as a test of climate models rather than a definitive proof of global warming.

103. Walter Munk to Carl L. Williams and Members, California Coastal Commission, June 9, 1995, ATOC AC32 11: ATOC, California Coastal Commission Public Hearing, June 15, 1995, Agenda, SIOA.

104. On the no effects claim, see Andrew Forbes to Jeffrey Benoit, January 13, 1995, ATOC AC32 2: California Coastal Commissions (Jan–May 1995), SIOA. For copies of public comments, see ATOC AC32 11: numerous folders, SIOA. In comparison with the flood of negative public comments, relatively few letters were received in support of the project, and many of these were hedged by conditions. The Monterey Bay Chapter of the American Cetacean Society noted that its members were divided, even polarized, and so they had decided to focus their comments solely on the MMRP component. If the concerns raised by the Center for Marine Conservation were addressed—particularly the issue of how and by whom a decision to terminate the project might be made—as well as additional concerns outlined in their own four-page letter about the research protocols for whales—then they would support the project. John Pearse, Professor Emeritus of Biology at UC Santa Cruz, had been a member of the NRC Committee on Low-Frequency Sound and Marine Mammals, which had concluded that scientific data on the question were "extremely limited and cannot constitute the basis for informed prediction or evaluation of the effects of intense low-frequency sounds on any marine species." For this reason, he felt that "the proposed ATOC program in general, and the associated Marine Mammal Research Program in particular, represents a tremendous opportunity to obtain much of the information need to better protect both marine mammals and their food sources. ... Without such information, we have no basis for protecting marine mammals, and other forms of marine life, from anthropogenic sound." A rare expression of unqualified support from someone with a strong interest in the Monterey Bay Sanctuary came from Stephen K. Webster, Director of Education at the Monterey Bay Aquarium and a member-at-large of the Monterey Bay National Marine Sanctuary Advisory Council. Webster felt that ATOC had been misrepresented by the press and misunderstood by the public, that it "never deserved the play in received in the public press and resulting debate," and that "the combination of irresponsible 'tabloid' journalism, misinformed self-proclaimed 'experts,' and issue-hungry, fund-raising fringe environmentalists has elevated this issue to a level of controversy it never deserved, and still doesn't." The research was of high priority, he felt, and the Sanctuary should support and facilitate research within its boundaries. The MMRP was needed to provide "accurate, objective, and scientific supportable conclusions," and to allow an informed decision about the future of ATOC. He concluded: "The balance between 'heat' and 'light' in the public discussion to date has been strongly skewed in the direction of heat. It is time the balance shifted to the right—that is, in the direction of 'light.'" Support was also expressed by James Barry, an assistant scientist at the Monterey Bay Aquarium Research Institute, who also criticized the public discussion as based on emotion rather than fact: "It is ludicrous to think that uneducated, but well-meaning, citizens unfamiliar with ocean physics, sound transmission in water, behavior and physiology of marine mammals, fishes and invertebrates, and other information relevant to the ATOC proposal, should outweigh the highly informed and carefully considered views of experts in these fields. ... [T]he opinions of ill-informed conservationists are unwarranted and probably incorrect. In contrast, the opinions and advise [sic] of highly informed conservationists, like the scientists proposing this study, are based on fact rather than emotion, and should be heeded. Although public opinion is impor-

tant, it should carry little weight for advisory committee." Legally, the NMFS was required to consider public opinion, well informed or not.

105. Center for Marine Conservations, Comments of the Center for Marine Conservations on the Draft EIS/EIR, ATOC AC32 11: MMRP Draft EIS/EIR, volume 1, January 1995, comment received January 31, 1995 and responses C38 to C-61, SIOA.

106. Ibid.

107. Rodney M. Fujita, Comment C-24, SIO ATOC AC32 11: MMRP Draft EIS/EIR, volume 1, January 1995, comment received and responses C-1 to C-27, SIOA.

108. Lawrence D. Six, Pacific Fishery Management Council, Comment C-55, ATOC AC32 11: MMRP Draft EIS/EIR, volume 1, January 1995, comment received and responses C38 to C-61, SIOA.

109. Munk et al., "The Heard Island Feasibility Test."

110. Kauai Friends of the Environment, Comment C-56, ATOC AC32 11: MMRP Draft EIS/EIR, volume 1, January 1995, comment received and responses C38 to C-61, SIOA.

111. Statement by Robert M. White, March 22, 1994, ATOC AC32 11: California Coastal Commission Public Hearing, June 15, 1995, Agenda. See also *Ocean Science News*, March 22, 1994, in ATOC AC32 2: News clippings, March 1994–April 1994, SIOA.

112. Edward Frieman, "Dear SIO Associate," 29 March 1994, ODR MC77 3: Director's Office, Correspondence, March 1994, SIOA.

113. Edward Frieman, "Draft Letter to The Honorable <Field>, 18 March 1994," SIO ODR 3: Director's Office, Correspondence, March 1994, SIOA.

114. Ibid.

115. Ibid.

116. Rodney M. Fujita, Comment C-24, SIO ATOC AC32 11 MMRP Draft EIS/EIR, volume 1, January 1995, comment received and responses C-1 to C-27, SIOA. (In the original, the emphasized word was underscored.) For an analysis of these vast economic and political interests, see Naomi Oreskes and Erik M. Conway, *Merchants of Doubt: How a Handful of Scientists Obscured the Truth on Issues from Tobacco Smoke to Global Warming* (Bloomsbury Press, 2010). Fujita's analysis was crucial to my own dawning realization of the motivations behind attacks on climate science.

117. Debby Molina to Whom it May concern, Comment C-52, ATOC AC32 11: MMRP Draft EIS/EIR, volume 1, January 1995, comment received and responses C38 to C-61, SIOA; Debby Molina, 13 June 1995, ATOC AC32 2: California Coastal Commission [June–November 1995], SIOA.

118. Debby Molina, to Whom it May concern, Comment C-52, ATOC AC32 11: MMRP Draft EIS/EIR, volume 1, January 1995, comment received and responses C-1 to C-37, SIOA.

119. Derek J. Cole, Comment c-40, ATOC AC32 11: MMRP Draft EIS/EIR, volume 1, January 1995, comment received and responses C-38 to C-61, SIOA.

120. Which indeed, scientists did, not only in the nineteenth century, but earlier too. In the seventeenth century, Stephen Hales stuck glass tubes in dogs' arteries while researching blood pressure, see G. E. Burgett, "Stephen Hays (1677–1761)," *Annals of Medical History* 7, no. 2 (1925): 1–20. Such experiments were routine in the eighteenth and nineteenth centuries and continued in the twentieth, when laboratory animals were routinely "sacrificed" in scientific experiments. On the history of animal experimentation, see Anita Guerrini, *Experimenting with Humans and Animals: From Galen to Animal Rights* (Johns Hopkins University Press, 2003). For comments in the ATOC archival record on this point, see Elaine Sohier, Comment C-35, ATOC AC32 11: MMRP Draft EIS/EIR, volume 1, January 1995, comment received and responses C-1 to C-37; Peter Molitor, Comment C-57, ATOC AC32 11 MMRP Draft EIS/EIR, volume 1, January 1995, comment received and responses C38 to C-61. See also B. R. Harms, Comment C-19, ATOC AC32 11 MMRP Draft EIS/EIR, volume 1, January 1995, comment received and responses C-1 to C-27, SIOA.

121. Bob DeBolt, Comment C-37, ATOC AC32 11: MMRP Draft EIS/EIR, volume 1, January 1995, comment received and responses C-1 to C-37, SIOA.

122. "United Nations Framework Convention on Climate Change," http://unfccc.int/2860.php; Intergovernmental Panel on Climate Change, "The IPCC Second Assessment Synthesis of Scientific-Technical Information relevant to Interpreting Article 2 of the UN framework Convention on Climate Change," http://www.ipcc.ch/pdf/climate-changes-1995/2nd-assessment-synthesis.pdf. See also Oreskes and Conway, *Merchants of Doubt*, chapter 6.

123. Debby Molina to Whom it May concern, Comment c-52, ATOC AC32 11: MMRP Draft EIS/EIR, volume 1, January 1995, comment received and responses C38 to C-61, SIOA.

124. Edwards, *A Vast Machine*; David H. DeVorkin, *Science with a Vengeance: How the Military Created the Us Space Sciences after World War II* (Springer-Verlag, 1992); DeVorkin, "The Military Origin of the Space Sciences in the American V-2 Era," in *National Military Establishments and the Advancement of Science and Technology: Studies in the 20th Century History*, ed. Paul Forman and José M. Sánchez-Ron (Kluwer, 1996); Benjamin Wilson and David Kaiser, "Calculating Times: Radar, Ballistic Missiles, and Einstein's Relativity," this volume; Mittman, *Reel Nature*; Daston, Lorraine and Gregg Mittman, eds., *Thinking with Animals: New Perspectives on Anthropomorphism* (Columbia University Press, 2005).

125. "Outcry Could Give Scientists a Lesson in Public Relations," *Seattle Times*, April 26, 1994, ATOC AC32 2: News clippings, March–April 1994, SIOA.

126. Stanley Flatté, personal communication at conference on Oceanography: The Making of Science, People, Institutions, and Discovery, Scripps Institution of Oceanography, La Jolla, January 2000; Roger Dashen, Walter Munk, Kenneth Watson, Frederick Zachariasen, and Stanley Flatté, eds., *Sound Transmission through a Fluctuating Ocean* (Cambridge University Press, 1979).

127. Matt Hellman and Patti Kirby, Comment C-54, ATOC AC32 11: MMRP Draft EIS/EIR, volume 1, January 1995, comment received and responses C38 to C-61, SIOA.

128. Sarah Miquibas, handwritten letter in response to DEIS for Kauai ATOC program, ATOC AC32 11: DLNR comments to EIS, March–May 1995, SIOA.

129. David Seielstad, February 15, 1995, written response to the DEIS for Kauai ATOC program, ATOC AC32 11: DLNR comments to EIS, March–May 1995, SIOA.

130. The official website of the Pacific Missile Range Facility (http://www.cnic.navy.mil) touts this naval facility as the "world's largest instrumented multi-environment range capable of supporting surface, subsurface, air, and space operations simultaneously."

131. David Seielstad, February 15, 1995, written response to the DEIS, for Kauai ATOC program, ATOC AC32 11: DLNR comments to EIS, March–May 1995, SIOA.

132. Natalie De Pasquale, written response to DEIS for Kauai ATOC program, ATOC AC32 11: MMRP draft EIS /EIR, volume I, comments received and responses January 1995 C38–61, comment C-41, January 24, 1995, SIOA.

133. Dan Overmyer, written response to DEIS for Kauai ATOC program, ATOC AC32 11: MMRP draft EIS /EIR, volume I, comments received and responses January 1995 C38–61, comment C-43, January 27, 1995, SIOA.

134. Terry Jackson, written response to DEIS for Kauai ATOC program, ATOC AC32 11: MMRP draft EIS /EIR, volume I, comments received and responses Jan 1995 C38–61, comment C-42, January 24, 1995, SIOA. (In the original, the emphasized word was underscored.)

135. Matt Hellman and Patti Kirby, Comment C-54, ATOC AC32 11: MMRP Draft EIS/EIR, volume 1, January 1995, comment received and responses C38 to C-61, SIOA.

136. Michelle Waters, Comment C-36, ATOC AC32 11: MMRP Draft EIS/EIR, volume 1, January 1995, comment received and responses C-1 to C-37, SIOA.

137. Postcard from Ronald Peet and Sandra Castro of Seaside, California (1–7-95, ATOC AC32 11: MMRP draft EIS /EIR, volume I, comments received and responses Jan 1995, C1–C37, SIOA). (In the original, the emphasized word was underscored.)

138. California Coastal Commission, Revised Staff report and recommendation, June 15, 1995, ATOC AC32 2: California Coastal Commissions April=Sept 1994 [sic], quoting from Sylvia Earle's comments on the Draft EIS/R, SIOA.

139. California Coastal Commission, Revised Staff report and recommendation, June 15, 1995, ATOC AC32 2: California Coastal Commissions April=Sept 1994 [sic], SIOA.

140. Peter Douglas to ATOC, FILE CC-110–94, Attachment 3, Hearing Modifications, ATOC AC32 2: California Coastal Commission, June–November 1995; Executive Summary, June 15, 1995, California Coastal Commission, Revised Staff report and recommendation, ATOC AC32 2: California Coastal Commissions April=Sept 1994 [sic], SIOA.

141. Executive Summary, June 15, 1995, California Coastal Commission, Revised Staff report and recommendation, SIO ATOC AC32 2: California Coastal Commissions April=Sept 1994 [sic]; Andrew Forbes to Peter Douglas and Tami Grove, 8 May 1995, ATOC AC32 2: California Coastal Commissions Jan–May 1995, SIOA.

142. Walter Munk to Carl L. Williams and Members, California Coastal Commission, June 9, 1995, ATOC AC32 11: ATOC, California Coastal Commission Public Hearing, June 15, 1995, Agenda, SIOA.

143. Sylvia Earle evidently attended numerous meetings on the project, including two with the NRDC, one with Sierra Club Legal Defense, one with Save our Shores, one with Friends of the Sea Otters, and two with the ATOC project team (one on May 13, 1994 at the San Francisco Airport Hilton, one on July, 19, 1994 at the California Academy of Sciences). Scoping Process Summary, June 15, 1995, California Coastal Commission, Revised Staff report and recommendation, ATOC AC32 2: California Coastal Commissions April=Sept 1994 [sic], SIOA.

144. National Research Council, Committee to Review Results of ATOC's Marine Mammal Research, *Marine Mammals and Low-Frequency Sound: Progress since 1994* (National Academy Press, 2000), 3.

145. Ibid., 73.

146. Other data increasingly showed that the world was warming overall. See Houghton et al., *Climate Change 2001: The Scientific Basis*.

147. National Research Council, *Marine Mammals and Low Frequency Sound*, 5–6.

148. Ibid., 6–7.

149. Ibid., 82.

150. Scripps Institution of Oceanography, "ATOC Source Recovery Operation Update, August 9, 2000" (https://scripps.ucsd.edu/news/2760).

151. "Judge Orders Halt to Sonic Blasts," Associated Press, October 29, 2002, ATOC AC32 6: Publications Folder 1993–2002, SIOA; David Malakoff, "Suit Ties Whale Deaths to Research Cruise," *Science* 298, no. 5594 (2002): 722–723.

152. A. Bowles, M. Smultea, B. Wursig, D. DeMaster, and D. Palka, "Relative Abundance and Behavior of Marine Mammals Exposed to Transmissions from the Heard Island Feasibility Test," *Journal of the Acoustic Society of America* 96, no. 4 (1994): 2469–2484, on 2482–2483 (emphasis added).

153. Jacob Hamblin, "The Navy's "Sophisticated" Pursuit of Science," *ISIS* 93 (2002): 1–27; Hamblin, *Oceanographers and the Cold War: Disciples of Marine Science* (University of Washington Press, 2005).

154. Naomi Oreskes, *The Rejection of Continental Drift: Theory and Method in American Earth Science* (Oxford University Press, 1999).

155. Mott T. Greene, *Alfred Wegener: Life and Scientific Work* (Johns Hopkins University Press, forthcoming).

156. On the question of shifting regimes of accountability see Helga Nowotny, Peter Scott, and Michael Gibbons, *Re-Thinking Science: Knowledge and the Public in an Age of Uncertainty* (Polity, 2011).

157. Naomi Rose, telephone interview with author, 2001.

158. On the construction of ignorance, see Robert Proctor and Londa Schiebinger, eds., *Agnotology: The Making and Unmaking of Ignorance* (Stanford University Press, 2008).

6 Fighting Each Other: The N-1, Soviet Big Science, and the Cold War at Home

Asif Siddiqi

In August of 1989, a few months before the fall of the Berlin Wall, the official newspaper of the Soviet government, *Izvestiia*, published a long essay by Sergei Leskov titled "How We Didn't We Land on the Moon."[1] Leskov, *Izvestiia*'s science journalist, had been trying to publish the piece for some time, but Glavlit, the Soviet Union's censorship agency, had repeatedly rejected his appeals. Later he recalled that "even in 1989, when there were no limits to *glasnost'*, it was such a great effort to publish the essay."[2] When it finally appeared in print, with the personal permission of a top-ranked minister, the essay caused a minor sensation. In the piece, Leskov mentioned a rocket that few Soviet citizens had ever heard of (the N-I) and a program that had never been officially acknowledged (a 4.5-billion-ruble project to land a Soviet cosmonaut on the moon in the 1960s).[3] For more than twenty years, the effort had been whitewashed out of history; save for the occasional rumor and the speculations of a few Western observers, there had been no indication that one of the Soviet Union's largest, complex, and most expensive engineering projects of the Cold War had collapsed in a series of rocket explosions in the late 1960s and the early 1970s. The Soviet project had been hidden so well that some saw Neil Armstrong's triumphant step on the moon in 1969 as a pyrrhic victory. For example, in 1974 the American newscaster Walter Cronkite commented "It turned out that the Russians were never in the race at all."[4]

After Leskov's piece appeared in *Izvestiia*, more and more articles added to this recovered history. People whose names had been classified granted interviews, and journalists, given free rein, were able to put flesh on a skeletal tale that seemed to symbolize the institutional dysfunction of late-period Soviet science.[5] Managerial gridlock, technological limitations, and economic shortages had plagued the N-I project from the very beginning. But as journalists, historians, and participants reflected on the reasons for the catastrophic failure of the project, they kept returning to a central episode in the narrative: a clash of personalities that all claimed doomed the project at its very inception. Sergei Korolev, the famous "chief designer" of the Soviet Union's spaceships and Valentin Glushko, the chief designer of its rocket engines, had almost come to blows over the selection of propellants for the N-I and eventually ceased communicating with each other. Korolev was left to guide the N-I project to success without Glushko. Despite the best efforts of thousands of engineers, and just as Glushko had warned, the N-I program—a quintessential yet largely unknown exemplar of Soviet big science and technology—eventually collapsed in a pile of rubble.

Big Science in the Soviet Context

Since the early 1990s, historians have devoted considerable attention to the fate of "big science" during the Cold War.[6] Having emerged out of interwar research and development into a full-blown phenomenon during World War II, such large-scale government-sponsored projects typically involved money, manpower, monumental machines, and often the military. In revisiting the Cold War, historians found that big science, and scientific practice in general, was hard to divorce from the forces, stresses, and demands of the national-security state. Scholars argued that scientific practice, at the institutional, cultural, and epistemological levels, thrived on instrumental, overlapping, and symbiotic relationships with high politics. Big science, because it was funded by the state, took on features that reflected the state's priorities. The possibility that Cold War imperatives altered the direction of particular disciplines was highlighted most famously in Paul Forman's meditation on how military patronage shifted scientific priorities in the United States from theoretical to applied physics.[7]

In the Soviet case, the notion of big science has meant different things to different people, but two central defining assumptions guided scholars working in the pre-archival period: the scale of the effort and the pervasive role of the state, or, as the historian Loren Graham has noted, "its bigness and high degree of government centralization."[8] In other words, the scale of Soviet science during the Cold War and its seemingly close and almost indistinguishable alignment with state sponsorship and priorities underscored the notion that big science and Soviet science were synonymous concepts. In defining what was meant by "big," Graham added that, "Soviet science was 'big' in several different ways: large in numbers of researchers, highly centralized in organization, and dominated by powerful leaders."[9]

Beyond scale and sponsorship, historians discerned other features of Soviet big science. Already by the early 1930s, the three major constituent elements of Soviet science were firmly set in place. These—the university system, the Soviet Academy of Sciences, and the industrial ministry system—represented three points of a pyramidal structure that employed hundreds of thousands of scientists, engineers, technicians, and workers at its peak in the 1970s. This tripartite system inherited traits from pre-Revolutionary Russian science. Alexei Kojevnikov identifies, in particular, the formation of research institutes separate from higher education and the emphasis on applied over basic research as embryonic and ultimately enduring features of the Soviet scientific system that first emerged during the 1910s.[10] These peculiarities became more evident after the Revolution when leading Bol'sheviks fully embraced a more utilitarian approach to science and technology. To the extent that applied science efforts translated to "technologies for the masses" (to use inspirational parlance from the 1930s), Soviet science became closely intertwined with what some have called "gigantomania"—a penchant for the monumental in many infrastructural and industrial projects.[11] According to this interpretation, Stalinist ideologues (and their successors) saw science and technology as most effective when a utilitarian ethos was combined with ostentatious and awesome exhibitions; in other words, science and technology had to both serve *and* represent the nation. This combination of size, science, and spectacle was most obviously embodied in such projects as the Moscow Metro, the Dneprostroi Dam (and hydroelectric station), the trans-Siberian railroad, and the Tu-144 supersonic transport.

In reflecting upon Forman's claim about the Cold War altering the balance between fundamental and applied science, in the Soviet context, the problem might be more accurately characterized as an appropriate distribution between theory and praxis. Marxists would have articulated this relationship as a demand that the production of scientific knowledge be closely connected to the economic, industrial, and *practical* needs of society. In Stalinist times, this requirement was frequently articulated and manifested in the priorities of the Soviet scientific establishment.[12] One of the fundamental campaigns of Stalinist science was to reinforce the link between scientific practice and the real needs of Soviet society, a quest made much more urgent during World War II. In one sense, the postwar development of the atomic bomb—perhaps the most expensive Soviet scientific project ever, facilitated as it was by a web of institutions spanning the Academy of Sciences system, the defense industrial ministries, and the security services—can be seen as emanating from this mapping of theory with praxis.[13]

The nuclear project also established a precedent for postwar Soviet big science in fortifying the deep connection between science and military requirements. The alignment between science and defense in the Soviet context was difficult to ignore; during the postwar era, the lion's share of state investment in science and engineering was devoted not to the Academy of Sciences or the universities but to the industrial ministry system dominated by the nine ministries that made up the core of the Soviet military-industrial complex.[14] By 1990, 87 percent of the Soviet R&D budget was allocated for the industrial network, most of it for military needs, leaving the remainder for the Academy of Sciences and the universities.[15] Through institutional connections or by research priorities, Soviet science during the Cold War era was deeply enmeshed with the military-industrial complex. Science and defense (with some exceptions) coexisted as one, as the "normative" state of Soviet science. Here, interrogating whether military imperatives altered the priorities (and nature) of Soviet science during the Cold War promises few insights—the answer would unequivocally be affirmative. But priorities don't tell the whole story; what other factors distinguished Soviet science during the Cold War from its predecessors? For example, did civilian imperatives, particularly the demand to *display* or "civilianize" certain science projects that were military in nature (and thus secret) reinforce certain ideological and functional characteristics of Soviet science during the Cold War?

These questions framed around the tension between the military and the civilian (and between secrecy and publicity) lead us to other seeming dichotomies relevant to the broader context of Soviet science in the post-Stalin era. The conflicting demands of theory and praxis, for example, were loosely manifested in a battle between two competing constituencies, the first comprising scientists invested in the basic sciences (particularly physics) who had accrued the perquisites of state patronage and desired a science that was "detached" from the practicalities of the day and the second comprising engineers (especially missile designers) who emerged in the late 1950s as a powerful bloc of specialists in what Russians understood as the "technical sciences" (*tekhnicheskie nauki*)—generally fields that Westerners would consider applied sciences or engineering.[16] Here we see the mutable boundaries between science and engineering, distinctions frequently lost to official Soviet spokespersons who advertised, for example, the successes of Sputnik and Gagarin as successes of "Soviet science" rather than "Soviet engineering" or "Soviet industry." In this context, it was not a little ironic that the principal body associated with Soviet science, the Academy of Sciences, was hardly involved in either Sputnik or the launch of the first human in space, Iurii Gagarin.[17] Yet the Soviet engineers who directed the space program not only embraced this conflation between science and engineering but actively encouraged it, even though they had largely been educated in entirely different institutions than pure scientists. In the early 1960s, the rocket engineers assumed for themselves the mantle of the public notion of "Soviet science," a role held for more than a decade by Soviet physicists.

The N-I rocket program, one of the largest science and technology projects implemented during the post-Stalin era, carried within it all these conflicting (and conflated) tensions: between fundamental and applied science, science and engineering, civilian and military imperatives, display value and maintaining secrecy. In each case, the program was never entirely one or the other, but usually a mix of both. Such ambiguities destabilize the conceptual framework of historians such as Loren Graham and Paul Josephson, who, in many ways, exchanged idealized features of the Soviet *state* with those of Soviet *science*. By focusing exclusively on those aspects identified with the centralized state, they missed important phenomena—among them the popular and populist campaigns for science and, in the case of big science, the messy complexities and ambivalences that subvert Western stereotypes of orthodoxy, centralization, and lack of innovation. In this chapter, I explore all these complexities and ambiguities through one critical episode in the early history of the N-I project: the selection of propellants and rocket engines for the rocket. In this debate, the two principal actors in the Soviet space program, members of a new and powerful constituency of missile engineers who had become influential stakeholders in the system of Soviet science, found themselves on opposing sides. The result was a project that perfectly embodied the contradictions and heterogeneity of Soviet science during the Cold War.

The Rise of the Space "Scientists"

By the mid 1950s, Soviet physicists—particularly, nuclear physicists—had acquired, in the words of David Holloway, "unprecedented authority among the political leaders."[18] Soviet physicists' link to state power was underscored during Nikita Khrushchev's visit to Britain in 1956 when he introduced to Winston Churchill "Academician Kurchatov, who makes our hydrogen bomb."[19] The physicists also enjoyed a public role, fostering public interest in the possible uses of atomic energy for civilian purposes and reinforcing the notion that nuclear power was a panacea for a whole host of social ills.[20] Of course, the community of nuclear physicists did not act as one, nor did they share identical goals for the future of Soviet physics, but their influence was evidenced by the disproportionate power welded by the Division of Physico-Mathematical Sciences, the Academy section to which physicists belonged.

Both nuclear physicists and missile engineers took part in designing strategic weapons, but the missile engineers had little or no clout until the mid 1950s; their handiwork up until then—short-range missiles derived from the German V2—had been less than impressive. The first sign that rockets might have strategic uses appeared in 1953 when Sergei Korolev and his team in the northern Moscow suburb of Kaliningrad began test-firing a missile capable of flying 1,200 kilometers, just far enough to reach Great Britain. By early 1956, Korolev's engineers had modified this rocket, now known as the R-5M, and made it ready to carry a nuclear warhead. First launched on February 20, 1956, the missile flew 1,190 kilometers in a little over 10 minutes and deposited its 20-kiloton bomb over its target area in the Semipalatinsk range, where it exploded in a spellbinding inferno.[21] It was the first such missile test in the history of nuclear weapons. This naked display of power, spearheaded by Marshall Georgii Zhukov and leading nuclear physicists, was a watershed moment for the rocket designers, for it brought them, for the first time, squarely into the sights of top Party and government leaders. For nearly a decade, the missile engineers had been considered junior members in the pantheon of Soviet weapons makers. But by cooperating with famous nuclear project managers such as Igor' Kurchatov and Avramii Zaveniagin on this experiment, missile designers managed to equalize the power relationship with the nuclear empire.

Remembering the initial collaboration with the high-profile nuclear physicists, one of Korolev's senior test engineers noted:

> At the start of this work Sergey Pavlovich [Korolev] gathered the project leaders to make a speech concerning the program. This was a meeting before the start of work with the atomic people. ... The first thing he said was that we ought to be very careful in our activities ... because they had been spoilt, first, due to publicity and second, because they considered themselves superior to everybody else ... after developing the atomic bomb. ... S. P. Korolev said that at least in the beginning we should pander to them. But pander very precisely and carefully such that in the end we would prove to them that we were in the driver's seat and they were merely passengers.[22]

The success of the R-5M test swiveled the center of gravity of influence away from the nuclear elite for the first time since they began their work in 1945. After 1956, missile designers, especially Sergei Korolev, began to have increased access to the top levels of the Kremlin. This was reflected both in symbolic and practical terms. A week after the nuclear test, Nikita Khrushchev, Nikolai Bulganin, Viacheslav Molotov, and several other Politburo members graced Korolev's design bureau with their presence, a rare honor accorded to few design organizations.[23] In his memoirs, Khrushchev conceded that the visitors were bewildered by the rocket, "walked around [it] like peasants at a bazaar ready to buy some calico, poking it and tugging to test its strength," but noted that "the leadership was soon filled with confidence in [Korolev]."[24] On April 20, the Supreme Soviet bestowed on three nuclear scientists, Andrei Sakharov, Iulii Khariton, and Iakov Zeldovich, the USSR's highest civilian honor, "Hero of Socialist Labor." For the first time missile designers were among the honored: they included the six main chief designers involved with the R-5M project, Sergei Korolev, Valentin Glushko, Nikolai Piliugin, Mikhail Riazanskii, Viktor Kuznetsov, and Vladimir Barmin, and Korolev's right-hand man, Vasilii Mishin. Many other junior designers in the missile industry were simultaneously given less prestigious but notable national awards. These events significantly elevated the authority of missile designers, especially Sergei Korolev, within the Soviet defense industry. "From then on," Nikita Khrushchev's son Sergei has written, "[Korolev] could phone Father directly, bypassing numerous bureaucratic obstacles."[25] This newfound authority, established on the basis of missile development, would prove critical in firmly integrating two different aspirations among the missile designers—the job of designing powerful missiles for the Soviet armed forces, and the dream of breaching the cosmos. To realize this connection, the line to the Kremlin was one of paramount importance.

Besides access to the top of the Party and government structure, the missile designers also began to make inroads into the apex of the Academy of Sciences. Traditionally, Academy members—particularly theoretical physicists—had been hostile to scholars from the technical fields, including electrical, mechanical, chemical, and aeronautical engineering.²⁶ Established academicians had a point: few of the leading "chief designers" from the defense industry had higher degrees, such as Candidate of (Technical) Sciences, and fewer had Doctorates of Sciences. Almost all had specialized degrees from technical schools such as the Bauman Moscow Higher Technical School. Additionally, most of the chief designers in charge of the key organizations involved in missile development had been born in the five-year period between 1907 and 1912, putting them in the demographic educated during and after the "Great Break" (roughly 1928–29), when educational reforms fundamentally transformed the curriculum to a more practical bent.²⁷ Many of the first generation of nuclear physicists, by contrast, were at least five or six years older and educated *before* the Bol'shevization of Soviet education, and thus more theoretically inclined than their junior colleagues.²⁸ Barring rare exceptions, the missile designers represented an entirely different academic sensibility and generation than the nuclear physicists, who were educated abroad or at Moscow's most elite universities.

The launch of the first ICBMs and Sputniks in 1957 provided a further boost to the fortunes of these missile designers in the Academy system. In October, despite the objection of a number of academicians, Korolev was awarded a "doctor of technical sciences" without having written a dissertation (or indeed published a single scientific paper). In December, two months after the first Sputnik, Nikita Khrushchev signed an order giving free dachas to the six members of the missile program's Council of Chief Designers.²⁹ The realignment culminated in 1958 with the unprecedented election of thirteen leading rocket designers into the Academy, either as full members or as (junior) corresponding members; all were voted into the now-growing Department of Technical Sciences.³⁰ Membership in the Academy had many material benefits but also represented public recognition from their peers in the world of basic sciences of the value of their intellectual and practical work. There were further additions through the 1960s as the Department of Technical Sciences surged with rocket designers and other professional designers from the defense industry, who were seen as interlopers by many specialists in the "pure" sciences.³¹ In July of 1963, Korolev was elected to the Presidium of the Academy, the organization's highest deliberative body.³²

No one person more expertly negotiated across the various divides of Soviet science—fundamental, applied, civilian, military—than Academy President Mstislav Keldysh, an applied mathematician by training.³³ Keldysh's stature steadily rose through the 1950s, largely because of his close working relationships with influential members of the scientific elite such as Kurchatov and Sakharov. With rising clout, Keldysh's portfolio diversified; by the mid 1950s, he was directly involved in thermonuclear weapons development, ICBM design, the intercontinental cruise missile project, and the development of supercomputers.³⁴ After becoming president of the Academy in 1961, Keldysh served as one of the most prominent public faces of Soviet science, even as a vast amount of his energy was, in fact, devoted to advising on the development of various Soviet armaments. By serving as the chairman of numerous "interdepartmental" review commissions tasked by Nikita Khrushchev or Leonid Brezhnev to evaluate important weapons systems, he influenced the outcome of many intractable conflicts between designers. Keldysh's personal opinion (or relationships) were thus important barometers of the direction of such massive Soviet scientific and technical projects as anti-ballistic missile systems, research on charged particle beams, high-speed computing, and, most important, the space program.

Scientific research constituted a very small portion of the early Soviet space program, especially in the 1960s. In fact, the effort was overwhelmingly dominated by military infrastructure, needs, and services. In the formative years, almost every single aspect of the program, from the smallest electronic component to the largest networked system, was produced *by* the Soviet defense industry. On the client side, the spacecraft and rockets were all produced *for* the Soviet military. And all of the infrastructure was operated by the armed forces. Dedicated scientific projects were extremely rare in the first decade of the Soviet space program, and even those had a strong military bent to them.[35]

The most prominent contracting organization in the Soviet space program—similar in many ways to a giant aerospace firm in the Western context—was the Experimental Design Bureau-1 (Opytno-konstruktorskoe biuro-1, abbreviated OKB-1), based in the northeastern Moscow suburb of Kaliningrad (or Podlipki) and headed by Sergei Korolev. In the late 1950s, OKB-1 had driven the agenda for the early Soviet space program benefiting from its leading role in developing Sputnik and the rocket that launched it. In subsequent years, OKB-1 created further Sputnik and Luna spacecraft, and by the early 1960s it enjoyed a dominant position within the emerging space program, thanks largely to Korolev's headstrong personality and unbridled ambitions. Although only OKB-1's space accomplishments were known to the outside world, the overwhelming bulk of its work was dedicated to developing military systems, particularly ballistic missiles and intelligence-gathering satellites. This preference for military systems, dictated largely by the military, clashed with Korolev's personal interest, which was increasingly drawn to the kind of space exploration that inspired science fiction buffs. Weaned on the ideas of the early-twentieth-century theoretician Konstantin Tsiolkovskii, Korolev's vision for the Soviet space program—much like Wernher von Braun's for the American program—saw it as expanding progressively from Earth orbit to the moon and eventually to the inner planets.[36]

Korolev's monopoly, both in developing missiles and exploring space, faced stiff competition in the early 1960s as other ambitious designers began to encroach on his domain. By the time of Gagarin's flight, in 1961, two other prominent designers, Vladimir Chelomei and Mikhail Iangel', challenged Korolev's monopoly and influence in the space arena.[37] For all three, work on civilian spacecraft was at best a luxury, allowed if their primary work on missiles was not impeded in any way. In this situation, the missile-designers-turned-spacecraft designers faced a conundrum. The most effective way for them to accrue publicity was to engage in space activities that resonated deeply with a newly proud and hopeful Soviet populace. Yet their bread and butter—their funding—came from the armed services, which resisted their penchant for wasting time on space-related activities.[38] This dilemma was central to the battle that tore the N-I program apart.

The Market for Innovation

The increased authority of missile designers in the wake of the space successes of the late 1950s gave them unprecedented influence on the direction of future space research, particularly because the upper management had less expertise in evaluating the technical efficacy of space-related proposals than in assessing missile-related ones. In the post-Sputnik era, the Communist Party and the government had overlapping structures to direct and manage the space program. The most important organ at the government level was the so-called Military-Industrial Commission (Voennaia-promyshlennaia komissiia, VPK), representing the various ministries and industrial branches responsible for building hardware. The commission, established in December 1957 in the wake of Sputnik, was tasked with "leadership and monitoring of work on the creation and quick introduction into production of rocket and reactive armaments and other forms of military technology, and also to coordinate this work between branches of industry independent of their branch affiliation."[39] The VPK was established to coordinate work on all Soviet military technology—not only rockets but also tanks, airplanes, guns, ships, and submarines—but its leaders were largely grizzled veterans from the missile industry who were on good terms with missile designers such as Korolev and Iangel' and more receptive to their proposals than, say, to a proposal from a submarine designer.[40] On the other hand, these industrial managers were more than a bit bewildered by all this talk of space exploration; they had only the barest level of expertise with which to compare a wildly ambitious Mars-exploration program using ion-engine-equipped winged spacecraft (as Chelomei proposed) or a modest and sober idea for a film-return reconnaissance satellite (as Korolev proposed). This combination of familiarity with missile designers and lack of knowledge about space systems produced a systemic problem: there was a welcoming environment for the missile designers to send up all sorts of outlandish ideas for approval, but a lack of expertise to evaluate their value.

Conventional wisdom has it that the Soviet defense industry operated in much the same way as the rest of the economy, i.e., that this was a centrally driven command economy with no market choices. Already during the Cold War, it was evident to some Western analysts that this was not so. "Competition," David Holloway noted in 1984, "has been a common, though by no means universal, practice in the development of new weapons, especially of aircraft and missiles. Two or more design bureaus might be given the same requirements and asked to produce designs: the Ministry of Defense then selects the best design for development. This gives the customer a degree of choice unusual in the Soviet economy."[41] Recent evidence confirms this view that a uniquely Soviet quasi-market competition existed at certain stages of weapons design as a result of practices that dated back to the 1930s.[42] Naturally, both the buyer and the sellers of weapons systems were owned by the state; yet, at key points in the research and development process, market behavior very similar to US weapons research and development was tolerated; this quasi-market emerged at the level where the clients (usually, a broad coalition of representatives from the military-industrial complex) had to arbitrate between multiple proposals for a new weapons system. In principle, this meant that the military would select a particular designer's idea from a pool of proposals sent up to the VPK, based on a fit with requirements for the weapon. In practice the process rarely operated as expected.[43] Instead, other more subjective factors intervened. Favoritism predicated on professional and personal networks was crucial in the process; Chief Designer Mikhail Iangel', for example, hailed from Dnepropetrovsk, the Ukrainian industrial city where Brezhnev had served as a regional Communist Party secretary. Designers, like American companies responding to a request for proposals, also wildly exaggerated the capabilities of their own systems and promised highly optimistic timetables. Most crucially, they would each invoke American superiority in a particular field and guarantee that they and they alone could counter the potential threat. To the designers, new projects guaranteed continuing funding, and if they expressed some outward camaraderie or publicly appealed to a common national purpose, at the design proposal level, they were deeply competitive and often hostile toward one other. Each major chief designer of a weapons system ruled over a fiefdom whose well-being (and often existence) depended on large and continuing contracts.

The result was a chaotic research and development process that belied the public image of a command economy pursuing a sustained and well-conceived path. In reality, the VPK was completely unprepared to handle the large influx of proposals about future plans and, often, based on lobbying from a particular designer, approved multiple proposals for the same requirement, fearful that they would be treading on the toes of powerful patrons in Party and/or government who supported these ambitious chief designers. This combination of increased authority due to the successes of the early space program, personal connections with senior VPK officials, the (mis)use of technical knowledge as leverage, and inefficient institutional mechanisms meant that bureaucratic chaos was the norm rather than the exception in implementing large-scale Soviet space projects. And as more and more ambitious chief designers entered the fray by the early 1960s, formulation of any long-range and sustained vision of the

Soviet space program became all but impossible as the process became mired in petty disagreements nearly impossible to arbitrate.

A Tale of Rocket Propellants

The idea for a "super rocket" for the Soviet space program emerged as a part of plans to augment the standard and moderately powerful R-7 that had lofted the early Sputniks into orbit. As early as 1956, Sergei Korolev had referred to an idea for a massive rocket with a launch mass of 1,100 tons.[44] Such preliminary studies culminated in an intense period of investigation in early 1960 to develop some requirements and basic design choices. At this point, neither Party nor military officials evinced much interest in this idea, the former seeing this as a potentially costly diversion from immediate needs and the latter believing that a heavy-lift rocket would not be militarily useful. A meeting between Khrushchev and the leading space designers in January of 1960 appears to have altered the landscape, with Khrushchev calling for more intense efforts to develop space projects to respond to what he saw as ambitious American plans.[45] At the same time, Soviet military planners found statements from important American officials such as Senator Lyndon B. Johnson, the Democratic Senate Majority Leader, and Herbert F. York, the director of defense research and engineering at the Department of Defense, as being belligerent and advocating increased militarization of space. As a result, in the first few months of 1960, Soviet space designers scrambled to come up with an appropriate response, a grand seven-year plan for space exploration that would emphasize military operations. The central point in this ambition would be the development of a super-rocket.

After an intense series of negotiations, the Party and the government approved a long-range program of research on space travel in June of 1960. The heart of this program was assigned to Korolev's OKB-1, which was to create "a new powerful rocket system with a launch mass of 1,000–2,000 tons" capable of putting 60 to 80 tons into Earth orbit and sending 20 to 40 tons on translunar and interplanetary trajectories. The main goal of such rockets would be to launch a "heavy interplanetary ship." According to the plan, by 1962 there would be a initial rocket known as the N-I, and by 1967, and a more powerful one, the N-II. In drafting the decree to ensure that it would be approved at the highest level, Korolev and his associates noted that such super-rockets could be used for launching "space battle stations" into orbit and used for all manner of military operations in space, including "monitoring space and destroying enemy ... satellites" and reconnaissance missions and even for hitting ground targets from space.[46] Tellingly, none of these ideas for military applications came from the military; high officials in the Strategic Rocket Forces had no idea why they needed such a powerful rocket, and had, in fact, stayed out of the discussions on its specifications. As was not uncommon for weapons projects on both sides of the Cold War divide, this was a case where the contractor spent an inordinate amount of time trying to convince a client why they needed something that barely interested them.

As money for the new super-rocket project started to flow in, there were a number of decisions to be made about its design. The most contentious of these centered on the engines, whose designers drew on the science of chemical propellants, dating back to the early twentieth century. In 1903, when the Russian theorist Konstantin Tsiolkovskii first mathematically substantiated the possibility of space exploration in a published essay, he noted that the most energetic rocket propellants would be a combination of liquid hydrogen (fuel) and liquid oxygen (oxidizer).[47]

A rocket engine's measure of efficiency, which depends on the characteristics of the chemicals in question, is typically indicated by a number ("specific impulse") which measures the change in momentum per unit amount of propellant used; the higher the specific impulse, the more efficient a rocket engine. For rockets launching objects into space, engineers naturally gravitated to engines that promised higher specific impulse ratings since such engines would require less propellant to attain a given momentum. Theorists considered liquid oxygen the best oxidizer, one that when combined with kerosene (or especially, liquid hydrogen) could produce very high specific impulse values. That made liquid oxygen the first choice for space launch vehicles in the early years of the space age. But high-energy propellants brought their own challenges: oxygen, for example, takes on a liquid state only at very low temperatures, from −223°C to −183°C. Thus, in order to keep oxygen in its liquid form in the tanks of rockets, engineers needed to deal with many technical challenges, such as developing special systems to store super-cooled (or cryogenic) liquid oxygen both on the ground and in the rocket. By increasing tank pressure, it was possible to bring up the boiling temperature of liquid oxygen, but very high chamber pressures raised their own challenges. Rockets with cryogenic propellants were also notoriously difficult to ready for firing, especially in the early years of the space age: in the case of early versions of the R-7 ICBM, it took as much as 20 hours to prepare it for launch, which made it practically useless for a surprise attack.

Non-cryogenic propellant combinations had their own advantages and liabilities. For example, when nitrogen tetroxide was used as an oxidizer and standard kerosene as a fuel, the combination was storable, implying that a rocket fueled with such propellants could be kept at the ready for a long time. For a military rocket, this was a crucial asset. Unlike liquid oxygen, nitrogen tetroxide remained in a liquid state at close to room temperature (from −11°C to 21.5°C), which made it easier to handle. Such combinations, however, had low specific impulse values and thus were not quite as efficient as cryogenic engines. Many storable propellants were also highly toxic. In 1960, a new Soviet ICBM, the R-16, had exploded on its launch pad and killed nearly

90 people, many of them through exposure to the highly toxic propellants.[48] Yet the singular advantage of being able to get keep a missile ready for launch on command kept military commanders coming back to such storable propellants as the most ideal for use in the Soviet offensive strategic force.

When Korolev's engineers first proposed engines for the N-I, they gravitated to cryogenic combinations, especially liquid oxygen and kerosene, which they had successfully used in the R-7, recently put on service duty as the Soviet Union's first intercontinental ballistic missile.[49] For future upper stages, they assumed that other high-energy propellants, including the liquid oxygen–liquid hydrogen combination and perhaps even nuclear rocket engines, would be used. As before, the powerful first-stage engines for the rockets would be developed under the tutelage of Valentin Glushko, the Soviet Union's preeminent rocket designer, who headed a large organization, OKB-456, based in Khimki, a suburb northwest of Moscow.

Korolev and Glushko, the two giants of the Soviet space industry, already had a long and storied relationship, one that had been marred for many years by the debate over propellants. They had met as young men in the early 1930s and worked together at a government-sponsored organization for rocket research, the Reactive Scientific-Research Institute (RNII), in the interwar years. Debates over the appropriate choice of propellants almost tore the institute apart; Glushko had staked out a clear position in favor of storable propellants, particularly nitric acid, because they did not require complicated ignition systems, were cheap to produce, and were easy to obtain in Leningrad, where he had served his apprenticeship. Others favored liquid oxygen. Many engineers left the institute in disgust when their favored propellant was privileged over another. These battles added poison to the traumas at the height of the Great Terror in the late 1930s when Korolev and Glushko were forced to denounce each other on trumped-up charges of sabotaging equipment.[50] Both spent time in the depths of the Gulag and worked together in a prison camp for engineers, where Korolev was Glushko's deputy. After the war, they helped Soviet teams scour through the detritus of German industry and then assumed leadership of separate design organizations, with Korolev, more influential, designing missiles, and Glushko producing engines for them.[51]

Perhaps because of their shared traumas, the two men remained on friendly and respectful terms through the years. This connection began to fray by the mid 1950s as several progressively bigger technical disagreements pulled them apart. The disputes, initially technical, became increasingly personal. First, there was Glushko's refusal to design verniers (small steering engines) for the main engines of the R-7 in the mid 1950s. Then there was Glushko's failure to deliver on time a particularly crucial upper-stage engine for an advanced rocket—a delay that stretched into several years, until Korolev abandoned the contract.[52] These small fissures widened further with a major conflict over engines for Korolev's first post-R-7 missile, the R-9 ICBM.

By this time, Korolev and Glushko had staked out clear positions on the choice of propellants, the former now favoring cryogenic propellants (including high-energy fuels, such as liquid hydrogen) and the latter continuing to support storables.

Beginning in the early 1950s, the Soviet military had demanded that Korolev design newer missiles using storable propellants, a demand that he had resisted. Late in the decade, he proposed a new rocket, the R-9, that would use liquid oxygen, and under severe pressure from Korolev, a number of chief designers reluctantly came out in favor of it.[53] After almost a year of discussion, the military grudgingly supported the project, but only if Korolev could guarantee high-speed launch operations.[54] Glushko, the only major rocket engine designer in the Soviet Union who could be counted on to design such powerful engines (approximately 144 tons of thrust at sea level), was tasked to build engines for the R-9; he did this reluctantly, since he had begun to turn his entire organization away from the tried and tested liquid oxygen–kerosene combination that had powered the earlier R-7 ICBM. He had technical reasons for doing so; in the early 1950s, his last attempt to build a high-thrust single-chamber liquid oxygen engine had ended in disaster as model after model exploded in ground-test stands due to high frequency oscillations in the combustion chamber.[55]

Korolev himself had little confidence that Glushko could overcome these problems. Resentful that Glushko had a near monopoly on rocket engine design in the Soviet Union, Korolev invited a number of "outsiders" to submit proposals for the liquid oxygen engines for the R-9. One of these was an organization based in the large industrial city of Kuibyshev, nearly 1,000 kilometers southeast of Moscow, on the banks of the Volga river close to Kazakhstan. Known by its cryptic name, OKB-276, the design bureau was headed by Chief Designer Nikolai Kuznetsov, who had no experience designing rocket engines; for nearly a decade he had led the design of turboprop engines, including the NK-12 engines that powered the famous Tupolev Tu-95 ("Bear") strategic bomber.[56] Kuznetsov's attention was drawn to missiles in the late 1950s, when Khrushchev, mesmerized by the power of rockets, had begun to limit work for firms in the Soviet aviation industry. Numerous aviation firms struggled to make ends meet by diversifying into other fields, such as the rocket and space industry. The Soviet premier reportedly suggested to Korolev that he invite some of these design bureaus to be subcontractors for the space program. A growing number of these aviation firms, hungering for contracts, quickly turned their attention to Korolev and other missile designers and began to solicit contracts. Kuznetsov's design bureau was one of them.[57]

Kuznetsov's foray into missiles cracked open the rift between Korolev and Glushko.[58] The Soviet leadership had originally approved the development of the new R-9 ICBM in May of 1959. Contracts were handed out, and Glushko began to develop a new and powerful liquid oxygen–kerosene engine. Lacking confidence in Glushko's ability to develop such an engine, Korolev, somewhat abruptly, at the end of the year, wrote a letter to Leonid Brezhnev, the Party curator in charge of the missile and space program, to eject Glushko from the R-9 missile program in favor of newcomer Kuznetsov. Korolev argued that Kuznetsov, despite his lack of experience in designing rocket engines, could produce a much better and more efficient engine in a shorter time; it didn't help that Glushko had repeatedly failed to deliver major contracts on time.[59] It was unprecedented and rare for a designer to demand that a government decision be revised, but Korolev's relationship with Glushko had soured by then and he was keen to break his professional relationship with his former colleague. Glushko was livid when he found out; he fired off a letter to the Military-Industrial Commission rejecting Korolev's plea. In the end, Korolev lost his gamble, and the ministry in charge of the program reiterated that Glushko's engines would remain as part of the R-9 missile. Korolev was forbidden to test any other engine in support of the ICBM.[60] The R-9 flew, albeit much later than had been planned, and with Glushko's engines, as originally intended. Yet the battle over this military missile undoubtedly darkened the relationship between the two men.

Glushko's Refusal

The battle over the R-9 was only a prelude. From late 1960 to the summer of 1962, there was a protracted conflict between Korolev and Glushko over propellants that effectively split the entire program into two. Glushko recognized that Korolev's N-I and N-II rockets would constitute the future of the Soviet space program, and he wanted to have major contracts for these rockets. But there was a problem: his opinions about rocket engine design had dramatically shifted between 1958 and 1961, and his change of heart put him directly at odds with Korolev. In the 1930s, Glushko's favored propellants had been storables, in particular nitric acid (as oxidizer) and kerosene (as fuel). After the discovery of the German V2 ballistic missile at the end of World War II, Glushko had abandoned storables and reoriented his work to the use of liquid oxygen and alcohol for about five years. Building on this experience, his organization had produced engines using liquid oxygen and kerosene for the first R-7 ICBM. This combination made it difficult to prepare the missile for launch (which made the military unhappy), but it did add a modicum of extra lifting power to the rocket (which made the space enthusiasts happy). But between 1958 and 1961, Glushko's thinking slowly migrated back to his earlier position on the use of oxidizers and fuels; he now rejected both liquid oxygen and kerosene.

First, he found a new fuel to replace kerosene. In 1949, the Leningrad-based State Institute for Applied Chemistry developed a new toxic compound, a kind of hydrazine fuel known as unsymmetrical dimethyl hydrazine (UDHM). According to Glushko's calculations, when UDMH was paired with liquid oxygen instead of the usual kerosene, one could potentially increase specific impulse values by approximately 4 percent. By the late 1950s, when, on assignment from Glushko, this institute developed an industrial base to mass produce UDMH, Glushko immediately latched on to it, determined to stop using kerosene and replace it with UDMH. He began building a series of liquid oxygen–UDMH engines, and in January of 1958 proposed to Korolev that the next ICBM should use this propellant combination.[61] From then on, Glushko's organization developed almost no rocket engine without UDMH as the fuel.

Second, he began to go a step further and replace the oxidizer, liquid oxygen. This came as no surprise to anyone who knew Glushko's history; he had a long-standing animus toward liquid oxygen that he had suspended only because the Germans had been using the substance in their V2. Although Glushko was a diehard space enthusiast (and thus would be expected to prefer oxygen), he was also a realist. In the early 1930s, when he was searching for an ideal combination of propellants, he gravitated to materials that were available from industry. But one important criteria for him was the problem of keeping rockets at a ready state. In 1936, bearing in mind that military rockets had to be ready to be launched immediately on command, he had written that "in terms of battle applications liquid oxygen [has] acute operational shortcomings." He added that "careful consideration of the properties of these materials shows that [liquid] oxygen is not the best oxidant and [liquid] hydrogen is simply not suitable for practical use."[62]

In replacing liquid oxygen, Glushko proposed tried and tested oxidizers such as nitric acid. His engineers began development of a series of engines using the nitric acid–UDMH combination in 1958 for new missiles developed for Chief Designer Mikhail Iangel', Korolev's primary competitor at the time.[63] Eventually, he found the ideal oxidizer, nitrogen tetroxide, which promised even better specific impulse ratings when combined with UDMH. By the end of 1960, his position had solidified: the best combination of propellants for future rockets and launch vehicles would be nitrogen tetroxide (as oxidizer) and UDHM (as fuel). In a letter to ministry bureaucrats and military officials in December 1960, he noted that the availability of factories producing nitrogen tetroxide in the USSR created favorable conditions for its use in rockets and that his design bureau had completely turned its attention to creating engines using this oxidizer. He added—using a common strategy to strengthen an argument—that the Americans were increasingly turning to the use of nitrogen tetroxide in their missiles.[64]

The evolution in his thinking that led Glushko to abandon liquid oxygen angered one constituency (Korolev) but pleased another. At the very same time that Glushko embraced storable propellants, the Soviet Strategic Rocket Forces was gearing up for a massive expansion, soliciting contracts from many different organizations to build new generations of intercontinental ballistic missiles. Almost no one in the military wanted liquid oxygen missiles; it was clear to most that if the Soviet Union were to have an effective ICBM force, it would need to have missiles that could be launched at a moment's notice. In the early 1960s, when the military handed out several contracts, Glushko's organization snapped up all the major slots for designing powerful first-stage engines for these rockets. All of them used the nitrogen tetroxide–UDMH combination, highly toxic to handle but much easier for operational use. With some logic, Glushko believed that he would maximize his resources if he could produce "dual-use" engines that could be used for both the "civilian" N-I and another military rocket.

At the very beginning of the process, when Korolev's engineers were busy conceptualizing the giant N-I rocket, they entertained Glushko's insistence that they consider storable propellants as a possible option for it.[65] By March of 1961, Glushko clearly and without equivocation informed Korolev that his organization, having done some serious research into possible combinations for propellants, strongly preferred nitrogen tetroxide and UDMH for the new super-rocket.[66] He offered two engines, known as the RD-253 and RD-254, for the N-I; simultaneously he offered these engines for use on a new proposed military rocket proposed by a competitor to Korolev, Vladimir Chelomei.

In 1961, Korolev's engineers did some intensive analysis of possible configurations of the N-I. In considering propellants, engineers performed comparative analyses of several combinations, some cryogenic (i.e., using liquid oxygen) and some storable. Increasingly, they came to the decision that cryogenic combinations would be ideally suitable for this rocket. Korolev had already handed out competitive contracts to several organizations in March of 1961 to produce engines: some contracts went to Glushko to produce his favored engines, while a parallel assignment disbursed enough money for Nikolai Kuznetsov, the aircraft engine designer in Kuibyshev, to begin work on several liquid oxygen–kerosene rocket engines.[67] As the year ended, engineers on both sides of the debate fully understood that, if at one point, Kuznetsov's engines represented an insurance policy for Korolev, by the end of 1961, they were Korolev's primary choice. But Glushko refused to back down. In late 1961, he fired off several letters to Korolev, to Academy of Sciences President Mstislav Keldysh, and to high officials in the Communist Party, pressuring them to make a decision in his favor.[68]

The Keldysh Commission

The crisis culminated in July 1962 when an "extraordinary commission" tasked by Nikita Khrushchev convened to examine the course of work on the N-I rocket. Headed by Keldysh, the commission included dozens of academics, military officers, scientists, and engineers.[69] Its goal was to review, over a period of two weeks, the documentation on the rocket that had been prepared under Korolev's tutelage, and ensure that the government approved the most optimal and efficient path of development. It was unusual for a technological system to be subjected to such scrutiny at the highest level, but the N-I was no ordinary technology; it was to be the most expensive single project in the history of the Soviet space program. The obvious important issue at hand was the selection of propellants for the N-I, a battle between Glushko's storable propellants and Korolev's cryogenic ones.[70]

The arguments from each side advocating for their particular propellants were generally grouped under four criteria: efficiency, cost, safety, and engine design and operation. Glushko argued his case in a series of letters to Korolev and others in late 1961 and early 1962. Korolev presented his case during the actual meetings of the Keldysh Commission in July. The most important issue here was efficiency, i.e., the ability of a certain propellant combination to lift a larger payload into orbit. Here, Glushko's argument was weak. He noted somewhat vaguely that "the payload mass inserted into orbit, is evidently less" when using liquid oxygen–kerosene because of the need to reduce the evaporation of oxygen, which would require special insulation material for the rocket tanks, thus making it heavier and thus less effective.[71] Korolev's engineers had a very strong case against this argument, since all their calculations showed that liquid oxygen–kerosene was much more efficient than storable pairs, despite any additional weight on the rocket. Perhaps sensing that his position would not fly with the Keldysh Commission, Glushko made a last-ditch argument: if Korolev's engineers calculated that their liquid oxygen–kerosene pair was more efficient, i.e., could lift more into orbit, it was simply because of "the particular design of the N-I launch vehicle [and] thus we can assume that the design layout of the N-I is not optimal for a heavy-class launch vehicle."[72] In other words, he tried to deflect attention to the design of the rocket rather than the propellant combination.

The second important factor was cost. Each side did extensive calculations on the use of their respective propellants. They produced wildly different numbers, then interpreted them with their own biases. Glushko noted that in 1962–63 nitrogen tetroxide and UDHM cost 55 rubles and 1,800 rubles per ton respectively, whereas liquid oxygen and kerosene cost 41 rubles and 39 rubles per ton respectively. He conceded that the latter pair was "8 times cheaper" than the former, but only "if you don't consider the cost of super-cooled oxygen." This was because of the perceived extra cost of complicated systems and processes designed to ensure storage of liquid oxygen in liquid form (at very low temperatures), both on the ground and on the rocket.[73] "With such an objective assessment of the actual cost of tons of supercooled oxygen," he added, "it inevitably turns out to be several times more expensive. ... "[74] For his cost estimates, Korolev added overhead costs for both liquid oxygen and nitrogen tetroxide but still had a stronger argument: nitrogen tetroxide (181.4 rubles/ton) and UDHM (2,142.6 rubles/ton) came out a poor second to liquid oxygen (110.2 rubles/ton) and kerosene (79.6 rubles/ton). Korolev noted that both liquid oxygen and kerosene had large production bases in Soviet industry (as did nitrogen tetroxide) and were used widely in the Soviet economy. But concerns about having to develop storage and cooling systems for liquid oxygen, which tended to evaporate easily, could be put to rest, since such systems had already been developed for a military missile—the R-9A ICBM. On the contrary, he argued, using nitrogen tetroxide would require special equipment for the rocket, since the substance retained its liquid form only between −11°C (12.2°F) and 21.5°C (70.7°F), a range that was far exceeded at the launch site in Kazakhstan; in winter, special heating equipment would be required, and in summer, the tank pressure would need to be increased to ensure a higher boiling point, requiring thicker and thus heavier propellant tanks. In a comparison of one-time capital investments in the development of the engines, liquid oxygen–kerosene would be less than half as expensive (8.1 million rubles vs. 18.9 million rubles). The costs for subsequent launches would also favor liquid oxygen–kerosene (0.25 million rubles vs. 2 million rubles).[75]

The third issue was safety. Korolev noted that both UDMH and nitrogen tetroxide were highly toxic compounds, thus requiring extra ground equipment to neutralize waste, ensure drainage, "de-gas" facilities, and sanitize tanks after prolonged exposure to propellants. Ground crews would also need special masks and suits for their own safety. The fact that these components ignite upon contact with each other (that is, are hypergolic) increased the demands on tightness of joints significantly. Liquid oxygen and kerosene, on the other hand, were both non-toxic.[76] Glushko conceded that his propellants were toxic but noted that there had been no cases of poisoning when launch-site rules of operation had been strictly followed. In fact, experience with different rocket engines on earlier missiles showed that there were no cases of leaks in storable-propellant engines as opposed to many cases of dangerous leaks of liquid oxygen. The latter were especially hazardous, Glushko argued, because even a single leak of liquid oxygen was very dangerous in view of its low boiling point and extreme volatility, whereas with storable propellants *both* components would have to leak to cause an explosion.

The fourth major issue under discussion was engine design and operation. Both sides had compelling arguments. Glushko noted that because nitrogen tetroxide and UDMH were self-igniting (hypergolic), engines using such propellants would not require special ignition devices to start up; all that was needed was to put the propellants in contact with each other. Such engines were by definition more reliable and relatively easier to control—especially when simultaneously firing 24 engines, as would be the case for the first stage of the N-I. Hypergolic propellants also fired with less delay time, igniting on command, a facility critical to the operation of upper stages. Finally, Glushko argued, it was well known that high-thrust liquid oxygen engines suffered from irregular combustion and were more subject to high-frequency oscillations. In liquid oxygen engines, there was also the need to protect combustion chambers and nozzle walls from overheating. Glushko's design bureau had already faced these problems in the early 1950s in the course of developing single-chamber cryogenic engines. None of these problems afflicted storable-propellant rocket engines.[77] Korolev's engineers had a convincing counterargument: yes, they conceded, "normal" liquid oxygen engines were susceptible to unstable combustion and sometimes even exploded into fragments because of the particular mix of liquid and gaseous compounds that formed at the entry point of the combustion chamber. But all of Glushko's arguments were invalidated because Korolev was advocating the use of a new type of cryogenic engine: what Soviet engineers called a "closed-circuit" engine, known in the West as a type of "staged-combustion" engine. Such engines maximized the use of propellants by minimizing gas losses that occurred when driving turbines. They were extremely efficient (with high specific impulse ratings), safe from the common destructive properties of high-thrust liquid oxygen engines, and highly innovative for the period. American engineers had avoided such designs, believing them to be beyond the reach of current technology. Korolev, having already developed small staged-combustion engines, believed that a bigger one might be possible; in 1959, his new comrade-in-arms, Kuznetsov, had begun development of several new staged-combustion liquid oxygen rocket engines.[78]

As was typical for the time, final arguments were couched in terms of what the United States was doing. Glushko noted that "the early versions of the Atlas and Titan intercontinental rockets developed by the US used [liquid] oxygen and kerosene as propellants," but that "now [they are] urgently moving to use [nitrogen tetroxide] with hydrazine." "In this case," he continued, [they] have in mind the possibility to ensure long-term (several years) service of a fueled rocket in a battle-ready state with [launch] preparation time down to 1 minute. For some years now, the second stages of all Thor and Atlas missiles have been using only nitric acid and nitrogen tetroxide as oxidizers with UDMH."[79] Korolev argued almost the opposite:

There is evidence that 95% of the work on [rocket engines] in the US is focused on the use of [liquid oxygen]. In 1960–61, the Rocketdyne-North American firm finished development of the H-1 and H-2 oxygen-kerosene engines with thrusts of 85 tons and 112 tons. ... The H-1 engine has fully passed ground testing ... and is now part of stage I of the Saturn rocket, which has successfully passed its first flight test. ... All together in the US there are 19 [rocket engines] (90%) with a thrust [range] of more than 7 tons that use [liquid] oxygen and only two engines (10%) that work on nitrogen tetroxide.[80]

The arguments went back and forth for days without much compromise, sometimes fracturing the modicum of unity among the other chief designers. The choice, as presented by the two leading parties, was between two engines, those of Glushko and Kuznetsov, with Korolev arguing for the latter. Commission members debated various technical, industrial, and organizational issues. Eventually, the Keldysh Commission voted unanimously to recommend, as Korolev had argued, that the N-I use Kuznetsov's liquid oxygen engines, adding in its official report that the N-I technical documentation fulfilled "high scientific technical standards" that had been originally demanded in the initial proposals."[81] The commission justified its decision in favor of liquid oxygen and kerosene on the bases of efficiency, cost, and safety. On all three points, they were convinced that, as Korolev had argued, Kuznetsov's engines would have better lifting characteristics, would be safer to use, could take advantage of existing systems, and be cheaper, having accepted Korolev's cost numbers over those of Glushko.[82]

Glushko was livid. Despite the commission's conclusion, he insisted on a total revision of the N-I design so it would use his storable-propellant engines, under development for at least a year by then. Several prominent designers and highly placed military officials tried in vain to convince him to participate, but he categorically refused to make liquid oxygen rocket engines for the project.[83] Eventually Nikita Khrushchev was drawn into the battle, but even he was unable to mediate. "Differences of opinion," he wrote in his memoirs,

> started to pull [Korolev and Glushko] apart and the two of them couldn't stand to work together. I even invited them to my dacha with their wives. I wanted them to make peace with each other, so that they could devote more of their knowledge to the good of the country, rather than dissipate their energy on fights over details. It seemed to me that they were both talented, each in his own field. But nothing came of our meeting. Later Korolev broke all ties with Glushko.[84]

As a result, the job of developing the N-I engines went to Nikolai Kuznetsov, a designer of jet engines for Soviet civil aviation. The largest and most ambitious rocket ever built in the Soviet Union would have engines designed by an organization that had never flown a single one.

After the Decision

After Glushko was officially divorced from the program, he made repeated attempts to undermine the N-I project—a tactic he had adopted even before the 1962 settlement. In 1960–61, for example, during the conception stage of the N-I, Glushko had tried several times to push through alternative ideas for a similar monster rocket, using as a justification the goal of "maintaining the priority of the Soviet Union in this area [of rocket design]."[85] Korolev, who sought to maintain a monopoly on the building of the next generation of Soviet launch vehicles, bluntly rejected all these interventions without seriously evaluating their value. Glushko was also sufficiently shrewd to have an insurance plan in case the N-I didn't work out: long before the final decision on the N-I propellants had been made, and unknown to Korolev, Glushko had approached Korolev's rival Mikhail Iangel' and proposed the use of the same engines he was planning to use on Korolev's rocket for a competing variant produced by Iangel'.[86] When that attempt failed, he tried again the following year with a new Iangel' rocket, the R-56, proposing it as a much better alternative to the N-I, one that would use his unused nitrogen tetroxide–UDMH engines from the N-I. He tried to appeal to higher goals, imploring that "further delay in the development of rockets with ... lifting capacity greater than the [American] Saturn I ... will exacerbate the lag of the Soviet Union in the development of rocket technology."[87] Glushko's stubbornness eventually brought him into conflict with Mstislav Keldysh. In late 1964, two years after the decision against Glushko, when he brought up the propellant issue once again at a meeting on the N-I, Keldysh replied sharply: "The question over propellant components must stop. ... It's now necessary to firmly reject everything that interferes with [our work]. ... The arguments over this issue are just a waste of time."[88]

Glushko didn't give up. In 1964–65, he insisted on a repeat study to evaluate the characteristics of an N-1 rocket with his engines replacing Kuznetsov's liquid oxygen ones. In early 1965, a review commission rejected Glushko's suggestion to rework the N-1—not surprising, since millions of rubles had already been spent on the design approved by the Soviet government.[89] A last-ditch effort to derail the N-I program coalesced in the mid 1960s when Glushko joined with another Korolev competitor, Vladimir Chelomei, and sent appeals to the Party and the government proposing a new rocket that, if given the appropriate funds, could beat the Americans to the moon. This new imagined super-rocket would use powerful storable-propellant rocket engines developed by Glushko.[90] Even as more than 500 organizations nationwide were fully engaged in producing the N-1 rocket, a government decree allocated funds to Chelomei and Glushko to move ahead with their proposal. Eventually, saner heads prevailed, and the idea was scuttled in 1968.[91] Through it all, Glushko sent off several missives to the Soviet government severely criticizing Kuznetsov's work on liquid oxygen engines for the N-I. After a ground test of Kuznetsov's NK-15 engine went awry, Glushko wrote: "You can see for yourselves that the engine is bad. It's not fit for work, and certainly not for installation on such a crucial piece of hardware like the N-I."[92]

How was Glushko able to refuse a state mandate to participate in the N-1 project? How was he able to decline Khrushchev's overtures at mediation? And later, how was he able to mount repeated challenges to Korolev's program when it had already acquired significant organizational inertia? Three factors loom large here, all rooted in the way in which Cold War pressures at the international level affected "local" decision making.

First, Glushko's hubris was undoubtedly reinforced by the elevated authority of space-program chief designers in the aftermath of the success of Sputnik. One way this individual agency was instrumentalized was cowing Party and government bureaucrats with explicit claims that Khrushchev or Brezhnev had personally sanctioned some or other project and therefore the ministry had to act on it. Glushko was not shy about using firm language; in one letter to Korolev insisting on the use of storable propellants for the N-I, he underscored that his organization had been given the obligation to develop powerful rocket engines by the "repeated, direct, and personal instruction of N. S. Khrushchev."[93] With such invocations, missile chief designers were able to push through many projects that duplicated the efforts of others. There are innumerable cases of competitive projects tailored for singular goals when, because the Party and the government structure were ineffective in curbing the power of chief designers, simultaneous and similar projects were adopted and funded. The most striking case of such redundancy and waste was the so-called little civil war of the late 1960s, when competing missile designers—Vladimir Chelomei and Mikhail Iangel'—waged a battle through their patrons in the power structure to gain contracts for the third generation of Soviet ICBMs. In the end, Brezhnev, unable to decide between different options, funded similar high-performance missiles from both parties, squandering billions of rubles.[94]

Second, the authority of chief designers was undoubtedly affected by the perception of work being done in the United States. In the post-Stalin era, when missile chief designers appealed for funding for their pet projects, they invariably cited superior or better-funded work ongoing in the West. For example, in the battle over propellants for the N-I rocket, both Korolev and Glushko repeatedly used information about American missiles. In January of 1961, at a meeting with representatives of the Ministry of Defense on the future of the N-1, Glushko noted that "on the basis of published information it's worth nothing that in the second variant of the Titan rocket, the Americans are using nitrogen tetroxide as oxidizer, and a mixture of 50% dimethyl hydrazine and 50% hydrazine as fuel"—that is, storable propellants.[95] Later, in July of 1962, during the Keldysh Commission's two-week-long deliberations on the design of the N-I, Korolev produced a series of lengthy technical considerations to substantiate his position on the appropriateness of cryogenic propellants, but then in his conclusions specifically invoked concurrent American work.[96] As in the case of the N-I, each side could always find relevant information about American work to support its case, a task made easier by the inability of high government officials to discern actual sanctioned work going on in the United States from the speculations of American journalists.

Finally, there was the role of the Soviet military. When chief designers proposed ostensibly civilian space projects, such as a moon landing, they often articulated their ideas so as to suggest that these projects had both civilian and military uses. Barring rare exceptions—principally lunar and deep space missions—all Soviet space projects of the 1960s were military in nature or derived from military projects. To attract the military's attention, Korolev desperately tried to justify the N-I on the grounds that the military might need it. But the rocket's initial lifting capacity of 75 tons and its use of cryogenic components ensured that the military would find little or no use for it. In a meeting held in September 1960 to discuss the N-1, Major General Aleksandr Mrykin, a senior official in charge of procurement for the Strategic Rocket Forces, came right to the point: "Permit me to raise the following questions: for what purpose [do we need] heavy spaceships [weighing 75 tons] and what military application are they for?"[97] Even though several government decrees instructed the military to prepare proposals for what they could do with the N-I, the appropriate department within the Strategic Rocket Forces never produced a requirement, leaving Korolev to make up wildly ambitious ideas that bordered on fantasy, such as an idea to deploy an "orbital belt" of hundreds of military satellites that could continuously monitor the enemy and defend any space-based or ground-based asset belonging to the Soviet Union.[98] Even Korolev himself was self-aware enough to see the absurdity of some of his ideas for military space activities. In early 1961, in a letter to a defense industrialist, he conceded that "some of the proposals, on first glance, may seem dubious or even somewhat fantastic. But … one should not draw any hasty conclusions."[99]

Chief designers such as Iangel' or Chelomei or Glushko who tailored their work to be more in tune with prevailing military imperatives than Korolev did, were more likely to benefit from generous funding from the military services. In this context, developing a rocket to land a cosmonaut on the moon was seen by many in the military as a worthless sideshow to the real goal of achieving strategic parity. This was strikingly underscored by two consecutive Soviet ministers of defense, Marshal Rodion Malinovskii and Marshal Andrei Grechko. "We cannot afford to and will not build super powerful space launch vehicles and make flights to the moon," Malinovskii told Air Force officials in January of 1965.[100] His successor, Grechko, was equally firm, responding to a request for help by telling an official "I won't give you personnel. I won't give you money. Do what you like but I won't raise this with the government. … And in general, I am against flights to the moon."[101]

Because the military were hostile toward the "civilian" space program, Glushko was able to fortify his position by noting correctly that any storable-propellant engines he built could be used (or at least the technology would be useful) for military programs, particularly ICBM programs. Since the military were the primary clients for all space projects, even ostensibly civilian ones, by catering to military needs Glushko could have the military ensure a steady stream of funding for his organization. This security added to Glushko's rising stature; by the late 1960s, he enjoyed enormous authority as the man who produced the heart of the Soviet strategic missile force: its rocket engines. This connection to Soviet military power gave him significant leeway to continuously try to intervene in the ongoing N-I project. Who would challenge him?

Conclusions

The July 1962 decision by the Keldysh Commission effectively fractured the space program into the Korolev and Glushko camps, destroying any semblance of unity that may have existed during the Sputnik days. Although the break between Korolev and Glushko was ostensibly over technical issues, the repercussions were far-reaching: the two giants of the Soviet space program would not live to cooperate on another project. Korolev turned his back on the most powerful and successful rocket engine designer in the country and went to work with an organization that had almost no experience in the field, the Kuznetsov design bureau. Glushko, meanwhile, lost his role in what was to be the most expansive and greatest project in the history of the Soviet space program. In the end, these decisions, in favor of Kuznetsov's innovative, efficient, and "civilian" engines instead of Glushko's conservative, relatively inefficient "military" engines, doomed the remainder of the N-I project.

Kuznetsov, an outsider in the Soviet space program, found it very difficult to gain access to facilities for ground testing of his rocket engines, essential to certify his engines as flight-worthy. The majority of facilities at the premier Soviet site for testing rocket engines was devoted to Glushko's storable-propellant engines (built for ICBMs), and the resources to build ground infrastructure for Kuznetsov's engines were meager and late. His engines, though highly efficient, took far too long to develop, and their development was marred by the decision not to construct a full-scale ground-test stand for the rocket's entire first stage.[102] When four consecutive launches of the N-I ended in explosions in the late 1960s and the early 1970s, few were surprised.[103] For the Soviet space program, the collapse of the N-I project signaled the end of the beginning of a dramatic road that began with Sputnik, and it was the most visible manifestation of the program's fall from grace.

In untangling the main characteristic threads of this exemplar of late-period Soviet big science, it is worth revisiting Loren Graham's characterizations: "The system emphasized quantity over quality, seniority over creativity, military security over domestic welfare, and orthodoxy over freedom."[104] In the case of the N-I project, these rationales (quantity, seniority, security, orthodoxy) can be found in places, but they are neither the most important nor the most definitive attributes. What we see, in fact, are features (risk-taking, competition, discord within the scientific community, variable expertise) that are direct outcomes of the ways in which national goals set in the context of the Cold War, trickled down, and seeded science and engineering with "local" rationales, choices, and contours. In the case of the N-I, the result was a program that embodied multiplicities instead of singularities. Contradiction, messiness, ambivalence, and ambiguity were the *normative* modes of work in the case of N-I, not anomalies. Such seemingly discrepant strains are clearly also evident in other contemporaneous examples of Soviet big science in the postwar era, such as the anti-ballistic-missile project, the development of particle beams, and the Mars-exploration efforts, in each of which there was intractable conflict among the major players.[105]

From a purely technical perspective, perhaps the most important conflict was the tension between a conservative choice and a risky one (one whose outcome Graham saw as always being "quantity over quality"). Glushko's engines were less efficient, technically conservative, and could draw on established military contracts; Kuznetsov's motors, on the other hand, were highly efficient, technically innovative, and lacked institutional backing. When Korolev insisted on the latter for his giant space rocket, he was in, essence, trying to force an innovative and "civilian" solution into a milieu where conservatism and "military" options were privileged. This is not to suggest that innovation was the more difficult choice and was doomed to failure because of bureaucratic resistance; on the contrary, as the evidence shows, the N-I project made a space for both innovation and conservatism to exist in a tenuous balance. In each of these projects, powerful actors within the scientific and engineering communities exerted authority in favor of conservative or innovative solutions, sometimes in conflict with each other—solutions whose measure of success often depended on the degree of their professional clout. In the case of the rocket and space program, Korolev belonged to a small but powerful group of missile designers who had acquired unprecedented power and influence by the early 1960s, benefiting from the Cold War-driven successes of Sputnik and the space program. Their authority, predicated on access to the top levers of the Party and the government, combined with the institutionally "normal" Soviet approach to competition in the defense industry and the uneven technical expertise of managers, created a climate for chaotic infighting that existing institutional mechanisms were unable to arbitrate.

The experience of the N-I project shows that in the Soviet Union, competition and competitive contracts were designed not to invigorate innovation but instead to minimize risk or the chance of failure. Here, at one level, the competition was between different technological options: storable versus cryogenic, gas generator versus staged combustion, nitrogen tetroxide versus liquid oxygen, and so on. But at a deeper level, this was a competition between rival organizations. To the extent that organizations in the Soviet defense industry were identified with their chief designers, this was also a competition between *individuals*. Each of the designers competing for a contract would emphasize how his project was guaranteed to succeed and others guaranteed to fail; we see this dynamic in Glushko's continuing attacks on Kuznetsov's engines, for example. The bogeyman of America played a not insignificant role. Designers such as Korolev and Glushko could repeatedly invoke threats of American superiority or the blessing of Party leaders to defend their positions, and bureaucrats were too afraid to refuse their demands for fear of increasing risk—or, worse, offending the patrons of powerful designers. It was precisely this tendency—the growing power of chief designers—that the Military-Industrial Commission tried to counter in 1966 by signing into law a decree stipulating that every new proposal on a weapons system should be preceded by a detailed technical substantiation of the idea in the form of an "advance plan" (*avant-proekt*) that would be circulated *before* any direct conversation with top leaders. An official history of the Soviet military-industrial complex dryly notes that this decision "played a large role … in eliminating excessive expenses in creating new long-term technologies."[106]

The other built-in tensions, those between civilian and military imperatives and between publicity and secrecy, were also in evidence at the beginning of the N-I program. For example, the seemingly arcane and technical debate over propellant selection for the N-I rocket was, at heart, an outcome of different demands: should Soviet rockets use propellants appropriate for "military" use, or propellants appropriate for "civilian" use? The former would be wrapped up in the secrecy of the Strategic Rocket Forces. The latter would be elevated to display as a triumph of Soviet socialism for all to see—the first landing of humans on the moon. As leading architects of Soviet big science at the height of the Cold War, Korolev and Glushko embodied these conflicting rationales, but in slightly different and ultimately crucial ways. Korolev had firmly embraced the imperative for an expansive Soviet space program but was also acutely aware that he needed to cater to the military to realize his cosmic aspirations. These opposing impulses were in conflict. One the one hand, he wrote to defense industrialists about the military operations (such as "super-reconnaissance") that would be possible with the N-1, and invited the military to stipulate technical specifications (particularly, the launch mass) so that Korolev's designers could begin work on the rocket.[107] Almost simultaneously, he instructed his own deputies to determine the launch mass of the N-I so that it could perform a number of "civilian" tasks, such as circling and landing on the moon.[108]

For Korolev, then, the goal was to create a rocket that could, first and foremost, perform civilian missions such as landing on the moon. He would draw from this technology to cater to military needs. For Glushko, the goal was to create engines for ICBMs. He would draw on this technology to create a civilian rocket, the N-I, that could perform space missions. The former sought, with his innovative use of liquid oxygen, to create a military big science out of a civilian one. The latter sought, with his conservative storable propellants, to create a civilian big science from a military one. Fundamentally, both were trying to eliminate the inherent ambiguities and contradictions of Soviet big science by creating what they thought were more efficient versions. Unsurprisingly both failed in this quest.

In recovering the early history of the N-I project, then, one sees Soviet big science largely operating in an environment driven by conflicts between state intervention and competition, between military requirements and civilian goals, and between secrecy and display value. This was not the big science that Capshew and Rader described as possessing a "high degree of organization and coordination," nor was it Graham's model of quantity, seniority, security, and orthodoxy.[109] And neither does it echo accounts of the atomic bomb project—with its almost limitless state resources, involvement of security services, lack of competition (at least until the late 1950s), and insulation of leading scientists from broader ideological pressures—which for many has served as a surrogate for reflexive generalizations about Soviet big science when in fact the nuclear project was the exception and not the rule. What we find in the case of the N-I is a big science that embodied a clash of forces, one determined by imperatives defined at the global level of the Cold War (such as military, secrecy, and publicity) and the other pushed by a host of contradictory forces defined by local processes (such as professional, technical, historical) within various communities. The clash of the global and local in all its myriad forms created the archetypical Soviet big science: big, yes, but very different from the nuclear project, and full of contradictions, ambiguities, and contingencies.

Epilogue

When the N-I program was suspended, in 1974, Glushko was appointed to head the organization that Korolev—now dead—had headed. In a move that shocked many, Glushko immediately proposed development of a series of huge "super rockets," all using liquid oxygen–kerosene engines, of the very same type he had so vehemently railed against a decade earlier. One of these rockets, the Energiia, was successfully launched twice in the late 1980s, but the program was eventually canceled for lack of money after the Soviet Union collapsed. In the 1990s, the engines that had powered Energiia were scaled down and sold to General Dynamics (later acquired by Lockheed Martin), which now uses them on the American Atlas III and Atlas V launch vehicles. Meanwhile, the storable-propellant engines that Glushko originally offered to the N-I are now regularly used on the Proton rocket operated by International Launch Services, a joint US-Russian company. Because the Proton and the Atlas V are competitors, Glushko's storable-propellant and liquid oxygen rocket engines continue to compete with each other in the global launch market.

Equally striking was the "second act" for the highly innovative liquid oxygen engines that Kuznetsov designed and built for the N-I. Kuznetsov's engineers persevered and eventually flight-certified the engine despite the cancellation of the N-I project. For nearly twenty years, managers preserved 150 of the engines in a storehouse, three dozen of which were bought by the American company Aerojet in the 1990s. In early 2013, the Orbital Sciences Corporation used two of those engines—brought out of storage after nearly 40 years—on its Antares rocket, which launched a number of satellites into Earth orbit. A year later, an Antares rocket delivered supplies to the International Space Station, where American and Russian astronauts are stationed on long tours. All these Russian engines, widely considered high-performance systems, represent the peculiar but continuing embodiment of the arguments that shaped the discussions in 1962 between Korolev and Glushko. In that sense, it may be still too early to say whose argument won out.

Notes

1. S. Leskov, "Kak my ne sletali na lunu," *Izvestiia*, August 19, 1989.

2. S. Leskov, *Kak my ne sletali na lunu* (Panorama, 1991), 4. Much later, Leskov claimed that he published the piece without the permission of the censors. See "Obozrevatel' 'Izvestii' Sergei Leskov nagrazhden 'Znakom Gagarina'," *Izvestiia*, February 22, 2006.

3. The rocket has been various called "N-1," "N1," and "N-I." For the purposes of this chapter, I use the latter which is the most common designation in government documents.

4. James E. Oberg, *Red Star in Orbit* (Random House, 1981), 113.

5. For early published accounts of the lunar program, see A. Tarasov, "Polety vo sne i nayiu," *Pravda*, October 20, 1989; M. Rebrov, "A delo bylo tak: trudnaia sud'ba proekta N-1," *Krasnaia zvezda*, January 13, 1990; V. P. Mishin, "Pochemu my ne sletali na lunu?," *Znanie: seriia kosmonavtika, astronomiia* no. 12 (1990): 3–43; I. B. Afanas'ev, "Neizvestnye korabli," *Znanie: seriia kosmonavtika, astronomiia* no. 12 (1991): 1–64; R. Dolgopiatov, B. Dorofeev, and S. Kriukov, "Proekt N-1," *Aviatsiia i kosmonavtika* no. 9 (1992): 34–37; I. Afanas'ev, "N-1: sovershenno sekretno," *Kryl'ia rodiny* no. 9 (1993): 13–16; no. 10 (1993): 1–4; no. 11 (1993): 4–5.

6. For useful literature, see John H. Capshew and Karen A. Rader, "Big Science: Price to the Present," in Arnold Thackray, ed., "Science After '40,' " *Osiris* 7 (1992): 3–25; Peter Galison and Bruce Hevly, eds., *Big Science: The Growth of Large Scale Research* (Stanford University Press, 1992); Gregory McLauchlan and Gregory Hooks, "Last of the Dinosaurs? Big Weapons, Big Science, and the American State from Hiroshima to the End of the Cold War," *Sociological Quarterly* 36, no. 4 (1995): 749–776; David Reynolds, "Science, Technology, and the Cold War," in *The Cambridge History of the Cold War*, volume 3: *Endings*, ed. Melvyn P. Leffler and Odd Arne Westad (Cambridge University Press, 2010).

7. Paul Forman, "Behind Quantum Electronics: National Security as Basis for Physical Research in the United States, 1940–1960," *Historical Studies in the Physical and Biological Sciences* 18 (1987): 149–229.

8. Loren R. Graham, "Big Science in the Last Years of the Big Soviet Union," in "Science after '40," *Osiris* 7 (1992): 49–71.

9. Graham, "Big Science in the Last Years of the Big Soviet Union."

10. Alexei Kojevnikov, "The Great War, the Russian Civil War, and the Invention of Big Science," *Science in Context* 15, no. 2 (2002): 239–275.

11. Paul R. Josephson, "'Projects of the Century' in Soviet History: Large-Scale Technologies from Lenin to Gorbachev," *Technology and Culture* 36, no. 3 (1995): 519–559.

12. Loren R. Graham, ed., *Science and the Soviet Social Order* (Harvard University Press, 1990); Graham, *Science in Russia and the Soviet Union: A Short History* (Cambridge University Press, 1993); Nikolai Krementsov, *Stalinist Science* (Princeton University Press, 1997).

13. David Holloway, *Stalin and the Bomb: The Soviet Union and Atomic Energy, 1939–1956* (Yale University Press, 1994).

14. For post-Cold War works on the Soviet military-industrial complex, see N. S. Simonov, *Voenno-promyshlennyi kompleks sssr v 1920–1950-e gody: tempy ekonomicheskogo rosta, struktura, organizatsiia proizvodstva i upravlenie* (ROSSPEN, 1996); I. V. Bystrova, *Voenno-promyshlennyi kompleks sssr v gody kholodnoi voiny (vtoraia polovina 40-kh—nachalo 60-kh godov)* (Institut rossiiskoi istorii RAN, 2000). For a participant account in English, see Sergei Khrushchev, "The Military-Industrial Complex, 1953–1964," in *Nikita Khrushchev*, ed. William Taubman, Sergei Khrushchev, and Abbott Gleason (Yale University Press, 2000).

15. Graham, "Big Science in the Last Years of the Big Soviet Union," 51.

16. The Russian word *nauka* has historically implied a meaning closer to that of the German word *Wissenschaft* (meaning "scholarship") than to that of the English word, "science," with which it is most literally associated. Thus, *nauka* was used in popular media rather generally (and often carelessly) to encompass practices that Westerners might often associate with engineering. One of the most popular science journals during the Soviet era, *Nauka i zhizn'* (*Science and Life*) featured many stories about technology and engineering.

17. Asif A. Siddiqi, *The Red Rockets' Glare: Spaceflight and the Soviet Imagination, 1857–1957* (Cambridge University Press, 2010).

18. Holloway, *Stalin and the Bomb*, 366.

19. Ibid., 360.

20. See Paul Josephson, "Rockets, Reactors and Soviet Culture," in *Science and the Soviet Social Order*, ed. Loren Graham (Harvard University Press, 1990).

21. Iu. P. Semenov, ed., *Raketno-kosmicheskaia korporatsiia "Energiia" imeni S. P. Koroleva* [RKK Energiia named after S. P. Korolev], 1996), 51–54.

22. Memoir of A. I. Ostashev in *Nachalo kosmicheskoi ery: vospominaniia veteranov raketno-kosmicheskoi tekhniki i kosmonavtiki: vyp. vtoroi*, ed. Iu. A. Mozzhorin (RNITsKD, 1994), 69.

23. Besides the above named, the entourage also included L. M. Kaganovich, N. K. Kirichenko, and M. G. Pervukhin. The visit took place on February 27, 1956. Sergei Khrushchev has a long description of this visit based on his personal recollections. See Sergey N. Khrushchev, *Nikita Khrushchev and the Creation of a Superpower* (Pennsylvania State University Press, 2000), 101–112.

24. N. S. Khrushchev, *Vospominaniia, Kn. chetvertaia: Vremia. Liudi. Vlast'* (Moskovskie novosti, 1999), 191.

25. Khrushchev, *Nikita Khrushchev and the Creation of a Superpower*, 106.

26. See the discussion in Konstantin Ivanov, "Science after Stalin: Forging a New Image of Soviet Science," *Science in Context* 15, no. 2 (2002): 317–338 (see especially pp. 330–331). See also Alexander Vucinich, *Empire of Knowledge: The Academy of Sciences of the USSR (1917–1970)* (University of California Press, 1984).

27. See Sheila Fitzpatrick, *Education and Social Mobility in the Soviet Union, 1921–1934* (Cambridge University Press, 1979); Michael David-Fox, *Revolution of the Higher Mind: Higher Learning Among the Bolsheviks* (Cornell University Press, 1997); Kendall E. Bailes, *Technology and Society Under Lenin and Stalin: The Origins of the Soviet Technical Intelligentsia, 1917–1941* (Princeton University Press, 1978). These designers included those born in 1907 (S. P. Korolev), 1908 (M. M. Bondariuk, V. P. Glushko, A. M. Isaev, A. M. Liul'ka, N. A. Piliugin, A. A. Raspletin, D. D. Sevruk), 1909 (S. M. Alekseev, V. P. Barmin, M. S. Riazanskii), 1911 (N. D. Kuznetsov, M. K. Iangel'), and 1912 (B. M. Konoplev, B. P. Zhukov).

28. For a useful analysis of some of the elite designers as a demographic within the Soviet military-industrial complex, see Julian Cooper, "The Elite of the Defence Industry Complex" in *Elites and Political Power in the USSR*, ed. David Lane (Elgar, 1988).

29. G. S. Vetrov, ed., *S. P. Korolev i ego delo: svet i teni v istorii kosmonavtiki* (Nauka, 1998), 668. The six were S. P. Korolev (overall missile), V. P. Glushko (rocket engines), N.A. Pilyugin (inertial guidance), M. S. Riazanskii (radio guidance), V. I. Kuznetsov (gyroscopes), and V. P. Barmin (launch complex).

30. They included active members (V. P. Glushko, S. P. Korolev, and G. I. Petrov) and corresponding members (V. P. Barmin, V. N. Chelomei, P. D. Grushin, G. V. Kisun'ko, V. I. Kuznetsov, S. A. Lavochkin, V. P. Mishin, N. A. Piliugin, A. A. Raspletin, and M. S. Riazanskii).

31. By 1970 there were at least fourteen new corresponding members (G. N. Babakin, A. F. Bogomolov, B. V. Bunkin, K. D. Bushuev, B. E. Chertok, O. G. Gazenko, D. E. Okhotsimskii, S. S. Lavrov, N. S. Lidorenko, V. P. Makeev, B. V. Raushenbakh, V. S. Semenikhin, V. S. Shpak, and B. P. Zhukov) and thirteen new academicians (V. P. Barmin, V. N. Chelomei, P. D. Grushin, M. K. Iangel', A. Iu. Ishlinskii, V. I. Kuznetsov, V. P. Mishin, V. V. Parin, B. N. Petrov, N. A. Piliugin, A. A. Raspletin, R. Z. Sagdeev, and S. N. Vernov) whose primary work was in the missile and space sector. In addition, there were at least eight aviation designers (N. D. Kuznetsov, G. P. Svishchev, S. V. Il'iushin, A. M. Liul'ka, A. A. Makarevskii, A. I. Mikoian, V. V. Struminskii, and S. K. Tumanskii) who were elected into the Academy who did contract work for the missile and space programs. Lists of new Academy members from 1958 to 1970 show that the missile and space designers dominated the new entrants of designers from the defense industry.

32. These advances into the Academy occurred at the very moment when the institution was seeking to divest itself of applied scientific work. With the support of Nikita Khrushchev, the Academy adopted an official policy in 1961 of retaining the focus of the Academy on "fundamental science" while ejecting more than fifty applied research institutions to industry. See Ivanov, "Science after Stalin"; Nicholas DeWitt, "Reorganization of Science and Research in the USSR," *Science* 133, no. 3469 (1961): 1981–1991; Alexander G. Korol, *Soviet Research and Development: Its Organization, Personnel, and Funds* (MIT Press, 1965).

33. For more on Keldysh, see A. V. Zabrodin, ed., *M. V. Keldysh: tvorcheskii portret po vospominaniiam sovremennikov* (Nauka, 2001).

34. For Keldysh's role in nuclear weapons and missile development in the 1950s, see Iu. A. Trutnev, "M. V. Keldysh i ego kollektiv v reshenii atomnoi problemy" and V. A. Avduevskii and T. M. Eneev, "O rabotakh M. V. Keldysha po raketostroeniiu i kosmonavtike," in *M. V. Keldysh*, 66–78.

35. Asif A. Siddiqi, "Soviet Space Power During the Cold War," in *Harnessing the Heavens: National Defense Through Space*, ed. Paul G. Gillespie and Grant T. Weller (US Air Force Academy, 2008).

36. For English-language biographies, see Michael J. Neufeld, *Von Braun: Dreamer of Space, Engineer of War* (Knopf, 2007); James Harford, *Korolev: How One Man Masterminded the Soviet Drive to Beat America to the Moon* (Wiley, 1997).

37. William P. Barry, The Missile Design Bureaux and Soviet Piloted Space Policy, 1953–1970, DPhil dissertation, Merton College, University of Oxford, 1995.

38. See Asif A. Siddiqi, *Challenge to Apollo: The Soviet Union and the Space Race, 1945–1974* (NASA, 2000).

39. Quoted from the draft decree on the creation of the Military-Industrial Commission (December 4, 1957), Russian State Archive of the Economy (RGAE), f. 4372, op. 76, d. 320, ll. 33–38 (see especially l. 36).

40. Among the veterans of the missile industry who ended up near the top of the VPK structure were D. F. Ustinov (VPK chairman from 1957 to 1963), L. V. Smirnov (chairman from 1963 to 1985), G. N. Pashkov, S. I. Vetoshkin, A. N. Shchukin, and A. A. Kosmodem'ianskii. For more on the Military-Industrial Commission, see I. V. Bystrova, "K 50-letiiu voenno-promyshlennoi komissii," *Voenno-promyshlennyi kur'er* no. 47 (December 5–11, 2007).

41. David Holloway, *The Soviet Union and the Arms Race*, second edition (Yale University Press, 1984), 142.

42. See, for example, Mark Harrison, "A Soviet Quasi-Market for Inventions: Jet Propulsion, 1932–1946," *Research in Economic History* 23 (2005): 1–59; Andrei Markevich and Mark Harrison, "Quality, Experience, and Monopoly: The Soviet Market for Weapons under Stalin," *Economic History Review* 59, no. 1 (2006): 113–142; Mark Harrison, ed., *Guns and Rubles: The Defense Industry in the Stalinist State* (Yale University Press, 2008), chapters 3, 6, and 8.

43. These proposals were usually in the form of an *avant proekt* (advance plan). For an older but still quite useful summary of the Soviet weapons R&D process, see Arthur J. Alexander, "Decision-Making in Soviet Weapons Procurement," Adelphi Papers 147–148 (1978–1979), 1–64. For a more recent one, see Barry, "The Missile Design Bureaux and Soviet Piloted Space Policy, 1953–1970," 66–80.

44. Vetrov, *S. P. Korolev i ego delo*, 664.

45. Boris Chertok, *Rockets and People*, volume II: *Creating a Rocket Industry*, ed., Asif A. Siddiqi (NASA, 2006), 545–554; Vetrov, *S. P. Korolev i ego delo*, 288–289.

46. The decree, issued on June 23, 1960, was titled "On the Creation of Powerful Carrier-Rockets, Satellites, Spaceships, and the Conquest of Space in 1960–67." It has been published as "O sozdanii moshchnykh raket-nositelei, sputnikov, kosmicheskikh korablei i osvoenii kosmicheskogo prostrantsva v 1960–1967 godakh" (June 23, 1960) in *Sovetskaia kosmicheskaia initsiativa v gosudarstvennykh dokumentakh, 1946–1964 gg.*, ed. Iu. M. Baturin (RTSoft, 2008), 96–100.

47. Other early pioneers, such as the American Robert Goddard, and the Romanian-German Hermann Oberth, also came to the same conclusions.

48. There are many accounts of the R-16 disaster in print. See Chertok, *Rockets and People*, volume II, 597–634; Asif A. Siddiqi, "Mourning Star: The Nedelin Disaster," *Quest* 3, no. 4 (1994): 38–47.

49. The R-7 ICBM was officially declared operational on January 20, 1960.

50. Asif A. Siddiqi, "The Rockets' Red Glare: Technology, Conflict, and Terror in the Soviet Union," *Technology and Culture* 44, no. 3 (2003): 470–501.

51. Siddiqi, *The Red Rockets' Glare*, chapters 5–7.

52. This was the RD-109 engine. See I. Afanas'ev, "Neizvestnyi dvigatel' zabytoi rakety," *Novosti kosmonavtiki* no. 1 (2006): 66–67.

53. In the spring of 1958, Korolev pressured five other chief designers (including Glushko) to sign off on a report in favor of liquid oxygen for his new R-9. See S. P. Korolev et al., "O perspektivakh razvitiia kislorodnykh raket" (April 18, 1958), in Vetrov, *S. P. Korolev i ego delo*, 249–252.

54. On March 2, 1959, Minister of Defense R. Ia. Malinovskii and his deputy M. I. Nedelin wrote a letter to Korolev and chairman of the Military-Industrial Commission D. F. Ustinov agreeing to support the R-9 project as articulated by Korolev. See Vetrov, *S. P. Korolev i ego delo*, 286, 672.

55. This was the RD-110 engine. I. Afanas'ev, "'Kopii' dvigatelei dlia 'semerki,'" *Novosti kosmonavtiki* no. 7 (2005): 67–69.

56. For more on Kuibyshev and Kuznetsov, see Robert MacGregor, "The Little Engine That Could," paper presented at History of Science Workshop, Princeton University, 2010.

57. Korolev also made contact with A. M. Liul'ka's OKB-165 and S. A. Kosberg's OKB-154 to produce rocket engines for his missiles and spacecraft. In 1959, the Soviet party and government issued a decree calling for aviation design bureaus to produce rocket engines. See "O privlecheniia aviatsionnykh motorostroitel'nykh OKB k razrabotke raketnykh dvigatelei" (June 16, 1959) in *Zadacha osoboi gosudarstvennoi vazhnosti: iz istorii sozdaniia raketno-iadernogo oruzhiia i Raketnykh voisk strategicheskogo naznacheniia (1945–1959 gg.): sbornik dokumentov*, ed. V. I. Ivkin and G. A. Sukhina (Rosspen, 2010).

58. The Soviet government approved development of Kuznetsov's first rocket engine, the NK-9 (35 tons thrust), on June 26, 1959. See S. N. Tresviatskii et al., "Kosmicheskie dvigateli SNTK imeni N. D. Kuznetsova," *Aerokosmicheskii obzor* no. 3 (2006): 108–109.

59. S. P. Korolev to L. I. Brezhnev, November 25, 1959, in Vetrov, *S. P. Korolev i ego delo*, 284–287. Korolev also wrote a letter to the Central Committee of the Communist Party on December 8, 1959. The R-9 variant with Kuznetsov's engines was known as the R-9M.

60. For Glushko's report on a comparison between his and Kuznetsov's engines, see V. P. Glushko, "Vyvody k dokladu na komissii 14.12.1959 g" (December 14, 1959), in *Izbrannye raboty akademika V. P. Glushko: chast' 1*, ed. V. S. Sudakov et al. (NPO Energomash, 2008). For Glushko's letter to the government and other designers rejecting Korolev's appeal, see V. P. Glushko to D. F. Ustinov et al., December 25, 1959, in *Izbrannye raboty akademika V. P. Glushko*, 143–150. On January 18, 1960, the "minister" in charge of the defense industry, K. N. Rudnev, informed Korolev that R-9 rocket project would proceed as originally conceived, with Glushko's engines. See Vetrov, *S. P. Korolev i ego delo*, 676.

61. V. P. Glushko to S. P. Korolev (January 3, 1958), in *Izbrannye raboty akademika V. P. Glushko*, 133–135. The first engine he produced using this combination (liquid oxygen–UDMH) was the RD-109 upper stage. Subsequently, he proposed using the RD-112 and the RD-113 in his R-20 "super rocket" in early 1960, and the RD-114 and the RD-115 in an early variant of the N-I rocket in late 1960.

62. These words are from a series of lectures Glushko gave at the N. E. Zhukovskii Air Force Academy in 1933 and 1934, which were later published in a monograph in 1936. See V. P. Glushko, "Zhidkoe toplivo dlia reaktivnykh dvigatelei" (1936) in V. P. Glushko, *Put' v raketnoi tekhnike: izbrannye trudy, 1924–1946* (Mashinostroenie, 1977), 231–330. Glushko had other technical reservations about the use of liquid oxygen, which included: the challenge of cooling as a result of the unusually high temperatures during combustion that threatened to melt the metal casing; and the challenge of creating stable combustion within the combustion chamber since liquid oxygen engines at high pressure are prone to extremely dangerous high-frequency oscillations that destroy engines. He had identified the cooling problem as a technical challenge already in 1932 when he was only 24 years old. See his "Otchet po opytam s reaktivnymi motorami, provedennymi po 1 sentiabria 1932 goda" (September 1, 1932) in Glushko, *Put' v raketnoi tekhnike*, 143–157.

63. These included engines for the following missiles: the R-14 (the RD-216 engine) and the R-16 (the RD-218 and RD-219 engines). For details, see Asif Siddiqi, "Rocket Engines from the Glushko Design Bureau: 1946–2000," *Journal of the British Interplanetary Society* 54 (2001): 311–334.

64. V. P. Glushko to N. P. Antonov, December 9, 1960, in *Izbrannye raboty akademika V. P. Glushko*, 185–189.

65. At first, at an important meeting in September of 1960, Glushko had insisted on the nitric acid–UDMH combination for the N-I. See "Vypiska iz protokola soveshchaniia glavnykh konstrukturov po nositeliiu N-I" (September 23, 1960), in Vetrov, *S. P. Korolev i ego delo*, 305–308. In 1960, Glushko began developing several engines with this combination, including the RD-224 engine (for the R-26 ICBM), and the RD-220, RD-221, RD-222, and RD-223 engines (for early

conceptions of the N-I). But later, at a meeting in January of 1961, he replaced nitric acid with nitrogen tetroxide, and insisted on the nitrogen tetroxide–UDMH combination for the N-I. See "Vypiska iz protokola rasshirennogo soveshchaniia glavnykh konstruktorov" (January 31, 1961) on pp. 319–323 in the same source. These new engines for the N-I were the RD-253 and RD-254.

66. V. P. Glushko to S. P. Korolev, March 18, 1961, in *Izbrannye raboty akademika V. P. Glushko*, 195–199.

67. For details on these contracts, see Siddiqi, *Challenge to Apollo*, 314–318.

68. V. P. Glushko to S. P. Korolev, November 10, 1961, in *Izbrannye raboty akademika V. P. Glushko*, 204–211. See also Glushko to D. F. Ustinov, November 14, 1961; Glushko to M. V. Keldysh, November 24, 1961, in the same volume (pp. 211–212).

69. Korolev notes that there were "seven Academicians, nine Corresponding Members [of the Academy of Sciences], representatives of the Ministry of Defense, many doctors and candidates of science, including the best specialists in engine design, propellants, combustion processes, etc." See S. P. Korolev, "Otsenka plana rabot OKB V. P. Glushko" (September 30, 1963) in Vetrov, *S. P. Korolev i ego delo*, 426–431 (see especially 429).

70. The commission examined four different pairs of propellants: liquid oxygen–kerosene, liquid oxygen–UDMH, nitrogen tetroxide–UDMH, and nitric acid–UDMH.

71. V. P. Glushko to S. P. Korolev, November 10, 1961, in *Izbrannye raboty akademika V. P. Glushko*, 204–211. Glushko sent similar letters to D. F. Ustinov (on November 14), M. V. Keldysh (on November 24), and I. D. Serbin (on November 29).

72. V. P. Glushko to B. A. Komissarov, February 19, 1962, in *Izbrannye raboty akademika V. P. Glushko*, 216–218.

73. These, according to Glushko, would include additional electrical power, the cost of refrigeration equipment, costs of their maintenance and depreciation, maintaining extra staff, storage tanks with special insulation at the launch site, etc.

74. Glushko to Korolev, November 10, 1961, 207. Later in the letter, he notes that "the cost of oxygen-kerosene propellants is not much less expensive than nitric tetroxide with [UDMH], considering the costs of manufacturing and operating storage tanks for super-cooled oxygen and units for super-cooling."

75. "Doklad o moshchnoi rakete-nositele N-I na zasedanii ekspertnoi komissii" (July 2–16, 1962) in Vetrov, *S. P. Korolev i ego delo*, 363–382.

76. "Doklad o moshchnoi ... ," 363–382.

77. Glushko to Korolev, November 10, 1961, 207–208.

78. Korolev's OKB-1 produced the S1.5400 engine for an upper-stage application in 1958–61, while Kuznetsov's OKB-276 started work on the NK-9, originally meant for an abandoned version of the R-9 known as the R-9M, in 1958. Korolev's arguments on staged-combustion engines are from "Doklad o moshchnoi ... ," 368–370; S. S. Kriukov, "N-1: Istoriia proektirovaniia, stroitel'stva, ispytanii," in *S. S. Kriukov: izbrannye raboty: iz lichnogo arkhiva*, ed. A. M. Pesliak

(Izd.-vo MGTU im. N. E. Baumana, 2010), 49–138 (see especially pp. 57–60). For more on Kuznetsov's NK-15 engine and staged combusion in general, see MacGregor, "The Little Engine That Could."

79. Glushko to Korolev, November 10, 1961, 209.

80. "Doklad o moshchnoi ... ," 369–370.

81. Kriukov, "N-1," 79–80.

82. The actual text of the decision notes the following advantages of the liquid oxygen–kerosene combination over storable propellants: (1) higher specific impulse; (2) lighter rocket; (3) higher payload to orbit; (4) safer; and (5) cheaper. See the excerpt from the commission's final decision quoted in Korolev, "Otsenka plana rabot OKB V. P. Glushko," 428–429.

83. G. Vetrov, "Trudnaia sud'ba rakety N-1," *Nauka i zhizn'* no. 5 (1994): 20–27.

84. Nikita Khrushchev, *Khrushchev Remembers: The Glasnost Tapes* (Little, Brown, 1990), 186.

85. Glushko describes his various attempts to propose alternatives to the N-I in a report in early 1962. See V. P. Glushko to D. F. Ustinov and L. V. Smirnov, March 12, 1962, in *Izbrannye raboty akademika V. P. Glushko*, 221–229.

86. This was Iangel's R-46 "super-rocket," an early competitive project to Korolev's N-I, which was never formally approved by the Soviet party and government. See V. P. Glushko to M. K. Iangel' (April 3, 1961) in *Izbrannye raboty akademika V. P. Glushko*, 199–202.

87. Glushko to Ustinov and Smirnov, March 12, 1962, in *Izbrannye raboty akademika V. P. Glushko*, 229. For an excellent discussion of the R-56 rocket, see Bart Hendrickx, "Heavy Launch Vehicles of the Yangel Design Bureau," *Space Chronicle JBIS* 63, Suppl. 2 (2010): 50–62 and 64, Suppl. 1 (2011).

88. "Protokol'naia zapis' vystupleniia na soveshchanii glavnykh konstruktorov o khode rabot po tiazhelomu nositeliiu N-I" (June 23, 1964), in Vetrov, *S. P. Korolev i ego delo*, 459.

89. On July 18, 1965, the leading research institute of the missile and space sector, NII-88, issued a report rejecting the proposal to replace Kuznetsov's engines on the N-I with those of Glushko. See Vetrov, *S. P. Korolev i ego delo*, 696.

90. For the initial proposal for this rocket, known as the UR-700, see V. N. Chelomei, V. P. Glushko, V. P. Barmin, and V. I. Kuznetsov, "Predlozhenie po sozdaniiu raketno-kosmicheskoi sistemi UR-700" (October 16, 1965) in *Izbrannye raboty akademika V. P. Glushko*, 288–293. See also "O provedenii rabot po raketno-kosmicheskoi sisteme UR-700-LK-700" (June 30, 1967), Russian State Archive of the Economy (RGAE), f. 4372, op. 81, d. 2519, ll. 125–129.

91. Siddiqi, *Challenge to Apollo*, 480–481, 538–546.

92. I. Afanas'ev, "N-1: sovershenno sekretno" [Part 2], *Kryl'ia rodiny* no. 11 (1993): 4–5.

93. V. P. Glushko to S. P. Korolev, November 10, 1961, in *Izbrannye raboty akademika V. P. Glushko*, 211.

94. These were Chelomei's UR-100N and Iangel's MR UR-100 missiles. For discussions of the "little civil war," see Boris Chertok, *Rockets and People*, volume III: *Hot Days of the Cold War*, ed. Asif A. Siddiqi (NASA, 2009), 148–154; N. A. Anfimov, ed., *Tak eto bylo ... : memuary Yu. A. Mozzhorin: Mozzhorin v vospominaniiakh sovremennikov* (ZAO Mezhdunarodnaia programma obrazovaniia, 2000), 144–188.

95. Vetrov, *S. P. Korolev i ego delo*, 321.

96. "Doklad o moshchnoi ... ," 369–370.

97. Vetrov, *S. P. Korolev i ego delo*, 307.

98. S. P. Korolev, "Dokladnaia zapiska o razvitii upravliaemykh chelovekom korablei-sputnikov i podgotovke neobkhodimykh kadrov spetsialistov dlia kosmicheskikh poletov" (April 20, 1962) in Vetrov, *S. P. Korolev i ego delo*, 360–363.

99. S. P. Korolev to K. N. Rudnev (January 15, 1961) in Vetrov, *S. P. Korolev i ego delo*, 316–319.

100. Siddiqi, *Challenge to Apollo*, 481.

101. N. P. Kamanin, *Skrytyi kosmos: kniga tret'ia, 1967–1968gg.* (OOO IID Novosti kosmonavtiki, 1999), 35.

102. For details on the static testing of the individual engines, see A. A. Makarov, ed., *Nazemnye ispytaniia raketno-kosmicheskoi tekhniki: opyt otrabotki raketnoi i raketno-kosmicheskoi tekhnik, 1949–1999 gg.* (Roskosmos/FGUP NII Khimmash, 2001).

103. For details, see Siddiqi, *Challenge to Apollo*, 679–684, 688–693, 701, 729–730, 754–756, 780, 818–824.

104. Graham, "Big Science in the Last Years of the Big Soviet Union," 51.

105. For the anti-ballistic missile program, see Mikhail Pervov, *"Annushki"—chasovye Moskvy: istoricheskii ocherk* (Stolichnaia entsiklopediia, 2010). For the particle beams program, see Peter J. Westwick, "'Space-Strike Weapons' and the Soviet Response to SDI," *Diplomatic History* 32, no. 5 (2008): 955–979. On Mars, see V. G. Perminov, *The Difficult Road to Mars: A Brief History of Mars Exploration in the Soviet Union* (NASA, 1999).

106. N. S. Stroyev, "Voennaia aviatsiia," in *Sovetskaia voennaia moshch' ot Stalina do Gorbacheva*, ed., A. V. Minaev (Voennyi parad, 1999), 280. This move to limit the chaos in the defense industry may have been part of a larger process of bureaucratic and industrial rationalization in the post-Khrushchev era connected with A. N. Kosygin's failed reforms to introduce quasi-market features into the command economy.

107. See the letter from Korolev to Rudnev (January 15, 1961) in Vetrov, *S. P. Korolev i ego delo*, 316–319.

108. S. P. Korolev to S. S. Kryukov, February 5, 1962, in Vetrov, *S. P. Korolev i ego delo*, 355–357.

109. Capshew and Rader, "Big Science," 10.

7 Embedding the National in the Global: US-French Relationships in Space Science and Rocketry in the 1960s

John Krige

One evening in March of 1959 a team of young French space scientists led by Jacques Blamont stared anxiously into the sky over the Sahara desert as their Véronique sounding rocket (a small launcher used to study the properties of the upper atmosphere) soared skyward from its ramp in Hammaguir. To their relief, after a few minutes a trail of bright yellow sodium vapor spewed from a small capsule carried in the nose cone. The trail gradually dispersed, blown by the prevailing winds. By visually tracking the dispersal of the sodium cloud, the French team gained new insights into the dynamic properties of the upper atmosphere.[1]

This seemingly minor scientific experiment was a major national event. The French press attended the launch and enthusiastically reported the spectacle. The prestigious French daily *Le Monde* devoted eight columns to the campaign. An enthusiastic journalist's report appeared on the first page of the newspaper *Combat* on March 11, 1959 under the title "I saw France's first artificial comet launched at 19h. 38."[2] Hundreds of baby girls born in France were immediately named Véronique. France had entered the space age.

But not without help. The sodium vapor capsules were cheap, robust, simple, proven devices based on an American design. Scientists at the Air Force Cambridge Research Laboratory near Boston had used them for similar experiments with an Aerobee sounding rocket. Blamont had collaborated with them in 1957 while at the University of Wisconsin, where he had built his own sodium vapor capsule. The launches were under the control of three Germans: Wolfgang Pilz, Nettersheim, and Karl Bringer. (Blamont has forgotten Nettersheim's first name.) All three had worked in the Nazi regime's missile program. They had relocated to France after the war, where they played a major role in French rocket/missile development.[3] The launching ramp was at a military base in Algeria, then engaged in a bitter war of independence against the metropolitan power. In short, the "French" sodium vapor experiment was launched from a colony that would soon pass out of French control, using a rocket developed by engineers from an erstwhile and hated enemy, while the cloud was ejected from a device that had been designed and developed in the United States. The sodium vapor that drifted across the Saharan sky at sunset wasn't a national product. It was a global hybrid.[4]

The French press was disturbed. Why, they asked Blamont, was German being spoken on the launching ramp? They had a point: the sodium vapor cloud drew on the financial, political, and military power of the French state. However, what the journalists couldn't stomach or say was that the expression of that power was possible only because the experiment was also embedded in a network of interconnections that overflowed territorial boundaries. Newspaper reports ignored global linkages when they took the steam out of nationalist narratives; any dilution of sovereignty had to be avoided if the sodium trail was to represent the resurgence of French *grandeur*.

The urge to nationalize Blamont's experiment was only to be expected. In the shadow of superpower rivalry expressed through competing "national" space achievements, the French media and the population could not but invest the spectacular trail of sodium and the rocket that launched it with national significance. The space race, and the arms race in general, structured the meaning of technological achievements in both the space and nuclear domains. National pride, nationalist ideology, and the legitimacy of a political system and its leaders were tied up with the successful testing of a bomb or the launch of a satellite: this was what great powers did. For France in particular, as MacDougall and Hecht have emphasized, space and nuclear technologies affirmed national sovereignty and self-respect, providing the platforms on which to construct a postwar identity for a people that was humiliated by defeat and occupation, threatened by American hegemony and coming to terms with the loss of its colonies.[5]

It is understandable that the media should promote a nationalist agenda. More pertinent here, this agenda has also structured the *historiography* of the nuclear and space; until recently it, too, has been bounded by the walls of national containers.[6] Now, with our imaginations liberated from the crushing binary logic of Cold War competition, historians of science and technology, like many others, are increasingly striving to rupture the national frame and to situate scientific and technological practices in a transnational or global framework. The nuclear and space occupy a very particular niche in such a project. The Cold War irreversibly politicized them; both were intimately tied to national security, and interstate rivalry. At the same time both were also embedded in global networks through which knowledge in all its forms circulated. As Itty Abraham put it, "No atomic program anywhere in the world has ever been purely indigenous." Andrew Rotter called the atomic bomb "The World's Bomb."[7] Asif Siddiqi recently emphasized that "*every* nation engaged in [ballistic missiles and space] technology has been a proliferator and has benefited from proliferation."[8] These interconnections are invisible in a historiography that is complicit with national narratives. To move beyond those confines it behooves us to study the history of space (and the nuclear) through a lens that describes the tissue of global linkages that make "national" space programs possible and that sustain them.

That global tissue is not spontaneously generated: it requires social work. Indeed, as Frederick Cooper warned us some time ago, "There are two problems with the concept of globalization, first the 'global' and then the '-ization.'"[9] The first suggests that a single system of linkages and connections has penetrated the entire globe. The second implies that this occurs by an ineluctable process that propels entire societies, come what may, toward an interdependent world. What this conceptualization overlooks is that the "global" is constructed by human agents who establish and sustain networks that tie them together in specific patterns of interdependence. If we unpack the global, unravel the networks of transnational relationships of inclusion (and exclusion) that constitute it, our attention is drawn as much to nodes and blockages as to movement, to regulation and control as much as to circulation and fluidity. The world, as Cooper puts it, is "a space where economic and political relations are very uneven; it is full of lumps, places where power coalesces surrounded by those where it does not, places where social relations become dense amid others that are diffuse."[10]

Charles Bright and Michael Geyer share Cooper's concerns. They "applaud" the current emphasis on global flows that spill beyond the boundaries of the national container, and that decenter America. But they are wary of much writing on globalization that "tends to presume the (relative) openness of the world and to become preoccupied with the (relative) ease and multidirectional complexities of flows. ... "[11] Against this Bright and Geyer insist on the need to analyze "the structured networks and webs through which interconnections are made and maintained—as well as contested and renegotiated."

This chapter elaborates these insights through an analysis of US-European relationships in space science and technology in the 1960s. It throws light on the practices of inclusion/exclusion that shaped the transnational flow of knowledge in the early Cold War. This network was constituted by actors on both sides of the Atlantic in an asymmetrical field of force, in which knowledge/power was concentrated on the American side. The analysis doesn't assume that international collaboration defined the norm of what was possible and desirable, and that ruptures and regulations were "externalities" that disturbed its otherwise smooth functioning. Rather it explores the conjuncture of multiple factors that made collaboration possible—and that also set limits on what was possible. Knowledge that is so closely tied up with national economic and military competitiveness can only flow across borders if the states concerned see good reason for it to do so. International collaboration transcends national boundaries, but it doesn't dissolve national interests. On the contrary, it is one strategy among others for pursuing national interests, at least in domains, such as space and the nuclear, that constituted the core of state power after World War II. Through sharing—or denying—knowledge that its allies wanted, the US sought to construct a regime of order which strengthened the Western alliance on the front lines of the Cold War without undermining American hegemony in the region.

In what follows, I first briefly describe the national and international ambitions of the National Aeronautics and Space Administration and of Western Europe, notably France, as regards the exploration and exploitation of space in the 1960s. This contextualization is crucial to my argument, for it emphasizes that the patterns of international collaboration in space science and technology are shaped by the foreign policies of governments, and cannot be understood apart from them. Geopolitical relationships of power are embedded in the global circulation of knowledge, even if they are sometimes implicit in the vectors, human and otherwise, that constitute the network through which knowledge flows. This introduction is followed by an analysis of the different strategies adopted by NASA to implement its mandate to collaborate internationally in space science and technology. The more generous approach adopted with French engineers who came to the United States to learn how to build their first satellite is contrasted with the narrower and evolving restrictions on technology transfer in the domain of rocketry. Frank Ninkovich has emphasized that one of the abiding themes of American foreign policy in the twentieth century has been the recognition that "the very forces that made progress possible—technology, trade, a global division of labor, and interdependence—also made possible the system's destruction if pushed in the wrong direction and not checked." "The greater the degree of integration," he goes on, "the more explosive would be the disintegration produced by a runaway modernity."[12] This chapter takes a transnational approach to the proliferation of space science and technology in the 1960s, showing how the regulation of sensitive knowledge flows in the early Cold War served as an instrument of American foreign policy. It simultaneously encouraged national space programs, enhanced European integration and interdependence, and tried to stop France's technological ambitions pushing the continent "in the wrong direction," so catalyzing the "disintegration" of America's grand design for postwar Europe.

The US-European Geopolitical Context for Space Collaboration

In the 1960s, the collaboration between NASA and Western Europe was embedded in the national ambitions of both partners. It was initially driven by NASA's mission, as specified in the 1958 Space Act that established the agency, to foster American space leadership and to promote international cooperation in space. This agenda was quickly implemented at a meeting of COSPAR (Committee on Space Research), set up by the International Committee of Scientific Unions to maintain the momentum of the IGY (International Geophysical Year, a major collaborative event that ended in December 1958).[13] At a meeting in The Hague in March of 1959, the American delegate, on behalf of NASA, offered to support the work of COSPAR by launching "worthy experiments proposed by scientists of other countries," either as "single experiments as part of a larger payload, or groups of experiments comprising complete payloads."[14] In the former case the proposer would be "invited to work in a United States laboratory on the construction, calibration, and installation of the necessary equipment in a US research vehicle." In the latter the United States was willing to "advise on the feasibility of proposed experiments, the design and construction of the payload package, and the necessary pre-flight environmental testing."

The United States' offer had an electrifying impact on those present. Arnold Frutkin, who was responsible for NASA's office of International Affairs for almost two decades, has written that "the future of international cooperation in space exploration was raised at a stroke from the token to the real."[15] NASA's prestige and desirability as an international partner of choice was also confirmed. While NASA was making a "purely technical proposal of an inherently generous character, without strings," the Soviet delegate was busy threatening to leave COSPAR if its membership wasn't more representative of the Eastern Bloc.[16] This "intrusion of politics into the meeting," as Frutkin puts it, was the first in a long-line of measures taken by the Soviet Union to derail the smooth functioning of the Committee. The effect, of course, was to draw the attention of those present to "the stark contrast between the US and Soviet space programs in openness and willingness to share with others."[17]

This first step toward international collaboration in space science laid the foundation for a vast program that has persisted for more than fifty years: NASA has entered into no less that 4,000 cooperative ventures in space science, in satellite applications (e.g. for weather forecasting) and in technology, notably the giant International Space Station.[18] The precise terms of these projects differ. In particular they are subject to restrictions on the sharing of sensitive knowledge when that may be to the detriment of American economic or military competitiveness. NASA's global aspirations are tempered by the need to protect key national interests in a scientific and technological domain that is of immense strategic significance.

NASA's (and the Department of State's) willingness to consider sharing potentially sensitive knowledge with Western Europe was of a piece with overall policy in the region. The Eisenhower, Kennedy, and Johnson administrations all sought to build a scientifically and technologically strong, united Europe that could contribute its share to the burden of defense on the front lines of the Cold War. This explains why in the mid 1960s President Johnson was perturbed by strident complaints in France, and to some extent Germany, that a "technological gap" had opened up between the two sides of the Atlantic. American business was accused of invading Europe and dominating key sectors of European industry.[19] The issue was famously highlighted with the publication of Jean-Jacques Servan-Schreiber's *Le défi americain* (*The American Challenge*) in 1967.[20] Some American commentators were quick to see European fears of being outstripped by American technological and managerial prowess as an extension of President Charles de Gaulle's hostility to American "domination." They placed the blame for Europe's relative "backwardness" squarely on the continent's own shoulders (as did Servan-Schreiber to some extent).[21] Others were more prudent. Indeed, The President took this matter so seriously that in November 1966 he personally signed (National Security Action Memorandum) NSAM357, instructing his science adviser, Donald Hornig, to set up an interdepartmental committee to look into "the increasing concern in Western Europe over possible disparities in advanced technology between the United States and Europe."[22]

In its preliminary report, the committee concluded that "the Technological Gap [was] mainly a political and psychological problem" but that it did have "some basis in actual disparities." These included "the demonstrated American superiority in sophisticated electronics, military technology and space systems." Particularly important were "the 'very high technology industries' (particularly computers, space communications, and aircraft) which provide a much greater military capability, are nationally prestigious, and are believed to be far-reaching in their economic, political and social implications."[23]

Frutkin forcefully endorsed the idea that investments in space were crucial national needs at a meeting of the American Academy of Political and Social Science in Philadelphia in April of 1966. The American space program, he said, had pushed established scientific and technical disciplines to probe new frontiers in a wide variety of fields from physics to geodesy, from materials to structures. "In fact," he insisted, "we may with increasing confidence say that the peculiar quality of space science and technology is its forcing function, its acceleration of joint progress in a wide range of disciplines."[24] Hence space research and development had contributed "significantly to the fundamental strength and viability of the United States in a world where economic and military security increasingly rest[ed] upon technology." Western Europe was spending only a small fraction of what the United States did on space, to their detriment. It was in America's interest to help them close the technological gap in the space sector: "What has stimulated, energized and advanced us, may well stimulate, energize and advance them," Frutkin suggested.[25]

For the Johnson administration, then, the technological gap, even if inflated in Europe, was a problem that had to be addressed. European scientific and technological strength was essential if capitalism was to compete successfully with the Soviet system. This Cold War agenda, and the relatively paltry investment in space in Western Europe, obliged NASA to step in if it could. As a CIA report put it, whatever measures the Europeans took to build their capability, "the assistance of the US—both officially and through unofficial commercial channels—has been, is, and will probably remain the critical factor in the success of any European space program in this decade."[26]

France was at the centerpiece of this agenda. An (internal?) report on the French scientific and technical system stressed the emphasis the French now placed on scientific and technological pre-eminence: "Behind all efforts to accelerate the growth of science and technology in France has been the belief that the influence of a modern power is in proportion to its scientific and industrial strength and excellence ... ," the report stated. "The battles of price in the conquest of markets are being replaced by battles of innovation, in which scientific and technological superiority is the only effective weapon."[27] Space science and space technology, in particular, were crucial to modernization. In another special report on the state of space programs in Western Europe, written in May of 1964, the CIA quoted General Aubinière, the director of the new French space agency CNES (Centre Nationale des Études Spatiales) who had welcomed Blamont to Hammaguir in 1959, as saying that "space technology touches so many disciplines that to neglect it would signify for our peoples, formerly masters of the world, a decadence and underdevelopment and an unacceptable economic servitude, no matter whence it comes."[28] The French and the Americans drew different conclusions from this basic credo, of course. For Aubinière, alert to the rapidly declining influence of his country in a world dominated by the superpowers, any modernizing, self-respecting nation had to invest in space to avoid becoming little more than a colonial chattel of either the United States or the Soviet Union.[29] For Aubinière national autonomy and technological sovereignty were at stake. For the Johnson administration, that was inspired neither by colonial nostalgia nor by Spenglerian pessimism, the strength and unity of the free world in the face of the Communist threat were the main concern. NASA's task was to promote space science and technology in Europe without unduly fueling "runaway modernity" in France. It was to lock Western Europe into the American sphere of influence without helping France acquire national technological capabilities that would lead to the "disintegration" of American hegemony in the region.

The considerations outlined above don't simply throw light on the different motives held by key social actors in the United States and Europe for collaborating with one another in space science and technology. They do more. They emphasize that the transnational circulation of knowledge in the Cold War (at least in sensitive domains) was embedded in foreign-policy concerns of *both partners* and cannot be understood apart from them. Washington's determination to lead the "free world" and to contain the spread of communism was expressed through promoting cooperative ventures in space in which it could lever its leadership to help build a robust scientific, technological, and industrial base in Europe in a sector deemed to be of importance to economic growth and military preparedness. Western European governments couldn't afford, for that very reason, to ignore space science and technology, on pain of becoming even more dependent on the superpower. They were willing to collaborate from a position of weakness in the short to medium term, in the hope of being more equal partners in the longer term. Among continental countries, de Gaulle's France stands out as both seeking to have access to the knowledge that America had to offer, and as being determined to retain the freedom of action that befitted its quest for *grandeur*. This is the foreign-policy context that framed cross-border knowledge flows between the United States and Western Europe in the period under consideration here.

Standardizing Technical and Managerial Practices: The "Americanization" of French Space Science in the Early 1960s

NASA, in making its offer at COSPAR in March of 1959, was inviting others to enter US laboratories and firms and to acquire the most advanced instruments and techniques then available. The French were quick to take advantage of this opportunity. Soon after having established CNES in 1961 they embarked on the development of their first "national' satellite," the FR1—with American help. About half a dozen young men were sent to NASA's Goddard Space Flight Center in Maryland to learn American technological and managerial practices. This was to mutual benefit. As one Goddard engineer, Gilbert Ousley, later remembered it, their extended stay

> was a great excuse for us to share technology and training but we also had a selfish purpose. It was to get young engineers that were experienced to participate in our program and later come back to France speaking the same terminology that NASA uses, that understood our review process and did not feel insulted by peers looking at what was being done, and making constructive criticism. So it worked out in NASA's interest, and that was one of the main reasons that we set up the training program for the French engineers.[30]

Ousley and his colleagues certainly won over their French visitors. Jean-Pierre Causse, FR1's project director, explained that, while working at Goddard, his young cohort learned

> a method that we made our own—professional, rigorous, systematic and uncompromising, a method that we imposed on ourselves and also on all our scientific and industrial partners. ... In our dealings with industry, inspired by NASA, we exercised direct responsibility for the integration of our satellite. ... We familiarized ourselves with concepts like 'Memorandum of Understanding,' of 'No exchange of funds,' of 'Design Review' etc. We tried them out and perfected their application in [the FR1 project] ... . Bravo NASA and the United States![31]

During their six-month visit then, NASA engineers in Maryland imparted their established procedures to the young men from CNES. They sought not only to train them as engineers with a shared perception of what counted as significant problems and how to solve them, but also to teach them how to manage a large project. They went even further: they sought to inculcate in them a non-hierarchical management style, to have them acquire new attitudes to authority in their daily behavior, in their gestures, in their emotional makeup, and in their relationships with industry. The appropriation and performance of NASA's "best practice"—the use of standardized experimental and managerial methods, techniques and protocols—was essential if CNES engineers wanted to contribute to shaping the contours of the international research frontier. It was the *sine qua non* of their entering into a successful international dialog with the world leader.

In the early 1960s the transnational, transatlantic circulation of knowledge of how to build a successful scientific satellite occurred in an asymmetric field of force in which NASA and the United States constituted a dominant pole. In a previous study I have explored how, in this situation, various other social actors—scientific statesmen, Foundations, the Department of State and the NATO Science Committee—instrumentalized American scientific leadership with a view to locking the Western European scientific community into the Atlantic alliance, transforming their practices and their ideological engagements in line with American interests in the region.[32] The engineers at Goddard were part of that same movement. They helped construct a space science community in CNES that shared their ways of doing things, that adopted their techniques and their terminology. From the United States' point of view, a collaborative effort under its tutelage was an opportunity to build a global community that spoke the same "language" and that was organized along the same lines as their American homologs, facilitating knowledge circulation, the penetration of new markets, and the consolidation of the Atlantic community. From the European point of view the standardization, the "Americanization" of technical and managerial practices enabled them to jump-start their exploration of space by creatively applying NASA's procedures to local needs.

Many (cultural) historians regard the concept of "Americanization" to be too general and totalizing to be of much analytical value. It has the advantage, however, of alerting us to the hold that "America" had, and still has, on the imagination of people all over the world. Already in the early 1960s its scientific and technological achievements in space, in particular, were so impressive that European partners gladly engaged with and adopted NASA's best practices: "Americanization" was not imposed, it was coproduced, shaping behaviors, even professional identities. This is not to say that its effects were uncritically absorbed. As social actors navigated between the attractions of the American way and its disruption of deeply ingrained customs, traditions, and values, they selectively appropriated, adapted, or simply rejected the model on offer to satisfy the specificities of particular situations. The global project to refashion the world in America's image not only had to deal with the agency of those whom it sought to transform. It was also constrained by the stamina of local cultures.

Regulating Flows of Sensitive Technology: Propulsion

The stories told by Ousley and Causse celebrate the open exchanges that marked the design, development, and construction of France's FR1 satellite. They don't speak of the barriers to knowledge sharing that were surely imposed when the satellite was integrated as the payload on an American launcher, a Delta rocket descended from a Thor intermediate-range ballistic missile. Indeed the knowledge sharing at Goddard took place in a privileged zone of free circulation that was carved out from a far broader domain structured by concerns for regulation and control.[33] Arnold Frutkin summed up the guiding principle defining this domain in one pithy phrase: there had to be "clean technological and managerial interfaces" between the knowledge contributed by each participant to an international project. When the possibility of technological collaboration arose, knowledge flows were closely regulated, and pressures for technological sharing had to contend with counterpressures for technological denial. Technological, more than scientific co-operation touched directly on commercial competitiveness and national security and was subject to their imperatives.

From the dawn of the nuclear age American authorities were determined not to give the French access to any scientific and technological knowledge that might encourage the proliferation of nuclear weapons or, later, of their delivery systems.[34] The nuclear scientists who fled occupied France weren't admitted to the Manhattan Project with their British counterparts; they worked at Chalk River in Canada instead. Relations remained strained after the war as long as Frédéric Joliot-Curie, an active member of the resistance, and a convinced communist, was the scientific director of the French Commissariat à l'Énergie Atomique. His departure in 1950, and Eisenhower's Atoms for Peace initiative in the mid 1950s improved the transatlantic flow of knowledge and materials between the two countries, though a morbid fear that left-wing scientists in France's atomic complex would leak secrets to the Soviet Union always lurked in the background. The United States deplored France's decision to develop a nuclear weapon in 1956. When Congress liberalized the highly restrictive requirements of the 1946 McMahon Act in June of 1958, it specifically excluded France by limiting the reciprocal exchange of restricted data on atomic weapons to those countries that had made "substantial progress" on their own at the time, i.e., the United Kingdom.[35] Charles de Gaulle, who came to power in 1958, tried and failed to include France, along with the United States and Britain in a triumvirate that would define nuclear strategy for Europe. Rebuffed, he struck off on his own. He made the bomb the centerpiece of his multi-pronged *force de frappe* that included nuclear-tipped missiles launched from silos and from submarines. France became the fourth nuclear power with a successful test of its weapon in the Sahara desert in February of 1960 and it refused to sign the 1963 treaty limiting all further nuclear testing to underground facilities. For de Gaulle, who would have had to test French weapons under US surveillance in the Nevada desert, this was an unacceptable violation of sovereignty imposed by the superpowers who wanted to maintain their nuclear monopoly and stabilize the international system under their control.

This determination to deny proliferation-related knowledge to France also shaped American policy regarding dual-use delivery systems in the 1960s. France's growing capacity in the domain of rockets/missiles was evident. In 1965 it became the third space power when it successfully launched its own satellite using a French-built Diamant rocket. The Diamant combined stages that had been developed previously as part of the "Precious Stone" series of sounding rockets and missiles (Emeraude, Topaze) with a state-of-the-art solid-fuel third stage derived directly from the military program.[36] The new Johnson administration was quick to stop any knowledge flows in this sector. In April of 1964 McGeorge Bundy signed off on NSAM 294, which contained this stipulation:

Given current French policy it continues to be in this government's interest not to contribute to or assist in the development of a French nuclear warhead capability or a French national strategic nuclear delivery capability. This includes exchanges of information and technology between the governments, sales of equipment, joint research and development activities, and exchanges between industrial and commercial organizations, either directly or through third parties, which would be reasonably likely to facilitate these efforts by significantly affecting timing, quality or costs or would identify the US as a major supplier or collaborator.[37]

Washington couldn't stop Paris from developing an independent nuclear deterrent and delivery system. But it could retard its progress by refusing to collaborate and, by denying it cutting edge science and technology, make it less effective and destabilizing than it might otherwise have been.

De Gaulle's determination to go it alone posed something of a dilemma for the administration. On the one hand they were keen to strengthen European science and technology, notably in the space sector. On the other hand, national-security policy dictated that France was not to be helped in acquiring a "strategic nuclear delivery capability." How could one support the first without fostering the second? The European Launcher Development Organization (ELDO), established in the early 1960s was a potentially fruitfully instrument to serve this dual role.[38] To explain how that was possible, a short detour is called for.

ELDO brought together five of the six founder members of the European Economic Community (Belgium, the Federal Republic of Germany, France, Italy, and the Netherlands) plus Britain and Australia. In 1962 they agreed to build a three-stage satellite launcher called Europa.[39] The first stage would be derived from Britain's Blue Streak intermediate-range ballistic missile, which would be stripped of its military characteristics. France would build the second stage, Germany the third. The other participants would provide the test satellite, telecommunications and ground equipment, while the launching base would be in Woomera, South Australia.

NASA and the Department of State welcomed the formation of ELDO. To quote an early position paper on the issue, technological assistance to ELDO was coherent with "our objective of an economically and politically integrated European Community with increasingly close ties to this country within an Atlantic community." In addition, by working with a multinational organization rather than making bilateral arrangements with individual states one could divert scarce resources in countries like France and Germany away from national military programs and so stunt autonomous missile developments.[40] ELDO thus promised to kill two birds with one stone: it would stimulate space research and development in Europe without providing technological support for strengthening parallel national rocket/missile programs. Arnold Frutkin explained the terms on which NASA could work with Europe in the field of rocketry on an official visit to Britain, France, and Germany in December 1962. Knowledge would not flow relatively freely, as it did in the field of scientific cooperation. Collaboration in the launch vehicle area was possible only "to a limited extent."[41] The European programs had to be directed to peaceful civilian applications, and of mutual technological interest. They also had to be multilateral, not bilateral. In other words, NASA would collaborate only through ELDO and would not make bilateral agreements with individual national authorities on the continent in the domain of rocketry.[42]

Even this approach was not without its risks. It was well known that ELDO lacked a strong centralized system of project management and control; its Secretariat had little authority over the people and firms developing the separate stages in Britain, France and Germany.[43] A report prepared by the CIA in May of 1964 confirmed the danger: "the organization has no enforcement machinery to police compliance, and the possibility is raised that ELDO might contribute to the spread of ballistic missile technology." Under these circumstances, how did Frutkin hope both to assist ELDO and to respect the injunctions of NSAM 294? He suggested that it could be done by distinguishing between the *kind* of technology that could be shared. In the extremely sensitive domain of propulsion he drew a sharp line between solid propellants like powders, on the one hand, and non-storable liquid propellants like liquid hydrogen and liquid oxygen, on the other. France, he noted was heavily committed to using solid fuels in its military missiles. Non-storable and liquid fuels were unlikely to contribute much to the propulsion technologies France was developing for its strategic delivery objectives.[44] In other words, by distinguishing between the military potential of solid (high security risk) and non-storable (low military interest) propellants, NASA and the Department of State thought they could safely offer assistance to ELDO in a technologically crucial domain without significantly assisting France's missile program.

Frutkin implemented this distinction when asked for help in 1965. The Member States of ELDO decided that their launcher should be upgraded to have a geostationary capability (the Europa II program). This required constructing a more powerful third stage than previously planned for. A senior engineer in the European organization, Bill Stephens, asked NASA if they would authorize discussions about some of "more fundamental problems" that American engineers had encountered "in designing, testing and launching liquid hydrogen/liquid oxygen upper stages." He also hoped that European firms could work along with American companies who had developed such propulsion systems. Stephen's requests could be met. Cryogenic rocket fuels were highly inflammable, difficult to handle, bulky, and not the fuel of choice for the French military. This was, then, "a valuable opportunity to advance our relationship with ELDO as a multilateral institution, to establish a ground for limiting or delaying assistance in the missile field to competing interests in Europe, and to establish a counterweight to national missile programs."[45] NASA could encourage European integration and help ELDO technologically, fostering transatlantic exchanges of people and ideas, as well as corporate linkages that would respect the regulatory constraints imposed by NSAM 294.

We see then that the sharing of sensitive, dual-use propulsion technology with Europe in the 1960s required drawing two interlocking distinctions, one institutional, the other technological. First a sharp distinction was drawn between national and multilateral programs—only the latter would be candidates for assistance. Then a distinction was drawn between the kind of technology that would be shared: only propulsion systems that burned non-storable, liquid fuels that posed a relatively minor proliferation danger. Institutional and technological barriers to knowledge circulation co-existed with channels along which knowledge could flow relatively freely. Those barriers weren't fixed and immutable, but were themselves historical products, contingent on a variety of technological, industrial and political factors, as we shall see immediately.

Expanding the Scope of Collaboration: Knowledge Sharing as an Incentive to Sustain ELDO[46]

On July 29, 1966, Walt W. Rostow, one of LBJ's two national-security advisers, signed off on NSAM 354. NSAM 354 was a response to a request from the Department of State stating it was "a matter of urgency that we clarify and define our policy with respect to the development of [ELDO] and the extent of US cooperation with ELDO's present and future programs." The Memorandum went on to note that it was "in the US interest to encourage the continued development of ELDO through US cooperation." It referred to the results of an ad hoc interagency working group that had prepared a statement "defining the nature and extent of US cooperation with ELDO which the US government is now prepared to extend." And it confirmed that this statement was to be "continually reviewed by the responsible agencies," above all, the Department of Defense and the Department of State, along with NASA, "to ensure that it is current and responsive in terms of developing strategies."[47] In other words, the United States was prepared to be far more flexible than before as regards knowledge sharing with ELDO, expanding the scope as needed to help sustain its continued development. No explicit reference was made to the dangers of knowledge leaking to the French national missile program: the priority now was to keep ELDO afloat.

This shift in policy was precipitated by the British government that, in February of 1966, informed its partners in ELDO of its intention to withdraw from the organization.[48] In their view ELDO wasn't likely to produce any worthwhile result. Development costs of the Europa II rocket had more than doubled from the initial estimate and no end to the upward spiral was in sight. The time to completion had slipped by 50 percent, from five to seven and a half years. Britain's first stage, Blue Streak, had been successfully commissioned in June of 1965, while the French and German stages were still under development. The British were therefore effectively subsidizing continental industries to produce a launcher that, in fact, would be obsolete technologically and commercially uncompetitive with the US Titan III rocket. To add insult to injury, in January of 1963, President de Gaulle had vetoed Britain's application to join the Common Market. For the United Kingdom, who was paying almost 38 percent of the ELDO budget, the original technological, industrial, and political rationale for launching the organization had evaporated.[49]

The timing of this move was deemed most unfortunate in Washington. Firstly, the European integration process was in a very brittle state at the time and even NATO seemed to be on the brink of fragmentation.[50] The French had precipitated a crisis in the European Economic Community (EEC) by boycotting the EEC's decision-making machinery so as to liberate the country from its "subordination" to Community institutions and the dilution of sovereignty that that entailed.[51] De Gaulle was also frustrated by the constraints on French military ambitions imposed by NATO. "The French have emphasized their dissatisfaction by becoming increasingly an obstructionist force in NATO," one American task force wrote, "equating integration with subordination."[52] In this inauspicious climate, everything possible had to be done to sustain the momentum for European unity. As Under Secretary of State George Ball emphasized, "The United States has a direct interest in the continuation of European integration. It is the most realistic means of achieving European political unity with all that that implies for our relations with Eastern Europe and the Soviet Union … and is the precondition for a Europe able to carry its proper share of responsibility for our common defense."[53] ELDO wasn't central to European integration. But at a time when the momentum of European unity was being challenged in France, Britain's threat to leave

ELDO risked being amplified by those who were increasingly hostile to supranational ventures in Europe.

The United States also feared that if ELDO were dissolved "France might devote more of its resources to a national, military-related program or that it might establish undesirable bilateral relationships for the construction of satellite launch vehicles."[54] The Soviet Union was the most obvious "undesirable" partner. In June of 1966 the French president, affirming his determination not to accept the logic of a bipolar world structured by the superpowers, made a highly successful official visit to Moscow. He saw a satellite launched from the aerospace base at Baikonur, and he endorsed a major agreement for cooperation in science, telecommunications and meteorology that made of space the "emblematic flagship" of Franco-Soviet collaboration.[55] Sir Solly Zuckerman, the Chief Scientific Adviser to the British government, was exhorted by the Department of State not to withdraw from the European launcher organization for fear that "the Soviets would move into the vacuum if ELDO collapsed."[56] The United States had to contain this threat and to ensure that European institutions emerged "from the present crisis with their prestige, power and potential for building a united Europe as little impaired as possible."[57]

In response to this emerging "crisis," officials in NASA and the Department of State were emphatic that the constraints imposed by NSAM 294 on sharing rocket and missile technology had to be reviewed. Richard Barnes, the Director of Frutkin's Cooperative Projects Division, wrote to the chairman of the NSAM 294 Review Group in the Department of State to help define guidelines for a less restrictive policy. The United States should refuse to a foreign power, he suggested, "*only* those *few critical* items which are clearly intended for use in a national program, would significantly and directly benefit that program in terms of time and quality or cost, and are unavailable in comparable substitute form elsewhere than the US."[58] Correlatively, the US should be willing to share items that were "of only marginal benefit to the national program" or "were available elsewhere than the US without undue difficulty or delay."[59] In short, Barnes wanted US policy to take into account the kind of technology at issue, its likely uses in practice, the global state of the market for the technology, and the importance of collaboration from a foreign-policy perspective. The last, along with US business interests, were not to be sacrificed on the altar of an overcautious, generalized reluctance to share technology just because it might encourage national programs which sections of the US administration disapproved of.

While Barnes was putting NASA's case to the Department of State, NASA Administrator James Webb was doing what he could to get the Department of Defense to support a more liberal approach. Writing to Secretary of Defense McNamara in April of 1966, Webb pointed out that although high-energy, cryogenic, or "non-storable" rocket fuels might conceivably be employed for military purposes, in practice they would probably not be deployed in that way. He argued that in any case the risks of technological leakage into the military program were outweighed by the benefits of promoting a civilian rocket being built by a multilateral organization: "Even in the case of France it seems likely that encouragement to proceed with upper stage hydrogen/oxygen systems now under development might divert money and people from a nuclear delivery program rather than contribute to that which is already under way using quite different technology." In a supportive reply, McNamara reassured the NASA Administrator that he strongly favored international cooperation in space and that he had directed the DoD staff "to be as liberal as possible regarding the release of space technology for payloads and other support items."[60]

Walt Rostow's NSAM 354 of July 1966, and the administration's concern to formally investigate the roots of the technological gap (NSAM 357) were responses to this changing situation. The Johnson administration had come to think that NSAM 294 on sharing strategic nuclear and missile technology with France was harming the United States' interests in Europe. It hoped that, by expanding the scope for knowledge sharing it could entice the British to stay in ELDO, and boost the organization's launcher program.

The new suggestions that emerged that summer from an interagency working group led by the Department of State proposed extensive US support in three categories: general, short-range, and long-range.[61] The first contained some standard items—training in technical management, facilitating export licenses, use of NASA test facilities—but also suggested that a technical office be established within NASA "specifically to serve in an expediting and assisting role for ELDO." Short-range help included "technical advice and assistance" in items like vehicle integration, stage separation, and synchronous orbit injection techniques, as well as the provision of unclassified flight hardware, notably the strapped-down "guidance" package used on the Scout rocket that had already been exported to Japan. Long-range assistance was focused on helping with a high-energy cryogenic upper stage of the rocket, as had been requested by Stephens on behalf of ELDO the year before. It was proposed that Europeans be given access to technological documentation and experience available in the Atlas-Centaur systems (that mated a ballistic missile with an upper stage that used liquid hydrogen and liquid oxygen). It was also suggested that ELDO's technical personnel "have intimate touch with the problems of systems design, integration, and program management of a high-energy upper [sic] such as the Centaur" and even that the United States consider "joint use of a high-energy upper stage developed in Europe."[62] In short, in mid 1966, the United States was considering making a substantial effort to help ELDO develop a powerful launcher by sharing state-of-the-art knowledge and experience in both technology and project management, as well as by facilitating the export of hardware.

It is difficult to evaluate the impact of these proposals. Certainly they helped persuade the British to remain a member state of ELDO: in September 1966 the U.K. agreed not to withdraw in return for their contribution to the budget being reduced from 38 percent to 27 percent.[63] Apart from that the archives are silent on whether or not engineers in ELDO took advantage of what the United States had to offer. In any event the organization collapsed in 1972, its launcher having failed to put even one satellite into orbit.

Concluding Remarks

The "globalization" of science and technology doesn't take place of its own accord. Sometimes it does indeed seem as if knowledge "simply travels by itself," as Jim Secord put it, since "the work that has gone into making this appear to be the case is so pervasive and institutionalized that it has become hard to see.[64] This chapter has emphasized that we should not be trapped by such appearances. Surely some information circulates unimpeded, as when Blamont learned how to build a sodium vapor capsule at the University of Wisconsin in 1957. However, knowledge transfer to a foreign national in a university setting is at one of extreme of a spectrum, of which classification and total exclusion is the other. In between lies a vast grey area of sensitive but unclassified knowledge whose 'export' is subject to ongoing negotiation and regulation.

Boundary work is part and parcel of the management of knowledge flows. These boundaries are constantly negotiated by diverse social actors invested in the system of knowledge transfer—scientists, engineers, firms, research-and-development agencies, various arms of the national-security state. They are not watertight: knowledge can seep through enclosing walls in many ways, from informal exchanges between scientists and engineers to espionage. They also shift over time to keep abreast of the research frontier and in response to changing institutional and political agendas. Like any "frontier," they are, as Charles Maier reminds us, "partially a virtual construction … as much a site of the demonstrative extent of power as a real barrier."[65] They are porous, but not infinitely so for then they, and those who manage them, would lose their meaning and their legitimacy. The social work that is done to make it appear that "knowledge simply travels by itself" through an unbounded global space, conceals the repeated display of state power in the implementation and enforcement of barriers, effaces the regulatory machinery that controls the movement of knowledge in all its forms across borders, and mystifies the political processes that ultimately determine the uneven, lumpy contours of the global.

"*Every* nation engaged in this technology [ballistic missiles and space technology] has been a proliferator, and has benefited from proliferation," writes Siddiqi. But no nation, as far as I know—and certainly not the United States—proliferates indiscriminately. There is the methodological rub. As the example of US policy in Europe shows, transatlantic collaboration in the preparation of scientific payloads was relatively unimpeded; the sharing of launcher/missile technology was carefully calibrated. It was driven by a foreign-policy agenda that supported European integration and that was strongly opposed to French autonomy. Scientific and technological exchange had to consolidate the former without catalyzing the latter, above all as regards de Gaulle's quest for an independent nuclear deterrent. The solution was to "proliferate" launcher technologies to ELDO without "proliferating" missile technologies to France. Drawing boundaries between space science and technology, between the national and the supranational, and between non-storable and storable engine fuels, NASA and the Department of State constructed one set of international scientific, technological, and institutional linkages while doing all they could to stop another taking shape. A transnational history that dissolves one boundary—that defined by the autonomous nation-state—must analyze another—that constituted by the regulations whereby an interconnected nation manages its linkages to serve its national ambitions and its foreign-policy goals.

The history of an "indigenous" launcher is a prime candidate for a nationalistic narrative. A government's prestige, access to markets, and military potential acquired by having independent access to space inspires nation-centered stories that occlude the multiple borrowings and transnational interactions that are sustained by interpersonal, inter-institutional and inter-firm relations and that are eventually built into the hardware. As Siddiqi rightly points out, a global history of rocketry punctures these national and nationalistic narratives, and exposes the "connections and transitions of technology transfer and knowledge production" that they render "invisible." That granted, the interconnections that a national history renders "invisible" are *not* invisible to the social agents that are engaged in international relations—senior government officials like McNamara and Webb, administrators like Frutkin and Barnes, scientists and engineers like Blamont. On the contrary, it is they who constitute those interconnections by their policies and practices. Bearers of state patronage, they orchestrate national and regional science and technology policy on an international stage, leaving an indelible "national" stamp on the transnational web of linkages and interconnections that they construct and maintain. In a "decentered" global history, Siddiqi suggests, a nation's space program would be "rendered as a more nebulous transnational process. ..." This injunction is misguided if it means that we must blur the role of national actors in writing transnational history. Our task rather is to recapture the not-so-nebulous web of international linkages that are mobilized by national actors to serve national and foreign-policy objectives. It is to embed the national into the transnational narrative—not as a bounded container but as one node in an interconnected global network. We must position our national actors on a global stage of competing and collaborating, actual and aspirant space powers, and follow them as they choose, or refuse, to establish links with institutions and individuals in other countries. A global space history must retain the national as a key analytic category—not as an autonomous but as an interdependent actor, whose scientific and technological practices are inspired by national interest and framed by foreign policy. It is that policy that determines, for a strategic sector like space, who a country will cooperate with, that defines the terms of the engagement, and that structures the channels through which knowledge will flow across borders—as well as which channels will be plugged to ensure that it doesn't. That French journalist had a point when he excitedly wrote, in a moment of nationalist fervor, that "I saw France's first artificial comet launched at 19h. 38." Jacques Blamont had a point when he decried "the highly exaggerated success of our pathetic sounding rockets"[66] that drew on so much external support. The challenge that this chapter addresses is to craft an historical narrative that does justice to the sentiments expressed by both.

Notes

1. The campaign is described by Jacques Blamont, "Origines et principes de la politique spatiale de la France," *Rayonnement du CNRS* 48 (juin 2008): 24–32; Jacques Blamont, "Les prèmieres éxperiences d'aéronomie en France," in *Actes. Première Rencontre de l'I.F.H.E. sur l'essor des recherches spatiales en France: Des prèmieres expériences scientifiques au premiers satellites, Paris, France, 24–25 octobre 2000* (ESA SP-472, 2001), 31–41.

2. The page is reproduced in Blamont, "Les prèmieres éxperiences d'aéronomie en France," 35.

3. Frances Durand-de Jongh, *De la fusée Véronique au lanceur Ariane. Une histoire d'hommes 1945–1979* (Stock, 1998), 24. Marie-France Ludmann-Obier also deals with this issue; see "Un aspect de la chasse aux cerveaux: les transferts de techniciens allemands en France; 1945–1949," *Relations internationales* 46 (été 1986): 195–208. The most comprehensive treatment is Michael Neufeld's in "The Nazi Aerospace Exodus: Towards a Global, Transnational History," *History and Technology* 28, no. 1 (2102): 49–67.

4. I have developed this notion in "Hybrid Knowledge: The Transnational Coproduction of the Gas Centrifuge for Uranium Enrichment in the 1960s," *British Journal for the History of Science* 45, no. 3 (2012): 337–357.

5. Gabrielle Hecht, *The Radiance of France: Nuclear Power and National Identity after World War II* (MIT Press, 2009); Walter A. McDougall, "Space-Age Europe: Gaullism, Euro-Gaullism, and the American Dilemma," *Technology and Culture* 26, no. 2 (1985): 179–203. See also Kristin Ross, *Fast Cars, Clean Bodies: Decolonization and the Reordering of French Culture* (MIT Press, 1998).

6. Asif Siddiqi, "Competing Technologies, National(ist) Narratives, and Universal Claims. Towards a Global History of Space Exploration," *Technology and Culture* 51, no. 2 (2010): 425–443.

7. Itty Abraham, "The Ambivalence of Nuclear Histories," in *Global Power Knowledge: Science and Technology in International Affairs*, ed. John Krige and Kai-Henrik Barth, *Osiris* 21 (2006): 49–65; Andrew J. Rotter, *Hiroshima: The World's Bomb* (Oxford University Press, 2008).

8. Siddiqi, "Competing Technologies," 437.

9. Frederick Cooper, *Colonialism in Question: Theory, Knowledge, History* (University of California Press, 2005), 91–112, at 91.

10. Ibid., 90–91.

11. Charles Bright and Michael Geyer, "Regimes of World Order. Global Integration and the Production of Difference in Twentieth-Century World History," in *Interactions: Transregional Perspectives on World History*, ed. Jerry H. Bentley, Renate Bridenthal and Anand A. Young (University of Hawaii Press, 2005), 202–237, at 204. See also Frederick Cooper, "What Is the Concept of Globalization Good For? An African Historian's Perspective," *African Affairs* 100 (2001): 189–213.

12. Frank Ninkovich, *The Wilsonian Century: US Foreign Policy Since 1900* (University of Chicago Press, 1999), 66.

13. On COSPAR see H. Newell, *Beyond the Atmosphere: Early Years of Space Science* (NASA, 1980), chapter 18. For a detailed account of the meeting in March of 1959, see H. Massey and H. O. Robins, *History of British Space Science* (Cambridge University Press, 1986), 58–62.

14. For NASA's initiative, see letter Newell to Glennan et al., March 12, 1959; for Porter's offer, see his letter to van de Hulst, March 14, 1959. Both letters are available in *Exploring the Unknown: Selected Documents in the History of the US Civil Space Program*, volume II: *External Relationships*, ed. John M. Logsdon (NASA, 1996),.

15. Arnold Frutkin, *International Cooperation in Space* (Prentice-Hall, 1966), 38.

16. Massey and Robins, but not Frutkin or Newell, see this objection as justified. They write that the composition of COSPAR was at this time "heavily weighted in favor of the West," and describe the efforts made to redress the imbalance, Massey and Robins, *History of British Space Science*, 59–61 and Annex 2.

17. Newell, *Beyond the Atmosphere*, 314.

18. John Krige, Angelina Long Callahan, and Ashok Maharaj, *NASA in the World: Fifty Years of International Collaboration in Space* (Palgrave Macmillan, 2013).

19. SC No. 00666/65B, "US Investments in Europe," CIA Special Report, April 16, 1965, folder Memos [2 of 2], Vol. II, 7/64–7/66, Country File, Europe, National Security File, box 163, Lyndon Baines Johnson Presidential Library, Austin (hereafter LBJ Library).

20. Jean-Jacques Servan-Schreiber, *Le défi americain* (Paris, 1967), published in English as *The American Challenge* with a foreword by Arthur Schlesinger Jr. (Atheneum, 1968).

21. See for example the lead article by Philip H. Abelson, "European Discontent with the 'Technology Gap'," *Science* 155, no. 3764 (1967). For a summary of Servan-Schreiber's argument see

"West Europeans Attribute Continuing Technological Lag Behind US to Inferior Management," *New York Times*, December 13, 1967. See also Henry R. Nau, "A Political Interpretation of the Technology Gap Dispute," *Orbis* 15, no. 2 (1971): 507–524.

22. NSAM 357, "The Technological Gap," November 25, 1966, available at http://www.lbjlib.utexas.edu. Hornig's official title was the Special Assistant to the President for Science and Technology.

23. "Preliminary Report on the Technological Gap Between US and Europe," attached to David Hornig's letter to the President, January 31, 1967, folder Technological Gap [1 of 2], Subject File, National Security File, box 46, LBJ Library.

24. Arnold W. Frutkin, "The United States Space Program and Its International Significance," *Annals of the American Academy of Political and Social Science* 366 (July 1966): 89–98, at 90. See also *Space Business Daily* 25, no. 35 (April 18, 1966): 285–286.

25. Frutkin, "The United States Space Program," 91, 92.

26. Central Intelligence Agency, Office of Current Intelligence, *Special Report: Western European Space Programs*, May 1964, 3, record no. 15707, International Cooperation and Foreign Countries, European Launcher Development Organization, folder ELDO, NASA Historical Reference Collection, Washington.

27. "Growth of Scientific and Technical Manpower in France," report attached to memo Frutkin to Webb and others, April 2, 1965, record no. 14596, International Cooperation with Foreign Countries, France, folder France Space General, NASA Historical Reference Collection, Washington.

28. CIA, *Special Report. Western European Space Programs*, 13.

29. See also the report "Growth of Scientific and Technical Manpower in France."

30. Gilbert Ousley, "French-American Space Relationship," in *Les relations franco-américaines dans le domaine spatial (1957–1975), Quatrième rencontre de l'IFHE, 8–9 décembre 2005* (IFHE, 2008), 80–95, at 83–84.

31. Jean-Pierre Causse, "Le Programme FR-1," in *Les relations franco-américaines*, 214–231, at 215, 216, 231 (my translation).

32. John Krige, *American Hegemony and the Postwar Reconstruction of Science in Europe* (MIT Press, 2006).

33. See also John Krige, "Building the Arsenal of Knowledge," *Centaurus* 52, no. 4 (2010): 280–296.

34. Frank Costigliola, *France and the United States: The Cold Alliance Since World War II* (Macmillan, 1992), 37–39.

35. John Baylis, "Exchanging Nuclear Secrets: Laying the Foundations of the Anglo-American Nuclear Relationship," *Diplomatic History* 25, no. 1 (2001): 33–61.

36. Diamant-A used a mixture of N_2O_4 /UDMH (storable liquid fuels) in its first stage and solid fuel in the second and third stages. Claude Carlier and Marcel Gilli, *Les trente premières années du CNES. L'Agence Françaises de l'Espace, 1962–1992* (La documentation Françaises/CNES, 1994), 143–145.

37. NSAM 294 is available at http://www.lbjlib.utexas.edu.

38. I have described the launch of ELDO in detail in John Krige and Arturo Russo, *A History of the European Space Agency, 1958–1987*, volume I: *The Story of ESRO and ELDO, 1958–1973* (ESA, SP-1235, 2000), chapter 3. See also Michelangelo De Maria and John Krige, "Early European Attempts in Launcher Technology," in *Choosing Big Technologies*, ed. John Krige (Harwood, 1993), 109–137; Also McDougall, "Space-Age Europe."

39. Krige and Russo, *A History of the European Space Agency, 1958–1987*, chapters 3 and 4.

40. "Draft US Position on Cooperation with Europe in the Development and Production of Space Launch Vehicles," attached to internal correspondence between Frutkin and Milton W. Rosen, October 15 and 30, 1962. Frutkin told Rosen that the "Draft" position developed by the Department of State was in fact now firm US policy, record no. 14548, International Cooperation and Foreign Countries, Europe, folder US-Europe 1965–1972, NASA Historical Reference Collection, Washington. The arguments here resonate with those used to justify sharing civilian nuclear energy with Euratom: see John Krige, "The Peaceful Atom as Political Weapon: Euratom and American Foreign Policy in the late 1950s," *Historical Studies in the Natural Sciences* 38, no. 1 (2008): 9–48.

41. Undated memo, "Trip to European Launcher Development Organization (ELDO) and Member Countries, France, England, and Germany (December 4–17, 1962)," Alfred M. Nelson, record no. 15707, International Cooperation and Foreign Countries, European Launcher Development Organization, folder ELDO, NASA Historical Reference Collection, Washington.

42. More precisely, as Frutkin put it, official US policy was that "Exchanges on a purely national basis will be discouraged or rerouted through ELDO," Memorandum for Rosen, October 30, 1962, record no. 14548.

43. Stephen B. Johnson, *The Secret of Apollo: Systems Management in American and European Space Programs* (Johns Hopkins University Press, 2002), chapter 6, describes the failure of ELDO's management system in detail.

44. Memo, Frutkin to Robert N. Margrave, Director, Office of Munitions Control, Department of State, "ELDO interest in high energy upper stages," June 6, 1965, record no. 14465, International Cooperation and Foreign Countries, International Cooperation, folder International policy manual material from Code I, NASA Historical Reference Collection, Washington.

45. Memo Frutkin to Margrave, June 6, 1965.

46. This section draws heavily on the story told in John Krige, "Technology, Foreign Policy and International Collaboration in Space," in Stephen J. Dick and Roger D. Launius, *Critical Issues in the History of Spaceflight* (NASA, 2006).

47. NSAM 354, "US Cooperation with the European Launcher Development Organization," July 29, 1974, available at http://www.lbjlib.utexas.edu.

48. For details on the crisis in Europe see Krige and Russo, *A History of the European Space Agency*, volume I, chapter 4.3.2.

49. Britain's partners didn't accept these arguments of course. They insisted that ELDO's aim was not to produce a competitor to American launchers in the immediate future. It was rather intended to build up the industrial infrastructure and experience in Europe which would enable its member states to develop their own launcher in the long-term, and so cooperate with the United States from a position of relative autonomy and technological and industrial strength.

50. For general accounts of the United States' relations with France during the Cold War, and with de Gaulle in particular, see Frank Costigliola, *France and the United States: The Cold Alliance Since World War II* (Twayne, 1992); Christian Nuenlist, Anna Lochner, and Garret Martin, eds., *Globalizing de Gaulle: International Perspectives on French Foreign Policies, 1958-1969* (Lexington Books, 2010); Robert O. Paxton and Nicholas Wahl, eds., *De Gaulle and the United States: A Centennial Reappraisal* (Berg, 1994); John Newhouse, *De Gaulle and the Anglo-Saxons* (Viking, 1970).

51. For the term 'subordination', see Ted Van Dyk to the Vice President, July 7, 1965, folder Germany Erhard Visit [12/65], 12/19–21/65, Germany, Country File Europe and USSR, National Security File, box 192, LBJ Library.

52. "France and NATO," position paper, September 25, 1965, folder Germany Erhard Visit [12/65] 12/19–21/65, Germany, Country File Europe and USSR, National Security File, box 192, LBJ Library.

53. Department of State to Amembassy Bonn 1209, outgoing telegram, November 18, 1965, signed [George] Ball, folder Germany Erhard Visit [12/65], 12/19–21/65, Germany, Country File Europe and USSR, National Security File, box 192, LBJ Library.

54. "US Cooperation with ELDO," position paper, July 21, 1966, folder Cooperation in Space—Working Group on Expanded International Cooperation in Space. ELDO #1 [2 of 2], Charles Johnson File, National Security Files, box 14, LBJ Library.

55. Marie-Pierre Rey, "De Gaulle, French Diplomacy and Franco-Soviet Relations as Seen from Moscow," in Nuenlist, Locher and Martin, *Globalizing De Gaulle*, at 35.

56. "Memorandum for the Files, Cooperation with ELDO," meeting with Zuckerman, May 6, 1966, folder Cooperation in Space—Working Group on Expanded International Cooperation in Space. ELDO #1 [2 of 2], Charles Johnson File, National Security Files, box 14, LBJ Library.

57. Department of State to Amembassy Bonn 1209, outgoing telegram, November 18, 1965, signed [George] Ball, folder Germany Erhard Visit [12/65], 12/19–21/65, Germany, Country File Europe and USSR, National Security File, box 192, LBJ Library.

58. Richard Barnes to Scott George, Chairman, NSAM 294 Review Group, Department of State, April 15, 1966, record no. 14459, International Cooperation and Foreign Countries,

International Cooperation, folder Miscellaneous Correspondence from Code I—International Relations 1958–1967, NASA Historical Reference Collection, Washington. The words italicized here were underscored in the original.

59. For example, an American company had recently been refused a license to assist France with the development of gyro technology even though gyros of comparable weight and performance were already available in France; this was incoherent.

60. Webb to McNamara, April 28, 1966, and reply, Bob [McNamara] to Jim [Webb], May 14, 1966, record no. 14459, International Cooperation and Foreign Countries, International Cooperation, folder Miscellaneous Correspondence from Code I—International Relations 1958–1967, NASA Historical Reference Collection, Washington.

61. This paragraph is derived from "Policy Concerning US Cooperation with the European Launcher Development Organization (ELDO)," attached to U. Alexis Johnson's "Memorandum," June 10, 1966, record no. 15707, International Cooperation and Foreign Countries, European Launcher Development Organization, folder ELDO, NASA Historical Reference Collection, Washington.

62. In summer 1965, ELDO had asked for help from NASA on "designing, testing and launching liquid hydrogen/liquid oxygen upper stages" (Frutkin to Margrave, June 6, 1965).

63. Krige and Russo, *A History of the European Space Agency, 1957–1985*, volume I., chapter 4.3.2.

64. Jim Secord, "Knowledge in Transit," *Isis* 95, no. 4 (2004): 654–672, at 670.

65. Charles S. Maier, *Among Empires: American Ascendancy and Its Predecessors* (Harvard University Press, 2006), 106.

66. Blamont, "Origines et principes," 27.

8 Bringing NASA Back to Earth: A Search for Relevance during the Cold War

Erik M. Conway

During the past 30 years, the National Aeronautics and Space Administration has become the largest funder of climate research in the United States, and a major player in the earth sciences more generally. When NASA was created after the launch of Sputnik, though, Congress gave the agency a very limited role in atmospheric science. Since then, as the Cold War waxed and waned, the agency has gained for itself a much broader "mission," to study the Earth, with profound consequences for the Earth and planetary sciences.

In 1989, the administration of George H. W. Bush approved the largest science project in US history: the Mission to Planet Earth. Designed as a $53 billion enterprise, this program was intended to study global change from a series of large orbital platforms administratively attached to Space Station *Freedom*. Now called the Earth Observing System and funded at a dramatically reduced level, it began its life in the late 1970s in a series of proposals aimed at expanding NASA's role in the earth sciences while also beginning a systematic study of climate-related processes. In one sense, the Mission to Planet Earth was a quintessential Cold War endeavor. In its original form, it was an example of engineering gigantism, a set of astronaut-tended platforms with dozens of instruments. The initiative also made NASA the largest funder of climate science in the United States.[1]

NASA is a creation of the Cold War. Founded after the Sputnik launch of October 1957, for its first few years the agency was aimed at developing orbital technologies, including remote sensing. President Eisenhower had not thought that men had much future role in space and opposed pursuing a men-to-the-moon program, a stance that was influenced by his scientific advisors.[2] That changed in May of 1961, when Vice President Lyndon Johnson convinced President John Kennedy to launch the Apollo expeditions to the moon.[3] That decision enshrined a tension within NASA between advocates of human space flight and the advocates of robotic science.[4] The Apollo decision also cemented a tendency toward the technological sublime in NASA—or, put more bluntly, toward technospectacle.[5]

That tendency toward technological spectaculars might imply that NASA's large role in the earth sciences has been turning Earth scientists into gadgeteers, the complaint Paul Forman has levied against Cold War physicists.[6] But in NASA, the opposite has been occurring. At first, the agency didn't have an expansive earth science program. The Space Act of 1958 didn't assign it one. Rather, like its predecessor agency, the National Advisory Committee on Aeronautics, it was given a narrower purview: "The expansion of human knowledge of phenomena in the atmosphere and space."[7] Two different research emphases derived from this mandate. Within NASA's science directorate, rocket-based stratospheric research imported from its predecessor continued. But its remote sensing activities took place in an Earth Observations technology program established within the "applications" portion of NASA's Office of Space Science and Applications, or OSSA. The Applications program, in turn, was intended to produce technologies useful to other government agencies or private industry. It wasn't specifically a science program. But by the mid 1980s, the Applications program had disappeared, replaced by a comprehensive geosciences program.

The disappearance of Applications and its replacement with an expansive earth science program reflects NASA's internal dynamics as it tried to remain relevant to its funder, Congress. Relatively early in NASA's existence, the Cold War competition with the Soviet Union lost its ability to command large funding levels, leaving the agency unable to finance its ambitious plans to build space stations and lunar bases. This also produced a crisis of relevance. What purpose was NASA to serve if it wasn't going to be funded for lunar or space colonization? The agency has still not solved that larger question. But in its effort to remain relevant in changed circumstances, it took on one new mission, ultimately called the Mission to Planet Earth. It was built on the remote sensing technologies developed by its Applications program, but with a focus on fundamental research instead.

A Short History of the NASA Applications Program

The Applications program's mode of operation was to develop Earth remote sensing technologies for use by other agencies.[8] During the 1960s, it had developed two series of Earth remote sensing satellites in partnership with other government agencies. The Nimbus series of weather satellites was developed to demonstrate the ability to improve weather forecasting with space data, and utilized instruments developed by NASA centers and by the predecessor to the National Oceanic and Atmospheric Administration, the Environmental Science Services Administration. The Nimbus program developed ultraviolet and infrared instruments for various kinds of atmospheric research; it also developed a set of microwave instruments for measurement of atmospheric temperature and for measuring sea ice.[9] Some of the instruments developed under the Nimbus program were eventually moved into the NOAA operational satellite series;

at least one microwave instrument migrated onto the Department of Defense's weather satellites.

The Applications program also developed a series of Earth resources satellites called Landsat, initially in partnership with the Department of the Interior. Landsat has had a very troubled history, struggling with national-security restrictions on the resolution of its imagery, with the availability of data processing and display technology for the imagery, and the challenge that still faces the system decades later: the inability to generate a sufficient user base to make it profitable. NASA and the US Geological Survey funded research on and with the Landsat datasets in order to create a user base that might enable the system to become at least self-supporting financially, if not actually profitable. But that strategy, while producing good scientific results, didn't create a large enough body of paying customers to continue the series. Repeated efforts to privatize and commercialize the system have not been successful, and NASA is funding the latest replacement effort, the Landsat Data Continuity Mission.[10]

Institutionally, during the 1960s, the Applications program was housed within the Office of Space Science and Applications.[11] The applications budget grew from $98 million in 1969 to $239 million in 1978.[12] Reflecting the elevated political stature, late in 1971, NASA removed the Applications program from OSSA and created a new directorate co-equal to the remaining Office of Space Science, the Office of Applications.[13]

President Nixon's 1972 approval of the Space Shuttle program further reinforced the trend toward Earth applications. The Shuttle's ability to reach only low Earth orbit impaired its utility for interplanetary launches—the Shuttle would have had to carry into orbit both the spacecraft and a high-energy (thus highly explosive) "upper stage" to thrust the spacecraft onto its interplanetary trajectory. Earth-orbiting satellites didn't need high-energy propulsion systems for their maneuvering; like the Shuttle's orbital maneuvering system, they use a fuel called hydrazine, which is toxic but not very explosive. Thus, with the Shuttle, interplanetary launches were riskier and more expensive than Earth-orbit missions.

In 1974, under pressure from its funding committee in Congress, NASA established a small stratospheric ozone-depletion research program within its science directorate. In part, this was a response to the Shuttle's own potential to deplete ozone, as well as ongoing interest in supersonic transports, which would inject potentially ozone-destroying oxides of nitrogen into the stratosphere. But the new initiative became politically salient very rapidly as the first of the American "ozone wars" broke out around chlorofluorocarbons, which at the time were used in aerosol sprays as well as refrigeration and air conditioning.[14] NASA defined a "pollution satellite" intended to examine this issue.[15] This was Nimbus 7, the last of the Applications program's series of meteorological research satellites. NASA also supported a host of balloon-borne and aircraft-borne instruments to further elaborate the chemistry of stratospheric ozone.

In 1976, Congress revised the Space Act to make study of the stratosphere a statutory responsibility, and in 1977 the program manager, Shelby Tilford, initiated studies for two more ozone research satellites, the Solar Mesosphere Explorer, to investigate ozone production in the mesosphere and upper stratosphere, and the Upper Atmosphere Research Satellite, aimed at the lower and middle stratosphere.[16]

The stratospheric research program drew in remote sensing specialists from both the Applications program's earlier efforts and some who had been working on planetary science during the previous decade. One scientist who moved between Mars and Earth was Crofton B. Farmer, whose stratospheric research became a central part of NASA's effort in the 1980s.[17] Others include Michael McElroy of Harvard University and Charles Barth of the University of Colorado, who practiced on Venus and Mars before beginning to work in photochemistry on Earth. Atmospheric modelers also found it relatively easy to move between planets, as it were.[18]

NASA's effort to bring remote sensing into stratospheric science was controversial. Planetary scientists had accepted remote sensing immediately—they had no other options, and many people calling themselves planetary scientists in the 1970s had come from astronomy, which had nothing but remote sensing. But Earth scientists had many other methodologies already. It didn't help that early remote sensing instruments didn't earn a great reputation for stability or accuracy, either. They tended to decay in space from radiation bombardment, and it proved difficult to correct for that decay. For example, during the effort to understand the Antarctic ozone hole discovered in 1985, there was a good deal of controversy surrounding the question of whether ground-based instruments or a Nimbus 7 instrument called SBUV/TOMS were providing more accurate data. Robert T. Watson, a NASA program manager, appointed an expert committee to review the question, and they sided with the ground-based data based on some evidence of instrument decay. Two years later, a scientist at the Goddard Space Flight Center in Maryland found a way of correcting the satellite data for on-orbit decay, without reference to the ground-based data.[19] The SBUV/TOMS instrument had been operating for twelve years by this time.

In part because of the problem of instrument decay, and also in part due simply to unfamiliarity with the technology among the geosciences, from the early 1970s on, NASA scientists had worked to improve remote sensing's credibility with the larger American scientific community. One way this was done by field experiments, which could bring NASA and outside scientists together physically, measuring the same thing with different techniques.[20] In addition to building credibility, these field experiments were also ways of recruiting new scientists (and graduate students in particular) into NASA-sponsored research.

In the process of building the stratospheric science program, the new program's managers also began to bring to the Applications program some of the other strictures of academic science, and that were already in use in NASA's space science directorate.

Donald Laurance of Langley Research Center implemented peer review of instrument proposals and opened proposing to researchers outside NASA. He, Ron Greenwood, the first of NASA's atmospheric science managers, and Greenwood's successor, Shelby Tilford, also began to demand publication in journals, trying to erase an old tradition—left over from the National Advisory Committee on Aeronautics days—of publishing internally (the so-called "gray literature").[21] The NACA had been the dominant research organization in aeronautics for decades, so its gray literature had very high standing in the aeronautics community.[22] But Greenwood and Tilford were trying to remake the Earth Applications program into an earth science program, which to them meant accommodating NASA as much as possible to the standard practices of the geosciences.

Despite the apparent political popularity of the Applications program in the mid 1970s, the basic Applications model of developing technology for others to use was faltering. It had worked well in the case of communications satellites, because, as with aviation, there was a ready commercial infrastructure that could apply NASA's research, but not for other parts of the Applications program. The endless problems with finding a paying customer to support the Landsat program, as Pamela Mack has documented, was one strike against the model.[23] Another was the difficulty NASA had getting oceanographers and the US Navy to accept, and pay for, ocean remote sensing.[24] Even the most apparently successful, the Nimbus experimental weather satellite program, ran aground at the end of the decade. The Nimbus satellite bus itself wasn't compatible with Shuttle deployment and needed to be replaced. By 1978, NOAA had also adopted some of the Nimbus-developed generation of remote sensing instruments into its own weather satellites, and when NASA sent tentative specifications for a set of next generation instruments to NOAA for verification, Tilford recalls, the agency would not accept them.[25] Nor did NOAA use the satellite data for its operational weather forecasting, although military weather forecast agencies did. Ultimately, the Operational Satellite Improvement Program the two agencies had pursued for the preceding two decades ended in 1982, and was not replaced by another formal mechanism for transferring NASA technology to NOAA.[26] The Applications program's success depended upon partnerships with other federal agencies and finding users willing and able to pay for sustaining operations, and these proved very difficult to forge and maintain.

Yet while the Applications program's long-standing efforts were ending, Congress was actually beginning to demand still more Earth-oriented work from NASA. In addition to the mandated upper atmosphere research program on ozone depletion, in 1978 Congress passed the National Climate Program Act, which required federal agencies to develop research programs to study what was then still often referred to as "inadvertent climate modification."[27] Tilford and David Atlas at Goddard Space Flight Center began to plan a comprehensive, space-based climate research program in response. It was not to be set up within the Applications program, though. NASA's senior leadership didn't think the Office of Applications had aggressive enough leadership to pursue the opportunities that had appeared.[28] Instead, like the Upper Atmosphere Research Program, it was to be established within the science directorate.

NASA's senior leaders recognized the popularity of applied research in Congress, but they also believed that the Applications program wasn't particularly effective. Outgoing NASA administrator James Fletcher told his successor in 1977 that he thought the Applications Program was the "'wave of the future' as far as NASA's public image is concerned. It is the most popular program (other than aeronautics) in the Congress, and as you begin to visit with community leaders, you will understand it is clearly the most popular program with them as well." He also explained, though, that since the weather satellites that had such a high public image weren't actually being used by the National Oceanic and Atmospheric Administration to drive weather forecast models, "to many [in NASA], weather satellites are mostly talk and not much show."[29] For this and other reasons Fletcher laid out, the Applications program didn't have broad support within NASA or with the very important Office of Management and Budget, regardless of its popularity in Congress.

NASA and the Problem of Relevance

NASA was a product of the Cold War, and all of its programs were affected as Cold War fears became less effective at fostering funding, but none more so than its human space-flight program. Like any public agency, NASA had to perform a function that its congressional funding committees thought was worth performing. It had to remain relevant. The increasing cost of the war in Vietnam helped undermine both congressional and public support for NASA's human component. Apollo had never been very popular with Congress, and attacks on its budget took place every year from 1963 on and expenditures fell after 1965. Although many believe that Apollo enjoyed broad public support, at the time polls asking whether Americans supported government funding of human trips to the moon generally found a minority of support. Even polls taken just as Apollo 11 was making its historic voyage found only 53 percent of the public thought Apollo was worth its price.[30] The former NASA Chief Historian Roger Launius concluded that "the political crisis that brought public support to the initial lunar landing decision was fleeting and within a short period the coalition that announced it had to reconsider their decision. It also suggests that the public was never enthusiastic about human lunar exploration, and especially about the costs associated with it."[31]

The Apollo program died in stages. In 1970, President Nixon, a supporter of balanced budgets despite the costs of the Cold and Vietnam wars, insisted "space expenditures must take their proper place within a rigorous system of national priorities."[32]

Polls showed a majority of the public believed that the Apollo program was too expensive and a large majority thought NASA received too much money.[33] And his administration imposed budget cuts on NASA that ended the moon program and refocused its efforts on low Earth orbit. The final Apollo mission then planned, Apollo 20, was canceled that January, and a few months later, production of the Saturn V was "suspended"—permanently, as it turned out. By the end of 1970, two more missions were canceled: Apollo 15 and Apollo 19. In 1971, there was consideration of canceling Apollo 16 and Apollo 17, but the Office of Management and Budget director intervened to preserve them.[34] One set of Apollo hardware from the canceled missions was used for the Skylab mission in 1973, which among other scientific activities carried out a number of Earth remote sensing experiments.[35] Another set was used for the final Apollo flight—the Apollo/Soyuz test project of July 1975, a visible symbol of détente.[36]

The decline in fortune for NASA's human space-flight program left agency leaders scraping for funds to sustain the Apollo successor program, the Space Shuttle. The agency's planetary science program was the most damaged. After a strong increase in funding for planetary exploration between 1970 and 1973, funding fell until 1978 and then roughly stabilized until 1985.

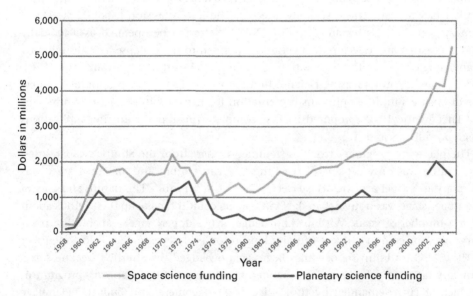

Figure 8.1
Funding for space science and planetary science in the years 1958–2005 (in fiscal year 2006 dollars). The gap in the planetary line is due to elimination of the separate budget line item for planetary missions during that period. Data used for 1958–1996 were from NASA's Historical Databook series; data used for 1997–2005 were from annual budget submissions.

There were no planetary mission "new starts" authorized between 1978 and 1984. The decline in funding for planetary missions wasn't a result of congressional action; indeed, on several occasions, the congressional funding committees criticized NASA for not proposing new planetary missions. In the fiscal year 1977 budget discussions, the House Committee on Science and Technology complained about the lack of new planetary mission proposals, stating "such a gap will likely create an inefficient employment of resources."[37] The following year, the House and Senate committees attempted to force NASA to start a Lunar Polar Orbiter mission by funding one over agency objections; NASA canceled it anyway, and in the fiscal year 1979 budget discussions the congressional committees demanded again to see it proposed.[38] And they complained: "The Committee is concerned by the lack of new starts in the Physics and Astronomy Program and Planetary Exploration Program for fiscal year 1980. The National Aeronautics and Space Act of 1958 mandate for the expansion of human knowledge of phenomena in the atmosphere and space dictates a commitment to program continuity. Without new and challenging initiatives, the currently healthy character of NASA's space science activities cannot be sustained."[39]

NASA Headquarters and the White House Office of Management and Budget, then, were responsible for the decline in planetary science's fortunes in the 1970s and the 1980s. The agency's focus on completing Shuttle led it to starve its scientific arm, with its planetary activities most strongly affected. The Jet Propulsion Laboratory, NASA's primary planetary center and the only NASA center to be operated as a federally funded research and development center, responded to the collapse in its funding by establishing energy and transportation research programs that ran through the 1970s, and by drawing on increasing Defense funding in the 1980s.[40] It also initiated a new line of business in developing instrumentation for Earth-orbiting scientific missions, and later in the 1970s proposed its first complete mission within the Applications program, called Seasat A.[41]

The planetary program's troubles stemmed in part from the Shuttle's capabilities. The Shuttle was only capable of reaching low Earth orbit, and an upper stage was necessary to propel a planetary spacecraft out of Earth orbit. The nature and cost of that upper stage was hotly debated; NASA leaders couldn't make up their minds about it for a number of years. Without knowledge of the upper stage's capabilities, planetary missions couldn't proceed in any kind of orderly development.[42] In September of 1981, NASA Administrator James Beggs acknowledged the Shuttle's weakness as a planetary launcher when he recommended termination of the planetary program in the face of cuts demanded by the Office of Management and Budget. "Planetary exploration is much more highly dependent on launch vehicles, and it is our opinion that the most important missions that can reasonably be done within the current launch vehicle capability have, more or less, been done."[43] "Full development" of the

Shuttle should be achieved before planetary exploration should be revived, Beggs had argued.

The first space shuttle, *Columbia*, was launched three years late, on April 12, 1981. Its first four flights were considered "engineering" flights, although beginning with the second flight, in November of 1981, it did carry Earth-observation experiments.[44] As a low-Earth-orbit launcher, it favored the kinds of Earth-oriented scientific and applied research that Congress had sought from the agency during the 1970s and that the Applications program existed to facilitate. Indeed, it could do little else, as the space station it was intended to serve had not been approved, and would not exist, as it turned out, until the late 1990s. Though its advocates had seen the Shuttle as the first piece of infrastructure necessary for sustained human exploration of the solar system, the reality that the other pieces of such an infrastructure (space station, orbit transfer vehicles, and so on) weren't built meant that the Shuttle confined human "exploration" to low Earth orbit. And from that vantage point, Earth and applied sciences were an obvious choice of justification. They solved, at least in part, the agency's relevance problem.

But it wasn't at all clear to NASA's leaders that the Applications program should play host to a re-commitment to Earth-oriented research. In the same memo Beggs had recommended termination of planetary exploration to OMB in, he had also recommended elimination of the Applications program.[45] Like Fletcher, Beggs believed that the leadership of the Applications program was weak, and a reformulation of its approach was necessary.

Constructing Earth System Science

NASA had recruited a small legion of scientists interested in planetary climate during its first decade of existence, many of them primarily involved in studies of Venus and Mars. By the late 1970s, scientists outside NASA had gotten interested in climate, too, in no small part because it had started to become a subject of policy debate. NASA's leaders chose to respond to the 1978 National Climate Program Act by beginning to plan for a comprehensive Earth-observation program to study climate processes. But in fact, Tilford and the deputies he had brought in, Robert Watson from the Jet Propulsion Lab and Dixon Butler from Goddard Space Flight Center, went far beyond what NASA had done under the Applications program.

At NASA, the comprehensive Earth-observation satellite system that entered its planning phase in the late 1970s had a number of different names, and assumed a number of different forms as it evolved. The first version, and certainly the simplest, was devised during 1980. David Atlas of Goddard Space Flight Center organized the effort, which was chaired by Verner Suomi, a meteorologist at the University of

Wisconsin. Suomi argued for gradually expanding the weather satellite program to include chemistry and climate instruments, with NASA developing and demonstrating the new instrument technologies and NOAA supporting and managing the expanded system over the long term.[46]

This initial plan went nowhere, perhaps due to the termination of the Applications program's partnership with NOAA. In 1981, Ronald Reagan had been inaugurated as president with a strong antipathy to government-provided services. His administration dismantled the public/private Comsat Corporation, attempted to privatize the weather satellite system, and also further commercialized the Landsat series of satellites.[47] In short, the sorts of things that NASA had been doing in its Applications program were not things the administration was willing to support. In response, late in 1981 NASA officials re-merged the Office of Applications with the Office of Space Science to reestablish the Office of Space Science and Applications. Reflecting the increasing interest in earth science, the old applications directorate took the name "Earth Science and Applications Division."[48]

Yet Tilford and, as it turned out, the political appointees Reagan brought into NASA, still wanted to expand the agency's role in earth science. The first initiative to gain real traction within NASA was called Global Habitability. NASA leaders spent the early 1980s trying to sell the Reagan White House on a grand space station endeavor, ultimately called Space Station *Freedom*.[49] It was designed to compete directly with the Soviet Union's space station, eventually launched in 1986 as *Mir*. But NASA's deputy administrator, Hans Mark, wanted to use the station project to further develop remote sensing technology, while its associate administrator for space science, Burt Edelson, wanted to use it to develop large space platform technologies for use by the telecommunications industry. They agreed to attach (organizationally, not physically) development of large remote sensing platforms to the space station project as part of its justification.[50]

The research program intended to justify the large space platforms was initially called Global Habitability. As the name suggested, it was aimed at investigating "long-term physical, chemical and biological trends and changes in the Earth's environment, including its atmosphere, land masses, and oceans." It would "specifically investigate the effects of natural and human activities on the Earth's environment," and "estimate the future effects on biological productivity and habitability of the Earth by man, by other species, and the effects on natural causes."[51] It was announced in 1982, at that year's UNISPACE meeting, receiving near-universal condemnation.[52] NASA had not told its European allies of the initiative, and the reality that the polar platforms would be collecting high-resolution imagery of every nation in the world didn't go over well either with the Warsaw Pact nations or with the non-aligned nations. Global Habitability, as a result, also didn't go very far. The one significant change to have come from the initiative was another change to the Space Act. Congress added the requirement to "materially contribute to ... expansion of human knowledge of the Earth," in 1984, justifying a truly expansive earth science program.[53]

The basic idea of a comprehensive Earth-observation system didn't die with Global Habitability. In part, the idea had momentum due to a pair of studies done in parallel with the Global Habitability initiative, the reports on which were titled "Global Change: Impacts on Habitability" and "Toward an International Geosphere-Biosphere Program."[54] Further, the manager of NASA's earth science program, Shelby Tilford, had created an advisory panel of scientists in 1982 to formulate a discipline around the idea of studying Earth as a dynamic system. Tilford thought this was "simply the next logical step." "We'd been trying to do things piecemeal, some atmospheric satellites, some ocean satellites, some of the solar-observation satellites," he continued. "But no one had sat down to figure out how all this fits together."[55] Wesley T. Huntress, who moved from JPL to NASA Headquarters to become Tilford's deputy, reflected later that NASA "was the right place to do it because we were the ones who were going to be studying the planet from a global perspective, looking at it at large scales and trying to understand how to put the smaller scales together."[56]

The committee chairman Tilford chose was Francis Bretherton, who had just stepped down as the director of the National Center for Atmospheric Research in Boulder, Colorado. It took Bretherton several years to wrangle an agreement out of his group. At its root, this was the product of classic disciplinary disputes over standards of evidence, measurement methodologies, and even timescales. The atmosphere changes far faster than does the solid earth, so an observing system suitable for atmospheric measurement wasn't necessarily suitable for solid earth sensing. This made it very difficult to produce agreement within the committee. As a result, the committee's report, Earth System Science: A Program for Global Change, wasn't released until 1986. The classic icon of this report is the "Bretherton diagram," a graphical exposition of the links between various parts of the Earth system.

The basic idea underlying Earth System Science was to study the interlinked processes involved in global change, encompassing a "variety of interrelated natural processes, including changes in the climate system, in solar processes, in the Earth's orbit, in volcanic processes, and in the distribution of biological species and land masses that may have been ongoing for centuries."[57] It wasn't strictly confined to climate change, because that was far too narrow a subject to interest the broad range of Earth scientists whose support NASA sought.

Earth System Science, a term coined by panel member Moustafa Chahine of the Jet Propulsion Laboratory, mapped closely onto James Lovelock's Gaia hypothesis.[58] Both Gaia and Earth System Science were based in a systems engineering view of the world (Lovelock had called it cybernetic engineering), a profession that specialized in (nonliving) feedback control systems. For Earth scientists, Lovelock and the Bretherton committee presented a view of the world that required examination of complex,

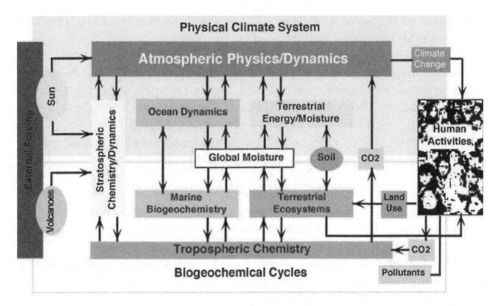

Figure 8.2
A simplified Bretherton diagram showing the links between various portions of the climate system. From *Earth System Science: An Overview* (NASA, 1988).

interlocking feedback loops. Understanding them required interdisciplinary research that the American scientific community wasn't set up to foster, and was therefore difficult to sustain.[59]

After the loss of *Challenger* in 1986, the entire Space Shuttle program was put under review. The Department of Defense had already begun questioning its reliance on a vehicle that was turning out to be unreliable and very expensive; two years before, the White House had approved an initiative to develop a new launch vehicle to replace it—though that effort came to naught.[60] In the shorter term, the Reagan administration decided to remove most commercial, military, and scientific payloads from the Space Shuttle and re-start production of expendable launch vehicles for them. The Shuttle's purpose became construction and servicing of Space Station *Freedom* and the lofting of the occasional "attached payload," generally scientific missions that returned to the Earth when the Shuttle did.[61]

As part of the recovery from the *Challenger* accident, astronaut Sally Ride was asked to chair a committee that was to propose new missions for NASA to undertake. In 1987, her committee rebranded the comprehensive Earth-observation concept that had been evolving in NASA headquarters as "Mission to Planet Earth," recycling the name from an 1985 editorial by Burt Edelson, and again tying it to Space Station

Freedom.⁶² The "Mission to Planet Earth" wasn't approved until 1989 for fiscal year 1990, but when it first gained the White House's assent it had a price tag of $53 billion for 15 years of data collection.⁶³ This gigantism brought it under fire from scientists inside and outside NASA. Many scientists believed it would consume all the nation's earth science funds for the next two decades, depriving more traditionally trained and motivated investigators of research funds. Both the director of NASA's Goddard Institute of Space Studies, climate modeler James Hansen, and Senator Al Gore Jr. were outspoken critics of the system's pacing and cost, for example.⁶⁴

The sudden dissolution of the Warsaw Pact and the Soviet Union between 1989 and 1991 then threw all the American "big science" programs into disarray. Reducing the enormous budget deficit produced by the expansion of military spending during the 1980s came into vogue. Historian Daniel Kevles has already examined the sudden debt reduction fervor's impact on the Superconducting Super Collider (SSC), canceled in 1993 while NASA's space station survived, but with substantial cuts.⁶⁵ Space Station *Freedom* was revamped as the much less ambitious International Space Station, which drew (heavily, as it turned out) on the resources of the former USSR. Kevles concluded that the SSC's demise represented the relative disestablishment of high-energy physics. "Physics in the United States," he wrote, "has been irreversibly incorporated into the conventional political process, making it a creature of political democracy, its fortunes, like those of other interest groups, contingent on the outcome of the fray."⁶⁶

So too the "Mission to Planet Earth." Between 1990 and 1995, it was downscaled greatly. The approved budget of $30 billion for the period 1990–2000 shrunk over five years to $6.5 billion.⁶⁷ There were a variety of reasons for its scaling back. In addition to criticism of its pacing and architecture from elected officials and scientists who actually supported NASA's efforts to expand the agency's role in earth science, the "Mission to Planet Earth" effort ran into ideological opposition from the American political right.⁶⁸ One consequence was that NASA sought greater foreign participation in the program, hoping to protect some of its ambitious goals by drawing on the technical skills and financial resources of European, Japanese, and South American allies. But 1995 was the last year of major cuts to the program; it stabilized thereafter, and continued to be funded at between $1.2 and $1.4 billion per year after 2000.

The "Mission to Planet Earth" brought an end to the old Applications directorate. Applications had been organizationally downgraded in 1984, having been merged into Shelby Tilford's Office of Earth Science and Applications. In 1993, that office was removed from the Office of Space Science at Headquarters and made its own directorate named Mission to Planet Earth.⁶⁹ The Mission to Planet Earth was equal in status to the Office of Space Science, reflecting its size and complexity as a program. But applications fell off the agency organization chart for the first time, drawing congressional criticism.⁷⁰ When the Mission to Planet Earth was renamed in 1999 the "Earth

Science Enterprise," the applications label didn't stage a return to the upper reaches of the agency organization charts.[71]

The reduction in status of applications within NASA reflected a sea change in what NASA leaders thought the agency should be: a doer of earth science research, not primarily a developer of useful products. While the Mission to Planet Earth did provide small amounts of money for applied research, funding, for example, development of a forest fire alert tool based on MODIS sensor imagery, its focus was not on applications, but on fundamental research.[72] In one sense, this was an obvious result of the mid-1980s decision to move applications into the space science arm of NASA; the internal culture of the space science directorate was likely to bestow greater status on basic research than on applied, particularly given the poor perception of the old Applications program that lingered. But it was also reinforced by the larger political culture of the 1990s, which held that "applied research" or "operational" functions were properly the province of private enterprise.

The Mission to Planet Earth's mission of fundamental research also kept Earth System Science alive. By the early 2000s, when the program's satellites began to reach orbit, NASA was routinely spending between $300 and $400 million on earth science research across nearly all the geoscience disciplines (plus the cost of the satellites, generally another $900 million to $1 billion per year). For comparison, the National Science Foundation's Geosciences Directorate budget was $478 million in fiscal year 1999, and $488 million in fiscal year 2000.[73] Thus, although NASA didn't become the largest funder of the geosciences in the United States (unless one adds the cost of the space hardware), the Mission to Planet Earth provided sufficient new resources to be influential within the geoscience community.

At the same time that NASA's expansion of its earth science endeavor began to take hold in the 1990s, planetary exploration was beginning a gradual revival. After a near-death experience in 1981, when the Office of Management and Budget had tried to eliminate planetary exploration entirely, White House policy reversed course. A new Mars mission, Mars Observer, and a Venus mission, Magellan, had been approved, and once the space shuttles began flying again after the 1986 *Challenger* explosion, the Galileo mission to Jupiter was launched as well.[74] The second NASA administrator chosen by George H. W. Bush, Daniel S. Goldin, enabled a renaissance in planetary exploration. Goldin also sought deep cuts in the NASA budget as part of a reform agenda. Under the rubric of 'faster better cheaper,' he and Wesley Huntress, who became associate administrator for space science under Goldin, hoped to accomplish far more planetary missions for less money.[75] For the 1990s, they succeeded, although the end of the decade also saw the end of the "faster, better, cheaper" approach and a return to a slower, more expensive planetary program.

The end of the Cold War competition with the Soviet Union, then, although it jeopardized NASA's space station and earth science dreams for a few years in the

1990s, didn't threaten the planetary program, another of the Cold War's legacies. Instead, something much more interesting happened. As NASA began funding both Earth and planetary science more fully in the 1990s, these disciplines grew together. There had been a small number of interplanetary, and interdisciplinary, papers published in American scientific journals for decades; for example, 53 papers in the Web of Science database from the period 1980–1990 drew explicit comparisons between Earth and Mars. During the 1990s, more than a thousand such papers were published, even though before 1997 no new data from Mars were available.[76] Though the causes of this publication explosion need further research, one suspects this enormous increase is as much due to the increasing availability of inexpensive computing, which made the old Mariner and Viking datasets much more accessible than they had been in the era of expensive mainframes, as it is to NASA program managers' stated desire to foster interdisciplinary study. A consequence of NASA's support for the twin disciplines of Earth and planetary sciences has been their increasing merger at the university level, and a blurring, if not an erasure, of disciplinary boundaries.[77]

Conclusion

Although the Cold War fostered NASA, that agency has actively sought new missions for itself since the need to compete with the Soviet Union lost its ability to command funding. W. Henry Lambright has called this willingness, if not eagerness, to adapt to new circumstances "administrative entrepreneurship."[78] NASA leaders used opportunities provided by the changing tenor of the Cold War and by the vagaries of American domestic politics to pursue scientific goals that they thought were important. NASA's component centers followed this lead. During the 1960s, for example, the Jet Propulsion Laboratory was entirely devoted to lunar and planetary exploration; by the late 2000s, about 15 percent of its budget consistently came from earth science missions.[79] As a consequence, NASA became a major funder of the geosciences as well as of the planetary sciences.

This story also helps to illustrate the limits of what Cold War politics could accomplish. Though NASA officials continued to use Cold War rhetoric in their public communications, the agency quickly lost the ability to draw large amounts of money from Congress for its Cold War mission. Unlike the defense-related scientific agencies, which also struggled financially during the 1970s but saw a strong funding revival during the 1980s, NASA never recovered its original post-Sputnik support. Instead, it became immersed in "normal" domestic politics. Normal politics required demonstrating the agency's utility to domestic constituencies, and as a result NASA focused more on Earth. As a consequence of this redirection of effort, it became a major actor in environmental controversies of the 1980s and the 1990s. Another consequence is that

NASA's science budget has gradually increased relative to the rest of the agency's budget, reflecting the success of its programmatic redirection.

But NASA experienced its funding crisis far earlier than is suggested in this volume by Naomi Oreskes, who asserts that physical oceanographers sought new funding opportunities after the sudden end of the Cold War reduced their traditional sources of military funding. For NASA, the search for renewed political relevance began in the 1970s. (In fact, NASA's first attempt to recruit physical oceanographers into its own constituency took place in the late 1970s.[80]) Thus, different US agencies, and the different pieces of the scientific enterprise that they supported, had divergent histories in the same period.

Disclaimer and Acknowledgment

The author is employed by the Jet Propulsion Laboratory, California Institute of Technology. All opinions and interpretations contained in this chapter are those of the author and do not represent JPL, Caltech, or NASA policy.

Notes

1. Of the $1.86 billion dollars spent by the United States on climate science in fiscal year 2005, NASA spent $1.24 billion. Of that sum, NASA spent $519 million on climate-related research, and $722 million on developing and procuring new climate-related space-based observation technologies. The next largest funder of climate science, the National Science Foundation, spent $198 million that year. The National Oceanic and Atmospheric Administration spent only $120 million (source): OCP2007-budget-table3.pdf at http://www.usgcrp.gov). The political controversies surrounding NASA's earth science activities are documented in a variety of histories of stratospheric ozone and climate science: at the diplomatic level, see Edward A. Parson, *Protecting the Ozone Layer: Science and Strategy* (Oxford University Press, 2003); at NASA's level, Erik M. Conway, *Atmospheric Science at NASA: A History* (Johns Hopkins University Press, 2008), especially chapters 6 and 9; Naomi Oreskes and Erik M. Conway, *Merchants of Doubt* (Bloomsbury USA, 2010); Mark Bowen, *Censoring Science: Inside the Political Attack on Dr. James Hansen and the Truth of Global Warming* (Dutton, 2008); also see James Hansen, *Storms of My Grandchildren: The Truth about the Coming Climate Catastrophe and Our Last Chance to Save Humanity* (Bloomsbury USA, 2009), especially chapters 2 and 3.

2. Zuoyue Wang, *In Sputnik's Shadow: The President's Science Advisory Committee and Cold War America* (Rutgers University Press, 2008), 97.

3. John M. Logsdon, "Project Apollo: Americans to the Moon," in *Exploring the Unknown*, volume VII: *Human Spaceflight: Projects Mercury, Gemini, and Apollo* (NASA, 2008), 395.

4. See chapter 1 of Roger Launius and Howard McCurdy, *Robots in Space: Technology, Evolution, and Interplanetary Travel* (Johns Hopkins University Press, 2008).

5. In chapter 9 of *American Technological Sublime* (MIT Press, 1994), David E. Nye presents Apollo as the epitome of the dynamic sublime.

6. Paul Forman, "Behind Quantum Electronics: National Security as Basis for Physical Research in the United States, 1940–1960," *Historical Studies in the Physical and Biological Sciences* 18, no. 1 (1987): 149–229.

7. "National Aeronautics and Space Act of 1958," Public Law 85-568, 72 Stat., 426, signed July 29, 1958. The unamended version is available at http://history.nasa.gov/spaceact.html.

8. For a discussion that emphasizes the technology developed by NASA and the agency's inability to recruit an adequate user base, see Pamela E. Mack and Ray A. Williamson, "Observing the Earth from Space," in *Exploring the Unknown: Selected Documents in the History of the US Civil Space Program*, volume III: *Using Space*, ed. John M. Logsdon et al. (NASA, 1998).

9. Erik M. Conway, *Atmospheric Science at NASA: A History* (Johns Hopkins University Press, 2008), 141–145.

10. Mack and Williamson, "Observing the Earth from Space."

11. Linda Newman Ezell, *NASA Historical Databook*, volume 3: *Programs and Projects, 1969–1978* (NASA, 1988), 237.

12. Ezell, *NASA Historical Databook*, volume 3: *Programs and Projects, 1969–1978*, 245.

13. Ibid., 243.

14. The first popular history of the campaign against CFC use, authored by Lydia Dotto and Harold Schiff, was titled *The Ozone War* (Doubleday, 1978).

15. Conway, *A History of Atmospheric Science at NASA*, chapter 6.

16. history.nasa.gov/spaceact-legishistory.pdf.

17. Conway, "Earth Science and Planetary Science: A Symbiotic Relationship?" in *NASA's First 50 Years: Historical Perspectives*, ed. Steven J. Dick (NASA, 2010).

18. Ibid., 572–579.

19. Robert Watson established a panel of scientists known as the Ozone Trends Panel to resolve the conflict between ground and satellite ozone data. See Edward Parson, *Protecting the Ozone Layer: Science and Strategy* (Oxford University Press, 2003), 153–155; J. R. Herman et al., "A New Self-Calibration Method Applied to TOMS and SBUV Backscattered Ultraviolet Data to Determine Long-Term Global Ozone Change," *Journal of Geophysical Research* 96, no. D4 (1991): 7531–7545.

20. Conway, *Atmospheric Science at NASA*, 122–154.

21. Tilford also credits Donald Hearth, Paul Holloway, and Donald Laurence at Langley Research Center for promoting journal publication within the former NACA centers: Shelby Tilford, interview with Conway, February 11, 2004, NASA History Office, Washington.

22. The NACA's role in aeronautics is subject of many histories. See Alex Roland, *Model Research* (NASA, 1985); James R. Hansen, *Engineer in Charge: A History of the Langley Aeronautical Laboratory, 1917–1958* (NASA, 1987); Roger Bilstein, *Orders of Magnitude: A History of the NACA and NASA, 1915–1990* (NASA, 1989); Richard P. Hallion, On *the Frontier: Flight Research at Dryden, 1946–1981* (NASA, 1984); Virginia P. Dawson, *Engines and Innovation: Lewis Laboratory and American Propulsion Technology* (NASA, 1991).

23. Pamela E. Mack, *Viewing the Earth: The Social Construction of the Landsat Satellite System* (MIT Press, 1990).

24. Erik M. Conway, "Drowning in data: Satellite oceanography and information overload in the Earth Sciences," *Historical Studies in the Physical and Biological Sciences* 37, no. 1 (2006): 127–151.

25. Shelby Tilford, interview with author, February 11, 2004.

26. Board on Atmospheric Sciences and Climate, National Research Council, *From Research to Operations in Weather Satellites and Numerical Weather Prediction: Crossing the Valley of Death* (National Academy Press, 2000).

27. Public Law 95-367, September 17, 1978, available at epw.senate.gov/ncpa.pdf.

28. James C. Fletcher to Bob Frosch, "Problems and Opportunities at NASA," May 9, 1977, reproduced in *Exploring the Unknown: Selected Documents in the History of the US Civil Space Program*, volume 1, ed. John M. Logsdon et al. (NASA, 1995), 712.

29. James C. Fletcher to Bob Frosch, "Problems and Opportunities at NASA," May 9, 1977, reproduced in *Exploring the Unknown: Selected Documents in the History of the US Civil Space Program*, volume 1, ed. John M. Logsdon et al. (NASA, 1995), 712.

30. Roger Launius, "Public Opinion Polls and Perceptions of US Human Spaceflight," *Space Policy* 19 (2003): 163–175.

31. Ibid, 168.

32. John M. Logsdon, "Project Apollo: Americans to the Moon," in *Exploring the Unknown: Selected Documents in the History of the US Civil Space Program*, volume VII, ed. John M. Logsdon (NASA, 2008), quoted from p. 436.

33. Roger Launius, "Public Opinion Polls and Perceptions of US Human Spaceflight," *Space Policy* 19 (2003): 163–175.

34. Logsdon, "Project Apollo," 437.

35. W. David Compton and Charles D. Benson, *Living and Working in Space: A History of Skylab* (NASA, 1983).

36. For a history of ASTP, see Edward Clinton Ezell and Linda Neuman Ezell, *The Partnership: A History of the Apollo-Soyuz Test Project* (NASA, 1978).

37. House of Representatives, Committee on Science and Technology, "Authorizing Appropriations to the National Aeronautics and Space Administration," Report 94.897, 94th Cong., 2nd Sess., reproduced in NASA, Chronological History Fiscal Year 1977 Budget Submission, August 23, 1977, 11.

38. House of Representatives, Committee on Science and Technology, "Authorizing Appropriations to the National Aeronautics and Space Administration," Report 95-67, 95th Cong., 1st Sess., reproduced in NASA, Chronological History Fiscal Year 1978 Budget Submission, 11; House of Representatives, Committee on Science and Technology, "Authorizing Appropriations to the National Aeronautics and Space Administration," Report 95-973, 95th Cong., 2nd Sess., reproduced in NASA, Chronological History Fiscal Year 1979 Budget Submission, August 31, 1979, 12.

39. House of Representatives, Committee on Science and Technology, "Authorizing Appropriations to the National Aeronautics and Space Administration," Report 96-52, 96th Cong., 1st Sess., reproduced in NASA, Chronological History Fiscal Year 1980 Budget Submission, October 9, 1980, 12.

40. On JPL's attempt to accommodate itself to solving domestic problems, see pp. 232–242 of Clayton Koppes, *JPL and the American Space Program* (Yale University Press, 1982). For another interpretation, see pp. 124–163 of Jennifer Light, *From Warfare to Welfare* (Johns Hopkins University Press, 2003).

41. Erik M. Conway, "Drowning in Data: Satellite Oceanography and information overload in the Earth Sciences," *Historical Studies in the Physical and Biological Sciences* (2006) 37, no. 1: 127–151.

42. Bruce C. Murray, *Journey into Space: The First Thirty Years of Space Exploration* (Norton, 1989), 203–219; Michael Melzer, *Mission to Jupiter: A History of the Galileo Project* (NASA, 2007), 37–51.

43. James M. Beggs to the Honorable David Stockman, September 29, 1981, reproduced in *Exploring the Unknown: Selected Documents in the History of the US Civil Space Program*, volume V: *Exploring the Cosmos*, ed. John M. Logsdon (NASA, 2001), 433.

44. C. Elachi et al., "Shuttle Imaging Radar Experiment," *Science*, December 3, 1982: 996–1003; "Serendipitous Discoveries," Jet Propulsion Laboratory video AVC-2008-156.

45. James M. Beggs to the Honorable David Stockman, September 29, 1981, reproduced in *Exploring the Unknown: Selected Documents in the History of the US Civil Space Program*, volume V: *Exploring the Cosmos*, ed. John M. Logsdon (NASA, 2001), 433.

46. Conway, *Atmospheric Science at NASA*, 213–219.

47. Hugh R. Slotten, "Satellite Communications, Globalization, and the Cold War," *Technology and Culture* 43, no. 2 (2002): 315–350; David J. Whalen, "For All Mankind: Societal Impact of Application Satellites," *Societal Impact of Spaceflight*, SP-2007-4801 (NASA, 2007), 289–312. On meteorological and Landsat activities, see Judy Rumerman, *NASA Historical Data Book*, volume 6: *NASA Space Applications, Aeronautics and Space Research and Technology, Tracking and Data*

Acquisition/Support Operations, Commercial Programs, and Resources, 1979–1988 (NASA, 1999), 11–21.

48. Judy A. Rumerman, *NASA Historical Data Book*, volume 5: *NASA Launch Systems, Space Transportation, Human Spaceflight, and Space Science, 1979–1988*, 370–371; *NASA Historical Data Book*, volume 6: *NASA Space Applications, Aeronautics and Space Research and Technology, Tracking and Data Acquisition/Support Operations, Commercial Programs, and Resources, 1979–1988* (NASA, 1999), 11–21.

49. On the process of selling the space station, see Howard McCurdy, *The Space Station Decision: Incremental Politics and Technological Choice* (Johns Hopkins University Press, 1990).

50. Conway, *Atmospheric Science at NASA*, 220, 232.

51. S. G. Tilford and J. R. Page, "Global Habitability and Earth Remote Sensing," *Philosophical Transactions of the Royal Society A* 312 (1984): 115–118.

52. Edward S. Goldstein, "NASA's Earth Science Program: The Space Agency's Mission to Our Home Planet," in *NASA's First 50 Years: Historical Perspectives*, ed. Steven J. Dick (NASA, 2010); William H. Lambright, "Administrative Entrepreneurship and Space Technology: The Ups and Downs of 'Mission to Planet Earth,'" *Public Administration Review* 54, no. 2 (1994), 97–104; Conway, *Atmospheric Science at NASA*, 219–225.

53. See National Aeronautics and Space Act of 1958, As Amended, August 25, 2008 (http://history.nasa.gov/spaceact-legishistory.pdf), p. 4.

54. Richard M. Goody, "Global Change: Impacts on Habitability," JPL D-95, 1982; National Research Council, "Toward an International Geosphere-Biosphere Program: A Study of Global Change," Report of a National Research Council Workshop, Woods Hole, Massachusetts, 1983. Also see Chunglin Kwa, "Local Ecologies and Global Science: Discourses and Strategies of the International Geosphere-Biosphere Programme," *Social Studies of Science* 35, no. 6 (2005): 923–950.

55. Shelby Tilford, interview with author, February 11, 2004.

56. Wesley T. Huntress, interview with author, March 17, 2008.

57. Federal Coordinating Council for Science, Engineering, and Technology, Committee on Earth Sciences, "Our Changing Planet: A US Strategy for Global Change Research," 1989 (http://www.usgcrp.gov/usgcrp/archives), 7.

58. Conway, *Atmospheric Science in NASA: A History*, 114–116; Lynn Margulis and J. E. Lovelock, "Biological Regulation of the Earth's Atmosphere," *Icarus* 21 (1974): 471–489; J. E. Lovelock, *Gaia: A New Look at Life On Earth* (Oxford University Press, 1974), 13–32.

59. Lovelock, *Gaia*, 48–63; Margulis and Lovelock, "Biological Regulation of the Earth's Atmosphere," 487. On the difficulty of sustaining interdisciplinary research before the space age, see Ronald E. Doel, *Solar System Astronomy in America: Communities, Patronage, and Interdisciplinary Research, 1920–1960* (Cambridge University Press, 1996).

60. Conway, *High Speed Dreams* (Johns Hopkins University Press, 2005), chapter 6. Also see Ivan Bekey, "Exploring Future Space Transportation Possibilities," in *Exploring the Unknown: Selected Documents in the History of the US Civil Space Program*, volume IV: *Accessing Space*, ed. John M. Logsdon et al. (NASA, 1999), 503–512.

61. John M. Logsdon and Craig Reed, "Commercializing Space Transportation," in *Exploring the Unknown*, volume IV: *Accessing Space*, ed. Logsdon et al., 414–415.

62. Sally K. Ride, "Leadership and America's Future in Space: A Report to the Administrator," NASA report TM-89638, 1987, 23–25.

63. Lennard Fisk, interview with author, April 3, 2008.

64. James Hansen, William Rossow, and Inez Fung, "The Missing Data on Global Climate Change," *Issues in Science and Technology* 7, no. 1 (1990): 62–69, at 69; US Senate, Subcommittee on Science, Technology, and Space, "NASA's Space Science Programs and the Mission to Planet Earth," 102nd Cong., 1st session, April 24, 1991, 1, 35; US Senate, Subcommittee on Science, Technology, and Space, "NASA's Earth Observing System," 102nd Congress, 2nd session, February 26, 1992, 2–5; George E. Brown to Edward Frieman, 15 May 1991, NASA-EOS Engineering Review Panel Correspondence, March 1991–June 1991 (folder 4), box 124, Edward Freiman Papers (MC 77), Scripps Institution of Oceanography archives, La Jolla, and attachment; Gary Taubes, "Earth Scientists Look NASA's Gift Horse in the Mouth," *Science* 259 (February 12, 1993): 912–914.

65. Daniel J. Kevles, *The Physicists: The History of a Scientific Community in Modern America* (Harvard University Press, 1995 edition), ix–xlii.

66. Ibid, xlii.

67. For a more detailed discussion, see chapter 8 of Conway, *Atmospheric Science at NASA*.

68. Conway, *Atmospheric Science at NASA*, chapter 8; Oreskes and Conway, *Merchants of Doubt* (Bloomsbury USA, 2010).

69. NASA Organization Chart, January 1, 1993 (http://history.nasa.gov/orgcharts/orgcharts.html).

70. Conway, Atmospheric Science at NASA, 263.

71. NASA Organization Chart, March 8, 1999 (http://history.nasa.gov/orgcharts/orgcharts.html).

72. C. O. Justice et al., "The MODIS Fire Products," *Remote Sensing of the Environment* 83 (2002): 244–262.

73. Budget data from http://www.nsf.gov/about/budget/.

74. For a summary, see Amy Paige Snyder, "NASA and Planetary Exploration," in *Exploring the Unknown*, volume V: *Exploring the Cosmos*, ed. John M. Logsdon et al. (NASA, 2001), 263–300.

75. On the revival of planetary exploration, see Peter Westwick, *Into the Black: JPL and the American Space Program, 1976–2004* (Yale University Press, 2008), 207–276. On the idea of "Faster Better Cheaper," see Howard E. McCurdy, *Faster Better Cheaper: Low-Cost Innovation in the US Space Program* (Johns Hopkins University Press, 2001).

76. Conway, "The International Geophysical Year and Planetary Science," in *Making Polar Science Global*, ed. Roger Launius (Palgrave, 2010).

77. Ibid.

78. W. Henry Lambright, "Administrative Entrepreneurship and Space Technology: The Ups and Downs of 'Mission to Planet Earth,'" *Public Administration Review* 54, no. 2 (1993): 97–104.

79. In 2009, the fraction was 15.4 percent. See 2009 JPL Annual Report at http://www.jpl.nasa.gov.

80. Erik M. Conway, "Drowning in Data: Satellite Oceanography and Information Overload in the Earth Sciences," *Historical Studies in the Physical and Biological Sciences* 37, no. 1: 127–151.

9 Calculating Times: Radar, Ballistic Missiles, and Einstein's Relativity

Benjamin Wilson and David Kaiser

A popular image persists of Albert Einstein as a loner, someone who avoided the hustle and bustle of everyday life in favor of quiet contemplation. Einstein did much to contribute to that image, famously telling a journalist that his ideal occupation would have been that of a lighthouse keeper, isolated from society. Yet Einstein was deeply engaged with politics throughout his life. An outspoken socialist and pacifist, he worked tirelessly for civil rights, for civilian control of atomic energy, and to correct the abuses of domestic anti-communism. Indeed, he was so active politically that the Federal Bureau of Investigation kept him under surveillance for decades, compiling a 2,000-page secret file on his political activities.[1]

Einstein's most enduring scientific legacy, the general theory of relativity—physicists' reigning explanation for gravity and the basis for nearly all our thinking about the cosmos—has likewise been cast as an austere temple standing aloof from the all-too-human dramas of political history. In the late 1940s, the physicist George Gamow captured this notion in one of his many cartoons. He sketched a "temple of relativity" (it bore more than passing resemblance to the Taj Mahal) gilded with Einstein's field equations. General relativity, Gamow tried to suggest, was the purest of the pure, separate even from the messiness of the rest of physics, not to mention the grubby worlds of lucre and power. Gravity might have exerted a pull on Newton's apple, but it avoided the lure of Eve's.[2]

Gamow's construction fares no better when held up to historical scrutiny than does Einstein's self-depiction. The general theory of relativity was never above, beyond, or outside of politics. It was completed in a time of war, late in 1915. Its earliest converts included a Russian mathematician interred in a German prisoner-of-war camp, a German astronomer interred in a Russian prisoner-of-war camp, and another German astronomer who passed his time while serving on the Russian front by finding the first exact solutions to Einstein's equations, only to succumb to disease a few weeks later.[3] A British astrophysicist-Quaker-pacifist-internationalist, Arthur Eddington, saw Einstein's relativity as a means of salvaging an international "brotherhood" of science that had been riven by bitter wartime nationalisms. Eddington's ambitious eclipse expedition in 1919 sought to test one of the extraordinary predictions of Einstein's relativity: that a massive object, such as the sun, could bend the path of starlight. His results, announced almost exactly a year after the Armistice, propelled Einstein and his general theory to instant fame.[4] Just a year later, some of the earliest converts to German National Socialism targeted Einstein's theory when they launched their *deutsche Physik* campaign against "Jewish physics."[5]

Not long after Eddington's eclipse observations were announced, Einstein was greeted with a tickertape parade when he arrived in New York. Despite the hoopla, general relativity suffered a curious fate during the middle decades of the twentieth century: the topic all but vanished from the research agendas and teaching plans of most physicists. Einstein's achievement was a mathematical masterpiece, to be sure, but by the early 1930s the vast majority of physicists all over the world had turned away from the topic. Soon after consolidating their power, the Nazis formally banned the teaching of relativity throughout the Third Reich, squelching the world's most active research community on the topic and scattering many of relativity's leading exponents.[6] Beyond Germany, physicists all over the world shifted their focus away from general relativity. Some complained that the theory lacked adequate experimental support; others chafed at the recondite mathematics; most felt the tug of quantum theory and soon that of nuclear physics. Through the 1950s and into the 1960s, graduate-level courses on general relativity couldn't be found in any of the major physics departments in the United States, and in only a few in Europe. Problems on general relativity were absent from the general examinations of American candidates for PhDs in physics.[7]

During the 1960s, small pockets of interest in general relativity began to coalesce. On many physicists' telling, this "renaissance of relativity" sprang from a series of astronomical discoveries. Quasars, pulsars, and the cosmic microwave background radiation all seemed to require that physicists devote serious attention to subtle features of general relativity. The story has thus been cast in a familiar idiom of stubborn or surprising data forcing a new kind of theoretical engagement.[8] But where did those potent data points come from?

One of the most sensitive (and most heralded) tests of Einstein's general relativity to date was hatched in the 1960s at Lincoln Laboratory, an Air Force-funded defense laboratory operated under the auspices of the Massachusetts Institute of Technology. A team of physicists, astronomers, computer programmers, and radar gadgeteers led by a young researcher named Irwin I. Shapiro used Lincoln Lab's powerful radar apparatus—designed, built, and paid for to enable the Air Force to track Soviet ballistic missiles, as well as to aid other military research in communications and surveillance—to bounce radar signals off Venus and Mercury. The researchers then measured how long it took the echoes to return to Earth, testing a relativistic prediction that such signals would be effectively slowed by the curvature of spacetime near the sun. In the thick of the Cold War, using some of the United States' most advanced defense technologies, Shapiro and his collaborators verified Einstein's seemingly unworldly theory to an astonishing degree of accuracy.

This chapter tells the story of Shapiro's so-called time-delay test. By exploring the experiment's origins and location within a defense laboratory, we want to consider the roles played by Cold War priorities in the development of a seemingly pure and esoteric science. At the very least, the support of Air Force funding and the use of its hardware were indispensable to the success of Shapiro's time-delay test. But more than that, the methods that Lincoln Laboratory researchers pursued, the kinds of questions they asked, and the peculiar skills and attitudes they brought to the time-delay test were all deeply influenced by the technical and intellectual setting of their laboratory. In short, the time-delay test cannot be extracted cleanly from its several contexts. General relativity—ethereal and austere as it may often appear to be—was enmeshed in the political and technological environments of the Cold War.

Lincoln Laboratory and Ballistic Missile Defense

Radar and the Massachusetts Institute of Technology have had a long and close relationship, beginning with the decision by the National Defense Research Committee in October 1940 to create a laboratory at MIT for the development of microwave radar technology. The Radiation Laboratory (or "Rad Lab") grew from an initial staff of roughly thirty physicists to almost 4,000 personnel in 1945, managing industrial contracts totaling $1.5 billion (around $19.5 billion in 2014 dollars). In terms of overall spending and staff size, it was a larger undertaking than the Manhattan Project.[9]

The impact of the Radiation Laboratory on MIT's postwar relationship with the US military was profound. Near the war's end, several MIT officials made plans to channel some of the Rad Lab's best personnel and equipment into a peacetime laboratory for advanced electronics research. With the Rad Lab's official demobilization at the end of 1945, the armed forces, too, jumped at the chance to extend their relationship with MIT, sponsoring its work on advanced electronics. The new Research Laboratory of Electronics (RLE), buoyed by $600,000 from the military's Joint Services Electronics Program (nearly $8 million in 2014 dollars), initially employed seventeen members of MIT's physics and electrical engineering faculties and several graduate students.[10] One of the earliest proposals involved consulting on a classified Navy air-to-air-missile program called Project Meteor.[11] By 1950, as the Korean War re-intensified American efforts to mobilize its scientific and technological resources, the RLE had taken on three major projects (including work on radar and communications systems), each funded by a separate branch of the military.[12]

During the winter of 1950–51, MIT and the Air Force negotiated a contract to set up a new laboratory devoted to air defense research under the name Project Lincoln.[13] Its classified work—initially divided among five technical divisions, each working on a separate system, such as "Communications and Components" and "Weapons"—grew out of and greatly exceeded that of the RLE. By 1952, Project Lincoln had transformed almost seamlessly into Lincoln Laboratory, with a staff of nearly 2,000 and a budget of nearly $12 million (more than $100 million in 2014 dollars); a year later, it was receiving more than $18 million (nearly $160 million in 2014 dollars) from the military. In 1954, having outgrown its MIT campus home in Cambridge, Lincoln Lab departed for newly constructed facilities at Hanscom Air Force Base in Boston's western suburbs, near the town of Lexington. Though Lincoln Laboratory's physical separation from MIT was complete, its intellectual and social ties remained strong, and it kept MIT bound to the interests of the military as tightly as ever.[14]

In the early 1950s, air defense research at MIT concentrated on systems designed to counter a nuclear attack by manned bombers. In the most plausible scenario, a Soviet bomber would approach the continental United States from the north, flying low and fast, unnoticed by the United States' sparse array of World War II-era "Ground Control Intercept" radars (originally developed to detect large formations of bombers).[15] Project Lincoln's earliest charge was to develop a system to track and disable a Soviet nuclear bomber at any altitude in American airspace. A second task soon emerged: to develop a system to provide early detection and warning of such an attack. These two tasks led, respectively, to the Semi-Automatic Ground Environment (SAGE) and the Distant Early Warning (DEW) Line. SAGE involved a nationwide network of radars and anti-aircraft weapons, coordinated by a high-speed digital computer whose design was based on the Whirlwind machine under development at MIT.[16] The DEW Line consisted of a series of radar installations across the desolate Arctic expanse from Alaska to Greenland. The DEW Line would give at least three hours' warning of a bomber attack from over the North Pole; SAGE would take care of the rest.[17]

But as the technological characteristics and the potency of the Soviet threat evolved, so too did American efforts to confront them. SAGE and DEW, despite their sophistication and the stream of human, material, and financial resources poured into them (SAGE alone cost some $8 billion, nearly $70 billion in 2014 dollars), were practically obsolete by the mid 1950s.[18] With the advent of the intercontinental ballistic missile (ICBM), a nuclear warhead could be launched from well inside the Soviet Union and sent hurtling toward an American city. During the re-entry phase of its trajectory, it would be traveling at several times the speed of sound. SAGE, a system designed to track and destroy relatively slow-moving bombers, had no chance of intercepting an ICBM. The severe calculus of nuclear retaliation (including the possibility of "launch on warning") required precise tracking of approaching enemy missiles—not only to provide a desperate last-minute warning to a doomed metropolitan area, but to arm and deploy a retaliatory strike.[19]

The development of the so-called Ballistic Missile Early Warning System (BMEWS) began at Lincoln Laboratory in 1955. Each BMEWS station's "pencil-beam" missile-tracking radar was based on Lincoln Laboratory's prototype, built on Millstone Hill near Westford, Massachusetts (figure 9.1). The Millstone radar—the first to use real-time digital data processing, firing its radar pulses with a single klystron transmitter operating with an average power of 60 kilowatts—was completed in 1957, just in time to track Sputnik I.[20] In May of 1960, the first BMEWS station (in Thule, Greenland) welcomed reporters to watch as it "sent two radar beams, one atop the other, over the North Pole, 931 miles away, and across most of the Soviet Union." All quiet on the Eastern front, the *Washington Post* assured. "No rocket launchings were detected in the sweep," but should they ever be, "the electronic gear which is the brains of BMEWS instantaneously pinpoints both the launching site and the impact area, with eyelash accuracy."[21] By 1964, when the billion-dollar project was complete, three

Figure 9.1
Lincoln Laboratory's Millstone radar facility, a prototype missile-tracking radar for the Ballistic Missile Early Warning System (BMEWS). Reprinted with permission of MIT Lincoln Laboratory, Lexington, Massachusetts.

networked stations (one in Thule, one in Alaska, and one in Yorkshire, England) searched the heavens round the clock.[22]

Within Lincoln Lab's Radars and Weapons Division, Group 312 (Systems Research) was tasked with creating the design concept for the BMEWS radar configuration. The earliest classified report on the proposed system was written by Gordon Pettengill and Daniel Dustin.[23] Pettengill, an MIT graduate who wrote his Berkeley PhD dissertation on the scattering of high-energy protons, was hired at Lincoln in 1954, excited by the chance to apply his lifelong interest in radio electronics to problems outside the overworked field of nuclear physics. Dustin, after graduating with a master's degree from MIT's Department of Business and Engineering Administration in 1949, worked for the Research and Development Board (the government's topmost advisory body for R&D spending) before being hired at Lincoln Lab in 1953.[24]

The two young researchers contemplated possible BMEWS radar configurations—the designs and spatial arrangements of radars that would maximize both the amount of warning time BMEWS could provide, and the accuracy of its missile trajectory predictions. Their recommendation was that the Air Force adopt a "pencil-beam scanning system" for BMEWS. "In the pencil-beam scanning type of system," they advised, "two or more observations separated in space and time provide data to allow prediction of the missile's path." Multiple pencil-beam (that is, narrow-beam) radars separated by thousands of miles, each sweeping the horizon back and forth at fixed angles of elevation over the location of a possible launch site, would provide data on the range, velocity, and direction of a ballistic missile early in its flight. Depending on how steep or shallow the missile's orbital trajectory was, the configuration Pettengill and Dustin designed could provide from seven minutes to more than thirty minutes of warning time, with accuracies ranging from "good" for steep trajectories to "fair" or "poor" for shallow ones.[25]

For a more precise analysis of how to predict missile trajectories from radar tracking data, Pettengill and Dustin turned to one of their colleagues, a recent physics PhD named Irwin I. Shapiro. Shapiro had completed his undergraduate studies at Cornell in 1950 with a degree in mathematics and had immediately entered graduate school in physics at Harvard. Hoping to avoid the draft, he took up a position at Lincoln Laboratory in June of 1954, still a year away from finishing his doctoral thesis in theoretical nuclear physics. Shapiro was hired into Group 23, a radar development team in Division 2 (Aircraft Control and Warning), knowing "nothing whatever about radar at that time." Initially, Shapiro tackled theoretical problems related to the radar detection of objects in messy "clutter environments"—situations in which the radar echoes from an object of interest (such as a satellite or a ballistic missile) were mixed in randomly with echoes from upper atmospheric phenomena, the surface of the Earth itself, or nearby objects (such as the debris shed by a warhead's re-entry vehicle). Boiling the complex problem down analytically, Shapiro discovered that with a simple mathematical scheme the interesting radar reflection signal could be disentangled from a jumble of uninteresting signals by finely discriminating the frequency shifts or "Doppler signatures" of each individual reflection.[26]

In the summer of 1955, just as work on BMEWS was gearing up, Shapiro focused on a problem of critical importance: how to calculate the trajectory of a ballistic missile and the coordinates of its destination from radar tracking data. By 1956 he had joined the Systems Research group developing BMEWS, and he began to tackle two important questions: what sort of information would be sufficient to determine where a ballistic missile would end up (and when), and how that information could be acquired from radar tracking signals.[27] Shapiro's final report was published early in 1957. In its opening section he succinctly laid out the difficult problem before him: "In general, the totality of different radar measurements provides redundant data, i.e., provides more data than ... necessary to dynamically specify a trajectory. For measurements corrupted by noise, these data will be inconsistent," leading to "different dynamical trajectory predictions."[28] Troublingly, many possible missile trajectories could be gleaned from a single set of data. But a missile doesn't follow many trajectories; it follows only one. The only way to choose the real trajectory—and fast—was to rely on a sophisticated, statistically weighted decision-making algorithm. Shapiro used the remainder of his hefty report—175 pages filled with intricate calculations and technical diagrams—to develop just such a procedure.

Soon Shapiro's calculations were entered into Lincoln's missile-tracking computer systems by programmers in the Systems Research group. Algorithms such as "Preliminary Estimation of Radar Trajectory" took in radar tracking data (encoded in binary format on magnetic tape) and spat out IBM punched cards, converted by an automatic plotter to graphs of the elevation and range of the missile. In the end, a few computer printouts neatly summarized a detailed itinerary for nuclear obliteration.[29]

Curiously enough for a project of such deep importance for national defense, Shapiro's technical report was never classified; indeed, in 1958 it was published by McGraw-Hill.[30] The text, of obvious interest to Soviet engineers and government officials, was even translated into Russian.[31]

Irwin Shapiro and the Time-Delay Test

In the late spring of "1960, 1961, or 1962" (when interviewed in 1993 he wasn't able to recall the exact year), Shapiro attended a classified briefing on Air Force-funded research at MIT.[32] One of the speakers was George W. Stroke, a young electrical engineer at MIT's Research Laboratory of Electronics. In the late 1950s, Stroke had assisted in the RLE's work on the Polaris project, the Navy's effort to develop a submarine-launched ballistic missile. Stroke helped perfect Polaris's optical alignment, orienting the missile's inertial guidance system with reference to the North Star. By 1959, he was starting up a new set of experiments at the RLE, making high-precision interferometric measurements of the velocity of light.[33]

In the course of his light-velocity work, Stroke had stumbled across a curious prediction of general relativity, one that had long been recognized by cognoscenti but wasn't well known outside the small community of relativity specialists. Gravity doesn't just alter the path of a light ray, he reported at the military briefing; it alters its speed, too. "The velocity of light, measured by the same magnitude c independently of the state of motion of the frame in which the measurement is being carried out, should depend on the gravitational potential … of the field in which it is being measured," he wrote in 1959. Gravity slowed light down. And though the effect would be far too small to measure with even the most sensitive laboratory experiment, the speed of light would be "smaller by 2 parts in a million on the [surface of the] sun, as compared with measurements on earth."[34]

Stroke's comments struck Shapiro as surprising. Shapiro had acquired only a fleeting familiarity with general relativity during his graduate school days—Harvard didn't offer a single course on the subject at the time.[35] It isn't surprising that he had never heard about gravity's effects on light. More surprising is that he *did*, apparently, know about Einstein's explanation of the anomalous precession of Mercury's perihelion—one of the predictions that had won general relativity wide acclaim when it was first published. In 1959, it had occurred to Shapiro that perhaps Lincoln's radar techniques could be used to measure Mercury's precession rate with better accuracy than traditional optical astronomy could achieve.[36] Soon after hearing Stroke's talk, Shapiro began to wonder whether radar might be used to measure the slowing down of light by gravity. After all, detecting and extracting minute radar signals from complicated backgrounds was what BMEWS had been designed to do.[37]

Shapiro's "time-delay test" didn't spring unbidden from the intellectual ether. Certainly nothing in general relativity itself, and nothing in Einstein's writings, suggested a plausible way to test the time-delay phenomenon. Rather, Shapiro's experiment emerged as a natural product of the environment in which he was immersed at Lincoln Laboratory—its complex ecology of Cold War defense hardware of radar, electronics, and computing power and the technical know-how of the personnel, stimulated by their growing interest in radar astronomy and their access to Air Force funding. This interplay of technology and science, embodied in the skills and interests of Lincoln's technical staff, was exemplified by a series of attempts to reflect radar pulses from the planet Venus in the late 1950s and the early 1960s.

During the spring and summer of 1957, several factors converged to launch Lincoln Lab's venture into "planetary radar astronomy" (the use of radar to determine the location, range, and motion of another planet).[38] Plans for a radar-bounce experiment with Venus using the BMEWS prototype radar on Millstone Hill emerged in the course of a lively lunchtime conversation in the Lincoln Lab cafeteria. Robert Price and Paul

Green, two electrical engineers who had earned doctorates in MIT's Research Laboratory of Electronics, had been working on secure, unjammable long-distance radio communications technology at Lincoln. Together they had invented a system for encoding and extracting meaningful communications signals embedded in random noise.[39] Before joining Lincoln Lab, Price had gone to Australia's University of Sydney on a Fulbright fellowship to work with one of the world's first radio astronomy groups. He had returned to the United States with a book on radio astronomy that discussed some of the goals and challenges facing the new field.[40] Noting earlier successful attempts to bounce radar beams off of the moon (a US Signal Corps officer had received reflections from the moon as early as 1946), the text predicted that a future radar system might be powerful enough to send signals to Venus. Though neither Price nor Green was a radar engineer, they knew that the Millstone radar was set to begin operation in the next few months. On cafeteria napkins, Green later recalled, they calculated the estimated strength of signal returns from Venus at its point of closest approach to the Earth (called "inferior conjunction" by astronomers). To their disappointment, they found that, although Millstone was powerful enough to reach the planet with its pulses, its receiving apparatus wasn't sufficiently sensitive to distinguish reflected signals from random electrical background noise.[41]

Soon, though, Robert Kingston (another MIT-trained electrical engineer, specializing in solid-state devices) joined the discussion. He informed Price and Green that Millstone was about to become a far more sensitive instrument. Kingston had been developing a microwave-amplifying radar receiver (a "maser") that could be tuned precisely to the frequency at which the Venus experiment would have the best chance of success. And when Gordon Pettengill (who had earlier considered the best radar configurations for BMEWS, and was now working on readying the Millstone radar for its debut in the fall) also expressed interest in the test, the group resolved to use Millstone's powerful pulses to track Venus during its next inferior conjunction, which would occur the following February.[42]

"We dropped everything else and started spending the government's money without telling anybody," Green later insisted.[43] Though the young researchers probably didn't shirk *all* their official duties (Green was brought before the Lincoln Laboratory Steering Committee to justify the effort he and his co-workers were devoting to the Venus detection), soon enough the lab's administration saw the experiment as an opportunity to justify its work to its military patrons and the public at the same time. "It wasn't just an idle technology stunt," Green recalled. "The Space Program was about to be launched, and no one knew the size of the solar system at all accurately. … Everyone needed to know what the Astronomical Unit was, the yardstick of the solar system."[44] By accurately determining the distance between Earth and Venus at inferior conjunction—gathering precisely the kind of radar ranging data that Shapiro's missile trajectory algorithm had called for—the engineers could improve the estimate of the astronomical unit (the mean distance between Earth and the sun) by at least an order of magnitude. As Green pointed out, vast improvements in scientists' knowledge of the solar system's dimensions would be absolutely necessary for the planned space missions of the 1960s.

On February 10 and 12, 1958, the Millstone radar sent pulses with an average power of 256 kilowatts toward Venus in five separate trials. Using skills and techniques they had honed in their work finding small but interesting signals buried in random noise, Price and Green spent the next year refining and analyzing their collected data. Laboratory officials waited impatiently for news, hoping for a dramatic breakthrough at a time when the United States seemed to fall behind the Soviet Union in the race for technological superiority. Despite significant misgivings about the lack of useful data (only two of the five runs showed possible echo signals, and even those candidate echoes were less than fully convincing), Price and Green decided to declare the experiment a success. They published their data in the March 20, 1959 issue of the journal *Science*.[45] President Eisenhower telegrammed congratulations, and the *New York Times* and the *Boston Herald* lauded the test on their front pages.[46]

By the end of March, Lincoln Lab's director, Carl Overhage, had written to Lieutenant General Roscoe C. Wilson of the Air Force expressing his pride in the Lab's "achievement in long-range, interplanetary communication." (Wilson, chairman of the Joint Services Advisory Committee, was Overhage's immediate military overseer.) The experiment would have consequences for the Lab's work on defense technology, too, Overhage assured Wilson. It had "conclusively demonstrated" that radar detection and tracking measurements should be recorded on magnetic tape, and that digital computers were indispensable for processing such data. "This recording technique appears to be especially promising for those experiments in which *a priori* knowledge is lacking in several respects and only a brief period is available for observation."[47] The implications of Overhage's comments would not have been lost on Lieutenant General Wilson: in the tracking of ballistic missiles and satellites, "*a priori* knowledge" of the tracked object's range and location was slim, and only a brief window of time was available to radar operators.

The following September, during Venus's next inferior conjunction, radars at the California Institute of Technology's Jet Propulsion Laboratory (JPL) and the University of Manchester's Jodrell Bank facility failed to reproduce Lincoln's first detection results. Indeed, the Lincoln researchers failed to reproduce their own results that month, even with an upgraded Millstone radar. In March of 1961, however, a team at JPL succeeded in making the first unambiguous real-time radar detection of Venus. The Lincoln team, led by Gordon Pettengill and taking advantage of yet another round of improvements to the Millstone equipment, made unambiguous contact with Venus just after their JPL competitors. Though Lincoln had been disappointed by the false positive in 1959, the initial push to perform the Venus detection experiment had ushered in a new era in astronomy, opening up the solar system as a laboratory for experiments with radar. Ultimately, it offered Irwin Shapiro the techniques he would need to test general relativity.[48]

In 1963 Shapiro was transferred to Group 37 (Space Physics) in Lincoln Lab's Division 3 (Radio Physics). Now in the division primarily responsible for developing the new field of planetary radar astronomy at Lincoln, he began to work closely with the radar specialists—Price, Green, Pettengill, and others—who were continually perfecting their Venus tracking and ranging experiments.[49]

The time-delay test of general relativity developed naturally in Shapiro's newfound intellectual and technological environment. Perhaps, Shapiro would soon realize, the relativity experiment—which he had first dreamed up after hearing George Stroke's intriguing classified briefing—would now be possible with Lincoln's advanced radars. He would pass powerful radar signals through the intense gravitational field of the sun, bounce them off of Venus as his colleagues had done, and detect their echoes back at Lincoln. If Einstein and his general theory were right, the signals would take just a little too long to return home, relative to radar echoes sent when the sun was nowhere near the path to Venus. Solar gravity would delay them.[50]

But delay them by how much? Shapiro knew, from Stroke's briefing and a quick calculation, that the extra delay produced by a Venus reflection would be on the order of a few tens of microseconds; however, he had never calculated the relativistic time-delay effect with any precision. General relativity was outside Shapiro's expertise, and the full calculation required considerable familiarity with the theory. More important, Shapiro had remained convinced—despite Millstone's success in contacting Venus, and despite the relative ease with which the Arecibo dish (an enormous structure carved into the Puerto Rican jungle) could be converted to radar operation—that no functioning radar facility was capable of detecting and timing the echo with an uncertainty of only a few microseconds. In May of 1963, at the International Astronomical Union's symposium in Paris, Shapiro described the outlines of his time-delay test and his preliminary findings, which suggested that the experiment (and a more detailed calculation) would have to wait for a better radar.[51]

According to Shapiro, everything changed abruptly in the fall of 1964, when the construction of the most sensitive radar facility in the United States—called Haystack—was completed under the auspices of Lincoln Laboratory. Shapiro later claimed to have experienced something close to an epiphany during a cocktail-party conversation with Stanley Deser, a specialist in general relativity at Brandeis University. "Suddenly it occurred to me, while I was talking, that Haystack was coming on line, and that Haystack might have enough capability to do the experiment. ... I went home and quickly did some back-of-the-envelope calculations which showed that we could do the experiment. ... Now it became something real, instead of purely theoretical."[52] With Peter Bergmann's classic text on general relativity in hand, working feverishly

Figure 9.2
Irwin I. Shapiro. Reprinted with permission of MIT Lincoln Laboratory, Lexington, Massachusetts.

for two days at a card table set up in his bedroom at home (while, incidentally, his wife was in the hospital, having just given birth to their son), Shapiro churned out the first rigorous calculation of the time-delay effect.[53]

There is little doubt that the opening of the Haystack facility had a decisive influence on Shapiro's thinking. As Lincoln Laboratory's director William H. Radford would later explain to General Bernard A. Schriever of the Air Force, the design of the time-delay test was not only made possible by the development of hardware sophisticated enough to carry it out, but was actively *motivated* by it. "Prior to the existence of our new Haystack facility, this test was so far beyond the performance capability of any existing radar that it had not even seemed worthwhile to carry through the calculations on which the test is based," Radford wrote, echoing Shapiro. "Thus the recognition and formulation of the test may be regarded as an unforeseen benefit from the Haystack development effort."[54] But other aspects of Shapiro's work at Lincoln Laboratory between 1963 and 1965 point to a somewhat earlier and more gradual development of his thoughts on the possibility of testing general relativity with radar, and they make especially clear the time-delay experiment's conceptual and technological debts to Shapiro's defense research.

One of Shapiro's most important Lincoln Laboratory projects from this period was the creation of a sophisticated computer program for calculating and refining the "ephemerides"—the positions in the sky, as a function of time—of the moon and planets. The Planetary Ephemeris Program (PEP) grew directly out of Shapiro's work on the prediction of missile trajectories. As input, PEP was designed to take a variety of radar and optical observations of an object's position in the sky, use them as initial conditions for the complicated set of differential equations (derived from classical celestial mechanics) that described the object's orbit, and then integrate them numerically. As output, it converted the stream of numbers produced by the integration into accurate tables of the object's ephemeris, which could then act as pointing instructions for the computers that aimed the Millstone and Haystack radars. The more observational input, the better. With more and more "redundant" data telling PEP where an orbiting object was in the sky at a given point in time, the better the estimate that PEP could make of the object's position at earlier and later times. The statistical procedure behind PEP was exactly the kind that Shapiro had developed for his BMEWS missile-tracking project.[55]

Shapiro's text on missile tracking became a kind of recipe book for a major technical advance in astronomy: the generation of precise digital ephemerides. Shapiro later learned that his book had been required reading for members of the computer programming and radar tracking team when the Jet Propulsion Laboratory was developing its own ephemeris program, in the early 1960s.[56] As Lincoln Laboratory began to pursue its own radar astronomy program, Shapiro and a young researcher named Michael Ash began to develop the Planetary Ephemeris Program in earnest.

Ash, a recent Princeton mathematics PhD and an expert on computer programming, had been brought to Lincoln Laboratory in 1963 as an intern in the Space Physics Group, at that time concentrating on developing Lincoln's series of communications satellites. Under Shapiro's supervision, Ash began preparing the thousands of PEP Fortran statements that would be programmed on Lincoln's IBM 360 computer. In addition to that work, tables of optical ephemerides—published by the US Naval Observatory and the Royal Greenwich Observatory, some from as early as the eighteenth century—had to be converted into machine-readable format.[57] Soon William Boyd Smith, an MIT-trained electrical engineer who had helped with the magnetic tape recording instrumentation for Lincoln's first Venus radar ranging tests, was enlisted to help Ash with the numerical integration routines. It "had to be exquisitely accurate," Smith recalled. "We had to do nine-body integrations of the Sun, Moon, and planets back to 1750, and we had to do it with an accuracy over that entire period of, say, a tenth of a second of arc."[58]

In May of 1964, with work proceeding apace, Ash was hired as a permanent Lincoln staff member. His group leader wrote to the Lab's personnel department about Ash's continuing tasks assisting "the Laboratory's interplanetary radar experiments." He went on to indicate that Ash would "consider the special and general relativistic effects on the echo delay and frequency shift of a light signal sent from earth, reflected from another planet, and detected back on earth." The computer work would "be useful in generating more accurate planetary ephemerides"; and it would test "certain predictions of the theory of general relativity."[59] Shapiro rapidly hired several people to assist Ash in the development of the ephemeris program, even asking Gerald Dinneen (a member of the Lincoln Steering Committee) to help pay one programmer an extra two weeks' salary to finish coding his segment of the program.[60]

Shapiro's preparations for the time-delay test thus had begun long before the Haystack radar was up and running, and long before he had formally announced his intentions—either in publication or to Lincoln Lab's management. For many months, he and his colleagues had been tweaking the computational apparatus needed to find a small target in distant space, aim and fire a radar signal at it, and listen carefully for the echo.

There was a crucial difference, however, between the experiment Shapiro was imagining in late 1964 and the Lincoln Venus echo tests of 1958–1961. Instead of bouncing pulses off of the planet at inferior conjunction, when it lay directly between the sun and Earth, Shapiro's test would contact Mercury or Venus at "superior conjunction," when either planet lay on the far side of the sun from Earth. (Mercury was an attractive target because of its more frequent superior conjunctions, every three to four months. See figure 9.3.) Because the signals would therefore pass closer to the sun, the spacetime through which they traversed would be more strongly curved by the

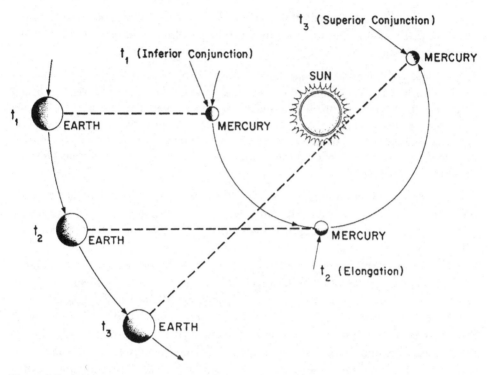

Figure 9.3
The solar-system geometry of Irwin Shapiro's proposed time-delay test of general relativity as represented in a 1965 report by the head of Shapiro's division at Lincoln Laboratory. From James W. Meyer, "Program Description—Radar Observations of the Planets," MIT Lincoln Laboratory Technical Note 1965-35, 1965. Reproduced with permission of MIT Lincoln Laboratory, Lexington, Massachusetts.

sun's mass, and thus the time-delay effect would be much larger. But they would have to be considerably more powerful than in previous experiments, and their echoes would have to be received with far greater sensitivity. Not only would the radar signals in Shapiro's proposed test travel a much longer distance (now roughly 520 million kilometers there and back in the case of Venus, requiring almost a half-hour of round-trip travel time); they would also have to pass through the fury of the solar corona—twice—without being hopelessly scattered or distorted.[61]

Shapiro described his calculation and his proposed experiment in a Lincoln Laboratory technical report published in December 1964, casting it as the latest in a series of "classical" tests of Einstein's general relativity. "Only three tests have been made since Einstein's theory was given its definitive form in 1916," he wrote, "all having been suggested in his original papers."[62] But the techniques of planetary radar astronomy, then being elevated to a high art at Lincoln Laboratory, would offer the means for a "fourth test":

> This test involves measuring the time delays between transmitting radar pulses toward either Venus or Mercury and detecting their echoes. These measurements must be taken at different relative orientations of the earth, the sun, and the target planet, with the most crucial ones being those near superior conjunction when the radar waves pass closest to the sun. For such configurations ... predictions based on general relativity indicate that the time delays will be increased by as much as 200 [microseconds] because of the influence of the sun's gravitational field on the speed of radio wave propagation.[63]

Shapiro titled his internal Lincoln Lab technical report "Effects of General Relativity on Interplanetary Time-Delay Measurements," thus fitting his proposal squarely within the program of radar ranging experiments that had become a fixture at Lincoln Lab since 1958.

The version of the paper Shapiro shipped off to the prestigious journal *Physical Review Letters* in early November, however, announced his experiment's place in the decades-old tradition of relativity research, successor to the three "classical" tests. Simply titled "Fourth Test of General Relativity," it was published in the issue dated December 28, 1964.[64] A press release issued the same day by Lincoln Laboratory hailed the experiment's important contribution to pure science, calling Shapiro's idea "the first practicable new test to be devised since the theory was published almost fifty years ago." Shapiro's experiment promised "a more subtle test of the theory, as it relates to the influence of gravity on light, than [did] the 'gravitational red-shift' test."[65] Under the headline "New Test Proposed for Einstein Theory," the *New York Times* similarly reported that the experiment "proposed tonight by a 35-year-old scientist" was to "determine whether the measured time delay does or does not vary exactly as it should according to Dr. Einstein's theory. The three other ways of testing the Einstein theory all have had larger margins of error."[66] Shapiro's time-delay test, should it be carried out, would join distinguished company.

The Haystack Radar

Shapiro's "fourth test" of general relativity was not only nourished but actively shaped from the start by its defense lab setting. The lines of force pointed in the opposite direction, too. Thanks to Shapiro's determined advocacy and the enthusiasm of his colleagues, the experiment had a strong effect on the activities and goals of Lincoln Laboratory's Radio Physics Division in the second half of the 1960s. These conduits of influence between the time-delay test and its context—linkages of money, materiel, and manpower—were nowhere more evident than in the development of the only instrument that was then capable of performing Shapiro's experiment: Lincoln's Haystack radar facility.

As Shapiro observed in his original technical report, Haystack boasted two closely related design characteristics that would allow it to handle the time-delay test. The more important of these was the high electromagnetic frequency of its transmissions. Operating at approximately 8,000 MHz (in the "X band" of the microwave spectrum), Haystack's radar signals could easily penetrate the solar corona—the sun's intensely hot and violent atmosphere of electrons and ions—without being absorbed, distorted, or delayed. Other facilities, such as the Arecibo Ionospheric Observatory, operated at much lower frequencies (for example, 430 MHz). At those frequencies, the time delay produced by the solar corona alone might be three times as large as the relativistic time delay produced by the sun's gravity. Because the exact amount of delay caused by the corona could only, in principle, be known with poor precision, the error in the corona-induced delay would make a careful measurement of the relativistic delay virtually impossible.[67] Related to the frequency issue was the exceptional narrowness of Haystack's "pencil beam"—as with all radar propagation, the higher the frequency, the narrower the beam. With a focused beam, Haystack could pass its transmissions much closer to the edge of the sun (and therefore through a much stronger gravitational field) than rival antennas, thereby lengthening the expected relativistic time-delay. Arecibo, which could fire its beam within roughly one degree of the sun, was no match for Haystack, whose beam was no wider than a few hundredths of a degree.[68]

In November of 1964, just as he was sending off his report and article for publication, Shapiro touted Haystack's distinct technical advantages in a memo to several members of Lincoln Laboratory's upper management (including James Meyer, the leader of Shapiro's own Radio Physics Division). Formally proposing a "fourth test of general relativity" to Lincoln's administration for the first time, Shapiro began by emphasizing the experiment's historic character: "Lincoln Laboratory, through proper use of the Haystack radar, has a unique opportunity to test an interesting consequence of general relativity. It is the first measurable effect to have been proposed since Einstein's classic paper of 1916." Moreover, Shapiro insisted, Lincoln had the chance to be *first*. The solar corona would prevent the Arecibo radar from performing the test, and "other program commitments will most likely prevent JPL from undertaking this experiment in the near future. The uniqueness of Haystack's contribution therefore seems clear."[69]

Indeed, Haystack seemed especially suitable for the job. As Shapiro told his superiors: "A 'raw power' analysis, assuming a 100 kw [kilowatt] tube [transmitter], an overall system noise temperature of 100°K, and planetary characteristics extrapolated from the best data available, indicates that Haystack is capable of measuring the general relativistic effect to an accuracy of 5 to 10%. ... (This is a considerably higher accuracy than has been achieved in the 'bending-of-light' experiment.)" Shapiro soon cut to the chase: if the Laboratory administration were to make a commitment to the experiment right away, one of Shapiro's colleagues at Haystack had informed him, "the necessary instrumentation could be constructed in time for the next favorable planetary configuration ... for Mercury on 12 June 1965." Without mentioning whether upgrades to Haystack might be necessary, Shapiro assured his superiors that the time-delay test was far from ungrounded technically or theoretically. "I have discussed this experiment with several of my friends who are specialists in relativity theory," Shapiro insisted, "and they spontaneously offered to write a supporting letter should the Laboratory seek outside support for this experiment."[70]

Shapiro's colleagues in Radio Physics would soon make it clear that Haystack *was* in need of significant—and costly—upgrades if it was going to be able to precisely time echoes from Mercury or Venus as the planets raced around the far side of the sun. The source of funds for the upgrades was another matter. Would the Air Force, which had funded Haystack's construction and paid for its maintenance, underwrite the radar's tune-up, or would the Lab have to look for "outside support"? How would the upgrades be justified? (Was general relativity the kind of thing the Air Force was in the business of testing?) In any event, Lincoln's researchers understood that Haystack (more specifically, its modular system of interchangeable transmitting and receiving electronics) could be tuned for one application (such as planetary radar astronomy) without permanently disabling the radar's other functions (missile and satellite tracking). It had been designed that way.[71]

The Haystack radar was conceived as a dual-use technology *par excellence*. Its design was first proposed in 1958, soon after the Millstone radar came online as a prototype missile tracker for BMEWS. As described by Lincoln Lab's director Carl Overhage in the summer of 1960, just a few months after the Lab began awarding procurement contracts to industrial firms for Haystack's construction, the dual scientific and technological benefits of the Haystack facility (especially in radio and radar astronomy, and various aspects of atmospheric studies) were mixed and numerous. The anticipated sensitivity of Haystack's antenna would improve knowledge of interplanetary distances, planetary rotation rates, and surface characteristics, as well as the "influence of atmospheric irregularities on realized antenna gain and narrow-beam pointing accuracy." But ultimately, Overhage argued, Haystack might make important contributions to defense: "These studies may have important consequences to the design of future radars for such purposes as missile discrimination or space-vehicle detection, where extremely high resolution and high effective radiated power are essential."[72]

The Haystack radar was designed by a Lincoln Laboratory team led by Herbert G. Weiss, an MIT-trained electrical engineer and a veteran of the wartime Radiation Laboratory. Weiss was experienced, having headed up the earlier Millstone radar design team. He was appointed chief designer for Haystack in November of 1959. Initial cost projections were at roughly $5 million (more than $40 million in 2014 dollars), though the Electronic Systems Division of the Air Force Systems Command—the ultimate source of funding for Lincoln Laboratory's "General Research" projects, including the Haystack construction—didn't approve the design until cost estimates were lowered to around $2 million.[73]

The new facility was to be a "ground station for space communications, radar and radio astronomy research," according to Weiss. "Communication experiments utilizing the moon, Echo-type balloons, the West Ford dipole belt, and other passive reflectors will receive early attention. … The Haystack system will also be used to explore the effects of the atmosphere and weather upon microwave communications at 8 [GHz] and at higher frequencies."[74] Indeed, Haystack was designed to transmit and receive in the X band precisely because those were the frequencies used by Air Force communications systems. (The ambitious and controversial Project West Ford, on which Irwin Shapiro worked for much of the early 1960s, was the Air Force's plan to orbit a ring of nearly 500 million inch-long, hair-thin copper wires high above the Earth's surface to act as an invulnerable, passive reflector for X-band communications signals.)[75] Though the facility would contribute to astronomy ("strong radar echoes" off Venus might be "detectable for many months of the year"), it would also profit space exploration and national defense by "obtaining radar back-scatter echoes from a medium-sized satellite" or scrutinizing "very small targets at ranges of several thousand miles."[76] MIT's *Technology Review* emphasized Haystack's dual benefits in 1962, reporting that, while Haystack was "the only US system presently planned that will be able to track a so-called 'stationary' communications satellite in a 24-hour equatorial orbit more than 22,000 miles above the earth," it would also, to the benefit of basic science, "extend our interplanetary radar contacts to Mars, Mercury, and Jupiter—as well as Venus."[77]

Haystack's huge reflector dish, 120 feet in diameter, was made of an extremely thin aluminum sheet, laid over a pattern of aluminum bars forming 96 honeycomb-shaped panels. The aluminum skin was stretched over its frame so tightly that at any point the reflector surface deviated from a perfect paraboloid by less than a tenth of an inch, the measure of its so-called "surface tolerance." Emanating from close to the center of the reflector dish were four rods on which a secondary, convex, hyperbola-shaped reflector was mounted, facing back toward the main dish. This reflector was about nine feet in diameter and surfaced by an aluminum sheet of even finer tolerance. When radar signals approached the main reflector dish from the sky at an angle parallel with the dish's central axis of symmetry, they would be reflected by the aluminum sheet toward the secondary reflector. From there, the signal would reflect again back toward the center of the main dish in a focused beam, where it would be received by a complicated cluster of electronic equipment known as the "target assembly."[78]

The radar's antenna structure—the reflectors and the trusses on which they were mounted, as well as the nearly 7,000-pound "plug-in equipment box" containing the target assembly and the antenna's first transmitter, which delivered an average 100 kilowatts of power—was encased within a 150-foot-diameter radome to shield the entire system from the weather. (The sun's heat might warp the reflector's aluminum surface, and even the smallest accumulation of ice and snow on the dish during the harsh New England winter would render the antenna inoperable at X-band frequencies, where exceptional surface tolerances were necessary.) The radome, originally procured by the Air Force for an Arctic radar station and designed to withstand winds of 130 miles per hour, was made of a frame of interlocking steel beams forming the triangular panels of a geodesic sphere. Each of the 930 panels was covered by a sheet of white fiberglass, less than a tenth of an inch thick, chosen for its transparency to X-band microwave radiation. Most important for tracking experiments, the Haystack antenna could be steered digitally. The whole apparatus rested on a bearing and control system weighing almost nineteen tons, floating on a nearly frictionless film of oil 0.005 inches thick, allowing the dish to be rotated smoothly around its central vertical axis, as well as tilted up or down to face targets higher or lower in the sky, all to accuracies of a fraction of a degree.[79]

To direct Haystack's extremely focused transmission beam, the steering system was interfaced with a state-of-the-art Sperry Rand Univac 490 digital computer, housed in an adjacent control room. Brightly lit, with polished floors, whirring tape reels along one wall and a small console station tucked into a corner, the control room was spare and ultramodern.[80] The "remote teletype console" allowed the antenna's position to be controlled in real time by a person sitting at a keyboard. "Software programs have been prepared which permit an operator to request that the antenna point at such targets as the moon, the planets, and known satellites by simply typing the name of the object in clear text," Weiss explained in his final design report. "The computer responds in clear text via the teleprinter and, if necessary, will ask the operator for additional pertinent information."[81]

Construction crews began pouring the one-story concrete foundation for Haystack in the fall of 1960 at the facility's new home, about a half mile up the road from the Millstone radar.[82] In December of that year, the Air Force awarded the Ohio division of North American Aviation a contract to fabricate and install the Haystack antenna system. As construction proceeded in the next four years, Lincoln Laboratory continued to pour funds into Haystack through its Air Force contract. By the spring of 1964, almost $850,000 had been shelled out to North American (more than $6.4 million in 2014 dollars); a memory upgrade for Haystack's Univac 490 computer alone cost the Air Force $152,000.[83] On October 8, 1964, with construction finally (and expensively) complete, Haystack was formally dedicated on Millstone Hill.[84]

Figure 9.4
Lincoln Laboratory's Haystack radar facility, completed in 1964. Reproduced with permission of MIT Lincoln Laboratory, Lexington, Massachusetts.

According to some Lincoln Lab scientists and engineers, obtaining government funds for research projects of all kinds was less a matter of proving military relevance, or of navigating tricky bureaucratic terrain, than of capitalizing on personal reputation within social networks in and around the military administration. Herbert Weiss recalled that proposing the multi-million-dollar Haystack radar simply depended on having "enough credibility then with the Defense Department ... that I could go in and make a pitch for a system and say, Look, we've really got to push the state of the art for space communications and radar. The kind of money we were talking about then was very small; a few million dollars a year, for three years or four years. ... Haystack was sold as an R&D exercise to try to push the state of the art."[85] Irwin Shapiro, similarly, said that securing support for subsequent upgrades to the Haystack radar—upgrades necessary for experiments in planetary radar astronomy, and for his time-delay test of relativity—was a matter of knowing the right people. Recognizing that a more powerful transmitter and sensitive receiver would be needed to reach Venus at superior conjunction, Shapiro apparently spoke with Lincoln Laboratory

Director William Radford. Radford, impressed (if somewhat baffled) by the proposal to test relativity with radar, called up the Harvard physicist Edward Purcell, who, already familiar with Shapiro's work on Project West Ford, vouched for Shapiro's qualifications. All that remained, according to Shapiro's recollection, was for Radford to notify his contact at the Rome Air Development Center, the Air Force's R&D lab in Rome, New York. With remarkable swiftness, the upgrade was funded with around $500,000 from the Air Force (around $3.8 million in 2014 dollars).[86]

In the case of Shapiro's time-delay test, a far more complicated picture unfolds when the actual uncertainty, sheer scale, and enormous cost of the hardware upgrade are taken into account. The Haystack facility had been designed and built during the boom years of defense spending after Sputnik. Upgrades to Haystack, on the other hand—prerequisite to fulfilling Shapiro's longed-for test—were negotiated in a fast-changing political environment. As early as January of 1965, the Department of Defense announced major job cuts at the Lawrence Livermore Laboratory in California. The following year, the Department of Defense released its first report from Project Hindsight, an in-house accounting of the benefits to military priorities and preparedness that had accrued in two decades of spending on basic science. For the first time since World War II, military bureaucrats began to second-guess the Cold War template of funding non-mission-oriented research in hopes of advancing military goals.[87] Military planners' own concerns about the effectiveness of defense spending for basic research were soon amplified by nationwide campus protests against the Vietnam War, in which calls rang out for the Department of Defense to get out of the business of higher education.[88] Lincoln Laboratory, having depended on Air Force money from the beginning, was especially sensitive to possible fluctuations in its funding stream. Lincoln administrators needed to adapt their strategies quickly to appeal to civilian as well as military sponsors. Suddenly the presence of Shapiro's time-delay test on the laboratory's roster assumed new significance.

In January of 1965, less than a month after Shapiro published his "fourth test" technical report and article, Paul Sebring (the leader of the Surveillance Techniques group and the manager of the Haystack facility) wrote to James W. Meyer, the head of Lincoln's Radio Physics Division, on the subject of NASA funding for both the Millstone and the Haystack radar. The Millstone radar, Sebring noted, had already provided valuable assistance to NASA's early communications satellite program, precisely tracking various satellites and communications balloons. But if Millstone was a valuable instrument, Haystack would be even more so. Millstone could track "a 1 sq. meter target at nearly 5,000 nautical miles." Haystack—if its 100-kilowatt-average-power transmitter were to be replaced with one delivering an average power of 500 kilowatts, and it were to be given a few "receiver improvements"—could easily spot a one-square-meter target at 10,000 nautical miles. "NASA may feel it beneficial to establish an arrangement with the Laboratory to support the Station to the extent of their anticipated requirements each fiscal year," Sebring wrote. "Once a channel were opened, monies could be adjusted each year to the assistance required." While the Millstone radar could be opened up for NASA use "up to several hundred hours per year," he added, "development of a supertracker facility at Haystack could be accelerated by NASA support. NASA may find such a capability of interest."[89] The Lincoln Lab researchers certainly hoped so, Irwin Shapiro not least among them.

What is especially striking about the costly plan to revamp Lincoln's most sophisticated radar only a few months after first switching it on is the degree to which Shapiro's esoteric experiment became the centerpiece of the Laboratory's efforts. While Sebring diligently outlined a host of NASA-relevant tasks that an upgraded Haystack radar could accomplish (mapping structural features and surface characteristics of the moon in fine detail, mapping the surface of Venus and determining the composition of its atmosphere, performing similar measurements on Mercury and Mars and refining the orbital parameters of the planets), the main reason for the upgrades, he admitted, was Shapiro's time-delay test. The trick was wording the proposal to NASA in a way that brought everything together. Sebring advised Meyer that "while the most important result of a 10-db increase in Haystack capability will be the ability to perform the fourth test of general relativity made against Venus or Mercury at superior conjunction," it was surely worth emphasizing the "many other experiments that will be greatly improved in quality."[90]

Shapiro made his own plug for NASA sponsorship the next day. He provided James Meyer, the head of Lincoln's Radio Physics Division, with a list of scientific experiments (including most prominently his time-delay test) that could be conducted only if Haystack received its expensive upgrades.[91] Meyer, in turn, prepared a draft proposal and budget estimate for "Deep Space Radar Measurements" for NASA. The new plug-in radar equipment box alone, including the 500-kilowatt transmitter (a significant improvement over the existing 100-kilowatt transmitter), would initially cost $675,000 (about $5 million in 2014 dollars); an upgrade to the radar-computer interface would set NASA back an extra $10,000; various architectural upgrades to the antenna structure would cost some $45,000. And in 1966, before the upgrades would be complete, an additional $300,000 would be needed to get the whole thing properly calibrated and tested. "The staff and their support ... will be provided by our General Research Program supported by the United States Air Force. The funds requested here are only those required to procure the necessary instrumentation to carry out the experiments described."[92] Meyer quickly followed up with a formal proposal to NASA for "Radar Observations of the Planets," again listing "the conduct of a 4th test of general relativity" as the first item of business.[93]

Despite some expressions of interest, the NASA money never materialized—at least not in time for the first iterations of the time-delay test in 1966 and 1967. As the proposals went higher and higher up Lincoln's and NASA's respective chains of

Figure 9.5
One of Haystack's modular plug-in equipment boxes, including the radar's powerful transmitter. Reproduced with permission of MIT Lincoln Laboratory, Lexington, Massachusetts.

command, NASA officials began to balk at the price tag. They expressed regret that neither the agency's Space Science and Applications division nor its Planetary Astronomy division had funds of such magnitude to spare (unlike, for example, NASA's Manned Space Flight division).[94]

At the eleventh hour, the Air Force stepped in with money to improve the Haystack radar. Exactly why it did so remains unclear. Irwin Shapiro and other scientist-participants (including Gordon Pettengill) would later recall that their arguments about the inherent scientific value of radar astronomy and the "fourth test" were compelling, setting off a chain of persuasion upward through Laboratory and Air Force command.[95] Of course, the Air Force might well have had its own reasons for seeking improved transmitter power: as Laboratory staff and management had consistently argued for both Millstone and Haystack radars, improved performance for tracking the moon and the planets implied improved performance for tracking missiles and satellites. Though Shapiro and his Lincoln colleagues had to sweat through the uncertainties that spring, by August of 1965, with the new Air Force funding in hand,

Lincoln could begin awarding contracts to outside companies to start working on the upgrades to Haystack.[96]

The Time-Delay Test and the Fate of the Haystack Radar

The announcement of plans to conduct the "fourth test" of Einstein's general relativity made waves far beyond Lincoln Laboratory. Enthusiastic letters began to arrive at Shapiro's office from all over the United States. To one eager inquirer from Norfolk, Virginia, Shapiro suggested grabbing a copy of Peter Bergmann's general relativity text to get a better handle on the subject. To another would-be relativity researcher from Quincy, Massachusetts, who wrote up his own "theory and proposed test" for Shapiro's approval, Shapiro was sorry to say "you did not analyze your experiment in sufficient detail," and "your analysis of Einstein's predictions are [sic] seriously in error." Shapiro regretted that a Chicago high school student's request that he "explain the entire theory of relativity ... in a letter" was "not possible to fulfill." And to one resident of Boston's tony Beacon Hill, who had sent a check to support the "fourth test" upgrades, Shapiro replied that, though he wasn't able to accept her generous donation, "every effort [was] being made to carry out the relativity experiment as soon as possible."[97]

A few of Shapiro's colleagues in physics also expressed interest in the upcoming test, but with varying degrees of enthusiasm. Some pressed him to clarify details of his calculation; his 1964 article in *Physical Review Letters* had been much briefer than the lengthy treatment in his Lincoln Laboratory technical report.[98] Others granted the importance of the proposed test, but disagreed that it should be counted the "fourth test"; Leonard Schiff of Stanford University, for example, had proposed a distinct experimental test of general relativity a few years earlier.[99] Still others engaged Shapiro in a priority dispute. Duane Muhleman of the Jet Propulsion Laboratory had written up a proposal for a time-delay test slightly before Shapiro's was in print but after Shapiro had discussed his idea at an international astronomy conference that both had attended. A crucial difference between the rival proposals, as Shapiro pointed out to his JPL colleague in increasingly heated correspondence, was that Muhleman's version involved radar propagation during a planet's inferior conjunction rather than its superior conjunction.[100]

While trying to keep up with the correspondence, Irwin Shapiro and Michael Ash were working frantically to complete their Planetary Ephemeris Program, incorporating more and more radar and optical observations of Mercury and Venus (away from superior conjunction) to make the program's orbital predictions more accurate. "Good news: The program ran successfully," Shapiro announced to Gordon Pettengill in fall 1965; still, there was more work to do debugging "parts of the program, statement by statement."[101] Shapiro badgered Richard Goldstein of the Jet Propulsion Laboratory to publish his Venus ranging data. "More than four and a half years have elapsed since JPL first made radar contact with Venus, yet no data at all has been published." Shapiro would not be delayed. If Goldstein's supervisor wouldn't allow him the time to prepare data tables, Shapiro would be happy to write a letter demanding the raw data himself.[102] Ash, Shapiro's right-hand-programmer, continued to pester the Naval Observatory with letters requesting optical ephemeris data for the sun, the moon, and the planets on punched cards to be fed into the IBM 360.[103]

By the summer of 1966, Shapiro was complaining to the Lincoln personnel department that it had become "quite clear that we are severely programmer limited in our attempt to analyze the interplanetary radar observations for the new test of general relativity." PEP's calculations had to be so accurate that precise optical observations of the *outer* planets were needed to account for their slight gravitational perturbations of the orbits of Earth, Mercury, and Venus. The need for manpower was dire. Though Shapiro expected to be able to complete at least a "semiquantitative test" of Einstein's theory to an error of perhaps 35 percent, "our results may fall almost an order of magnitude lower in test accuracy simply because of our inability to handle the programmer chores."[104] To Haystack's director, Paul Sebring, he wrote that Haystack had "the potential to provide the most accurate tests yet made of Einstein's predictions" and that nailing down the orbital parameters of Mercury and Venus was absolutely crucial: "general relativity literally hangs on it." But the researchers needed help finishing the computer work, including "approximately $5,000 to finish putting all of these [optical ephemeris] data on punched cards. The laboratory's card-punch operators have been shown to be incapable of handling this work on finite time scale."[105] Nearly forty separate tasks remained before PEP would be ready for the "fourth test," Shapiro explained in a separate memo.[106] Exhausted but still pushing hard, he wrote to an MIT colleague in early 1966 that preparing for the time-delay test was occupying an "inordinate fraction" of his time.[107]

Meanwhile, Shapiro's Lincoln Laboratory co-workers were engaged in an all-hands-on-deck effort to get Haystack and its planetary radar equipment box ready as soon as possible. Early in 1966, J. S. Arthur, deputy director for Haystack operations, laid out a schedule for radar activity during the coming year. In an accompanying memo to the research staff, Arthur outlined a few major tasks in the months ahead, including continuing Haystack's normal communications research program, and "integration of the 500 [kilowatt] Planetary Radar system." For the time being the facility was focused largely on radar mapping of the moon's surface for NASA's upcoming Apollo missions. But the relativity experiment loomed large. "It should be stated at this point that *a major goal* has been established toward a '4th test experiment', utilizing the planet Venus at the November [1966] superior conjunction." The new transmitter and receiver, then being constructed at Lincoln Lab in Lexington, were to undergo high-power tests and installation at Haystack in the summer. It was absolutely essential that the transmitter/receiver upgrade be kept on a tight schedule, Arthur continued; there was no room for "any shift in priority."[108]

Four days later, Gordon Pettengill pressed Paul Sebring to complete the upgrades right away. Preparing Haystack for its November debut with Venus was "worthy of unusual effort," Pettengill told Sebring. Venus' "angular approach" in November would place it far enough behind the sun (from Earth's viewpoint) that the time-delay test could be carried out, but not so far behind that radio noise from the sun itself would drown out the signal.[109] Sebring immediately penned a memo to the entire Haystack team, adding Pettengill's letter as an attachment. Sebring announced that the November time-delay test had become priority number one in the Radio Physics Division. Even the scheduled re-rigging of the antenna (in order to improve surface tolerance of the dish) was to be postponed until after the experiment. "No lessening of the importance with which we view [re-rigging] is to be inferred from this rescheduling," Sebring wrote. The time-delay test, until further notice, was simply more important.[110]

Preparations had reached fever pitch by May of 1966. Sebring gathered all the available manpower for preparations for the time-delay test, reapportioning it as he saw fit. The time-delay test, he wrote to the Haystack engineering team, had now "assumed priority equal to or above system operations at Haystack and Millstone, in order to free appropriate people." The day-to-day ranging, mapping, and communications research at Haystack—in other words, the work done at the bidding of the Air Force and of NASA—would be subordinated to relativity. He went on to instruct Louis Rainville, Haystack's test director, to drop his tasks on the lunar mapping project and assist Melvin Stone, chief project engineer in charge of the construction and testing of the new transmitter. Sebring even recommended that technicians be encouraged to work overtime on Saturdays, and wondered whether one shop foreman might be retained on the upgrade project for an extra couple of months to assist his teammates.[111]

The time-delay experiment, Sebring reminded Division 3, "is of the greatest scientific interest and seems certain to attract wide attention to Lincoln Laboratory ... if it succeeds." Only an intense effort would ensure Lincoln's "being first to make the proposed test." November 9, the date on which Venus would slip briefly behind the sun, was quickly approaching.[112]

Haystack was ready just in time. Data were collected during Venus's November 1966 superior conjunction and during the superior conjunctions of Mercury in January, May, and August of 1967. Early in March of 1968, Shapiro made the first public release of preliminary "fourth test" results at a meeting of the American Physical Society in Boston. The experiment, he announced, had confirmed general relativity, if only within fairly generous margins of error. Less than a week later, he sent a paper to *Physical Review Letters*.[113] Published in May, the paper explained that the experiment had been made possible by an "intensive program" to outfit Haystack with a new transmitter and receiver. Considering each experimental run separately, Shapiro and his co-authors (including Gordon Pettengill, Michael Ash, William Boyd Smith, and three of the Haystack engineers who had instrumented its upgrades) concluded that the "most reliable of these data agree, on average, with the excess-delay predictions of general relativity to well within the experimental uncertainty of ±20%."[114]

In March of 1968, Shapiro formally presented Lincoln Laboratory's director Milton U. Clauser with the first preprint of the *Physical Review Letters* paper. "Although the accuracy attained was not as high as we had hoped"—20 percent uncertainty hardly amounted to razor-sharp scrutiny or conclusive confirmation of Einstein's theory—"it nonetheless represents a substantial achievement in a very difficult experiment," he told Clauser. Moreover, "despite the unfortunate predilection of the popular press to present the result as a one-man show, the experiment was successful only because of the skill and spirit of a large number of Lincoln personnel." With understated admiration for his co-workers, Shapiro reminded Clauser that the design and construction of the new plug-in equipment box for Haystack had been a "massive effort." The grueling pace of work during May of 1967 ("unbroken except for Mother's Day"), when Haystack had tracked Mercury through its superior conjunction daily from 4:30 a.m. to 7 p.m., "served to demonstrate poignantly the dedication of the staff."[115]

No eye-catching headline heralded the findings, as had happened after Arthur Eddington's announcement in 1919, even though the news was the same: Einstein's relativity had held up. Instead, the scientific success of Irwin Shapiro and his colleagues was bittersweet. In 1968, before Shapiro had finished writing up the results of his "fourth test," ever clearer signs of a big shift in the US military's support of basic science could be detected throughout Lincoln Laboratory. For at least a year, high-level officials of the Department of Defense and the Air Force, and administrators at Lincoln Lab and at MIT, had asked searching questions about Lincoln Lab's future. The Lab, it had become clear, occupied a precarious position. As the Vietnam War raged, critics called ever more loudly for the bond between the national-security state and America's universities—held together by the clasp of the Cold War for more than twenty years—to be permanently broken. Lincoln Lab was caught in the center of the debate, and the Haystack radar would be among the first vestiges of its military-academic mission to be amputated by severe federal budget readjustments.[116]

In late summer 1967, Secretary of Defense Robert McNamara, MIT's president Howard Johnson, and the former MIT administrators and Presidential Science Advisors James R. Killian and Jerome Wiesner held a telephone conference addressing "long-range questions affecting Lincoln Laboratory in its service to the country, specifically the Department of Defense," and its "continuing operation as a part of MIT." In a letter to McNamara written after the conversation, Johnson summed up Lincoln's dual orientation. Its mission-oriented work was "aimed at assisting the military establishment in finding technical solutions to ... problems in ballistic missile penetration and ballistic missile defense, in space communications," and other areas. More basic research—buoyed by the General Research Program, which had fostered the construction of the Haystack radar—allowed "the Laboratory to look beyond its immediate tasks." However, Johnson continued, recent congressional budget cuts to Federal Contract Research Centers, of which Lincoln was a prominent example, had "opened the question of how MIT should plan for the future operation of Lincoln."[117]

Just as Johnson was writing to McNamara, a Defense Science Board task force commissioned by the Director of Defense Research and Engineering had recommended that each of the Defense Department's Federal Contract Research Centers (FCRCs) provide a statement of its "interpretation of the primary purposes and current technical objectives" of its work.[118] Only a month after the latest round of data collection for Shapiro's relativity experiment had taken place at Haystack, Lincoln Lab was being asked to explicitly justify its research pursuits to the Department of Defense. Howard Johnson's official statement to the Assistant Secretary of the Air Force included a tidy list of the Lab's accomplishments in radar and communications, but he was "deeply concerned" about Department of Defense budget cuts and Lincoln's "unfortunate" classification as an FCRC. Lincoln had "no financial cushioning and relies almost completely on annual appropriations from the Air Force and ARPA," the Department of Defense's Advanced Research Projects Agency. Budget cutbacks didn't curtail only military work at Lincoln; its "general research effort," too, had suffered deeply for years. Of what continuing value would Lincoln Lab be to MIT if its personnel and material assets were so easily buffeted by the changing winds of national defense?[119]

No matter how MIT, Lincoln Lab, and the Air Force might choose to handle them, the changes issued from Washington were sweeping and rapid. In 1969, Senate Majority Leader Mike Mansfield successfully tacked a rider (the so-called Mansfield Amendment) onto the military authorization bill requiring all research funded by the Department of Defense to have a "direct or apparent relationship" to military operations.[120] In November of that year, Milton Clauser told Howard Johnson that staff reductions and the "additional unplanned deletion of $450,000 from [Lincoln's] budget for rising overhead costs"—more than $2.8 million in 2014 dollars—would be "difficult to accommodate." Only a week after violent clashes between student protesters and riot police in front of MIT's military-funded Instrumentation Laboratory, Clauser wondered in frustration whether it was "inappropriate to suggest that the activities on the campus contributing to the rising overhead be given some scrutiny?"[121]

Just a few days later, Haystack's director, Paul Sebring, notified Clauser that negotiations for the transfer of the Haystack facility from the Air Force to civilian control were now proceeding at top speed. A year before, the Northeast Radio Observatory

Corporation (NEROC), a nonprofit consortium of New England research institutions, had submitted a proposal to the National Science Foundation for a phased-in, three-year transfer of control of Haystack from the Air Force. An advisory panel of prominent astrophysicists led by Robert Dicke of Princeton University recommended that the NSF approve the request (though, in Sebring's opinion, the panel had underappreciated Haystack's special "sophistication and broad application").[122] As the NSF delayed its final decision and NEROC made a parallel funding pitch to NASA, the Air Force forcibly set its own time line. In February of 1970, Grant Hansen, Assistant Secretary of the Air Force for R&D, warned Lincoln that by the following July, if Haystack had not been adopted by another government agency, it would be declared surplus and its computer apparatus scavenged for other military uses.[123]

In the summer of 1970, after last-minute guarantees from the NSF and NASA to help absorb Haystack's $1.5-million-per-year operating costs (more than $9 million per year in 2014 dollars), the Air Force handed Haystack over to MIT and NEROC. When Haystack's transmitter failed several times in the early 1970s, NASA—no longer in need of lunar radar maps for its Apollo missions—was, once again, unwilling to fund replacements. Haystack's last radar transmission—a ranging signal to Mercury—was sent in March of 1974.[124]

Conclusion

For nearly thirty years scholars have interrogated the interplay of scientists and military patrons during the Cold War. A debate about who set scientists' intellectual agendas has raged. What effects, if any, did abundant military spending have on the range of topics pursued or the styles of scientific inquiry? Some commentators have argued that the Cold War produced dangerous distortions of proper science.[125] Others have countered that scientists have never existed outside of society—for example, Galileo navigated patronage relationships no less complicated than those of the American physicists in the nuclear age.[126] Still other scholars have focused on aspects other than funding and patronage, highlighting the entanglement of personnel, training, instruments, and foreign relations. Yet even these cases have focused overwhelmingly on examples of science and technology that were of obvious military interest, including nuclear physics, materials science, electronics, computing, and the geosciences.[127]

Episodes like the general-relativistic time-delay test expand the pattern, exemplifying just how extensively the new template had reached. Despite the frequent portrayal of Einstein's general relativity as an austere and otherworldly temple, physicists' Cold War engagement with Einstein's work bore most of the marks we have come to expect of the era: seemingly limitless funding (at least for a time), sprawling machinery, soaring technological hubris, and a close mapping of personnel, tools, techniques, and skills between scientific practice and military priorities.[128] The long duration of the time-delay test, from conception in the late 1950s to completion in the early 1970s, further allows us to map the rocky transition from post-Sputnik boom to stagflation bust. "Cold War science" was no more a static entity than the planet Venus; both became moving targets.

For much of 1969 (when Haystack was still a military instrument), until nearly the end of 1970 (when it was officially a tool of civilian science), Haystack once again bounced radar echoes off of Venus for roughly 300 days on either side of superior conjunction. Using data from the Arecibo facility (far from superior conjunction), Shapiro's group combined about 1,700 time-delay measurements to reduce their earlier uncertainty estimates from 20 percent to less than 5 percent. Einstein was still right, and the Haystack team had now completed the most sensitive test of general relativity yet. The paper published in *Physical Review Letters* in May of 1971 was the culmination of Shapiro and colleagues' work testing relativity with the techniques of planetary radar astronomy. With little mention of the instruments involved, the group cast its "new radar result" as simply the latest refinement of a series of tests of Einstein's theory. By itself, the paper seemed the purest of pure science—an otherworldly theory, subject to a rather esoteric experiment.[129]

The original 1968 paper, however, was far more evocative of the time-delay experiment's deep and tangled roots in Cold War defense priorities. The bulk of the article was given, as the authors put it, "to a more detailed discussion of ... the novel experimental techniques" demanded by their test, including everything from the technical features of Haystack's transmitter to various methods of computer-aided signal processing. At times, they recounted, the radar echo had been "as small as 10^{-21} [watts], i.e., about 10^{27} times weaker than the transmitted signal power"; only clever design and exquisite instrumentation had made the experiment possible.[130] To produce, detect, and meaningfully interpret so tiny a return signal, Lincoln Laboratory's staff had drawn on resources that weren't available, or even imaginable, beyond the walls of their defense-lab home.

Amid the controversy as to whether MIT should divest its Air Force laboratory at Lincoln or spin off the Haystack radar to civilian overseers, MIT President Howard Johnson pointedly reminded the Air Force's Director of Laboratories that "the new test of Einstein's theory of general relativity [had] evolved from the Laboratory's intensive efforts in advanced radar technology."[131] Indeed it had. With technical and intellectual skills shaped by years of work on radar communications and satellite-tracking and missile-tracking projects, and equipped with multi-million-dollar Air Force-funded hardware, Lincoln Laboratory's researchers had put an extraordinarily subtle physical effect—and Einstein's most celebrated theory—to the test. Neither "pure" nor "applied," the relativistic time-delay test was Cold War science through and through.

Acknowledgments

We are grateful to archivists at the Institute Archives of the Massachusetts Institute of Technology, especially Nora Murphy, and at the Lincoln Laboratory archives, especially Tamar Granovsky and Nora Zaldivar. We also acknowledge Jimena Canales, Michael Gordin, Christopher McDonald, Suman Seth, Rebecca Slayton, and Dan Volmar for helpful comments on an earlier draft.

Notes

In the notes that follow, MITA means MIT Institute Archives and MITLL means MIT Lincoln Laboratory. In accordance with the Lincoln Laboratory's request, we cite archival materials from that lab's collections without specifying box and folder numbers.

1. Fred Jerome, *The Einstein File: J. Edgar Hoover's Secret War Against the World's Most Famous Scientist* (St. Martin's Press, 2002); David E. Rowe and Robert Schulmann, eds., *Einstein on Politics: His Private Thoughts and Public Stands on Nationalism, Zionism, War, Peace, and the Bomb* (Princeton University Press, 2007).

2. George Gamow, *One, Two Three ... Infinity* (New York, Viking, 1947), chapter 5.

3. Klaus Hentschel, *The Einstein Tower: An Intertexture of Dynamic Construction, Relativity Theory, and Astronomy* (Stanford University Press, 1997), 23; David Rowe, "Making mathematics in an oral culture: Göttingen in the era of Klein and Hilbert," *Science in Context* 17 (2004): 85–129, on 114–115; and Leo Corry, *David Hilbert and the Axiomatization of Physics* (Kluwer, 2004), 322, 364.

4. John Earman and Clark Glymour, "Relativity and eclipses: The British eclipse expeditions of 1919 and their predecessors," *Historical Studies in the Physical Sciences* 11 (1980): 49–85; Alistair Sponsel, "Constructing a 'revolution in science': The campaign to promote a favourable reception for the 1919 solar eclipse experiments," *British Journal for the History of Science* 35 (2002): 439–467; Matthew Stanley, "'An expedition to heal the wounds of war': The 1919 eclipse expedition and Eddington as Quaker adventurer," *Isis* 94 (2003): 57–89; Matthew Stanley, *Practical Mystic: Religion, Science, and A. S. Eddington* (University of Chicago Press, 2007), chapters 3 and 4.

5. Hubert Goenner, "The reaction to relativity theory, I: The anti-Einstein campaign in Germany in 1920," *Science in Context* 6 (1993): 107–133; Jeroen van Dongen, "Reactionaries and Einstein's fame: 'German scientists for the preservation of pure science,' relativity and the Bad Nauheim conference," *Physics in Perspective* 9 (2007): 212–230; Milena Wazeck, *Einsteins Gegner: Die öffentliche Kontroverse um die Relativitätstheorie in den 1920er Jahren* (Campus, 2009).

6. Alan Beyerchen, *Scientists Under Hitler: Politics and the Physics Community in the Third Reich* (Yale University Press, 1977).

7. Jean Einstaedt, "The low water mark of general relativity, 1925–1955," in *Einstein and the History of General Relativity*, ed. Don Howard and John Stachel (Birkhäuser, 1989), 277–292; David

Kaiser, "A psi is just a psi? Pedagogy, practice, and the reconstitution of general relativity, 1942–1975," *Studies in History and Philosophy of Modern Physics* 29 (1998): 321–338; Jean Eisenstaedt, *The Curious History of Relativity: How Einstein's Theory of Gravity Was Lost and Found Again* (Princeton University Press, 2006).

8. See esp. Clifford Will, *Was Einstein Right? Putting General Relativity to the Test* (Basic Books, 1986).

9. See Stuart W. Leslie, *The Cold War and American Science: The Military-Industrial-Academic Complex at MIT and Stanford* (Columbia University Press, 1993), especially 20–25; Michael Aaron Dennis, "'Our first line of defense': Two university laboratories in the postwar American state," *Isis* 85 (1994): 427–455; Deborah Douglas, "MIT and War," in *Becoming MIT: Moments of Decision*, ed. David Kaiser (MIT Press, 2010).

10. Leslie, *Cold War and American Science*, 25.

11. Ibid., 27.

12. Ibid., 31.

13. Ibid., 32–35; Eva Freeman, ed., *MIT Lincoln Laboratory: Technology in the National Interest* (MIT Lincoln Laboratory, 1995), 6; David Kaiser, "Elephant on the Charles: Postwar growing pains," in *Becoming MIT: Moments of Decision*, ed. David Kaiser (MIT Press, 2010), 108–109.

14. See Leslie, *Cold War and American Science*, 32–41; Freeman, *MIT Lincoln Laboratory*, 7–13.

15. Freeman, *MIT Lincoln Laboratory*, 3.

16. On the history of Project Whirlwind and SAGE see, e.g., Freeman, *MIT Lincoln Laboratory*, 15–33; Paul N. Edwards, *The Closed World: Computers and the Politics of Discourse in Cold War America* (MIT Press, 1996), chapter 3; Atsushi Akera, *Calculating a Natural World: Scientists, Engineers, and Computers During the Rise of US Cold War Research* (MIT Press, 2007), chapter 5. Edwards writes that SAGE, though terribly costly and technologically obsolete before it was even finished, was a paradigmatic example of the military's growing "hope of enclosing the awesome chaos of modern warfare (not only nuclear but 'conventional') within the bubble worlds of automatic, rationalized systems" (110). See also Donald L. Clark, "Early advances in radar technology for aircraft detection," *Lincoln Laboratory Journal* 12, no. 2 (2000): 167–180, especially 167–168.

17. Freeman, *MIT Lincoln Laboratory*, 35–36.

18. Leslie, *Cold War and American Science*, 35.

19. See, e.g., General Bernard A. Schriever of the Air Force (later chief military custodian of Lincoln Laboratory) to Secrety of the Air Force Eugene Zuckert, 13 August 1962, in Library of Congress, Papers of General Curtis LeMay, Box 141, Air Force Systems Command (AFSC) 1962. This and similar documents related to the history of early missile-detection and the policy of "launch on warning" may be found at http://www.gwu.edu/~nsarchiv/NSAEBB/NSAEBB43/. See also Donald MacKenzie, *Inventing Accuracy: A Historical Sociology of Nuclear Missile Guidance* (MIT Press, 1990).

20. Freeman, *MIT Lincoln Laboratory*, 47–49.

21. Warren Rogers Jr., "US tries huge radar on Russia," *Washington Post*, May 17, 1960.

22. Leslie, *Cold War and American Science*, 35; Freeman, *MIT Lincoln Laboratory*, 49. On the BMEWS installation at Fylingdales in Yorkshire, England, see Graham Spinardi, "Golfballs on the moor: Building the Fylingdales Ballistic Missile Early Warning System," *Contemporary British History* 21, no. 1 (2007): 87–110.

23. G. H. Pettengill and D. E. Dustin, "A comparison of selected ICBM early-warning radar configurations," MIT Lincoln Laboratory Technical Report 127, 1956.

24. For background on Pettengill, see the citation for the 1997 Whitten Medal of the American Geophysical Union: "Pettengill receives the Whitten Medal," *Eos* 78, no. 39 (1997): 419–420. On Dustin, see his obituary in the online edition of *Bates Magazine* (http://www.bates.edu). Dustin's master's thesis (An Investigation of the Environment for Scientific Research in Industry, in the University, and in Government) is in MIT's Hayden Library.

25. Pettengill and Dustin, "Comparison," 13–15.

26. Irwin I. Shapiro, interview by Andrew J. Butrica, September 30, 1993. (All oral history interviews conducted by Andrew J. Butrica cited in this chapter are courtesy of the NASA History Office.) See also "Interview with Irwin Shapiro," MITLL, 5. On radar in clutter environments, see Donald L. Clark, "Early advances," 172–174.

27. In an interview, Shapiro summarized the problem as follows: "You have this radar in, say, Alaska, and a missile comes over the pole. Where is it going? Is it going to New York? Is it going to San Francisco? Is it going to Chicago? Grand Forks, North Dakota? Where is it going?" See Irwin I. Shapiro, interview by Andrew J. Butrica, May 4, 1994, 1–2. Shapiro's group leader, David Falkoff, had earlier written a brief report summarizing the basic properties of ballistic trajectories. Shapiro's work expanded this simple theoretical framework to the context of radar observation. See D. L. Falkoff and E. C. Lerner, "Characteristics of trajectories," MIT Lincoln Laboratory Group Report No. 47.3 (January 1956), GR-47-3, MITLL; and Shapiro, interview by Butrica, September 30, 1993, 2.

28. I. I. Shapiro, "The prediction of ballistic missile trajectories from radar observations," MIT Lincoln Laboratory Technical Report 129, 1957.

29. Kent Kresa and Peter A. Willmann, Three Degrees-of-Freedom Trajectory Analysis Program, MIT Lincoln Laboratory Group Report 312G-5 (1962); Kent Kresa and Menasha Tausner, Programs for Calculating a Predicted Radar Trajectory, MIT Lincoln Laboratory Group Report 312G-13, 1962.

30. Irwin I. Shapiro, *The Prediction of Ballistic Missile Trajectories from Radar Observations* (McGraw-Hill, 1958).

31. Irwin I. Shapiro, interview by Andrew J. Butrica, September 30, 1993.

32. Irwin I. Shapiro, interview by Andrew J. Butrica, October 1, 1993.

33. On Stroke's background at the RLE, see George Wilhelm Stroke: An Interview Conducted by William Aspray, available at http://www.ieeeghn.org.

34. The quotation is from George W. Stroke, Light, MIT Research Laboratory of Electronics Technical Report 348, January 9, 1959, 19. Most experts on general relativity at the time agreed that gravity should affect the speed (and not just the direction and frequency) of light's propagation; see, e.g., Albert Einstein, "Die Grundlage der allgemeinen Relativitätstheorie," *Annalen der Physik* 49 (1916): 769–822, as translated and reprinted in Albert Einstein et al., *The Principle of Relativity* (Methuen, 1923), 109–164, on 114–115, 162–163; Albert Einstein, *Relativity: The Special and General Theory* (Holt, 1920), 111; Albert Einstein, *The Meaning of Relativity* (Princeton University Press, 1956 [1922]), 92–93; Wolfgang Pauli, *Theory of Relativity* (Dover, 1958 [1921]), 143–144, 154, 160; Hermann Weyl, *Space-Time-Matter* (Dover, 1922), 224, 252–255; A. S. Eddington, *The Mathematical Theory of Relativity*, second edition (Cambridge University Press, 1924), 93–94; Max Born, *Einstein's Theory of Relativity* (Dover, 1962 [1924]), 356–358. Even a modern textbook analyzes Shapiro's time-delay test in terms of variable speed of light: M. V. Berry, *Principles of Cosmology and Gravitation* (Hilger, 1989), 88–90. Physicists today often describe the effect differently: all *local* measurements of the speed of light conducted by freely falling (inertial) observers will agree, and hence one may conclude that the speed of light is constant even in general relativity. On the other hand, measurements of speed presuppose the establishment of a coordinate system. Having established a coordinate system extended in spacetime beyond one's infinitesimally small region, even freely falling observers would see the speed of a light ray vary as it traveled through a large region of gravitationally warped spacetime. That is, in any given coordinate system, the speed of propagation along a null geodesic, dr/dt, will depend on the spacetime curvature.

35. Kaiser, "A psi is just a psi?" 322.

36. Shapiro, interview by Butrica, October 1, 1993.

37. Ibid.

38. For accounts of the Venus detection experiment, see Andrew J. Butrica, "In conjunction with Venus," *IEEE Spectrum* 34 (December 1997): 31–38; Andrew J. Butrica, *To See the Unseen: A History of Planetary Radar Astronomy* (NASA History Office, 1996), 27–36.

39. Butrica, "In conjunction with Venus," 32. The system Price and Green worked on was known as Nomac, which stood for Noise Modulation and Control; it was an early version of what would later be widely known as "spread spectrum" technology. Specifically, Price and Green invented "Rake," a receiver that successfully combined a series of signals to enhance the communications channel above the background noise. See Freeman, *MIT Lincoln Laboratory*, 51–54.

40. The book probably was *Radio Astronomy* by Ronald L. Bracewell and Joseph L. Pawsey, two Australian physicists working in the Radiophysics Laboratory of the Australian government's Commonwealth Scientific and Industrial Research Organization, published by Clarendon in 1955. On the proposed radar detection of Venus, see p. 304ff.

41. Butrica, "In conjunction with Venus," 32.

42. Ibid.

43. "Dr. Paul Green: An Interview Conducted by David Hochfelder," IEEE History Center, October 15, 1999 (available at http://www.ieeeghn.org).

44. Paul Green, interview by David Hochfelder.

45. Butrica, "In conjunction with Venus," 32. For the actual report, see R. Price, P. E. Green Jr., T. J. Goblick Jr., R. H. Kingston, L. G. Kraft, G. H. Pettengill, R. Silver, and W. B. Smith, "Radar echoes from Venus," *Science* 129 (March 20, 1959): 751–753.

46. Carl F. J. Overhage to Lieutenant General Roscoe C. Wilson, 24 March 1959, Box 76, Folder "Lincoln Laboratory" 4/4, AC134, MITA; *New York Times*, March 20, 1959; Noah Gordon, "First signals bounced off Venus," *Boston Herald*, March 20, 1959, in Box 76, Folder "Lincoln Laboratory" 4/4, AC134, MITA.

47. Carl F. J. Overhage to Lieutenant General Roscoe C. Wilson, 24 March 1959, Box 76, Folder "Lincoln Laboratory" 4/4, AC134, MITA.

48. Butrica, *To See the Unseen*, 30–42. RCA's Missile and Surface Radar Division in Moorsetown, New Jersey (the prime contractor for the BMEWS missile-tracking radars) succeeded in detecting Venus in March and April 1961, as did Jodrell Bank in the UK, and a Soviet team under Vladimir Kotelnikov at the Long-Distance Space Communication Center in the Crimea. Each group (including Lincoln and JPL) used its time-delay values for Venus to determine an estimate of the astronomical unit, and the general agreement among the various groups' results was cause for a revision in the value accepted by the International Astronomical Union. See Butrica, *To See the Unseen*, 44–46. Bernard Lovell, director of the Jodrell Bank radio telescope facility, also promoted Jodrell Bank as a possible BMEWS site. In fact, the facility was used as an interim BMEWS station until September of 1963, when the installation at Fylingdales was ready. See Graham Spinardi, "Science, technology, and the Cold War: The military uses of the Jodrell Bank radio telescope," *Cold War History* 6, no. 3 (2006): 279–300.

49. Shapiro, interview by Butrica, September 30, 1993.

50. Shapiro, interview by Butrica, October 1, 1993.

51. Ibid.; J. Kovalevsky, ed., *The System of Astronomical Constants: International Astronomical Union Symposium 21* (Gauthier-Villars, 1963); Irwin Shapiro to Joseph Weber, 22 April 1965, MITLL. Shapiro informed Weber that he had first done "the necessary rough calculations in the summer of 1962."

52. Shapiro, interview by Butrica, October 1, 1993.

53. Peter G. Bergmann, *Introduction to the Theory of Relativity* (Prentice-Hall, 1942); Shapiro, interview by Butrica, October 1, 1993.

54. W. H. Radford to General B. A. Schriever, 1 March 1965, Box 75, Folder "Lincoln Laboratory" 1/4, AC134, MITA.

55. M. E. Ash, "Generation of planetary ephemerides on an electronic computer," MIT Lincoln Laboratory Technical Report 391, 1965, 1-3; Ash, "Generation of the lunar ephemeris on an electronic computer," MIT Lincoln Laboratory Technical Report 400, 1965, 1-2. Shapiro discussed the basic outlines of the Planetary Ephemeris Program in an oral history interview conducted by Andrew J. Butrica on September 30, 1993. Butrica also discusses the program on pages 123-126 of *To See the Unseen*. In one of his oral history interviews, Shapiro made the connection between his calculations of ballistic missile trajectories and the design of the Planetary Ephemeris Program even more explicit: "Because I had worked on that ballistic missile problem, I was approached by those people at Lincoln Laboratory who were planning the first Venus radar experiment. They felt that they needed some expertise in celestial mechanics. That is how I first became involved in radar astronomy, and that led to PEP. It was that chain of events." See Irwin I. Shapiro, interview by Andrew J. Butrica, May 4, 1994.

56. Shapiro, interview by Butrica, May 4, 1994.

57. On Ash's background, see the Shapiro, interview by Butrica, September 30, 1993. Shapiro and Ash made written requests to several observatories around the world, including some in Japan and South Africa, for tables of centuries-old optical ephemerides. See, e.g., Irwin I. Shapiro to Dr. R. L. Duncombe, US Naval Observatory, 17 June 1965, MITLL; Irwin I. Shapiro to Astronomer Royal, Royal Greenwich Observatory, August 16, 1965, MITLL.

58. William Boyd Smith, interview by Andrew J. Butrica, September 29, 1993.

59. H. Sherman to Sidney Myers, 4 May 1964, MITLL. See also H. Sherman to Mrs. E. Simmons, 5 June 1964, MITLL.

60. Shapiro to G. P. Dinneen, 24 November 1964, MITLL; Shapiro, interview by Butrica, September 30, 1993. On Dinneen's career, see *The Global Agenda for American Engineering: Proceedings of a Symposium Held in Honor of Gerald P. Dinneen* (National Academy Press, 1996), 1-3.

61. Shapiro, interview by Butrica, October 1, 1993.

62. I. I. Shapiro, Effects of General Relativity on Interplanetary Time-Delay Measurements, MIT Lincoln Laboratory Technical Report 368, 1964, 1. As Shapiro (and most other physicists) viewed the history of gravity research, three essentially distinct kinds of experimental tests of general relativity had been proposed before his time-delay experiment. These 'classical' tests included the bending of starlight by the sun (as first observed by Eddington in 1919), the observation of the anomalous precession of Mercury's perihelion (well known among astronomers since Urbain Le Verrier first observed it in the middle of the nineteenth century), and the "redshift,"—the decrease in the frequency of light as it travels from regions of stronger gravitational field to regions of weaker gravitational field (measured precisely in atomic experiments at Harvard University by Robert Pound and Glen Rebka in 1959). Shapiro's proposed test of general relativity was thus the "fourth" in a series.

It is not obvious that the redshift experiment constitutes a clear test of general relativity, since the redshift of light is not a consequence of Einstein's theory specifically, but of something called the "equivalence principle" generally. (The equivalence principle says, simply, that an object's inertial mass and its gravitational mass are equivalent.) Many rival theories of gravity incorporate the equivalence principle, but they differ in the quantitative details of their predictions of how spacetime is curved in the presence of a given distribution of matter and energy. On the equivalence principle and the history of 'classical' tests of general relativity, see chapter 3–5 of Will, *Was Einstein Right?*

63. Shapiro, "Effects of general relativity," 1.

64. Irwin I. Shapiro, "Fourth test of general relativity," *Physical Review Letters* 13 (December 28, 1964): 789–791.

65. Allen S. Richmond to Julius A. Stratton, 22 December 1964, Box 76, Folder "Lincoln Laboratory" 2/4, AC134, MITA.

66. "New test proposed for Einstein theory," *New York Times*, December 28, 1964.

67. Shapiro, "Effects of general relativity," 21–22. Mcps is a measure of frequency (millions of cycles per second). In the International System of Units (commonly called SI), frequency is measured in units of hertz, abbreviated Hz; 1 Hz is equivalent to one cycle per second. Thus, 1,000 MHz (megahertz) is equivalent to 1,000 Mcps.

68. Shapiro, "Effects of general relativity," 21; Herbert G. Weiss, "The Haystack experimental facility," MIT Lincoln Laboratory Technical Report 365, 1964, iii.

69. Shapiro to J. Arthur, W. Davenport, B. Lax, J. Meyer, and P. Sebring, 10 November 1964, MITLL.

70. Ibid.

71. The Air Force did fund research on general relativity at the time, but almost exclusively small projects in theoretical physics, rather than major equipment upgrades, as in the case of Haystack. See Joshua N. Goldberg, "US Air Force support of general relativity: 1956–1972," in *Studies in the History of General Relativity*, ed. Jean Eisenstaedt and A. J. Kox (Birkhäuser, 1992).

72. Carl F. J. Overhage to Lieutenant General Roscoe C. Wilson, 30 August 1960, 76:1, AC134, MITA. A Lincoln Laboratory information pamphlet similarly played up Haystack's unrivaled resolving power, noting that it was "designed to detect a one-square-meter target at a range of 20,000 nautical miles, and a target only 0.0001 square meter in cross-section at 2,000 miles." Pamphlet located in Box 76, "Lincoln Laboratory" Folder 2/4, AC134, MITA.

73. John V. Harrington, "The Haystack Hill station," MIT Lincoln Laboratory Technical Memorandum 78, 1959; Butrica, *To See the Unseen*, 65.

74. Herbert G. Weiss, "The Haystack experimental facility," MIT Lincoln Laboratory Technical Report 365, 1964, iii.

75. For background on Project West Ford, see Freeman, *MIT Lincoln Laboratory*, 65–66, and William W. Ward and Franklin W. Floyd, "Thirty years of research and development in space communications at Lincoln Laboratory" in *Beyond the Ionosphere: The Development of Satellite Communications*, ed. Andrew J. Butrica (NASA History Office, 1997), 79–81. For a discussion of

the West Ford controversy, see Edward M. Purcell, "The case For the 'needles' experiment," *New Scientist* 13 (February 1, 1962): 245–247. For details concerning Shapiro's work on West Ford, see R. W. Parkinson, H. M. Jones, and I. I. Shapiro, "Effects of solar radiation on earth satellite orbits," *Science* 131 (March 25, 1960): 920–921; Irwin I. Shapiro and Harrison M. Jones, "Lifetimes of orbiting dipoles," *Science* 134, October 6, 1961: 973–979. Shapiro, in a later interview, hinted that Haystack's connection to Project West Ford (and its needle-like orbiting copper dipoles) went even as far as its name. Receiving signals from the diffuse dipole belt was apparently as difficult as finding a needle in a haystack. See "Interview with Irwin Shapiro," p. 7, MITLL.

76. Weiss, "Haystack," 55.

77. *Technology Review* 64 (January 1962), 17, Box 76, Folder "Lincoln Laboratory" 3/4, AC134, MITA.

78. Weiss, "Haystack," esp. 1–6. Such a configuration—a secondary hyperbolic reflector mounted on a larger parabolic reflector—was said to be "Cassegrainian," inspired by a standard optical telescope design traditionally associated with the seventeenth-century priest Laurent Cassegrain.

79. Information on the radome construction is included on the back cover of *Technology Review* 64 (January 1962), Box 76, Folder "Lincoln Laboratory" 3/4, AC134, MITA. The thickness of the "bearing and control" system's oil film is mentioned in Weiss, "Haystack," 11.

80. For an image of the Univac 490 computer control room, see Weiss, "Haystack," 30.

81. Weiss, "Haystack," 27.

82. Carl F. J. Overhage to Lieutenant General Roscoe C. Wilson, 21 November 1960, Box 76, Folder "Lincoln Laboratory" 4/4, AC134, MITA.

83. Lincoln Laboratory Procurement Information, Serial No. 301, Purchase Order No. B-00375, May 1, 1964 (North American Aviation, Inc.); Serial No. 252, Purchase Order No. BB-112, 18 June 1963 (Univac Division, Sperry Rand Corporation), MITLL.

84. "Haystack facility dedication ceremony," Box 76, Folder "Lincoln Laboratory" 2/4, AC134, MITA.

85. Weiss, interview by Butrica, September 29, 1993, MITLL, 13.

86. "Interview with Irwin Shapiro,"MITLL, 8.

87. Anon., "1,000 atomic workers face loss of jobs," *Los Angeles Times*, January 7, 1965; C. W. Sherwin and R. S. Isenson, First Interim Report on Project Hindsight, Office of the Director of Defense Research and Engineering, June 1966. See also Karl Kreilkamp, "*Hindsight* and the real world of science policy," *Science Studies* 1 (1971): 43–66.

88. See Daniel Kevles, *The Physicists: The History of a Scientific Community in Modern America*, 3rd ed. (Harvard University Press, 1995 [1978]), chapters 24 and 25; Roger Geiger, *Research and Relevant Knowledge: American Research Universities since World War II* (Oxford University Press, 1993), chapters 8 and 9; Leslie, *Cold War and American Science*, chapter 9; Matthew Wisnioski, "Inside

'the System': Engineers, scientists, and the boundaries of social protest in the long 1960s," *History and Technology* 19 (2003): 313–333; Kelly Moore, *Disrupting Science: Social Movements, American Scientists, and the Politics of the Military, 1945–1975* (Princeton University Press, 2008), chapters 5 and 6.

89. P. B. Sebring to J. W. Meyer, January 12, 1965, MITLL.

90. Ibid.

91. I. I. Shapiro to J. Meyer, January 13, 1965, MITLL.

92. J. W. Meyer, "Work statement and budget estimate for deep space radar measurements," January 27, 1965, MITLL.

93. J. W. Meyer, ed., "Program description: Radar observations of the planets," January 25, 1965, MITLL. Meyer's proposal was also published in an internal Lincoln Laboratory Technical Note: see J. W. Meyer, "Program description—Radar observations of the planets," MIT Lincoln Laboratory Technical Note 1965-35, 1965. For cost estimates, see "Table V" ("Personnel, costs, & time, preparation for 4th test of general relativity") in "Budgetary memorandum," (undated), probably prepared by Paul B. Sebring for James W. Meyer, MITLL.

94. Homer E. Newell to Dr. W. H. Radford, 1 March 1965; Paul B. Sebring to Dr. William Brunk, 20 April 1965; J. W. Meyer to W. H. Radford, 24 May 1965, all in MITLL.

95. Pettengill described the unusual funding situation at Haystack in the mid 1960s as follows: "We had money from NASA to do some lunar work, and we had special money that came in a peculiar way to support the fourth test of General Relativity. That money paid for the transmitter, which became the backbone of the Haystack radar facility." See Pettengill, interview by Butrica, September 28, 1993. Pettengill also recalled that radar astronomy at Lincoln Laboratory, more generally, had always been regarded as perfectly consistent with (and fostered by) military work: "At Millstone [in contrast to the Arecibo facility in Puerto Rico], radar astronomy was never considered the prime mission. It was an allowable activity. It was even encouraged. It was certainly funded, but you did not hire people specifically to do that work. It was to be kept in its proper place. It was never put that way, but you knew it. Your prime involvement was with the military work, the classified work. You could never let that lapse. However, as long as you did your homework, you could do this other work, which was more fun. It was okay, but it was never a prime objective. It was tolerated." See Pettengill, interview by Butrica, September 28, 1993.

96. Lincoln Laboratory Procurement Information, Serial No. 413, Purchase Order No. BB-232, August 24, 1965 (Lehigh Design Company), MITLL.

97. Shapiro to Wilbert F. Buie, January 25, 1965; Shapiro to Emory Ladner, February 26, 1965; Shapiro to Robert N. White, December 7, 1966; Shapiro to Cordelia Galt, undated, MITLL.

98. Shapiro to George H. Brigman, February 26, 1965; Shapiro to Steven Weinberg, March 4, 1965; Shapiro to Joseph Weber, 22 April 1965; Shapiro to B. Bertotti, 21 December 1965; Shapiro to James C. W. Scott, 22 August 1967, MITLL.

99. P. G. Bergmann, review of "Fourth test of general relativity" by Irwin I. Shapiro, *Mathematical Reviews* 30 (December 1965); Irwin I. Shapiro to P.G. Bergmann, January 12, 1966; Shapiro to L. I. Schiff, 14 May 1965, 21 July 1965, 29 October 1965, and 12 November 1965, all in MITLL. Leonard Schiff's proposal was first published as Schiff, "Possible new experimental test of general relativity theory," *Physical Review Letters* 4 (March 1, 1960): 215–217. Schiff teamed up with Stanford experimentalists William Fairbank and Robert Cannon, experts on low-temperature physics and sensitive gyroscopic instrumentation, to propose an orbiting gyroscope experiment to NASA in 1961. (A similar calculation and experiment had been outlined by MIT scientist George Pugh in 1959, while working for the Department of Defense's Weapons Systems Evaluation Group.) Though the proposal received initial funding, the challenging experiment would fail to get off the ground in time for Schiff to see actual results (he passed away in 1971). A more recent incarnation of the gyroscope experiment—NASA's "Gravity Probe B"—was flown in 2004 and 2005. See Will, *Was Einstein Right*, chapter 11. For the later history of (and latest news on) the gyroscope test, see Stanford's Gravity Probe B website: "Gravity Probe B: Testing Einstein's Universe" (available at http://einstein.stanford.edu/).

100. Duane O. Muhleman and Paul Reichley, "Effects of general relativity on planetary radar distance measurements," *Jet Propulsion Laboratory Space Programs Summary* 4, no. 37–29 (1964): 239–241; D. O. Muhleman and I. D. Johnston, "Radio propagation in the solar gravitational field," *Physical Review Letters* 17 (August 22, 1968): 455–458. See also Shapiro to Duane O. Muhleman, 17 December 1964, 14 May 1965, 2 September 1966, and 26 September 1966, MITLL. Shapiro would later call the episode "one of the major annoyances of my professional life." "I felt he had taken my idea, and he didn't give me any credit for it. ... He would never admit it. He would never admit it." See Shapiro, interview by Butrica, October 1, 1993, 12–16.

101. Shapiro to Pettengill, 13 September 1965, MITLL.

102. Irwin I. Shapiro to R. M. Goldstein, 17 November 1965, MITLL. Still sore from the ordeal with Muhleman, Shapiro aimed a quick shot at the rival laboratory: "From experience I know that JPL people don't like to write letters, but I would really appreciate your answer."

103. See, e.g., Michael E. Ash to A. N. Adams, 26 November 1965; Michael E. Ash to Superintendent, US Naval Observatory, 6 December 1965; Michael E. Ash to R. L. Duncombe, 15 March 1966, MITLL. See also Michael E. Ash to R. L. Duncombe, 5 August 1966, MITLL; Michael E. Ash to Superintendent, US Naval Observatory, 28 April 1967, MITLL.

104. I. Shapiro to H. Sherman, 22 July 1966, MITLL.

105. I. I. Shapiro to P. B. Sebring, 3 March 1967, MITLL.

106. Irwin I. Shapiro, "Schedule for program completion and checkout for planetary ephemerides," 7 April 1966, MITLL.

107. I. I. Shapiro to Reginald E. Newell, 16 February 1966, MITLL.

108. J.S. Arthur to S.H. Dodd et al., 4 March 1966, MITLL.

109. G. H. Pettengill to P. B. Sebring, 8 March 1966, MITLL.

110. P. B. Sebring to J. S. Arthur et al., 9 March 1966, 1966, MITLL.

111. P. B. Sebring to J. S. Arthur et al., 17 May 1966, MITLL.

112. P. B. Sebring to J. S. Arthur et al., 9 March 1966, MITLL. Our reconstruction of the experiment is inspired by similar close studies of the usually unseen labor behind laboratory work. See, e.g., Steven Shapin, "The invisible technician," *American Scientist* 77 (November–December 1989): 554–563; Peter Galison, *Image and Logic: A Material Culture of Microphysics* (University of Chicago Press, 1997); Harry Collins, *Gravity's Shadow: The Search for Gravitational Waves* (University of Chicago Press, 2004).

113. Irwin I. Shapiro to Samuel Goudsmit, 7 March 1968, MITLL. Leonard Schiff acted as referee for the paper. Shapiro and Schiff proceeded to debate, among other things, the effect of Venus's surface topography on the time-delay measurements, the question whether Shapiro had "stolen" the form of the "generalized metric" from Schiff and Ross's analysis (in relativity, the "metric" is the mathematical object that allows one to calculate distances between different points in four-dimensional spacetime), and the sticky issue of whether to credit Muhleman and Reichley with an independent proposal of the experiment. (It was a non-issue for Shapiro, of course. "I hope you agree with me that no one should be given credit for an idea that wasn't his," he wrote Schiff. When Schiff reported to Shapiro that Muhleman claimed their results were "*completely* independent," Shapiro actually sent copies of his entire personal correspondence with Muhleman for Schiff to read for himself. "I don't ordinarily distribute copies of my correspondence with others," he added, "but since my honesty seems to be in question, I have made an exception for this case.") See Irwin Shapiro to Leonard I. Schiff, 29 May 1968; and Irwin I. Shapiro to Leonard I. Schiff, 10 July 1968, MITLL.

114. Irwin I. Shapiro, Gordon H. Pettengill, Michael E. Ash, Melvin L. Stone, William B. Smith, Richard P. Ingalls, and Richard A. Brockelman, "Fourth test of general relativity: Preliminary results," *Physical Review Letters* 20 (May 27, 1968): 1265–1269, on 1265.

115. I. I. Shapiro to M. U. Clauser, 19 March 1968, MITLL.

116. On the dramatic transformations in the science-state relationship in the late 1960s, see, e.g., Kevles, *The Physicists*, 406–415; Leslie, *Cold War and American Science*, chapter 9; and Kelly Moore, *Disrupting Science: Social Movements, American Scientists, and the Politics of the Military, 1945–1975* (Princeton University Press, 2008), chapter 5.

117. Howard W. Johnson to Robert S. McNamara, 22 September 1967, Box 203, Folder 13, AC118, MITA.

118. Frank R. Haggerty to Milton U. Clauser, 26 September 1967, Box 203, Folder 12, AC118, MITA.

119. Howard W. Johnson to Alexander H. Flax, 31 October 1967, Box 203, Folder 12, AC118, MITA.

120. As Daniel Kevles notes, although the amendment was removed the following year, "its statement of congressional intent had a lasting impact, especially in the defense bureaucracy's

interpretation of its latitude in research funding." See Kevles, *The Physicists*, esp. 414–415. Part of the amendment is quoted in Butrica, *To See the Unseen*, 80.

121. Milton U. Clauser to Howard W. Johnson, 13 November 1969, Box 203, Folder 12, AC118, MITA. Johnson mentioned the news coverage of campus protests at MIT to Brigadier General Raymond A. Gilbert of the Air Force Systems Command, assuring him that the "events as reported by the news media have seldom been presented in the proper perspective." That said, Johnson admitted, the 1969 "period of critical self-examination as to the basic nature and purpose of the Institute" would have fundamental consequences for the way MIT balanced "advancing knowledge" with "rendering service to the community and nation." See Howard W. Johnson to Raymond A. Gilbert, 13 February 1970, Box 203, Folder 12, AC118, MITA. On protests against the Vietnam War at MIT, see Kenneth Hoffman et al., *Creative Renewal in a Time of Crisis: Report of the Commission on MIT Education* (MIT Press, 1970); Dorothy Nelkin, *The University and Military Research: Moral Politics at MIT* (Cornell University Press, 1972); David Kaiser, "Elephant on the Charles: Postwar growing pains," in Kaiser, *Becoming MIT*; Stuart W. Leslie, "'Time of troubles' for the special laboratories," in Kaiser, *Becoming MIT*.

122. P. B. Sebring to M. U. Clauser, 21 November 1969, contained as an attachment to Milton U. Clauser to Jerome B. Wiesner and Edward M. Purcell, 25 November 1969, Box 203, Folder 12, AC118, MITA.

123. Howard W. Johnson to Robert C. Seamans Jr., 21 May 1970, Box 203, Folder 11, AC118, MITA.

124. Howard Johnson attributed the transfer of Haystack directly to "the constraints imposed by Section 203 of the Fiscal Year 1970 Military Procurement Authorization Act" (the Mansfield Amendment) in a letter to Secretary of the Air Force Robert Seamans Jr. See Howard W. Johnson to Robert C. Seamans Jr., 21 May 1970, Box 203, Folder 11, AC118, MITA. The yearly operating cost estimate for Haystack is found in P. B. Sebring to M. U. Clauser, 21 November 1969, contained as an attachment to Milton U. Clauser to Jerome B. Wiesner and Edward M. Purcell, 25 November 1969, Box 203, Folder 12, AC118, MITA. Andrew Butrica reports the last Haystack radar transmission recorded in the logbooks of Haystack engineers Richard Ingalls and Alan Rogers, in *To See the Unseen*, 83.

125. See esp. Paul Forman, "Behind quantum electronics: National security as basis for physical research in the United States, 1940–1960," *Historical Studies in the Physical Sciences* 18 (1987): 149–229; Leslie, *Cold War and American Science*.

126. See esp. Daniel Kevles, "Cold War and hot physics: Science, security, and the American state, 1945–1956," *Historical Studies in the Physical Sciences* 20 (1990): 239–264.

127. See esp. Forman, "Behind quantum electronics"; Leslie, *The Cold War and American Science*; Galison, *Image and Logic*; Spencer Weart, "Global warming, Cold War, and the evolution of research plans," *Historical Studies in the Physical and Biological Sciences* 27 (1997): 319–356; Edward Jones-Imhotep, "Disciplining technology: Electronic reliability, Cold-War military culture, and the topside ionogram," *History and Technology* 17 (2001): 125–175; David Kaiser, "Cold War requisitions, scientific manpower, and the production of American physicists after World

War II," *Historical Studies in the Physical and Biological Sciences* 33 (2002): 131–159; Naomi Oreskes, "A context of motivation: US Navy oceanographic research and the discovery of sea-floor hydrothermal vents," *Social Studies of Science* 33 (2003): 697–742; Jacob Hamblin, *Oceanographers and the Cold War: Disciples of Marine Science* (University of Washington Press, 2005); John Krige, *American Hegemony and the Postwar Reconstruction of Science in Europe* (MIT Press, 2006); and Akera, *Calculating a Natural World*.

128. See Cyrus C. M. Mody, "How I learned to stop worrying and love the bomb, the nuclear reactor, the computer, ham radio, and recombinant DNA," *Historical Studies in the Natural Sciences* 38 (2008): 451–461.

129. Irwin I. Shapiro, Michael E. Ash, Richard P. Ingalls, William B. Smith, Donald B. Campbell, Rolf B. Dyce, Raymond F. Jurgens, and Gordon H. Pettengill, "Fourth test of general relativity: New radar result," *Physical Review Letters* 26 (May 3, 1971): 1132–1135.

130. Irwin I. Shapiro et al., "Fourth test of general relativity: Preliminary results," 1265.

131. Howard W. Johnson to General Raymond A. Gilbert, 18 November 1968, Box 203, Folder 12, AC118, MITA.

10 Defining (Scientific) Direction: Soviet Nuclear Physics and Reactor Engineering during the Cold War

Sonja D. Schmid

Like no other science, nuclear physics and its related scientific and engineering disciplines shaped, and were shaped by, the Cold War. In particular, what was considered physics, and what was valuable about "fundamental research" more broadly, became subjects of intense debate in the early 1950s. These dynamics took on a distinct character in the post-Stalin Soviet Union, where internal ideological debates increasingly confronted renewed international exchange among nuclear scientists and engineers. Encouraged by their peers' recognition for their achievements, Soviet nuclear specialists started using the words for "fundamental" and "applied" in ever more sophisticated ways. Most remarkably, and against considerable odds, Soviet scientists succeeded in defending the place of "fundamental" science, even in very "applied" areas such as the nuclear power industry.

In this chapter, I show that in the Soviet context technical choices, specifically in the case of nuclear reactor physicists and engineers, grew out of earlier conceptual debates about the nature of Soviet science and its place in Soviet society. Technical designs, I argue, emerged hand in hand with a uniquely Soviet institutional landscape, which in turn reflected the ongoing boundary work on the role of "fundamental" and "applied" science.[1]

After Stalin's death, and in the context of increasing international discussion, Soviet nuclear experts gradually learned to use new arguments at home that utilized science's claim to universal truth as a powerful rhetorical resource. Ultimately, a specific way of understanding the distinction between "fundamental" and "applied" science shaped the organizational structures of the emerging nuclear industry. These structures in turn influenced who could claim to do "fundamental" nuclear science, "applied" nuclear science, or both.[2] The industry's institutions took shape in combination with technical choices, which in turn reflected the delineations between "fundamental" and "applied" and between international science and Soviet science, and reinforced the institutional expression of these distinctions.

Soviet Science and the Beginnings of Soviet Nuclear Physics

What was unique about Soviet science? According to Sergei Vavilov, president of the Soviet Academy of Sciences from 1945 to 1951, science in the Soviet Union was thoroughly democratic: it was science of the people and for the people. To serve the people, Soviet science was planned just like the rest of the economy; its programs were oriented toward practical outcomes, and its goals were authorized by the Communist Party.[3] Furthermore, since science was the declared "tool of Communist social design," only the Soviet kind of society was able to sustain legitimate and authentic science.[4] The objective knowledge produced by science, in turn, justified the Soviet social order.[5]

In the West, of course, Cold War propaganda was quick to emphasize the allegedly biased nature of science in the Soviet Union—in contrast, presumably, to its pristine character in American-style capitalist democracies. Many Western analysts discarded Soviet science as hopelessly "subdued" by political authorities, and by Stalin's personal interventions in particular. The most widely referenced episode that supposedly confirms this view is the so-called Lysenko affair.[6] As Alexei Kojevnikov put it, innumerable references to this episode connected "the failures and problems of Soviet science and technology to the pernicious influences of politics and ideology, while refusing to see the very same forces at work in the cases of achievements and triumphs."[7] And of achievements and triumphs there were many.

The Soviet Union's test of a fission device in 1949 caught Americans by surprise, and its first test of a fusion weapon, in 1952, came years earlier than American scientists had expected. Somewhat belatedly, in 1955, the world found out at the first UN Conference on Peaceful Uses of Atomic Energy in Geneva that the Soviet Union had also pioneered "peaceful" applications of nuclear energy by starting up the world's first nuclear power plant the preceding summer.[8] Two years later, in 1957, the Soviet Union launched the first artificial satellite, and in 1961 it sent the first human into space.

How was this possible? How could a country that had within the past fifty years gone through the political turmoil of empire collapse, back-to-back revolutions, and civil war, a country whose population had suffered devastating famine during the forced collectivization of agriculture and horrendous loss of life during World War II, succeed in cutting-edge science and highly innovative technology? Students of Soviet science have pointed to the new institutional system of research institutes, created after the 1917 October Revolution, that emphasized the *outcomes* of scientific research. True to the early Bolshevik motto "Science in the service of Socialist construction," the young polity claimed science as its foundation.[9]

The discipline of nuclear physics was forming just as science, technological modernization, and state patronage became explicitly linked within newly founded Soviet research institutes. In 1922, the Radium Institute was established in Leningrad (then

Petrograd). Among the other research institutions created around that time that would later become famous for their accomplishments were the Ukrainian Physico-Technical Institute (UFTI) and its Leningrad equivalent, the Leningrad Physico-Technical Institute (LFTI). LFTI's leader, Abram Ioffe, trained an entire cohort of scientists who would become the country's nuclear elite.[10] In 1932, Ioffe set up two experimental laboratories to explore the atomic nucleus, one under the direction of Igor Kurchatov, who later became the leader of the Soviet atomic bomb project.[11] Before World War II, Soviet publications on nuclear research drew international attention, in particular the 1939–1940 papers by Zeldovich and Khariton that advanced sophisticated analyses of the uranium chain reaction.[12] In 1939 and 1940, the Soviet Academy of Sciences absorbed both the Radium Institute and LFTI; UFTI was integrated into the Ukrainian Academy of Sciences. A "Uranium Commission" set up under the Academy's auspices in 1940 began prospecting for uranium deposits. Hitler's attack, however, prompted most Soviet nuclear scientists to abandon their research and join the war effort.[13] Only when it became clear that the Americans were working on a nuclear weapon did Stalin revive nuclear research.[14]

In 1943, Igor Kurchatov was appointed scientific director of Laboratory No. 2, the top-secret heart of the Soviet nuclear weapons project. Although this institution was nominally under the Academy of Sciences, its sole task was to produce an atomic bomb. Resources for Kurchatov's team grew considerably once the United States had detonated their first nuclear weapon in 1945. After the bombing of Hiroshima and Nagasaki, Stalin put Lavrentii Beria, deputy chairman of the Council of Ministers and former chief of the secret police, in charge of directing the Soviet effort to create a nuclear weapon.[15] Under Beria's ruthlessly effective leadership, the newly created Special Committee on the Atomic Bomb turned a small research undertaking into an enormous industrial project. By the end of 1946 the Soviet Union's first research reactor was up and running, and in 1948 its first large-scale reactor started producing plutonium.[16] A year later, on August 29, 1949, the Soviet Union tested a plutonium bomb.

But Stalin didn't just appoint leaders and provide resources for the atomic bomb project. His preferred research strategy was assigning parallel teams to work on the same problem, and often one team wasn't aware of the other's existence.[17] In addition, Stalin personally intervened in debates about scientific disciplines as diverse as physics, biology, and linguistics.[18] In the late 1940s and the early 1950s, several scientific disciplines were subjected to "conferences" (or councils)—meticulously planned, well-rehearsed events that involved staged confrontations between scientists and party ideologues, self-critical confessions, and conclusions that not only authoritatively settled the matter but also set the standard for future research.[19] Such a "conference" on physics was planned for 1949. It emerged from a heated debate between physicists at the Moscow State University and Academy physicists. The university physicists argued that Academy physicists were corrupted by idealist Western science, and later accused physicists of anti-Soviet "cosmopolitanism."[20] The Academy physicists, however, were the ones involved with the atomic project, and Kurchatov turned out to be a skillful spokesman for them. The physics "conference" was canceled—perhaps, as Ethan Pollock has suggested, "because so many prominent physicists were unwilling to compromise and admit to supporting idealist theories or behaving in an unpatriotic manner," or perhaps, as Alexei Kojevnikov has argued, because the Academy's president never handed in his final presentation, upon which all responses were supposed to be based.[21] Another explanation, put forth by David Holloway, credits the significance of the atomic project, as well as Kurchatov's brokering, for the successful protection of physicists (and physics) from ideological intervention.[22]

With or without a physics conference, Stalin's personal involvement fundamentally reshaped Soviet science, including physics. He consolidated rigid hierarchical structures in scientific organizations, established the party-state apparatus at the core of the scientific community, and put this apparatus in control of research agendas, personnel appointments, international scholarly exchanges, and overall science policy.[23] The specific state-science relations established during the Stalin era became more firmly engrained in the post-Stalin period, and had a lot in common with what has been characterized as "big science."[24] Soviet-style "big science" integrated science and the state as interdependent components of one complex system.[25]

And yet, the relationship between scientists and the Party-state apparatus was anything but straightforward. Most important, the state was not the all-powerful oppressor, and scientists the victims. As Nikolai Krementsov has observed, nowhere else in the world had a regime come up with a more repressive apparatus to control science, but, by the same token, nowhere else in the world had scientists been as creative as in the Soviet Union in responding to these pressures, and in finding ways around state power.[26] Soviet scientists in all disciplines developed sophisticated strategies to avoid, elude, and even exploit this system of control—strategies that, paradoxically, reinforced the symbiotic relationship between science and the state by buttressing the rhetoric of "scientific neutrality," and weakening the influence of state ideology.[27]

As scientists joined the highest state agencies in growing numbers, the balance of power shifted and party ideologues gradually lost their ability to brutally enforce political loyalty. Thus, in the wake of the atomic bomb project, the state's attention slowly but surely returned to the practical outcomes of scientific research.[28] Moreover, as Konstantin Ivanov has shown, scientists (especially nuclear physicists) used "their increased political capital and social status to push for a major change in the political organization and management of science" and eventually, after Stalin's death, "succeeded in imposing their own specific agenda on the Communist Party leaders."[29] This success resulted in a role reversal. Previously, scientists had been called to operate in tune with the state's ideology; now, ideologists increasingly had to orient themselves toward the latest scientific achievements.[30] In the process, Ivanov argues, scientists became insiders, "conscious and loyal participants in the Soviet polity."[31]

Post-Stalin Reforms and the Return of Soviet Science to the World

Soviet science in general is often equated with Stalinist science. Stalin's undeniable effect on Soviet science notwithstanding, this simplification neglects the important changes that occurred after his death in 1953.[32] These changes involved the shift of authority from philosophers and Party ideologues to scientists, discussed above. But the post-Stalin reformers accomplished even more: they achieved nothing short of the rehabilitation of "fundamental science" and the liberation of science from the demand of applicability.[33] Political reforms had become possible only because scientists, especially physicists, had earned tremendous clout in the successful nuclear bomb project. Despite the fact that both "fundamental" science and "applied" engineering had contributed tremendously to the success of the bomb project, physics set a precedent for other disciplines: the post-Stalin Soviet state ultimately allowed *more*, not less, fundamental science.[34]

After Stalin's death, Soviet scientists increasingly demanded space for fundamental science. Academy physicists in particular pushed for research that wasn't solely defined by its applicability—a criterion that most engineers still defended as critical.[35] Alexandr Nesmeyanov, who had succeeded Vavilov as Academy president in 1951, tried to sit out this smoldering conflict, but when Mstislav Keldysh was elected president of the Academy, in 1961, he transferred most engineering departments from the Academy into industry. Keldysh justified his decision with reference to the very distinction between fundamental and applied sciences, but one could argue that it was in fact his bold managerial maneuver that gave this distinction institutional legitimacy in the first place.[36]

The distinction between "applied" and "fundamental" research contributed to an image of fundamental research that would be able to make unique contributions only when sheltered from the day-to-day pressures of mundane production processes. By so distinguishing fundamental from applied research, the Academy's president and Soviet scientists engaged in what Thomas Gieryn has termed "boundary work." According to Gieryn, boundary work is a rhetorical strategy that scientists use to make science "look empirical or theoretical, pure or applied," depending on "which characteristics best achieve the demarcation in a way that justifies scientists' claims to authority or resources."[37] That does not necessarily mean that the actors involved didn't understand the interplay between "fundamental" and "applied" research; it may mean, instead, that they were keenly aware of the power of this rhetorical strategy and how it could be used to convince different audiences. Exactly how a scientist will depict science at any given time depends on the "cultural repertoires" available and on the scientist's specific goal.[38] What counts as "science," or what kind of research counts as "fundamental" (as opposed to "applied"), can appear quite confusing until this context becomes clear.[39]

The successful test of a nuclear weapon gave Soviet physicists a new set of rhetorical resources with which to expand their authority vis-à-vis party ideologues, to monopolize their role as the creators of the Soviet atomic bomb, and to protect the extraordinary privileges granted to them during the bomb project. They used their prestige and social status to transform the political management of science. The volatile relationship between scientists and the state gave way to an equilibrium that provided physicists with "almost unlimited resources and the state with the ultimate tokens of Cold War politics—nuclear weapons, missiles, and spacecraft."[40] By reasserting the legitimacy of fundamental science and thus demarcating the "proper" place of science in Soviet society, nuclear physicists effectively renegotiated their own role in the Communist project. Not only did they transform physics; they also redefined the relationship between science and technology, that between research and industry, and ultimately that between science and the state.[41]

Basic or "fundamental" research had had a long tradition in Russia. During the early years of the Soviet Union, political leaders regarded that tradition with profound suspicion, as it seemed to suggest a separation of science from technical applicability.[42] The idea also was uncomfortably close to the notion of "pure science"—according to Lenin a trick label for what was really "bourgeois science," that is, science detached from labor, and elite scientists serving the interests of the powerful.[43] Despite the Soviet state's tenuous relationship with its scientists, the early political leaders knew that they depended on scientists and engineers for their project's success.[44]

Stalin's successors didn't embrace "fundamental science" overnight. Ivanov has shown vividly how mindful physicists had to be of what was politically possible at any particular time, especially during the power struggle that followed Stalin's death. Their goal was aided by the rapidly growing numbers of scientists and engineers: as training institutions multiplied, science became a mass profession, and scientists were no longer perceived as a suspicious elite. In the wake of the "Scientific-Technological Revolution," science was increasingly conceptualized as "part of the economic 'basis'—rather than ideological 'superstructure'—of contemporary society."[45]

Hand in hand with this rehabilitation of fundamental science went a gradual reinstatement of international scientific cooperation.[46] The Manhattan Project had triggered yet another period of militant Soviet nationalism and fierce isolationism—unless mutual espionage counts as "interaction."[47] By the same token, the United States' atomic monopoly had prompted Stalin to declare state support for scientific research a strategic priority.[48] Only in the context of the first successful Soviet nuclear test were Soviet scientists able to begin reaffirming the legitimacy of fundamental science, and to promote international scientific exchange and cooperation as something that was in the Soviet state's best interest.

For nuclear science and engineering, the pivotal event of renewed international exchange was the first International Conference on the Peaceful Uses of Atomic Energy, held at Geneva in 1955. This conference (which was followed by others in 1958, 1964, and 1971) was a direct outcome of President Dwight Eisenhower's "Atoms for Peace" speech to the UN General Assembly in December of 1953.[49] The announcement that a small nuclear power plant had been connected to the Soviet national grid the summer before and the revelation that the world's most powerful accelerator was under construction at Dubna convinced many American observers that "the Soviets were nipping at their heels."[50]

It is difficult to overemphasize the significance of the Geneva conference. The participating nations declassified a significant amount of technical information for it, and the papers that were presented provided insights into nuclear research all over the world. More important, it created the first opportunity in decades for Soviet scientists and engineers to mingle informally with their Western counterparts. The conference allowed, as John Krige has pointed out, an assessment of research directions in the other camp, while at the same time strengthening the coherence within each camp.[51] Moreover, Soviet scientists learned to use references to what the Americans (and the British, and the French) were doing to justify funding requests at home.[52]

Once they were back from Geneva, Soviet nuclear scientists and engineers launched an impressive promotional campaign to counterbalance the concerns about nuclear war and to emphasize the peaceful orientation of the Soviet program.[53] They created what Paul Josephson has called an "iconography" of nuclear power. The "icons" included gigantic nuclear power plants and nuclear-powered space rockets, icebreakers, and automobiles.[54] This campaign spanned magazines, museums, movies, and mundane artifacts (for example, tree ornaments for the New Year's celebration). Although it was successful in stimulating the public imagination, the percentage of nuclear energy relative to the country's actual and projected energy production remained small. Funding and material support for the nascent civilian nuclear industry stayed tight, despite the fact that Kurchatov and his soon-to-be successor, Anatolii Aleksandrov, personally intervened in favor of the new technology at the highest political, economic, and planning levels.[55] And although a nuclear-powered icebreaker (the *Lenin*) was launched in 1959 with great fanfare, nuclear-powered automobiles, trains, or airplanes never materialized.[56] The research that persisted despite scarce resources was devoted to reactors, accelerators, and fusion and was directly inspired and sustained by the Cold War rationale of proving that the Soviet system was technologically and morally superior.

Here, I use a subset of nuclear physicists—those involved in creating the nuclear power industry—to illustrate how the elusive distinction between fundamental and applied research played out in unexpected ways. The international discussion that scientists reignited in Geneva in 1955 informed the arguments Soviet nuclear experts could make at home.[57] For the first time since the end of Stalin's reign, these arguments invoked fundamental science and its claim to universality.

Nuclear Reactors in the Service of Soviet Electrification

As we have seen, Stalin's death inspired a major reorganization of Soviet science. Beginning in the early 1950s, scientists were able to renegotiate profoundly the relationship between science and industry. A space opened up for science and scientists that had previously been subject to political control, contingent upon applicability and Party loyalty. The distinction between "fundamental" and "applied" played an important but highly arbitrary and often confusing role in these negotiations. The way nuclear physics and engineering became institutionalized reflected this imagined demarcation between "fundamental" and "applied" science, and subsequently shaped how the nuclear industry was organized. Moreover, I argue, the specific division of labor that these institutional arrangements came to embody had palpable effects on *technical* decisions: Soviet decision makers selected reactor designs at least partly on the basis of which research institute and engineering bureau promoted them. But these research institutes and engineering bureaus themselves, with their specific strengths and areas of expertise, emerged directly from the boundary work over what constituted "fundamental" nuclear science and "applied" nuclear engineering. This boundary, especially when and where it was drawn to distinguish scientific research from practical applications, became critical when scientists and engineers proposed, challenged, and ultimately determined the feasibility and appeal of different options. The experts who emphasized the "fundamental" side of the divide designed and built a number of different reactors and, based on these small-scale prototypes, asserted the technical feasibility of a design. But scientists and engineers who worked in planning agencies or industrial organizations related "feasibility" primarily to economic and financial parameters. For them, a design was feasible only when it could rely on existing operational experience, a strong supply industry, or both as reliable indicators of potential profitability.

Institutional reform and technical choices were the two most determining factors in the creation of the Soviet nuclear industry, and they were tightly connected to the specific character of Soviet nuclear science (and engineering) itself. Previous accounts have tended to emphasize cultural characteristics of Soviet science and industry: the Stalinist legacy of megalomania, a specifically Soviet emphasis on narrow specialization in the training of engineers, and (especially after the Chernobyl accident) a lack of "safety culture" in the nuclear industry.[58] I would argue that the institutional set-up of the Soviet nuclear sector was at least as significant, because it shaped the technical decisions Soviet research centers put forth, which in turn influenced the ongoing institutional reform.[59] In other words, the decision about which reactor designs to standardize and mass produce structured the organization of tasks among different groups in the nuclear industry. Each of these groups developed particular ideas about what constituted legitimate scientific research. The demands of the emerging nuclear industry thus looped back into the development of nuclear science itself. And yet, nuclear scientists and engineers engaged in boundary work throughout to delineate and keep separate the directions of Soviet nuclear *research* and the institutional aspects of the nascent nuclear *industry*.[60]

Institutional Designs and the Role of the "Scientific Director"

According to Viktor Sidorenko, a leading nuclear expert and chronicler of the Soviet Union's nuclear energy program, theoretical science played a central role in the practical management of the nuclear industry. In this respect, Sidorenko claims, nuclear energy was distinct from other technological sectors, such as shipbuilding, aviation, and rocket design, in which "research" and "construction" were clearly defined, separate tasks.[61] In the nuclear industry, he argues, fundamental science was intimately involved in all technical, applied decisions. Sidorenko's assessment is, of course, biased, but the division of labor between the scientists who developed reactor designs and the engineers who constructed and operated power plants was in fact the nuclear industry's greatest liability. The quality of cooperation between research institutes (and researchers) and administrative entities often played a decisive role in determining success or failure of any particular project.

The ideas for several reactor designs, including the two designs that eventually prevailed, originated at the Institute of Atomic Energy (known today as Kurchatov Institute).[62] Researchers at the Institute of Atomic Energy set up a series of experimental facilities, including a test reactor and several "hot chambers" for materials science, a research reactor for neutron physics, and another research reactor for experiments in radiation safety. In particular, these researchers established and maintained a monopoly relating to processes inside various reactor cores. The most important limiting factor was access to computers: the weapons program controlled most of the early Soviet computing power, which led to a notorious lag in computer access for civilian nuclear researchers, even among elite nuclear scientists.[63]

Apparently, the Institute's scientific community constantly debated the relationship between fundamental and applied research. Sidorenko reports that the physicists doing "fundamental science" (*fiziki-fundamentalisty*) at the Institute of Atomic Energy worried that the reactor specialists (*reaktorshchiki*) might take over the Institute. And yet, with increasing resources devoted to fusion research, studies on thermal reactors decreased significantly. More and more, a number of newly created branch institutes outside of Moscow took on the "applied" part of the Institute's work, in addition to the secret nuclear weapons research institutes at Sarov (Arzamas-16) and the Semipalatinsk testing grounds.[64] And in 1979, the "scientific management of nuclear power plant operations," which included fundamental research relating to "already mastered systems," was transferred to the newly created Institute for the Operation of Nuclear Power Plants (VNIIAES) in Moscow[65]—the only nuclear research institute of its kind and caliber that existed within the Ministry of Energy and Electrification (Minenergo), a large, production-oriented authority very different from the secretive Ministry of Medium Machine Building (Sredmash). After 1966, these two ministries jointly managed the country's nuclear power plants, but the division of labor among them was subject to ongoing changes.[66]

The Ministry of Medium Machine Building (Sredmash) was essentially in charge of the Soviet nuclear weapons complex, while the Ministry of Energy and Electrification (Minenergo) built and ran all conventional power plants, in addition to maintaining the country's electricity grid. Initially, the construction of nuclear power plants mirrored the construction of conventional power plants, but to take nuclear reactors' "special status" into account, Sredmash authorities kept control over a number of sensitive tasks.[67] Beginning in the late 1960s, the nuclear industry started expanding aggressively. The emerging division of labor distinguished the role of the "scientific director" from that of the "chief design engineer" and that of the "chief project manager."[68] Minenergo and its organizations typically performed the role of project manager; it competed with Sredmash, whose organizations typically supplied the scientific director and answered for the design and construction of the reactor. Typically, the Institute of Atomic Energy took on the role of "scientific director" and coordinated the intense collaboration among the other groups. This coordination was intended to prevent decisions from being taken in isolation, and thus constituted an internal system of quality control. This peculiar division of labor reflected grave proliferation concerns and sincere attempts to ensure that the unique requirements of nuclear plants were being met. By the same token, however, it set up overlapping responsibilities and asymmetries in the access to and distribution of knowledge, and thereby reified an increasingly artificial boundary between "fundamental science" and "practical application." The "scientific director," in particular, was supposed to take charge of the "fundamental" science relevant to the operation of nuclear reactors. According to Sidorenko, the scientific director was responsible for addressing all new problems, ranging from those that required very theoretical research to those with very practical policy implications. Such problems might relate to neutron physics and chain reaction, removal of heat from a reactor's core, the effects of radiation on materials and related problems in materials science, the chemical characteristics of a coolant, and the characteristics of fission products under normal and accident conditions, but also to the construction and durability of nuclear fuel, the setting of limits on operational parameters, and the development of radiation-protection protocols.

Competent "scientific direction" was particularly important during a reactor's start-up, when system interactions were especially unpredictable and experience was crucial.[69] During a start-up, scientists from the Institute of Atomic Energy supported the plant's operating staff. A nuclear power plant's workforce often included a few highly qualified specialists who had received their training on nuclear submarines or at military facilities. An important part of their job at a civilian plant was to supervise their younger colleagues, who had joined the team without a comparable background and who received "training on the job." The central part of start-up work was the inspection and testing of all of the reactor's safety mechanisms, in an effort to demonstrate the plant's readiness for operation. The experts performing this task had to have a deep understanding of the reactor's operating processes, how to ensure the safety of machines and personnel, and how all the sophisticated technical equipment worked together.[70] At military facilities, once a reactor had been started up, the "scientific director" delegated all responsibilities to the facility's managers and retained only a peripheral role as expert consultant. At civilian plants, by contrast, the scientific director remained an important presence throughout the life of the reactor.[71] In other words, the negotiated definitions of what constituted the proper role of the "scientific director" reflected *institutional* turf issues, but they were couched in terms of "fundamental" versus "applied" science. The "scientific director's" ubiquitous jurisdiction thus paradoxically sharpened the boundary between what was framed as theoretical (or fundamental) and applied competencies, and exacerbated the competition between parties who were supposed to cooperate closely. One area in which this competition bore concrete results was reactor design.

Competing Technological Designs

Starting in the late 1940s, Soviet research institutions developed and tested a wide variety of reactor designs, yet in the end they standardized only two of them. Soviet physicists developed numerous reactor prototypes, which they set up as research reactors in laboratories, or built as larger experimental facilities in a designated city, Melekess, to gain experience with scaled-up reactors.[72] The first two industrial-scale nuclear reactors to begin operating were one near Voronezh in southern Russia and one in the Urals near Sverdlovsk (today Ekaterinburg). Each of them featured a unique design. Although these two pilot reactors began producing electricity in 1964, not until the late 1960s did planners decide which reactor designs would be standardized. No other government, including that of the United States, supported this kind of lavish nuclear research and development for such a long period; everywhere else, governments narrowed down the options quite a bit earlier.[73] Typically, they justified their decisions on economic grounds, so it is even more curious that Soviet economic planners, who claimed to be the most rational defenders of efficient state spending, agreed to an expensive strategy of funding multiple designs.

Both the small reactor at "The World's First Nuclear Power Plant" (that's its official name), which began operating in 1954, and the "Siberian nuclear power plant," a dual-use reactor at a classified site near the city of Tomsk that began producing plutonium and electricity in 1958, used graphite as moderator and water as coolant—a combination preferred for military reactors around the world.[74] One of the two pilot power stations, the Beloiarsk nuclear plant in the Urals, also relied on water and graphite for its reactor. However, according to Igor Kurchatov, Soviet research and development laboratories were actively pursuing as many as ten different reactor designs at the time.[75] By the late 1960s, when the first civilian reactors had been operating for only a few years, the Soviet Union narrowed down its efforts to two designs, one based on submarine-propulsion reactors and one based on plutonium-production reactors.

When we try to assess the design selection process today, our explanations tend to follow a functionalist model of innovation according to which the most efficient and economical designs prevailed. But the reactor prototypes built in laboratories and at testing facilities in the Soviet Union were all perfectly functional, or at least became so over time and with increased experience, and at the time no one could predict which design would prove most economical in the long run. Furthermore, some of the other design options that Soviet scientists originally pursued but later abandoned came to fruition in other parts of the world, once again contradicting the functionality argument. Also, if economics played such an important role, why did the Soviet Union select two designs, not (as France did) one "best" design?[76]

Perhaps not surprisingly, the initially diverse portfolio of nuclear reactor designs fell victim to the requirements of a streamlined production process in the mid to late 1960s. The Soviet industry still struggled with limited capacities to manufacture industrial-scale prototypes and, more important, switch to mass production.[77] The economy was recovering from the war slowly, and the arms race was draining scarce resources from other areas, notably the consumer-products sector.[78] And while nuclear scientists at large research institutes benefited from funding for nuclear weapons work, their peers in industry did not. "The electrification of the entire country" remained a state priority, but nuclear power plants had yet to prove their potential contribution to large-scale electricity generation. Meanwhile, the Soviet Union's East European satellite states, encouraged by the "Atoms for Peace" initiative, increasingly demanded Soviet nuclear assistance. They signed the first agreements for the delivery of Soviet research reactors in 1955, when reactor science and reactor engineering were still in their infancy. And soon agreements about the delivery of power reactors challenged the domestic industry's capacity to the breaking point.[79]

Scientists and planners followed international trends in reactor design preferences with great interest. During the 1950s, nuclear scientists still generally interpreted such trends in relation to the "big picture," which at the time pointed to the so-called plutonium economy, with breeder reactors ensuring a self-sustaining fuel-supply, until fusion reactors would become a reality. In the meantime, common trends in design choices seemed to indicate that reactor specialists around the world had come up with similar preferences independently.[80] Soviet scientists could argue that following international trends would allow joint improvements and help the Soviet nuclear industry accumulate operating experience at a much faster pace than it would on its own. But others emphasized differences and national variations in design ("the American reactor," "the French system," and so on) and used them as grounds for pushing a "uniquely Soviet" design.

More important than economic constraints and international trends, however, was the Soviet state's demand for military applications of nuclear energy. The two reactor designs that the Soviet Union eventually selected for mass production were not necessarily better, more efficient, or safer than the others, but they had the advantage of having predecessor designs in the military realm. The very first reactor, a graphite-water design, had delivered the material for the first Soviet plutonium bomb. Since then, many more of its kind had started operating in the Urals and Siberia. They increasingly served two purposes: they provided heat and electricity to the nearby secret cities, in addition to producing weapons-grade plutonium. Engineers later advanced the idea of "dual use" in the opposite direction as well: when the decorated engineer Nikolai Dollezhal proposed a power-reactor design based on military plutonium-production reactors, he argued that his team had modified the design to optimize electricity generation, but that if the need to produce extra plutonium should arise they could easily use them as back-up reactors for plutonium production. Not only could the RBMK (better known in the West as the "Chernobyl-type reactor") rely on two decades' experience with graphite-water models; it was a hybrid design that allowed—at least in principle—for operations beyond electric power generation. Soviet designers, acutely aware of that possibility, never allowed the RBMK to be exported, despite ample availability of parts and relative ease of assembly.

When scientists sought to determine which reactor design would work best for nuclear submarine propulsion, their top criterion was size. Adapting the design of the Siberian plutonium-production reactors nicknamed *krokodily* to the tight environment of a submarine proved daunting. At Obninsk (site of the World's First Nuclear Power Plant), they reduced the size of the core as much as they were able to, but that was still far too big for a submarine. The pressurized-water reactor, by contrast, lent itself to miniaturization. By increasing the fuel enrichment from about 4 percent to 90 percent, scientists managed to create a compact reactor. After its successful debut in 1957 in the first nuclear submarine, production of nuclear-powered attack submarines began in 1959.[81] By the early 1960s, when deliberations about choosing the most promising types of power reactors began, the pressurized-water design's advantage lay in its substantial operating experience. Nuclear scientists modified the design to allow the use of less enriched uranium and built progressively scaled-up models at the Novo-Voronezh site. Planners swiftly adopted the VVER (its Russian acronym) as one of the country's standard designs, and soon thereafter approved it for export.

Thus, it is clear that factors outside of "fundamental" science motivated the decision about which designs to standardize for Soviet power reactors. Firmly rooted in the "applied" science, such factors included industrial capacity, operating experience, and overall "fit" with existing structures, organizations, and personnel. Promoters of reactor designs emphasized international compatibility, Soviet uniqueness, or national security, depending on who they were trying to convince. Notably, the decision to adopt two designs, the VVER and the RBMK, looped back into the practice of "fundamental" science: as a direct consequence of this decision, research and development on alternative designs slowed down or stopped altogether, and entire subdivisions within the scientific research institutes dedicated their resources and expertise to improving and supporting the selected designs.[82]

How did the Cold War influence the idea of "scientific direction," and thus the boundary work involving "fundamental" and "applied" science? Two countertendencies operated at the same time in Soviet discourses on science: one emphasized Soviet independence, autarchy, and technical superiority; the other emphasized the fundamentally universal nature of science. The former trend catered to the political leadership by suggesting not only that the two sides in the Cold War were politically incompatible, but that ideological differences would ultimately produce distinct kinds of science. In the latter view, science *required* the free exchange of ideas, and therefore international cooperation. The definitions of "science" were very elastic: they could, but didn't necessarily have to, include technology. Often the audience had to figure out how exactly to interpret references to "fundamental" and "applied" science. Such rhetorical flexibility allowed Soviet scientists to make credible pitches for two very different reactor designs: while they promoted the pressurized-water type (VVER) as an "international" design, they represented the graphite-water model (RBMK) as an exclusively "Soviet" tradition. Ultimately, both arguments were successful: Soviet decision makers approved both designs for the Soviet nuclear industry.

The internationalist rhetoric enabled scientists from the Soviet Union to visit the United States and meet with their peers there, even during the Cold War. But those trips were short, and the information reported at home was fragmented and often miscellaneous. The construction of a Soviet-designed nuclear power plant at Loviisa, in Finland, was a different matter. There, cooperation involved the relocation of numerous Soviet specialists for several years and allowed an unprecedented level of mutual insight into work practices, safety principles, and the standards of professional conduct.[83] The Finnish plant was an example for the way that technical design choices led to a renegotiation of what constituted Soviet nuclear science. In 1965, when Finland announced a tender for its first nuclear power plant, the Soviet leadership instructed their nuclear specialists to win the competition. According to some Western sources, the Finns' decision to accept the Soviet bid was politically motivated, and was taken despite the fact that the proposal was found lacking on technical grounds.[84] Until the contract was signed, in 1969, Soviet nuclear specialists made great efforts to accommodate the demands of their Finish customers, who insisted on additional safety features. This was a tricky task, as Viktor Tatarnikov, an engineer who spent his entire career working in the nuclear industry, remembers:

> The first huge problem was the containment of an accident involving a breach of the reactor loop's hermetic sealing. Common practice abroad was to define the maximum possible accident involving hermetic sealing as the sudden, complete breach of the largest pipe in the most undesirable location (in our case a pipe with a 500 mm diameter). The Finnish specialists wanted this requirement to be met. In the USSR, breach of a 100 mm diameter pipe was assumed [as the maximum possible accident]. Our domestic engineering community supported the foreign standard, but our administration refused to accept it. We had to substantiate at home that it was valuable to do it like everyone else, and argued the opposite when talking to the Finns—that their requirements were exaggerated.[85]

Ultimately, the Soviet specialists modified the original design significantly, and the Finns insisted on purchasing important instrumentation and control components from German and American companies. From the Finns' diligence, Soviet specialists learned firsthand about current industrial standards and safety requirements in the West. In terms of reactor safety, then, the Loviisa experience was a turning point for Soviet nuclear science and engineering. The experience led to the creation of the first Soviet normative safety regulations (*Obshchie pravila iadernoi bezopasnosti*) just two years after the start of the collaboration with the Finns; it also prompted discussions about the best format to organize independent regulatory supervision for the nuclear industry within the specific context of the Soviet command-administrative economy.[86] The Soviet-Finnish cooperation provides an instance where new kinds of international cooperation led not only to the adjustment of some technical parameters, but to the restructuring of an entire regulatory apparatus.[87]

Conclusion

In 1949, an "All-Union Council of Physicists" was supposed to show the world, but especially physicists at home in the Soviet Union, how truly Soviet physics should be done. Rooted in an institutional conflict between Moscow University and the Academy of Sciences, the meeting would have addressed the question of what made physics in the Soviet Union distinctly Soviet. Despite careful planning and rehearsing, the meeting never came about, in part because the country's top physicists were busy building a nuclear weapon. Any questions about whether Einstein's theory of relativity was compatible with dialectical materialism, or which group of physicists should be calling the shots, were answered on August 29, 1949. However, the success of the atomic bomb project didn't affect only physicists; it also affected the status of Soviet science and scientists in general. The Academy of Sciences and its budget expanded dramatically during the postwar years, and Party ideologues no longer had the authority to tell physicists—and by extension, scientists—what they should or should not be doing.[88] With increasing confidence, Soviet nuclear physicists began emphasizing a universalist concept of science. The respect they had earned at Geneva for their civilian nuclear applications encouraged them to re-engage with the international scientific community, in addition to satisfying domestic authorities. That was a balancing act, as the two audiences often had widely differing expectations.[89]

Soviet nuclear physicists and engineers developed an impressive portfolio of operational reactor designs, and began training a workforce in ways consistent with uniquely Soviet ideas about the role of human operators in complex technological systems. They also created institutions that distributed expertise and accountability in accordance with the distinct Soviet political and economic context. The reactor designs selected for the Soviet nuclear industry reflected the dual vision of an "international" design and a uniquely "Soviet" model. Both projects found resonance among Soviet decision makers. These sophisticated machines owe their development to the generous support of research and design institutes; institutes that had been set up and strengthened with the clout gained from the atomic bomb project. Soviet nuclear physicists solidified a system of research institutes quite independent from direct political control—a system that emphasized fundamental research over teaching and that linked large, interdisciplinary research projects to the development of new technologies for military or civilian purposes only in a second step.[90] By doing so, these nuclear scientists rehabilitated a field that had been more vulnerable to political whim than its "applied" cousins.

Doing "fundamental science" in the Soviet Union (and calling it that) thus became a possibility only in the context of the successful nuclear weapons project, when physicists could claim renewed authority and autonomy. In the years after the detonation of the first Soviet atom bomb, scientists gradually re-established the legitimacy of "fundamental" science. For the first time in 30 years they could talk openly about engaging in "fundamental science" without the fear of getting purged. The number of university graduates dramatically increased and transformed science into a mass profession.[91]

To make sure this hard-won space for scientific research for the sake of scientific research didn't vanish again as it had before, Soviet nuclear scientists pursued a double strategy. First, their public image campaign for peaceful uses of nuclear energy linked their research agenda to concrete applications and the public good. This campaign also contributed to yet another Cold War competition, that for political and economic influence in Central and Eastern Europe. The status of individual scientists involved in this campaign, and the reputation of the institutes they were affiliated with, often rubbed off on the science they performed, and this helped them transform the value attributed to "fundamental" science into a positive one. Simultaneously, Soviet nuclear experts set up organizational structures anchoring fundamental science firmly in the country's institutional landscape. They did so first and foremost by singling out and transferring much of what counted as "applied" to other institutions. Such institutional maneuvers entailed a clear division of labor, and the scientists were careful to establish positions that would enable them to supervise the outsourced "applications" of their research.

The drawing of a boundary between "fundamental" and "applied" science allowed Soviet nuclear specialists to have their cake and eat it too: they used the political clout inherited from the weapons project to claim independence from "political" (applied) affairs. At the same time, they could make a case for their indispensability in all applied matters of, for example, nuclear energy. The flexibility with which they were able to re-assign responsibilities from "fundamental" to "applied" experts, or share competencies among them, indicates that these categories were in some sense artificial. But the institutional re-organizations reified the boundary between these categories, and the resulting institutional arrangement affected what kind of research was possible at what institution, and determined which technical decision, for example the choice of a particular reactor design, made sense. The rhetorical demarcation between "fundamental" and "applied" science thus materialized itself in specific organizational arrangements, which in turn shaped the kind of research deemed appropriate, and the kinds of applications regarded as desirable. Consequently, nuclear experts justified proposals for particular research projects, reactor designs, and training curricula with reference to the institutional expertise—the same expertise they had carved out in the process of demarcating "fundamental" from "applied" science.

The boundary work nuclear specialists engaged in involved not only the distinction between fundamental and applied research, but also between military and civilian, and between science and technology, research and engineering. Nuclear power plants turned out to be particularly strange animals that required competent "applied" science just as much as insights from "fundamental" research. Nuclear experts continued to grapple with this problem, as debates about where to draw these boundaries and how to legitimize them continued throughout the Cold War. The perpetual restructuring of the nuclear industry peaked in the frantic reorganization of jurisdiction and redistribution of responsibilities after the Chernobyl disaster.[92] Kojevnikov is certainly correct when he writes that behind the official rhetoric of ideological opposition between the two superpowers "one finds mutual learning and uncoordinated two-way borrowing."[93] But the obsession with separating "fundamental" research from "applied" science and engineering also stands in contrast to a growing trend in the West, where starting in the 1960s a group of scholars engaged in "science studies" began questioning these boundaries. For Soviet scientists, acknowledging that science was deeply immersed in and responsive to society's needs and values was tantamount to endorsing a quintessentially Marxist view.[94] By contrast, positing a clear boundary between fundamental and applied science provided them with a maximum of rhetorical flexibility: they could invoke the distinction to defend their research agendas against the occasional onslaught of Party ideologues, and at the same time justify the institutionalization of fundamental research to guarantee optimal applications.

Notes

1. Thomas Gieryn first applied the term "boundary work" to science and technology studies; it originally referred to a rhetorical style used by scientists to describe science to the larger public and to protect their professional autonomy (Gieryn, "Boundary-Work and the Demarcation of Science from Non-Science: Strains and Interests in Professional Ideologies of Scientists," *American Sociological Review* 48, no. 6 (1983): 782). Scientists perform "boundary work," for example, when they demarcate the production of scientific knowledge by scientists from the consumption of its products by non-scientists (ibid., 789). Demarcation, then, "is as much a practical problem for scientists as [it is] an analytical problem for sociologists and philosophers" (ibid., 792).

2. Coincidentally, at the same time that Soviet scientists, especially physicists, re-discovered the rhetorical power of distinguishing between fundamental and applied science, a similarly radical change of course took place on the other side of the Iron Curtain (Konstantin Ivanov, "Science after Stalin: Forging a New Image of Soviet Science," *Science in Context* 15, no. 2 (2002): 317–338, at 335).

3. Ibid., 317–318; Alexei B. Kojevnikov, *Stalin's Great Science: The Times and Adventures of Soviet Physicists* (Imperial College Press, 2004), xii.

4. Alexander Vucinich, *Empire of Knowledge: The Academy of Sciences of the USSR (1917–1970)* (University of California Press, 1984), 42.

5. This is analogous to the role science was to play in Western societies: as the objective foundation for liberal democracy. See, e.g., Yaron Ezrahi, *The Descent of Icarus: Science and the Transformation of Contemporary Democracy* (Harvard University Press, 1990); Michael Polanyi, "The Republic of Science," *Minerva* 1 (1962): 54–73; Steven Shapin and Simon Schaffer, *Leviathan and the Air-Pump: Hobbes, Boyle, and the Experimental Life* (Princeton University Press, 1985), 343.

6. David Joravsky, *The Lysenko Affair* (University of Chicago Press, 1970); Nils Roll-Hansen, "Wishful Science: The Persistence of T. D. Lysenko's Agrobiology in the Politics of Science," *Osiris*

23 (2008): 166–88; Kirill O. Rossianov, "Stalin as Lysenko's Editor: Reshaping Political Discourse in Soviet Science," *Configurations* 1, no. 3 (1993): 439–456.

7. Kojevnikov, *Stalin's Great Science*, xii.

8. David Holloway, *Stalin and the Bomb: The Soviet Union and Atomic Energy 1939–1956* (Yale University Press, 1994), 347.

9. Science "constituted one of the central elements of the Soviet polity, being an important creator as well as a creation of Soviet civilization" (Kojevnikov, *Stalin's Great Science*, xi–xv).

10. Among the many others who worked at LFTI was Iakov Frenkel, who received the Nobel Prize in physics in 1958. See Paul R. Josephson, *Physics and Politics in Revolutionary Russia* (University of California Press, 1991).

11. Kojevnikov, *Stalin's Great Science*, 129.

12. In the years 1939–1941, Zeldovich and Khariton wrote ground-breaking papers on the conditions under which a nuclear chain reaction would become possible (Holloway, *Stalin and the Bomb*, 51–56; Kojevnikov, *Stalin's Great Science*, 132–134).

13. Holloway, *Stalin and the Bomb*, 75.

14. The Soviet physicist Georgii Flerov had noticed the absence of scholarly publications on nuclear fission and alerted Stalin (Holloway, *Stalin and the Bomb*, 78).

15. Ibid., 129; Arkadii Kruglov, *The History of the Soviet Atomic Industry* (Taylor & Francis, 2002), especially 77; Amy Knight, *Beria: Stalin's First Lieutenant* (Princeton University Press, 1993).

16. The reactor, nicknamed Annushka and located at the secret Cheliabinsk-40 nuclear site, was started up in June of 1948.

17. This practice resulted in gross overspending of resources, but it also improved the chance of success and reduced the time needed to accomplish the set goals (Kojevnikov, *Stalin's Great Science*, 148–151). See also Asif Siddiqi's chapter in this volume and his book *The Red Rockets' Glare: Spaceflight and the Soviet Imagination, 1857–1957* (Cambridge University Press, 2010); Kruglov, *History of the Soviet Atomic Industry*; Ethan Pollock, *Stalin and the Soviet Science Wars* (Princeton University Press, 2006).

18. Krementsov, *Stalinist Science*; Pollock, *Stalin and the Soviet Science Wars*.

19. The results of Stalin's intervention were extremely inconsistent. In linguistics, the internationally accepted Indo-European approach was rehabilitated and "a maverick homegrown rival" (Nikolai Iakovlevich Marr) defeated. The debate in biology had the opposite outcome, as the often-cited Lysenko affair illustrates (Kojevnikov, *Stalin's Great Science*, 239–240). See also Pollock, *Stalin and the Soviet Science Wars*; Krementsov, *Stalinist Science*.

20. In the parlance of the time, this was code for an anti-Semitic attack. Pollock, *Stalin and the Soviet Science Wars*, especially chapter 4.

21. Ibid., 89; Kojevnikov, *Stalin's Great Science*.

22. David Holloway has argued that the atomic bomb not only saved Soviet physics from Stalinist prosecution, but also allowed Soviet physics institutes to protect a measure of "civil society," as well as other disciplines, most prominently genetics ("Physics, the State, and Civil Society in the Soviet Union," *Historical Studies in the Physical and Biological Sciences* 30, no. 1, 1999: 173–193). Kojevnikov, by contrast, dismisses the idea that "the bomb saved Soviet physics" as a story that lacks historical plausibility (*Stalin's Great Science*, 218). Kojevniko argues that in 1949 neither Kurchatov nor Beria could tell the political leaders that they had to pick physics (that is, the bomb) *or* ideology: communists believed "that true ideology and true pragmatism were in perfect agreement" (Vladimir P. Vizgin and Gennadii E. Gorelik. "The Reception of the Theory of Relativity in Russia and the USSR," in *The Comparative Reception of Relativity*, ed. Thomas F. Glick; Reidel, 1987; David Joravsky, "The Stalinist Mentality and the Higher Learning," *Slavic Review* 42, no. 4, 1983: 575–600; Kojevnikov, *Stalin's Great Science*, 218–221).

23. Krementsov, *Stalinist Science*, 8.

24. Kojevnikov, *Stalin's Great Science*, especially chapter 2; Peter Galison and Bruce Hevly, eds., *Big Science: The Growth of Large-Scale Research* (Stanford University Press, 1992); Nikolai L. Krementsov, "Russian Science in the Twentieth Century," in *Science in the Twentieth Century*, ed. John Krige and Dominique Pestre (Harwood, 1997).

25. Krementsov, "Russian Science in the Twentieth Century," 777.

26. Ibid., 778.

27. Krementsov, *Stalinist Science*; Slava Gerovitch, *From Newspeak to Cyberspeak: A History of Soviet Cybernetics* (MIT Press, 2002). See also Alexei B. Kojevnikov, "Introduction: A New History of Russian Science," *Science in Context* 15, no. 2 (2002): 177–182, and the 2008 special issue of *Osiris* on the Russian and Soviet technical intelligentsia—especially Alexei B. Kojevnikov, "The Phenomenon of Soviet Science," *Osiris* 23, no. 1 (2008): 115–135.

28. Krementsov, *Stalinist Science*, 97.

29. Ivanov, "Science after Stalin," 318.

30. Ibid., 329.

31. Ibid., 318. See also *Osiris* 2008. Another change brought about by post-Stalin reforms in science was the *de facto* abandonment of the "planning" principle, even though it remained formally in force. Aleksandr Nesmeyanov, who served as president of the Soviet Academy of Sciences from 1951 to 1961, argued that it was by definition impossible to define the results of fundamental research in advance. And although submitting plans for research "continued as a regular ritual even in the most abstract fields of science," "the practice of approving and enforcing such plans became increasingly pro-forma" (Ivanov, "Science after Stalin," 334).

32. Holloway, *Stalin and the Bomb*; Kojevnikov, *Stalin's Great Science*; Ivanov, "Science after Stalin"; Krementsov, *Stalinist Science*.

33. The term "pure science" was still considered bourgeois, but "fundamental science" could be made acceptable (Ivanov, "Science after Stalin," 326).

34. See also Siddiqi, this volume. In contrast to the US context, where, as Rebecca Schwartz has argued, the atomic bomb was predominantly represented as "the intellectual achievement of a group of brilliant physicists, rather than as a practical achievement of design, engineering, and management" (Rebecca P. Schwartz, The Making of the History of the Atomic Bomb: Henry Dewolf Smyth and the Historiography of the Manhattan Project, PhD thesis, Princeton University, 2008, 3), the Soviet atomic project was publicly celebrated as a feat of management and engineering. But the official narrative also acknowledged that the brilliance of Soviet scientists had contributed to this success, which in turn allowed them to claim a modicum of autonomy.

35. Ivanov, "Science after Stalin," 331; Kojevnikov, *Stalin's Great Science*, 226.

36. As Siddiqi points out in this volume, the picture wasn't clear: at the same time as Keldysh transferred engineering departments to industry, numerous academicians joined the Academy's Department of Technical Sciences, especially in areas such as aviation, space, control systems, and radar.

37. Gieryn, "Boundary-Work," 781.

38. Gieryn distinguishes three such goals: expansion of authority or expertise claimed by other professions or occupations, monopolization of professional authority and resources, and protection of autonomy over professional activities (Gieryn, "Boundary-Work," 791–792). In later work, Gieryn adds "expulsion" (of scientists who violate the rules) to these goals (Gieryn, *Cultural Boundaries of Science: Credibility on the Line* (University of Chicago Press, 1999)).

39. For example, as Sabine Clarke has shown, the multiple meanings of "fundamental research" allowed early-twentieth-century British science administrators to reassure simultaneously at least three communities: scientists (that the government would not dictate the nature of their research), industry (that researchers would not be receiving funds solely to satisfy their scientific curiosity), and the government (that scientific research would benefit the state) (Sabine Clarke, "Pure Science with a Practical Aim: The Meanings of Fundamental Research in Britain, circa 1916–1950." *Isis* 101, no. 2 (2010): 285–311; here especially 285–288, 295–296).

40. Krementsov, *Stalinist Science*, 290.

41. Kojevnikov, *Stalin's Great Science*, 127, 156; Ivanov, "Science after Stalin."

42. See, e.g., Bailes 1978 on the conflict over "bourgeois" and "red" specialists.

43. Marx 1975, 422 (cited in J. W. Grove, "Science as Technology: Aspects of a Potent Myth," *Minerva* 18, no. 2 (1980): 297–298); Hessen 1971; Loren R. Graham, *Science and Philosophy in the Soviet Union* (Knopf, 1972), especially chapter 2; Rosalind J. Marsh, *Soviet Fiction since Stalin: Science, Politics and Literature* (Croom Helm, 1986), especially 81–82.

44. E.g., Holloway, *Stalin and the Bomb*, Krementsov, *Stalinist Science*, Pollock, *Stalin and the Soviet Science Wars*.

45. Ivanov, "Science after Stalin," 334–5. See also the chapter by Aronova in this volume.

46. Ivanov, "Science after Stalin," 321.

47. "Soviet intelligence officers code-named their file on the Manhattan Project ENORMOZ (in Cyrillic transliteration), which probably reflected their astonishment at the scale of the industrial enterprise." (Kojevnikov, *Stalin's Great Science*, 139. See also Lev D. Riabev, ed., *Atomnyi proekt SSSR: dokumenty i materialy*. Vol. 1, 1938–1945, part 1 (Nauka—Fizmatlit, Ministerstvo Rossiiskoi Federatsii po atomnoi energii, Rossiiskaia akademiia nauk, 1998), 345). On the role of espionage for the Soviet nuclear project, see, e.g., Kojevnikov, *Stalin's Great Science*; Holloway, *Stalin and the Bomb*.

48. Krementsov "Russian Science in the Twentieth Century," 787–788.

49. Eisenhower's proposal, of course, was itself a reaction to the first Soviet atomic bomb and originally aimed at limiting the Soviets' capacity to enlarge their arsenal (see, e.g., Leonard Weiss, "Atoms for Peace," *Bulletin of the Atomic Scientists* 59, no. 6 (2003): 34–44).

50. Peter J. Westwick, *The National Labs: Science in an American System, 1947–1974* (Harvard University Press, 2003), 161. On Geneva see, e.g., Laura Fermi, *Atoms for the World; United States Participation in the Conference on the Peaceful Uses of Atomic Energy* (University of Chicago Press, 1957).

51. John Krige, "Atoms for Peace, Scientific Internationalism, and Scientific Intelligence, *Osiris* 21 (2006): 161–181. According to Peter Westwick, the conference thus "combined the rhetoric of international cooperation with the reality of Cold War competition" (Westwick, *National Labs*, 161).

52. Peter Westwick relates the following anecdote: "According to a member of the congressional Joint Committee who talked to a Soviet physicist at Dubna: 'The Dubna Laboratory asked our group when we were there 2 years ago how we got the money to build our accelerators. We told them the legislative process of getting money on our program. He said, 'That is not the way I understand.' He said, 'I understand you get it by saying the Russians have a 10 million electron volt synchrotron and we need a 20 billion electron [volt] synchrotron and that is how you get your money.' I said, 'There may be something to it.' I said, 'How do you get your money?' He said, 'The same way.' [FN: Rep. Melvin Price in JCAE, 86th Cong., 1st sess., *Stanford Linear Electron Accelerator*, 36.]" (Westwick, *National Labs*, 168–169).

53. Sonja D. Schmid, "Shaping the Soviet Experience of the Atomic Age: Nuclear Topics in *Ogonyok*, 1945–1965," in *The Nuclear Age in Popular Media: A Transnational History, 1945–1965*, ed. Dick van Lente (Palgrave McMillan, 2012).

54. Paul R. Josephson, "Atomic-Powered Communism: Nuclear Culture in the Postwar USSR," *Slavic Review* 55, no. 2 (1996): 297–324; Josephson, *Red Atom: Russia's Nuclear Power Program from Stalin to Today* (Freeman, 1999).

55. Sonja D. Schmid, *Producing Power: The Pre-Chernobyl History of the Soviet Nuclear Industry* (MIT Press, forthcoming).

56. The icebreaker *Lenin* was launched on December 5, 1957 and sailed down the Neva for its maiden voyage on September 15, 1959, with crowds waving from the river banks and cameras rolling (Global Security.org, Project 92M Lenin, at http://www.globalsecurity.org).

57. John Krige, "The Peaceful Atom as Political Weapon: Euratom and American Foreign Policy in the Late 1950s," *Historical Studies in the Natural Sciences* 38, no. 1 (2008): 11; Richard G. Hewlett and Jack M. Holl, *Atoms for Peace and War, 1953–1961: Eisenhower and the Atomic Energy Commission* (University of California Press, 1989).

58. Loren R. Graham, *Between Science and Values* (Columbia University Press, 1981); Graham, *Science in Russia and the Soviet Union: A Short History* (Cambridge University Press, 1993); Graham, *The Ghost of the Executed Engineer: Technology and the Fall of the Soviet Union* (Harvard University Press, 1993); Graham, *What Have We Learned About Science and Technology from the Russian Experience?* (Stanford University Press, 1998); Josephson, *Red Atom*; Viktor A. Sidorenko, "Nuclear Power in the Soviet Union and in Russia," *Nuclear Engineering and Design* 173 (1997): 3–20; Grigori Medvedev, *The Truth About Chernobyl* (Basic Books, 1989); Klaus Gestwa, "Herrschaft und Technik in der spät- und poststalinistischen Sowjetunion. Machtverhältnisse auf den 'Großbauten des Kommunismus', 19481964," *Osteuropa* 51, no. 2 (2001): 171–197.

59. Sheila Jasanoff calls this process "co-production" ("Science, Politics, and the Renegotiation of Expertise at EPA," *Osiris* 7 (1992): 195–217). See also Jasanoff, ed., *States of Knowledge: The Co-Production of Science and Social Order* (Routledge, 2004); Shapin and Schaffer, *Leviathan and the Air-Pump*.

60. Limited archival access still plagues this field. Notable exceptions are the document collections edited by former minister Lev Riabev (*Atomnyi proekt SSSR*), and the useful, although more anecdotal, series on nuclear reactor research and engineering edited by former deputy minister Viktor Sidorenko (*Isotoriia atomnoi energetiki*). On the Soviet space program, by contrast, we already have outstanding institutional histories: Asif Siddiqi, *Sputnik and the Soviet Space Challenge* (University Press of Florida, 2003); Siddiqi, *The Soviet Space Race with Apollo* (University Press of Florida, 2003); Siddiqi, *The Red Rockets' Glare*; Slava Gerovitch, "Creating Memories: Myth, Identity, and Culture in the Russian Space Age," in *Remembering the Space Age*, ed. Steven J. Dick (NASA History Division, 2008); Gerovitch, "Stalin's Rocket Designers' Leap into Space: The Technical Intelligentsia Faces the Thaw," *Osiris* 23 (2008): 189–209.

61. Viktor A. Sidorenko, "Nauchnoe rukovodstvo v atomnoi energetike," in *Istoriia VVER*, volume 2: *Istoriia atomnoi energetiki Sovetskogo Soiuza i Rossii*, ed. Sidorenko (IzdAt, 2002a), 5. My translations throughout.

62. In addition to these two designs, the VVER and the RBMK, the VK (a boiling vessel type) and the AST (district heating plants, *atomnye stantsii teplosnabzheniia*) also originated in the Institute of Atomic Energy (renamed Kurchatov Institute after Kurchatov's death in 1960). Breeder reactors, by contrast, were designed at Obninsk (FEI). The Institute of Atomic Energy's singular role was increasingly complemented, if not challenged, by other Institutes, such as the Bochvar Institute for Inorganic Materials (VNIINM), the Karpov Institute for Physical Chemistry (*Fiziko-khimicheskii Institut, FKhI*), and the Moscow Power Engineering Institute (*Moskovskii energeticheskii institut, MEI*), which specialized in secondary-loop water chemistry (Sidorenko, *Nauchnoe rukovodstvo*, 24).

63. Ibid., 11–15.

64. Ibid., 22–23.

65. Ibid., 23. Some of the research was delegated even further: the fuel fabrication facility in Elektrostal' received equipment so they themselves could perform the necessary testing and quality control of the active zone, before it was delivered to a nuclear power plant (ibid., 24).

66. Sonja D. Schmid, "Organizational Culture and Professional Identities in the Soviet Nuclear Power Industry," *Osiris* 23 (2008): 82–111.

67. These tasks included the design and manufacturing of nuclear reactors, and the design, production, and handling of nuclear fuel (Viktor A. Sidorenko, "Upravlenie atomnoi energetikoi," in *Istoriia atomnoi energetiki Sovetskogo Soiuza i Rossii*, volume 1, ed. Sidorenko (IzdAt, 2001). These tasks were outsourced to VNIIAES thirteen years later, in 1979.

68. The terms are mine; they are liberal translations from the Russian terms to make their tasks clear and distinct. "Scientific direction" refers to *nauchnoe rukovodstvo*, "chief design engineer" to *glavnyi konstruktor*, and "chief project manager" to *general'nyi proektirovshchik*. This three-part division of labor is not unique to the nuclear industry; in fact, the nuclear industry adopted it from other branches of industry, where it constituted the standard model.

69. The start-up process involves two main stages: that of physical start-up, when the reactor first goes critical, and, once physical start-up has been completed successfully, that of power start-up, when the nuclear station gets connected to the power grid.

70. At the Institute of Atomic Energy, the central figure among these inspectors was Alevtina Petrovna Pankratova. An eminent expert, she was the Institute's only physicist who oversaw the start-ups of every single VVER (Sidorenko, *Nauchnoe rukovodstvo*, 18–19).

71. Ibid., 22.

72. In 1956, construction of the "Scientific Research Institute for Nuclear Reactors" began. In 1972, Melekess was renamed Dimitrovgrad in honor of the ninetieth birthday of the Bulgarian communist Georgi Dimitrov.

73. Krige, "The Peaceful Atom as Political Weapon."

74. L. A. Alekhin and G. V. Kiselev, "Istoriia sozdaniia pervogo v SSSR i v mire dvukhtselevogo uran-grafitovogo reaktora EI-2 dlia odnovremennogo proizvodstva oruzheinogo plutoniia i elektroenergii," *Istoriia nauki i tekhniki* 12 (2003): 2–35.

75. Igor V. Kurchatov, "Nekotorye voprosy razvitiia atomnoi energetiki v SSSR," *Pravda*, May 20, 1956; Igor V. Kurchatov, "Rech' tovarishcha I. V. Kurchatova (Akademiia nauk SSSR)," *Pravda*, February 22,1956; V. V. Goncharov, "Pervyi period razvitiia atomnoi energetiki v SSSR," in *Istoriia atomnoi energetiki Sovetskogo Soiuza i Rossii*, Volume 1, ed. Sidorenko (IzdAt, 2001).

76. Gabrielle Hecht, *The Radiance of France: Nuclear Power and National Identity after World War II* (MIT Press, 1998). The latter point seems at first surprising for a planned economy, but actually shows analogous features to the bomb project, where Stalin reportedly had two teams work on the same project in parallel.

77. One particular challenge was (and remains) the manufacture of reactor vessels: they have to withstand not only high pressure and normal wear but also the effects of sustained ionizing radiation.

78. David R. Stone, *Hammer and Rifle: The Militarization of the Soviet Union, 1926–1933* (University Press of Kansas, 2000); Ruth Oldenziel and Karin Zachmann, eds., *Cold War Kitchen: Americanization, Technology, and European Users* (MIT Press, 2009).

79. Sonja D. Schmid, "Nuclear Colonization? Soviet Technopolitics in the Second World," in *Technopolitics and Imperialism in the Global Cold War*, ed. Gabrielle Hecht (MIT Press, 2011).

80. Such preferences were, of course, shaped by international exchange: the Geneva conferences, in particular, allowed international connections to develop in the first place (see, e.g., International Conference on the Peaceful Uses of Atomic Energy, Geneva, Delegation from the United States, *The International Conference on the Peaceful Uses of Atomic Energy, Geneva, Switzerland, August 8–20, 1955; Report, with Appendices and Selected Documents* 1956; International Conference on the Peaceful Uses of Atomic Energy, *Proceedings of the Second United Nations International Conference on the Peaceful Uses of Atomic Energy (2d, Held in Geneva, 1 September–13 September 1958)* (United Nations, 1958); International Conference on the Peaceful Uses of Atomic Energy, *Proceedings of the Third International Conference on the Peaceful Uses of Atomic Energy (3rd: 1964: Geneva)*, United Nations. Document; a/Conf. 28/1 (United Nations, 1965); United States, Delegation to the International Conference on the Peaceful Uses of Atomic Energy, *The International Conference on the Peaceful Uses of Atomic Energy, Geneva, Switzerland, August 8–20, 1955; report, with appendices and selected documents* (US Atomic Energy Commission, 1955); United States, Delegation to the International Conference on the Peaceful Uses of Atomic Energy, *Geneva, 1958; a Report* (US Atomic Energy Commission, 1959).

81. The first nuclear submarine, K3 (renamed *Leninskii Komsomol* in 1962), was built at the Sevmash shipbuilding yard in Severodvinsk. Its reactor was started up on September 14, 1957. Put to sea in 1958, it was transferred to the Northern Fleet later that year. In 1959, it was fully integrated into the Northern Fleet. See Sevmash Press Service, "Nuclear fleet celebrates half a century," December 16, 2008.

82. Experts within the Institute of Atomic Energy (renamed the Kurchatov Institute in 1960) were responsible for both the RBMK design and the VVER design. The FEI (Fiziko-Energeticheskii Institut) in Obninsk was in charge of the fast-neutron design and some the early graphite-water designs.

83. Viktor P. Tatarnikov, "Atomnaia elektroenergetika (s VVER i drugimi reaktorami)," in *Istoriia VVER*, volume 2: *Istoriia atomnoi energetiki Sovetskogo Soiuza i Rossii*, ed. Sidorenko (IzdAt, 2002).

84. Markku Lehtonen and Mari Martiskainen, "Governance of the 'Nuclear Revival' in Finland, France and the UK—Framings, Actor Strategies and Policies," presented at Sussex Energy Group conference on Energy Transitions IN AN Interdependent World, University of Sussex, 2010.

85. Tatarnikov, "Atomnaia elektroenergetika," 356.

86. Sidorenko, *Nauchnoe rukovodstvo*, 21. In contrast to the United States, where it made sense to establish the Nuclear Regulatory Commission independent from the state and the industry, the Soviet context lacked a space for an organization beyond state control; see B. G. Gordon, ed., *Gosatomnadzoru Rossii—20 Let* (NTTs IaRB, 2003).

87. The experience with Western standards prompted changes in the technical designs themselves: for example, most VVER reactors built after the Finnish experience would be equipped with a containment building, something deemed superfluous until then. See, e.g., the assessment by the International Nuclear Safety Program, a DOE-sponsored initiative (1993–2003) to improve safety at Soviet-designed nuclear power plants: "In some respects this design is more forgiving than Western plant designs with ... vertical steam generators" (http://insp.pnnl.gov/-profiles-reactors-vver230.htm; historical reference maintained by Pacific Northwest National Laboratory). Unfortunately, the Soviet nuclear industry rarely implemented suggested design changes promptly, in part due to the sluggish planning system. See also Thomas Wellock, "The Children of Chernobyl: Engineers and the Campaign for Safety in Soviet-Designed Reactors in Central and Eastern Europe," *History and Technology* 29 (2013): 3–32.

88. Kojevnikov, *Stalin's Great Science*, 303–304; Vucinich, *Empire of Knowledge*.

89. Kojevnikov, *Stalin's Great Science*, 304.

90. Ibid., 23–24. The Soviet budget for R&D in science and technology was smaller in absolute numbers than that of other nations, but proportionally it was among the largest (ibid., 304).

91. Ivanov, "Science after Stalin," 334–335.

92. Schmid, *Producing Power* and "Organizational Culture."

93. Kojevnikov, *Stalin's Great Science*, 305.

94. Ivanov, "Science after Stalin," 335. Kojevnikov argues that Soviet Marxists held on to two views of science simultaneously: on the one hand, they thought of science as a human activity related to economic, political, and class interests like any other, but on the other they believed that it provided true knowledge about nature. Notably, they didn't see these two views in conflict: they simply reasoned that to find truth in and through science was a matter of having the *right* interests (Kojevnikov, *Stalin's Great Science*, 223).

11 The Cold War and the Reshaping of Transnational Science in China

Zuoyue Wang

In an article in the October 1967 issue of *Foreign Affairs*, former vice president Richard Nixon advocated a more active US foreign policy toward China, declaring "There is no place on this small planet for a billion of its potentially most able people to live in angry isolation."[1] During the next several years, the perception of a China insulated from the outside world would receive confirmation as the country spiraled further into the Cultural Revolution maelstrom at home and engaged in conflicts with both the United States and the Soviet Union abroad. In science, technology, and education, almost all universities and research institutes were shut down, international scientific interactions ceased, and importation of foreign journals and books stopped. Yet less than five years later Nixon, as president, would land in Beijing, and a new era in the history of the Cold War and in China's scientific relations with the rest of the world would begin.

In this chapter I will sketch, in very general terms, how the Cold War reshaped China's scientific enterprise, especially its transnational character. The Cold War will be understood here not as a straightforward bipolar US-Soviet competition, but rather as a series of triangular US-Soviet-Chinese geopolitical interactions with alternating periods of alliance and hostility as seen from the Chinese perspective.[2]

The dynamics of the Cold War conditioned China's choice of major partners in international scientific exchanges and shaped domestic scientific priorities and institutions, which served to alter the existing patterns of Chinese transnational scientific interactions, sometimes in surprising ways. Thus, chronologically, the chapter takes into account both the conventional periodization of the Cold War—from the late 1940s, as US-Soviet tension rose, to the early 1990s, when the Soviet Union fell apart—and two major events in the history of modern China: the establishment of the People's Republic of China (PRC) in 1949 after the Communist revolution and a series of dramatic changes in late 1980s and the early 1990s that included not only the government's crackdown on a pro-democracy movement at Tiananmen Square in 1989 but also the subsequent acceleration of reform and opening up. Although the chapter focuses on the period 1949–1989, it also examines the years before and after that period in order to illuminate the background and the legacy of the development of transnational science in China during the Cold War. Specifically, the chapter examines five periods: the half-century before the 1949 revolution, when the influence of the United States was dominant; the decade that followed it, shaped by an alliance with the Soviet Union; the 1960s, the era of self-reliance; the Nixonian exchange of the 1970s, which reopened US-China scientific interactions; and the reform era of opening to the outside world that began in the late 1970s.

Here the term *transnational science* refers to the movements of scientists, scientific institutions, practices, instruments, and ideologies across national boundaries and how such movements interacted with the indigenous traditions and contexts within any particular nation-state to shape and shift scientific developments within it and internationally. As such, the concept of the transnationalization of science includes both formal, state-sponsored international activities and the informal, private cross-national networking that scientists engaged in outside the framework of the nation-states. It also refers to aspects of scientists' activities as they confront and even challenge the authority of the nation-states.[3]

It should be noted that to characterize science in modern China as transnational does not mean that nationalism didn't play an important role.[4] Clearly, both the pre-1949 Nationalist government and its Communist successors sought foreign aid, including technological aid, for the purpose of fulfilling their own national developmental aspirations.[5] Even the Western-trained Chinese scientists held a strong sense of Chinese nationalism, often sharpened by the history of national humiliation at the hands of Western powers and Japan, and generations pursued various versions of the dream of "saving China through science."[6] The Cold War, however, accentuated the latent tensions between the national and the transnational for both the Chinese party state and the Chinese scientists: the former had to balance its need for technical manpower for national security and its political distrust of Western-trained scientists; the latter had to deal with the intensified role of the party state in their scientific and personal lives, and also to reconcile their Chinese nationalism and their transnational scientific background and ideals.

In this chapter all Chinese names, except for those of overseas Chinese, are rendered in *pinyin*, with the family name first and the given name second.

Americanization before the Cold War, ca. 1900–1949

In the case of science in China, transnationalism predated the Cold War. International politics and scientific currents defined the social and political context of the introduction of modern science in China from the late nineteenth century onward.[7] The Sino-Japanese war of 1895 led to the end of the traditional civil service examination system and the introduction of the modern educational system in 1905, with the inclusion, for the first time, of natural science as part of formal schooling. Both before and after this educational reform, missionaries, most of them from the United States, established schools in China in which some modern science was taught. The anti-foreign Boxer Rebellion of 1900 and the subsequent intervention by Western powers and Japan eventually led to the establishment of the Boxer indemnity fellowships, first and foremost by the United States and later, on a much smaller scale, by other countries, to sponsor Chinese students to study abroad. These fellowships were made possible by funds remitted to China when it became clear that the massive indemnities ($330 million) that China was forced to pay the foreign powers were often based on exaggerated claims. The Boxer fellowships also played an important role in the emergence of the US as a dominant influence in the development of science in China in the first half of the twentieth century and thus deserve closer examination.

As the historian Michael Hunt convincingly argued, even though the United States publicly touted its first partial remission of the Boxer indemnity as a gesture of good will in 1908, its original claim of $25 million, with 4 percent of interest amortized until 1940, had been widely recognized from the beginning as excessive and subsequently proved to be so. Furthermore, when the US later forced the Chinese government to use the remission to send Chinese students to the US, it acted out of self-interested calculations at least as much as out of altruism.[8] Such a move would enable the US to influence China and, as Edmund J. James, president of the University of Illinois, put it in a letter to President Theodore Roosevelt, even control it with "the intellectual and spiritual domination of its leaders."[9] Roosevelt largely agreed. On December 3, 1907, in a message to Congress, he declared:

> This Nation should help in every practicable way in the education of the Chinese people, so that the vast and populous Empire of China may gradually adapt itself to modern conditions. One way of doing this is by promoting the coming of Chinese students to this country and making it attractive to them to take courses at our universities and higher education institutions. Our educators should, so far as possible, take concerted action toward this end.[10]

Yet, even though the Boxer indemnity funds emerged from one of the most humiliating episodes in modern Chinese history, as a transnational institution it probably played a more important role than any other financial and educational programs in the making of modern science and especially in the training of the first generation of modern scientists in China. During the negotiations leading to the establishment of the Boxer fellowship program, the Chinese government and the US government both agreed to emphasize science and technology. The "Proposed Regulations for the Students to Be Sent to America" prepared by Yuan Shikai, head of the Chinese Foreign Ministry (*wai wu bu*) in late 1908 contained this stipulation:

> The aim in sending students abroad at this time is to obtain results in solid learning. Eighty per cent of those sent will specialize in industrial arts, agriculture, mechanical engineering, mining, physics and chemistry, railway engineering, architecture, banking, railway administration, and similar branches, and 20 per cent will specialize in law and the science of government.[11]

According to these regulations, the Chinese Foreign Ministry and one official from the American legation in China would together design "the detailed method of procedure" for implementing the program. In the end, an Office for Students Going to the United States was established jointly by the Chinese Foreign Ministry and the Chinese Ministry of Education (*xue bu*) to oversee the selection of 180 students in China and their distribution in the US for the period 1909–1911 (47 in 1909, 70 in 1910, and 73 in 1911).[12] In 1911, the Tsinghua (Qinghua) School was set up in Beijing under heavy American influence to take over the job of preparing the Boxer fellows before their departure for the US and to supervise their activities thereafter. In all, from 1909 to 1929, the Boxer fellowship program brought about 1,500 Chinese men to study in the US, a majority of them studying engineering, science, agriculture, and medicine.[13] In addition, there were 54 women selected from outside of Tsinghua through special national competitions before 1928, when the school began to admit women.[14]

The admission of women came as part of a major transformation of Tsinghua that year: it became an independent national university under control of the Ministry of Education (not the Foreign Ministry as before). The broader context was the movement to recover Chinese rights in education after the establishment of the Nationalist government in 1927. The new Tsinghua also stopped the practice of automatically sending all its graduates to the US. Many of them continued to go to American universities with the special Tsinghua funds drawn from the original American Boxer remission, but some now could choose to go to Europe on Boxer indemnity fellowships from countries there when they became available. Most of them also now went abroad as graduate students, instead of undergraduates as before the change.[15]

A second and final remission of excessive Boxer indemnity funds by the US in the mid 1920s led to the establishment of the autonomous China Foundation for the Promotion of Education and Culture. Governed by a board of ten Chinese and five American educational leaders, the foundation used its funds to support not only Tsinghua but also a number of other universities and research institutions, including the Science Society of China. The foundation also sponsored graduate studies in the United States by non-Tsinghua graduates.[16] Likewise, the Rockefeller Foundation financed the creation and operation of the Peking Union Medical College and funded research and teaching at many other Chinese universities in this period.[17]

Significantly, the US Boxer funds, through both Tsinghua and the China Foundation, and the Rockefeller influence not only put a heavy American accent on Chinese science and education but were also among the few continuous institutional threads in these areas in China through the turbulent first half of the twentieth century, from the Republican Revolution of 1911, through the ensuing era of warlord chaos to the establishment of the Nationalist government in 1927, and through the War of

Resistance against Japan (1937–1945). Together with American missionary universities in China, they helped explain why an overwhelming number of Chinese students pursuing scientific studies abroad went to the US, even though it was widely recognized, even in China, that the center of most fields of science, especially during the early twentieth century, was in Europe and not the US. A survey published in China in 2007 of Chinese scientists and engineers who returned to China after studying abroad during the period 1879–1949, for example, found that about two thirds (66.14 percent) of them had returned from the US.[18] (See table 11.1.)

Although the United States played the most active role in Chinese science and education in this period, it was by no means the only foreign influence. Chinese students took advantage of the returned Boxer funds from other countries, especially from Great Britain in the 1930s, for studying abroad there as well. Indeed, because of a higher standard of selection, a disproportionate number of the leaders of Chinese science emerged out of the dozens of Boxer fellows who went to Britain.[19] Even before Tsinghua became autonomous in 1928, its graduates, aware of the gap between American and European science and the fact that many American scientists themselves had studied in Europe, found their way there at some points in their careers. Physicists were especially eager to get the European exposure after their American education. For example, Ye Qisun, a founding figure of modern physics in China and a Boxer fellow, received his bachelor's degree from the University of Chicago in 1920 and his PhD from Harvard University in 1923, then went on a four-month tour of Europe, then returned, after a short stint at Southeastern University in Nanjing, to Tsinghua, where he remained for the rest of his career.[20] As professors, the returned students intentionally sent their own students to strategically selected Western institutions and scientific fields so as to give China a balanced coverage in science. Ye Qisun, for example, helped arrange for the first three graduates of Tsinghua's Physics Department to go to Germany, France, and the United States respectively for graduate studies.[21] Thus, by the 1930s, Tsinghua had Chinese faculty members who had trained at and returned from not only the United States but also England, Germany, and France, as Norbert Wiener of MIT observed when he served as a visiting professor in mathematics at Tsinghua.[22]

Perhaps no institution epitomized the transnational features of Chinese science and the American scientific influence in China better than the Science Society. Founded in 1914 by Chinese students studying science at Cornell University, it was the first comprehensive scientific organization in China. Its membership grew rapidly, especially after it moved its headquarters to China in 1918 as its founders finished their studies in the US and returned home.[23] Ironically, owing to a number of factors, including their sense of Chinese nationalism, few of the US-trained Chinese scientists left with the US-backed Nationalists when the latter retreated to Taiwan in the wake of their defeat by the Communists in 1949.

Table 11.1

Numbers of Chinese scientists and engineers who returned to China after studying abroad by periods of return and countries of study, 1879–1949.

	US	UK	Japan	Germany	France	Canada	Belgium	Russia	Switzerland	Austria	Others	Total	%
1879–1911	109	62	41	19	37	0	7	0	0	0	1	276	7.15
1912–1928	815	59	129	43	38	3	9	4	3	5	12	1120	29.02
1929–1937	424	72	80	104	58	2	6	4	9	8	26	793	20.54
1938–1945	227	85	31	64	13	7	7	6	2	5	25	472	12.23
1946–1949	978	113	4	18	9	22	3	3	7	1	41	1199	31.06
Total	2553	391	285	248	155	34	32	17	21	19	105	3860	100
Percentage	66.14	10.13	7.38	6.42	4.02	0.88	0.83	0.44	0.54	0.49	2.72		100

Source: Ma Zusheng (T. S. Ma), *Linian chuguo/huiguo keji renyuan zonglan (1840–1949)* (Social Sciences Academic Press, 2007), 4.

Transnational currents also left imprints on some of the political events that occurred during this period. The republican Xinhai Revolution of 1911 was launched with strong support from overseas Chinese, especially in the United States and Japan, and in turn led to the establishment of a relatively stable Nationalist government that provided support for science. The importation of communism from Russia in the 1910s and the 1920s sowed the seeds for a Communist science policy. The May Fourth Movement of 1919 carried the banners of highly Westernized science and of democracy. The Nationalist government, based in Nanjing, made strides in science, technology, and education in the golden decade of 1927–1937, establishing universities (most of them American-style) and also establishing the Academia Sinica (modeled after the Soviet Academy of Sciences and the Kaiser Wilhelm Society of Germany).

During the Nanjing decade (1927–1937), the Nationalists also initiated ambitious technological development projects supported by international aid. The 1937–1945 war with Japan, however, disrupted progress and institution-building in science and other areas, and the Civil War between the Communists and Nationalists, which took place within the context of the global Cold War, irreversibly changed the fate of all Chinese, including scientists. Disgusted with the corruption-ridden Nationalists, most Chinese scientists, even those who had been educated in the West, stayed on the mainland instead of fleeing with the Nationalists to Taiwan. Few among them probably foresaw the nearly complete cut-off of transnational scientific communication after 1949 and especially after the outbreak of the Korean War in 1950. Yet the enlightened science policy of the early PRC years under the leadership of Mao Zedong seemed to have vindicated the scientists' choice. Under the direct leadership of Premier Zhou Enlai, whose vision for a developmental state best matched the nationalist aspiration of the scientists, the government seemed to support science and to make good use of the mostly Western-trained scientists.

Sovietization and Countercurrents, 1949–1960

During the 1950s, the first decade of the existence of the PRC, a geopolitical alliance with the Soviet Union, first launched in part as a way to fend off a possible American intervention in China and later consolidated during the Korean War, resulted in a massive Soviet technological transfer to China.[24] With it also came a wholesale restructuring of Chinese science, technology, and education policy and institutions from ones shaped by Western, especially American, practices to ones dominated by those of the Soviet Union. Yet even during this period of Sino-Soviet alliance one finds strong countervailing forces that resisted "Sovietization." They derived from several sources: ideological and political differences between the Chinese Communist Party (especially Mao) and the Soviet leadership under Nikita Khrushchev; the different social and cultural contexts for technological development in the two countries; and the presence of Chinese scientists and engineers who were trained in the West, especially in the US. Surprising as it may seem, the post-World War II Americanization of international science didn't leave China untouched even at the height of mutual antagonism.[25]

In the 1950s, the Soviet Union helped China construct or equip more than 300 major industrial projects, ranging from military technology to power generation and chemical engineering, laying the foundation for China's industrial infrastructure for decades to come. It sent thousands of Soviet scientists and engineers to China as technical advisors, and trained tens of thousands of Chinese engineers and scientists in the Soviet Union.[26] It helped China formulate the Long-Term Plan for the Development of Science and Technology (1956–1967), which served as a blueprint for its nuclear and space programs as well as general scientific and technological developments. Indeed, it was V. A. Kovda, a Soviet soil scientist and chief advisor to the Chinese Academy of Sciences in 1954–55, who suggested that such a plan be drawn up in the first place.[27] In a reversal of the Americanization of the pre-1949 period, almost all Chinese universities were restructured in the direction of narrow technical training, and almost all Chinese scientists were required to learn Russian and follow Soviet scientific literature. In the elite Chinese Academy of Sciences, for example, a survey taken in 1954 found that 93.2 percent of all the staff, including the scientists, were learning Russian, 73.5 percent could already read Russian scientific literature, and 26.8 percent were able to translate Russian papers into Chinese.[28]

Did the Cold War-inspired Sovietization transform the content of science in China? The most notorious candidate for a confirmation of this effect was the introduction of Lysenkoism and the effective banning of Western genetics in the teaching and research in biology in China in the first half of the 1950s. Close examination of this history indicates that Chinese Lysenkoism had elements that could be traced back to Communist science policy debate in the pre-1949 period, with the resultant emphasis on applied research, practical learning, and anti-Westernism, and thus not an entirely Cold War phenomenon.[29] Nevertheless, Sino-Soviet geopolitical alliance played an important part in the establishment of Lysenkoism as an orthodox biological doctrine.

An illustration for this point came in the case of Hu Xiansu, an outspoken American-trained botanist who, in a 1955 textbook, criticized Lysenkoism as pseudo-science propped up by political forces. The book drew protests from Soviet advisors in the Chinese Ministry of Education, and the controversy escalated into an international political problem. At a high-level meeting of the Chinese party-state leadership on April 27, 1956, Lu Dingyi, the party's propaganda chief, acknowledged that Hu was right scientifically but explained "what we focused on was his political problem. He attacked the Soviet Union at the time, which made us very mad."[30] Hu's book was banned, but fortunately political changes both at home and in the Soviet Union and

Eastern Europe soon led to the implementation of a liberal "Hundred Flowers" campaign that helped bring back modern genetics.[31]

Yet even in the heyday of the Sino-Soviet alliance, Sovietization had its limits. As a Chinese nationalist, Mao never had an easy relationship with Stalin. His relationship with Stalin's successor Khrushchev was even more uneasy. At one point in 1958, Mao told the Soviet ambassador to China "You say that Europeans look down upon the Russians. I believe that some Russians look down upon the Chinese."[32] Political and ideological divisions eventually widened, resulting in a souring of bilateral relations, an end to Soviet technical assistance (including assistance in China's atomic bomb project), and an end to the sending of large number of Chinese students to the Soviet Union to study science and engineering by the early 1960s.

In the same period, the West continued to influence science in China, if only in subtle and unheralded ways. For one, even at the height of Chinese isolation from the West, it was impossible to cut off all scientific and technological connections. Strict export controls didn't, for example, prevent a small number of scientific instruments being smuggled from the US into China, often through Hong Kong.[33] Likewise, transnational scientific networking demonstrated remarkable durability. For example, the prominent marine botanist Zeng Chengkui (C. K. Tseng), who had been trained in the United States and had returned to China in 1947, was able to communicate with his former colleagues at Scripps Institution of Oceanography in San Diego in 1951, amidst the Korean War. He requested that they send him scientific reprints and help him realize his dream of building "an equivalent of the Scripps Institution plus Woods Hole Marine Biological Laboratory" in Qingdao. Indeed, in the 1950s and the early 1960s, Zeng could still receive scientific papers from his American colleagues at Scripps and elsewhere, while they themselves tried to keep abreast of Zeng's scientific publications through Russian translations.[34]

Perhaps more important, the 1950s witnessed the return of about a thousand Chinese scientists from the United States to mainland China. Even though both the Chinese government and the US government harbored suspicions of the American-trained Chinese scientists (the former because of their Western educational background, the latter because some of them were thought to be loyal to Communist China, especially during the Korean War), each side sought to recruit them in order to use their technical talents, to deny them to the other side, and to show the superiority of their side's political system. The US government, in a move that echoed elements of its Project Paperclip to capture German scientists and engineers before they fell into the Soviet hands after World War II, not only encouraged but in some cases forced Chinese scientists and engineers who had been "stranded" in the US after the Communist revolution and Korean War to stay in the US. Yet many in this group were determined to return to China, often driven by a strong sense of Chinese nationalism, by political sympathy toward the Chinese Communists, and by a desire to reunite with their families. Eventually, though about 4,000 of the estimated 5,000 Chinese students and scientists remained in the US (including Chen Ning Yang and Tsung Dao Lee, who would win the Nobel Prize in physics in 1957), about 1,200 fought the policy and returned to China in the 1950s.[35] They were joined in China by several hundred others who returned from Europe and Japan. According to an internal report, 1,954 Chinese students and scientists returned to mainland China in 1949–1958, of whom 1,244 (about 64 percent) returned from the United States.[36] (See table 11.2.) Of the 1,954, the number who specialized in natural sciences (presumably including engineering) was 1,117 (57 percent); the other two categories were "social sciences" and "unknown."[37] (See table 11.3.)

Table 11.2
Chinese students and scientists who returned from abroad in 1949–1958, by country of training.

	Number	Percentage
US	1244	75.2
UK	221	11.3
France	103	3.8
Japan	198	10.1
Other Countries	128	6.6
Unknown Countries	60	3.1
Total	1954	100

Source: Beijing Municipal Archives, Beijing, file no. 002-020-339 "Guojia kewei zhuanjiaju 1959.3.5 tongzhi" (circular from the Bureau on Experts of the National Science and Technology Commission dated March 5, 1959), attachment 1.

Table 11.3
Chinese students and scientists who returned from abroad in 1949–1958, by field of training.

	Number	Percentage
Natural Sciences	1117	57.2
Social Sciences	716	36.6
Unknown Fields	121	6.2
Total	1954	100

Source: Beijing Municipal Archives, Beijing, file no. 002-020-339 "Guojia kewei zhuanjiaju 1959.3.5 tongzhi" (circular from the Bureau on Experts of the National Science and Technology Commission dated March 5, 1959), attachment 1.

Among those who returned to China from the US was Qian Xuesen (Hsue Shen Tsien). His case exemplified the difficulties that faced transnational Chinese scientists who were caught up in the international politics of the Cold War. Qian was born in 1911 in China and studied railroad engineering at Jiaotong University in Shanghai in the 1930s. He then went to the United States to study aeronautics on a Boxer fellowship. During and after World War II, he had a fast-rising reputation as a research scientist at Caltech and in national defense circles. His phenomenal ascent in the United States ended abruptly in June of 1950, when the government revoked his security clearance on suspicion of his past membership in the US Communist Party. It was the time of the Korean War, McCarthyism, and persistent racial discrimination. At one point Qian was both to be deported as a subversive (by the Immigration and Naturalization Service) and detained as a denial to Communist China (by the Pentagon and the State Department). After several years of virtual house arrest, Qian and dozens of other Chinese scientists who wanted to return to China eventually were exchanged for American prisoners of war and civilians held in China. Tsien went on to direct the Chinese missile program.[38]

Qian's case illustrates not only how the Cold War affected science and scientists, but also the reverse. It is reported that Qian turned out to be so valuable to China that Premier Zhou Enlai commented at one point that the long negotiations with the United States at Geneva were worthwhile if only to get Qian back.[39] One can also make the argument that the return of the hundreds of Chinese scientists from the West in the early 1950s gave the Chinese party-state leadership the self-confidence to pursue a path that was increasingly independent of the Soviet Union. Zhang Jinfu, the party chief of the Chinese Academy of Sciences during the late 1950s, recalled that "Qian Xuesen knew that the key to missiles was the boosters, and so after he came back from the US, the Academy of Sciences decided to work on [its own] new boosters. We quickly succeeded because we walked on two legs. If we had not walked on two legs, we would have been at a dead end when the Soviets reneged on their agreements with us and stopped their assistance to us."[40]

The return of these Chinese students and scientists from the United States and Europe in the 1950s helped to counterbalance Sovietization by reinforcing a strong sense of Chinese nationalism and, at the same time, by discreetly seeking inspirations from the Western model that they had become familiar. Qian, for example, reportedly voiced his shock at "seeing so many Soviet experts" at the famed Institute of Military Engineering (Hajungong) of the People's Liberation Army in the northern city of Harbin up on his return in 1955. "Don't we Chinese know how to teach? What are all these foreigners invited here to do?" he asked, to the delight of General Chen Geng, the president of the institute, who praised his patriotism.[41] Apparently motivated by an unspoken attachment to his American alma mater and a dissatisfaction with the

Soviet-inspired separation of science and engineering in Chinese universities in the early 1950s, Qian subsequently helped establish the University of Science and Technology of China in the Chinese Academy of Sciences on the model of Caltech.[42]

These post-1949 returnees further heightened the prominence of American-educated scientists among the leadership of the Chinese scientific community. In the late 1980s, the Chinese historian of science Li Peishan examined information on the 877 prominent Chinese scientists included in a contemporary biographical dictionary and found that 662—that is, 75.5 percent—had received advanced education abroad. Those who had received such education in the United States numbered 393 (59 percent)—more than four times as many as had received in the next favorite destination, Great Britain (91, or 14 percent). Li's data also revealed that surprisingly few Japanese-trained and Soviet-trained scientists were listed—only 34 (5.1 percent) and 28 (4.2 percent), respectively.[43] (See table 11.4.) Even though Japan was a popular destination for Chinese students studying abroad in the early twentieth century, few of those who went there chose to specialize in science and technology. Most of those sent to the Soviet Union studied engineering, not science; those few who did study science would become prominent in their fields in large numbers only after the mid 1980s.

The dominance and the high quality of Western-trained Chinese scientists and engineers in Chinese science and technology didn't escape the attention of the Soviet scientific leadership involved in Sino-Soviet interactions. As early as 1956, when a

Table 11.4

Countries of training for leading Chinese scientists, engineers, and physicians included in a biographical dictionary in 1986.

	No. of Chinese scientists trained in country	Percentage of all (877)	Percentage of all who studied abroad (662)
US	393	44.8	59.4
China	215	24.5	n/a
UK	91	10.4	13.7
Germany	54	6.2	8.2
France	35	4.0	5.3
Japan	34	3.9	5.1
Soviet Union	28	3.2	4.2
Other Countries	27	3.1	4.1
Total	877	100	100

Source: Li Peishan, "The Status and Roles of Returned Students in Scientific and Technological Developments in Post-1949 China," *Journal of the Dialectics of Nature* 11, no. 4 (1989): 28–36.

high-level delegation of the Soviet Academy of Sciences visited China to assist in the making of its Twelve-Year Plan for Science and Technology, it became clear to many in the group that, owing to the return of eminent scientists from abroad, the Soviet side needed to send to China truly world-class experts to serve as advisors. Otherwise, inferior Soviet advisors would quickly expose their limits and bring embarrassment to themselves and to the Soviet Union. At a debriefing for the Presidium of the Soviet Academy of Sciences in Moscow in July 1956, Academician Aleksandr Mikhailov, a member of the delegation, explained:

> It should be noted that in the last two or three years many prominent [Chinese] scientists have returned home from abroad. Based on our encounters with these scholars in China, their abilities are undoubtedly equal to those of our own Soviet scientists and their academic works are familiar to our scientists. We can point to the example of the visit of the aerodynamicist Qian [Xuesen] in our country. Returning to China from the US about seven or eight months ago, he is now visiting the Soviet Union by invitation. He has delivered a series of lectures and visited many research institutes. He is a well-educated scientist with a broad vision. Such scientists are growing in number in China. ... [Thus] I believe that if we could not send high-level personnel in some fields [to China], it's best that we do not send anyone.[44]

As further evidence of the disparity between Soviet-trained and Western-trained scientists in China, Mikhailov noted that among researchers at the highest level in China 339 had been trained in "capitalist countries," 164 of them in the United States; only five had studied in the Soviet Union. "Obviously the influence on them mainly came from the capitalist countries," he lamented, "our influence is negligible."[45]

Comparing Western-trained and Soviet-trained scientists was complicated by the little-known fact that at least a few of those who had been sent to the Soviet Union in the 1950s had previously returned to China from the United States. For example, Tu Guangchi, a geochemist, received his PhD from the University of Minnesota in 1949, returned to China in 1950, then was sent to Moscow University for another degree (which he earned in 1954). Likewise, Zhang Li, who had returned to China after studying physics at Cornell University in 1948–49, was sent to Leningrad University to study with Vladimir Fock. And Yang Guanghua had returned to China in 1951 after receiving a PhD in chemical engineering from the University of Wisconsin before going to the Moscow Institute of Petroleum in 1956 for additional training (though not another degree). In all three of these cases, the returnees' advanced technical preparations enabled him to take advantage of further Soviet training, but political reliability probably was another factor: all three had joined the Communist Party before going to the Soviet Union. Zhang was allowed to continue studying theoretical physics, but Tu switched from his postdoc basic research on titanium dioxide at the University of Pennsylvania to the more applied subject of mineral deposit geology in Moscow. In any case, such double exposure made this select group of scientists truly transnational hybrids during the Cold War.[46]

One could argue, of course, that in many ways the Cold War hindered Chinese scientific development. If there had been no Sino-US conflicts, especially the Korea War, perhaps more of the Chinese students and scientists would have returned home from the US. The ban on scientific contact and the embargo on sales of scientific and technical equipment between the two countries hampered scientific development in China, and arguably in the US. One could also point to China's participation in and withdrawal from the International Geophysical Year (1957–58) as evidence of the negative effects of Cold War international politics on Chinese scientific development. In the early 1950s, when invited by the organizers to participate in the IGY, the Chinese Academy of Sciences agreed to join the global effort only after the Soviet Union said that it would do so and after receiving assurance from the organizers that Taiwan would not take part. Seeing an opportunity to benefit from the international collaboration and also to showcase China's scientific achievements, the Chinese academy and the national government poured resources into preparing for China's participation in the IGY, including the acquisition of advanced instrumentation and standards from the Soviet Union, which greatly helped to promote geophysical research in China. But only a few months before the IGY was to begin, Taiwan entered into the IGY with instigation by the US Department of State. That led mainland China to withdraw at the last minute in order to avoid the creation of a "Two Chinas" problem. The planned measurements and observations were apparently carried out, but follow-up collection and processing of data suffered, partly as a result of the lack of the pressure of international participation.[47]

Yet it is doubtful that, in the absence of geopolitical rivalry with the United States, the Soviet Union would have provided the kind and the amount of technological assistance to China that it did provide.[48] It was within the framework and on the basis of this Soviet-style technical and industrial infrastructure that many of the most notable achievements of China's Western-trained scientists and engineers were produced. As Chen Mengxiong, a prominent geologist who never went abroad to study but who had been trained by Western-educated Chinese geologists in the 1940s, observed later:

The biggest difference between American and Soviet styles of engineering geology was that the American way was more open and focused more on innovation, which were reflected in their flexible practices, while the Soviets put more emphasis on applications. ... I think that it was right to learn from the Soviet Union at the time, because its approach was easier to follow. The advantage of the Soviets was that they had a set of comprehensive procedures, which we did not have in the past.[49]

Chen hastened to add that "of course the Soviet approach was not supposed to be copied mechanically." "Indeed, after the end of the Cultural Revolution," he continued, "we carefully revised those old procedures in accordance with the actual conditions in China."⁵⁰ What resulted in China was an interesting transnational hybrid science and technology that reflected both Soviet and Western influences, which often were both conflicting and complementary.

Self-Reliance in the 1960s

After the breakup between China and the Soviet Union in the early 1960s, the Cold War turned a new chapter and China, as elaborated by Sigrid Schmalzer in her contribution to this collection, entered into an era during which self-reliance, especially in the areas of science and technology, took on a new prominence.⁵¹ The crowning achievement in this period in this regard was the successful testing of China's first atomic bomb in 1964. Although the bomb was built with no formal participation by any foreigner, it nevertheless had many transnational characteristics: much of the infrastructure for it, including the uranium-separation facilities, was built with Soviet technical assistance. Most of the leading scientists involved in the project had been trained in the United States and in Western Europe, and a number of them had returned to China after 1949.⁵² Likewise, most of the senior scientists involved in the successful artificial synthesis of crystalline bovine insulin, another example of self-reliant science in this period, had received training in the West, and the final (though unsuccessful) official Chinese nominee for the Nobel Prize for this work, Niu Jingyi, had returned to China in 1956 with a PhD from the University of Texas.⁵³

The breakdown of the Sino-Soviet alliance and the continued Sino-American hostility led to the formation of self-reliance science and technology policy, but also to greater attention to the possibility of scientific communications with Europe, Japan, and other parts of the world. Even before the breakup with the Soviet Union, the Chinese government had hedged its Soviet bet with backup technical connections with Europe. In August of 1957, for example, Liang Sili, who had received a PhD in engineering from the University of Cincinnati in 1949 and had returned to China the same year, was sent by the Chinese government to Switzerland as a member of a trade mission with the purpose of purchasing missiles and related equipment in the black market. Liang's group succeeded in buying theodolites and was on the verge of acquiring missiles when news came that the Soviet government had agreed to provide prototypes to China. Liang's group was then turned into a trade delegation to West Germany, where they visited Siemens factories.⁵⁴

China extended its scientific contacts with other parts of Europe and the world. In 1962, the Chinese Academy of Sciences invited the Danish nuclear physicist Aage Bohr for a five-week visit and lecture tour, which in turn led China to send several physicists to Bohr's Institute of Theoretical Physics in Copenhagen for one-year and two-year visits in the years 1963–1967.⁵⁵ In 1964, a Beijing Science Symposium attracted 367 scientists from 44 countries in Asia, Africa, Latin America, and the Pacific. Formally sponsored by the left-leaning Association of Scientific Workers and its Beijing branch, the conference was used by the Chinese government as a way to learn of new scientific developments and also to boost its claim as a leader of the anti-colonial cause. Two years later, Beijing hosted a Summer Symposium in Physics.[56] In 1964–65, the Chinese Academy of Sciences sent 49 graduate students to study in Great Britain, France, Sweden, Australia, Denmark, and Switzerland.[57]

As the principal organizer of China's modernization drive, Premier Zhou Enlai was the main advocate for international scientific and technological contact. In early 1966, on the eve of the Cultural Revolution, Zhou urged Chinese diplomats to learn enough about science and technology to be able to coordinate the process of absorbing scientific and technological information from the countries where they were stationed.[58] Thanks in part to Zhou's protection, a number of prominent Western-trained scientists survived the Cultural Revolution, and some of them (including Zhu Kezhen, vice president of the Chinese Academy of Sciences) were able to keep abreast of international scientific developments by reading the journals *Science* and *Nature*. Thus, they were ready to plan for a resumption of scientific research and education under Zhou once the worst of the chaos was over, in the early 1970s.[59]

Post-Nixon Transformations

Sino-US transnational movement of scientists benefited greatly from the Nixonian exchange of the 1970s. The Shanghai Communiqué, signed during Nixon's first trip to China in 1972, included these sentences:

The two sides agreed that it is desirable to broaden the understanding between the two peoples. To this end, they discussed specific areas in such fields as science, technology, culture, sports, and journalism, in which people-to-people contacts and exchanges would be mutually beneficial. Each side undertakes to facilitate the further development of such contacts and exchange.

Of course, Nixon's trip to China in 1972 was a calculated move by both sides in reshaping the geopolitical balance in the US-Soviet-China triangular relationship in the context of the Cold War. As with most other state-sponsored scientific international exchanges, the primary purpose for the states involved was politics, not science.[60] Yet it wasn't an accident that science and technology featured prominently in the list of areas for initial contacts. The United States had long used its strength in science and technology as a diplomatic tool and as a way to demonstrate the superiority of the American system and way of life.[61] The US government sought to increase its appeal in Asia by equating science and technology, especially Americanized versions of them, with modernity. The US government was not unaware of the military implications of scientific exchanges and technological transfers; its counterintelligence officials would always keep a close watch on the visiting Chinese scientists conducted under governmental agreements.[62] Policy makers decided, however, that the benefit to American national interest was worth the risk.

Even though they knew that scientific exchanges formed only part of the overall new Sino-American relations, most Chinese scientists in China and Chinese American scientists in the US were heartened by the new developments, and indeed they capitalized on them in their continued pursuit of professionalism and Chinese cultural nationalism. For American-trained Chinese scientists who were still undergoing humiliating physical reeducation, the Nixon visit brought relief overnight. Many Chinese scientists saw a decisive improvement in their political fortunes when a visiting American scientist proposed meetings with them. For example, after the American marine biologist and deep-sea diver Sylvia Earle was allowed to meet with him, Zeng was able to return to research. In 1975, he was even chosen as vice chairman of a major Chinese scientific delegation to visit the United States, where he met with his fellow Michigan alumnus Gerald Ford in the White House.[63]

For Chinese American scientists, the Nixon overture opened a new world of possibilities. The thousands of "stranded" students and scientists who decided to stay in the United States in the 1950s had been motivated by a confluence of factors, including fear of and uncertainty about the Chinese communists, disappointment with the Nationalists, the prospect for a better life in the US (especially with the repeal of the Chinese Exclusion Act during World War II), and plentiful job opportunities for scientists and engineers. Yet most of them also suffered, to varying degrees, McCarthyist political harassment, persistent racial discrimination, and a feeling of being isolated and marginalized professionally and socially. In the late 1950s and the 1960s their sense of identity crisis grew as they began to sink roots in America by becoming citizens and raising their American children. Chen Ning Yang, for example, worried that his father would never forgive him for giving up his Chinese citizenship. The act of finally becoming Americans also prompted them to seek to understand the bitter history of Chinese in the US.

Some Chinese American scientists became political activists, supporting the Civil Rights movement, joining protests against the Vietnam War in the 1960s, and leading the Defend Diaoyu movement in the early 1970s. The latter was an effort to protest against the US government for returning control of the disputed Diaoyu Islands to Japan rather than to China (or Taiwan), which had ceded them to Japan after the 1895 Sino-Japanese war and had expected to gain them back as part of the settlement after World War II. Many Chinese American scientists of the generation that saw China ravaged by Japanese invaders hoped to see their nation of origin emerge as a strong, prosperous, and modernized country. As long as China and the United States remained isolated Cold War rivals, it was dangerous to express such attachment to China. The Nixon trip changed everything. It promised a future in which Chinese Americans in general, and Chinese American scientists in particular, could become transnational agents of exchange, legitimately serving the interests of both countries and healing the schism in their identity and political loyalty.[64]

Chen Ning Yang, the first well-known Chinese American scientist to take advantage of the Nixon administration's relaxation of travel restriction, paid a visit to China in the summer of 1971. Yang had been able to keep in touch with his family in China throughout his years in the US, and had met with his parents in 1957 and 1960 in Geneva and in 1962 in Hong Kong. His father, Yang Wuzhi, a mathematics professor in Shanghai, had tried, with the tacit encouragement of the Chinese government, to convince his son to return to China, but his mother disagreed, citing the poor living conditions. ("I had to wait in a queue for two or three hours in the night to just buy some pieces of tofu.") Chen Ning Yang never seriously considered returning permanently to China once he was offered a permanent position by the Institute for Advanced Study at Princeton, but he had heeded his father's advice to avoid visiting Taiwan. During his trip in 1971, he met with Zhou Enlai and with many of his old teachers and classmates. He was carefully shielded from the dark side of the Cultural Revolution, and what he saw impressed him greatly. And he claimed to have been moved to tears when Deng Jiaxian, his childhood friend and fellow Boxer student in the US who returned to China in 1950 and who went on to play a prominent role in China's nuclear weapons program, confirmed to him that no foreigner had participated in that program.[65]

However, Chen Ning Yang's meeting with another old friend, Huang Kun, and his brother Huang Wan, a heart specialist, touched a sensitive spot. When Yang mentioned the desirability of sending Chinese students to the US for training, Huang Wan said that there was a question of "who would they serve" after they finished. Taking the comment personally,

Yang responded immediately that he had wondered whether people in China might think that it was selfish for him to stay abroad. He said that yes, it was most selfish [for him to do so], but matters were not that simple. While many [Chinese] people in the US never cared for China, he thought of the country a lot. Huang Kun expressed to him [Yang] that he was especially delighted that he could overcome the many obstacles to be the first to return for this tour.[66]

The sense of guilt felt by Yang and many other Chinese American scientists reinforced their Chinese cultural nationalism and motivated them to become active participants in US-China scientific exchanges. They introduced recent scientific advances to their Chinese colleagues who had been largely isolated from the international scientific community during the Cultural Revolution, established programs and institutional connections to bring Chinese scientists to the United States as visitors, and advised the Chinese government on specific scientific problems and on major issues of science policy. Chinese science and scientists reaped benefits from US-China scientific exchanges, especially those promoted by Chinese American scientists.

The resumption of formal diplomatic relations between China and the United States in the late 1970s marked another milestone in transnational Chinese science. It came in the aftermath of the death of Mao, the end of the Cultural Revolution, and the coming to power of Deng Xiaoping, who remained the pragmatic national leader for the next two decades, pushing aggressively for the Zhou-initiated modernization drive and market-oriented economic reform for China. The normalization of relations with the US in 1979 made Chinese science even more transnational, bringing numerous bilateral cooperative scientific and technological projects into existence and further enlarging the influence of Chinese American scientists on China's science and education policy.

Yet perhaps the most significant result of formal diplomatic relations was that they opened the way for Chinese students to study in the United States, something that had proved politically impossible even after Nixon's trip. The size of the new wave of Chinese students studying in the US, which still shows no sign of abating, has been unprecedented, as was the speed with which it developed. The impact of those students is still unfolding. A large portion of them have stayed in the US after completing their studies, especially since the Tiananmen Square incident of 1989, and a growing number of them have become true transnationals by holding academic positions on both sides of the Pacific or by playing the role of facilitators in the rapid transnational development of the booming Chinese economy.[67] Those who did return also began to rise in leadership positions in Chinese science and other areas.[68] According to a 2003 Xinhua News Agency report commemorating the thirtieth anniversary of the Nixon trip, about 10,000 researchers from the Chinese Academy of Sciences alone had studied or worked in the United States in the previous 20 years, and "most of the CAS's leading scientists had a US education background."[69]

The beginning of the Cold War had made a dramatic difference to science, scientists, and science policy in China and the United States in the late 1940s and the early 1950s. The end of the Cold War, which the 1989 Tiananmen Square incident may well have precipitated, didn't seem to have nearly as large an effect, at least not immediately. After a short post-Tiananmen hiatus of scientific and educational exchanges between China and the West (especially the US), both scientific and educational exchanges resumed and then accelerated in the early 1990s. The number of Chinese students coming to the US dropped again when the US government tightened visa requirements after the terrorist attacks of September 11, 2001, but the restriction was later relaxed after protests by American scientists. In 2010, Chinese students once again constituted the largest group of foreign students (127,628) in the United States.[70]

Conclusion

Did the Cold War transform science in China and make it more transnational? It certainly brought about a massive Sovietization of Chinese science, technology, and education in the 1950s, within the Cold War framework of a geopolitical alliance. With the continuation of earlier American influence, reinforced by the return of about 2,000 Western-trained scientists from the United States and Europe in the 1950s, the era of Sovietization saw the emergence of what might be called a transnational hybridization of science and technology in China. Various foreign elements encountered, clashed, and merged with local ones within a highly politicized local environment. A fundamental character of science and technology in China during the Cold War was the combination of a Soviet-style scientific, technological, and industrial infrastructure built on the Nationalist legacy and a scientific and engineering workforce that was dominated by people educated in United States and Europe. This hybridization continued even during the subsequent decade of self-reliance, when both Soviet and American contacts were officially cut off. Transnational exchanges resumed in the 1970s as Westernization once again gained momentum, especially after the end of the Cultural Revolution in 1976.

Indeed, transnational science became so entrenched in China that its scientific and educational exchanges with the West, especially the US, withstood major potential countercurrents such as the 1989 Tiananmen Square tragedy and even the end of the Cold War in the late 1980s and the early 1990s when many commentators saw an increased possibility of Sino-US hostility as China rose as a major power. Today, science in China is more transnational than ever, marked by especially strong scientific and technological ties with the US, which in turn has been stimulated by booming trade and fostered by a new generation of American-educated Chinese scientists. In a 2007 speech on the importance of international scientific cooperation, Bai Chunli, who had done postdoctoral work at Caltech in 1985–1987, said: "As illustrated by the development of the Chinese Academy of Sciences in the last half century, almost all major scientific and technological achievements contained contributions from international collaboration."[71]

Yet an emphasis on the transnational transformations of science and technology in China during the Cold War doesn't, as mentioned above, mean that nationalism disappeared from the scene. In fact, nationalism played a big part in the transnational dynamics. China's decision to "lean toward" the Soviet Union during the Cold War was driven at least as much by concerns about national security and sovereignty as by ideological commonality. It was the former that would later lead to the Sino-Soviet breakup. Similarly, a form of cultural nationalism that had derived from a deep awareness of China's long-time international humiliations powerfully motivated many American-educated Chinese scientists to return home in the 1950s against US policy and to persist in their pursuit of modernization through science and technology despite political persecutions under Mao. Even those who had decided to stay in the US in the 1950s shared this sense of Chinese cultural nationalism, which was reflected in a widespread guilty feeling of having abandoned one's homeland and which drove many of them to engage actively in promoting US-China scientific exchange in the post-Nixon period and even in playing prominent roles in Chinese science, technology, and education policy.

In the end, a set of dualities shaped and reshaped the development of science and technology in China during the Cold War: a political context that reflected both the Cold War and developmental aspirations of the Chinese Communist party-state; a scientific community dominated by people trained in the West, especially the United States, working on the foundation of a technical and industrial infrastructure built with Soviet assistance and guidance; and finally, a strong sense of Chinese cultural nationalism both contrasted and underlined Chinese scientists' international scientific ideals, diverse transnational scientific training and influences, and shifting political environments. Some of these factors and dynamics arose from the Cold War; others did not. But in the case of the latter, such as Chinese scientists' sense of nationalism, the Cold War certainly sharpened it in many ways. Thus, to the extent that the Cold War helped to frame the development of transnational scientific institutions, practices, and interactions, we might say that it transformed science in China.

Acknowledgments

I would thank Naomi Oreskes, Sigrid Schmalzer, David Kaiser, John Krige, and anonymous referees for helpful comments and feedback on this chapter, and Wang Yangzong, Zhang Baichun, Zhang Li, Zhang Jiuchen, Chen Zhi, Yin Xiaodong, Xiong Weimin, and Pan Tao for helpful discussions and assistance with research materials and interviews. This work was supported in part by the National Science Foundation under grant no. SES-1026879. Any opinions expressed in this material are those of the author and do not necessarily reflect the views of the NSF.

Notes

1. Richard M. Nixon, "Asia after Viet Nam," *Foreign Affairs* 46, no. 1 (1967): 111–125, on 121.

2. There was much more to the Cold War than the actions and interactions of the United States, the Soviet Union, and China, of course. The so-called Third World in Asia, Africa, and Latin America, for example, played a key part in the Cold War. See Odd Arne Westad, *The Global Cold War: Third World Interventions and the Making of Our Times* (Cambridge University Press, 2005).

3. For some recent examples of transnational approaches to the history of science and technology, see Erik van der Vleuten, "Toward a Transnational History of Technology: Meanings, Promises, Pitfalls," *Technology and Culture* 49 (2008): 974–994; Zuoyue Wang, "Transnational Science during the Cold War: The Case of Chinese/American Scientists," *Isis* 101, no. 2 (2010): 367–377; John Krige, "Building the Arsenal of Knowledge," *Centaurus* 52 (2010): 280–296.

4. For more on the issue of foreign technological aid and self-reliance, see Sigrid Schmalzer's chapter in this volume.

5. See, e.g., William C. Kirby, "Technocratic Organization and Technological Development in China: The Nationalist Experience and Legacy, 1928–1953," in *Science and Technology in Post-Mao China*, ed. Denis Fred Simon and Merle Goldman (Harvard University Council on East Asian Studies, 1989); Izabella Goikhman, "Soviet-Chinese Academic Interactions in the 1950s: Questioning the 'Impact-Response' Approach," in *China Learns from the Soviet Union, 1949–Present*, ed. Thomas P. Bernstein and Hua-Yu Li (Lexington Books, 2010).

6. See, e.g., Zuoyue Wang, "Saving China through Science: The Science Society of China, Scientific Nationalism, and Civil Society in Republican China," *Osiris* 17, (2002): 291–322; Grace Yen Shen, "Taking to the Field: Geological Fieldwork and National Identity in Republican China," *Osiris* 24 (2009): 231–252.

7. On a broad survey of international influences in China in this period, see William C. Kirby, "The Internationalization of China: Foreign Relations at Home and Abroad in the Republican Era," *China Quarterly*, June 1997: 433–458. See also Benjamin Elman, *A Cultural History of Modern Science in China* (Harvard University Press, 2006).

8. Michael H. Hunt, "The American Remission of the Boxer Indemnity: A Reappraisal," *Journal of Asian Studies* 31, no. 3 (1972): 539–559.

9. Hunt, "The American Remission," 550. On the establishment of the Boxer fellowships, see also Stacey Bieler, *"Patriots" or "Traitors"? A History of American-Educated Chinese Students* (Sharpe, 2004), 42–51. Liang Shiqiu, a well-known Chinese writer who had studied English literature in the US in the 1920s as a Boxer fellow himself, first read James' letter in 1977 and commented: "How can we not react to these statements with alarm, bitterness, and shame?" Quoted in Cheng Xinguo, *Gengkuan liuxue bainian* (a century of Boxer indemnity funds for studying abroad) (East Publishing Center, 2005), 17.

10. Theodore Roosevelt, "Seventh Annual Message," December 3, 1907, in Public Papers of the Presidents (available at www.presidency.ucsb.edu).

11. The "Regulations," undated, were attached to a letter from William W. Rockhill, American minister in China, to the US Secretary of State, October 31, 1908, in US State Department, *Papers Relating to the Foreign Relations of the United States with the Annual Message of the President to Congress*, December 8, 1908 (available at http://images.library.wisc.edu). It is not clear who mandated the emphasis on science and technology in the fellowship program, but such a pattern was consistent with the advocacy of Zhang Zhidong, head of the Ministry of Education, that "Chinese learning should remain the essence while Western learning serve to bring utility." He

also insisted that Boxer candidates be tested primarily in Chinese, not Western subjects during the selection process. Hunt, "The American Remission," 557.

12. Cheng Xinguo, *Gengkuan*, 19–25.

13. According to Y. C. Wang (*Chinese Intellectuals and the West, 1872–1949*, University of North Carolina Press, 1966, 111), in addition to the 180 sent in 1909–1911, a total of 1,268 Tsinghua students had studied in the US with Boxer fellowships from 1912 to 1929, so the total for 1909–1929 was 1,448. Note, however, that Bieler (*"Patriots,"* 381) listed only 1,119 students in the US as having come from Tsinghua from the beginning to 1953.

14. Zhu Junpeng, "Qinghua yuan nei diyipi ruxiao nusheng" (the first group of resident women students at Tsinghua), April 20, 2009 (http://xs.tsinghua.edu.cn/docinfo/board/boarddetail.jsp?columnId=00302&parentColumnId=003&itemSeq=5602). See also Cheng Xinguo, *Gengkuan*, 35; Wang, *Chinese Intellectuals*, 112.

15. Cheng Xinguo, *Gengkuan*.

16. Bieler, 79–81; Wang, "Saving China through Science."

17. On the Rockefeller Foundation in China, see Mary Brown Bullock, *An American Transplant: The Rockefeller Foundation and Peking Union Medical College* (University of California Press, 1980); William J. Hass, *China Voyager: Gist Gee's Life in Science* (Sharpe, 1996).

18. Ma Zusheng (T. S. Ma), *Linian chuguo/huiguo keji renyuan zonglan (1840–1949)* (a comprehensive survey of scientific and technical personnel studying abroad and back to China, 1840–1949) (Social Sciences Academic Press, 2007). Ma was a Tsinghua graduate who received his PhD in chemistry from the University of Chicago in 1938, and after teaching stints in China and New Zealand in the late 1940s, spent much of his career at the City University of New York. See Cheng Xiankang, "Daonian guoji zhuming huaxuejia Ma Zusheng xuezhang [in memory of Ma Zusheng, a senior alumnus and internationally renowned chemist]," June 24, 2008 (http://www.tsinghua.org.cn/alumni/infoSingleArticle.do?articleId=10015446&columnId=10015367).

19. Cheng Xinguo, *Gengkuan*, 72–88.

20. Yu Hao and Huang Yanfu, *Zhongguo keji de jishi: Ye Qisun he kexue dashimen* [*The Cornerstones of Chinese Science and Technology: Ye Qisun and Scientific Masters*] (Fudan University Press, 2000), 100–112.

21. Fan Dainian and Qi Fang, "Wang Ganchang xiansheng zhuanlue [A Biographical Sketch of Mr. Wang Ganchang]," in *Wang Ganchang he tade kexue gongxian* [*Wang Ganchang and His Scientific Contributions*], ed. Hu Jimin et al. (Science Press, 1987), 228.

22. Norbert Wiener, *I Am a Mathematician* (MIT Press, 1964), 186.

23. Wang, "Saving China."

24. See Baichun Zhang, Jiuchun Zhang, and Fang Yao, "Technology Transfer from the Soviet Union to the People's Republic of China, 1949–1966," *Comparative Technology Transfer and Society* 4, no. 2 (2006): 105–171; Zhang Baichun, Yao Fang, Zhang Jiuchun, and Jiang Long, *Sulian*

jishu xiang zhongguo de zhuanyi, 1949–1966 [*Technology Transfer from the Soviet Union to the People's Republic of China, 1949–1966*] (Shandong Education Press, 2004).

25. Wang, "Transnational Science."

26. Zhang et al., *Sulian jishu*. See also Wu Yan, ed., *Zhongsu liangguo kexueyuan kexue hezuo ziliao xuanji* [*A Selected Collection of Documents on the Scientific Cooperation between the Academies of Sciences OF China and the Soviet Union*] (Shandong Education Press, 2008). On the sending of Soviet technical advisors to China, see Shen Zhihua, *Sulian zhuanjia zai zhongguo, 1948–1960* [*Soviet Specialists in China, 1948–1960*] (Xinhua Press, 2009).

27. See Wu Yan, *Zhongsu liangguo*.

28. Zhang et al., *Sulian jishu*, 162.

29. See Laurence Schneider, *Biology and Revolution in Twentieth-Century China* (Rowman and Littlefield, 2003), 155–159, and Schmalzer's chapter in this collection.

30. Hu Zonggang, *Bugai yiwang de Hu Xiansu* [*The Hu Xiansu That Should Not Be Forgotten*] (Changjiang Wenyi Press, 2005), 166.

31. Schneider, *Biology and Revolution*, 155–159.

32. Mao Zedong, "Tong sulian zhuhua dashi youjin de tanhua [Talks with Yudin, Soviet Ambassador]," July 22, 1958 (http://www.mzx.cn/theory/html/mx07385.htm).

33. Tao Wenzhao, "Jinyun yu fan jinyun: wushi niandai zhongmei guanxi zhong de yichang yanzhong douzheng [Embargo and Anti-Embargo: A Serious Struggle in China-US Relations in the 1950s]," *Zhongguo shehui kexue* [*China Social Science*] 3 (1997): 179–193.

34. Peter Neushul and Zuoyue Wang, "Between the Devil and the Deep Sea: C. K. Tseng, Mariculture, and the Politics of Science in Modern China," *Isis* 91, no. 1 (2000): 59–88, on 74.

35. Wang, "Transnational Science."

36. Beijing Municipal Archives, Beijing, file no. 002–020–339 "Guojia kewei zhuanjiaju 1959.3.5 tongzhi [Circular from the Bureau on Experts of the National Science and Technology Commission dated March 5, 1959]," attachment 1.

37. Beijing Municipal Archives, Beijing, file no. 002–020–339. The report didn't give specific breakdowns of disciplinary distributions within each country of education.

38. See Iris Chang, *Thread of the Silkworm* (Basic Books, 1995).

39. See ibid., 190.

40. Zhang is quoted in Jin Chongjin et al., *Zhou Enlai zhuan, 1949–1976* [*A Biography of Zhou Enlai*] (Central Documentation Press, 1998), volume 1, 235.

41. Zhu Zhaoxiang, "Qian Xuesen xiansheng zai lixuesuo chuangjian de rizili [Mr. Qian Xuesen in the founding period of the Institute of Mechanics]," *Kexue shibao* [*Science Times*], December 7, 2005, as quoted in Ye Yonglie, *Qian Xuesen* (Shanghai Jiaotong University Press, 2010), 205.

42. Wang, "Transnational Science."

43. Li Peishan, "Guiguo liuxuesheng 1949 nian yihou zai zhongguo kexue jishuyfazhan zhong de diwei yu zuoyong [The Status and Roles of Returned Students in Scientific and Technological Developments in Post-1949 China]," *Ziran biezhengfa tongxun* [*Journal of the Dialectics of Nature*] 11, no. 4 (1989): 26–34. These statistics didn't include those Chinese scientists who remained abroad after they finished their studies but later played active roles in facilitating Sino-foreign scientific exchange. The dictionary Li used, *Zhongguo kexuejia cidian* [*Dictionary OF Chinese Scientists*] (Shandong Science and Technology Press, 1982–1986), included engineers under "technological science."

44. For a Chinese translation of selections of the Russian transcript of a session of the Presidium of the Soviet Academy of Sciences on July 6, 1956, see Wu Yan, *Zhongsu liangguo*, 224. The English translation from the Chinese presented here is my own.

45. Wu Yan, *Zhongsu liangguo*, 223. Several delegation members also made the point that in two or three years, with the return of Western-trained scientists and with China's growing research capabilities, Soviet advisors in China probably wouldn't be in a position to provide guidance in many fields. At this point it is not clear whether this belief figured into Khrushchev's decision to withdraw all Soviet advisors in 1960.

46. On Tu, see Tu Guangchi et al., *Tu Guangchi huiyi yu huiyi Tu Guangchi* [*Recollections by and of Tu Guanczhi*] (Hunan Education Press, 2010), 45–49. On Zhang Li, see interview with Zhang Li, Tsinghua University, Beijing, by Zuoyue Wang and Yin Xiaodong, December 17, 2011. On Yang, see "Yang Guanghua xiansheng shengping [Obituary of Mr. Yang Guanghua]" (http://www.cup.edu.cn/heavyoil/Bulletin/37957.htm). Tu had joined the Chinese Communist Party in 1949 while still in the US (*Tu Guangchi huiyi*, 55–56). Zhang joined in 1949 after his return to China. Yang joined in 1956.

47. Zuoyue Wang and Jiuchen Zhang, "China and the International Geophysical Year," in *Globalizing Polar Science: Reconsidering the Social and Intellectual Implications of the International Polar and Geophysical Years*, ed. Roger D. Launius, James R. Fleming, and David H. Devorkin (Palgrave Macmillan, 2010).

48. See, e.g., Shen Zhihua, *Sulian*.

49. Zhang Jiuchen (interviewer), "Wo suo jiechu de sulian zhuanjia—Chen Mengxiong fangtan [The Soviet Experts I Encountered—An Interview with Chen Mengxiong]," *Zhongguo kejishi zazhi* [*Chinese Journal for the History of Science and Technology*] 32, no. 2 (2011): 154–165, on 162.

50. Zhang Jiuchen, "Wo suo jiechu de sulian zhuanjia," 162.

51. Schmalzer, "Self-Reliant Science."

52. Wang, "Transnational Science," 373–374.

53. Xiong Weimin and Wang Kedi, *Hecheng yi ge danbaizhi: Jiejing niu yidaosu de rengong quan hecheng* [*Synthesize a Protein: The Story of the Total Synthesis of Crystalline Insulin Project in China*]

(Shandong Education Press, 2005), 17 and 121. On this project, see also Schmalzer's chapter in this book.

54. Guo Mei and Pang Ru, *Liang Sili zhuan* [*a biography of Liang Sili*] (Jiangsu People's Press, 2009), 41–42; Zuoyue Wang interview with Liang Sili, Beijing, July 14, 2011. See also Zhou Bolun, ed., *Nie Rongzhen nianpu* [*A Chronicle of the Life of Nie Rongzhen*] (People's Press, 1999), volume 1, 592 and 624. Marshal Nie was in charge of Chinese missile and atomic bomb programs in this period.

55. See Yin Xiaodong, "Danmai wulixuejia aoge boer laihua fangwen shimo jiqi yingxiang [Danish Physicist Aage Bohr's Visit to China and Its Impact]," *Ziran kexueshi yanjiu* 31, no. 3 (2012): 329–342. See also John Krige, "The Ford Foundation, European Physics, and the Cold War," *Historical Studies in the Physical and Biological Sciences* 29, no. 2 (1999): 333–361.

56. Xiong Weimin, "Zai kexue yu zhengzhi zhijian: 1964 nian de Beijing kexue taolunhui [Between Science and Politics: The 1964 Beijing Scientific Symposium—An Oral History Interview with Mr. Xue Pangao]," in *Xue Pangao wenji* [*Collected Papers of Xue Pangao*] (Division for the History of the Chinese Academy of Sciences of the Institute for the History of Natural Science, Chinese Academy of Sciences, 2008).

57. Qian Linzhao and Gu Yu, eds., *Zhongguo kexueyua* (Contemporary China Press, 1994), volume 1, 137.

58. Zuoyue Wang, "US-China Scientific Exchange: A Case Study of State-Sponsored Scientific Internationalism during the Cold War and Beyond," *Historical Studies in the Physical and Biological Sciences* 30, part. 1 (1999): 249–277.

59. See, e.g., Zhu Kezhen and Wu Youxun to Zhou Enlai (draft), September 26, 1968, in *Zhu Kezhen quanji* [*Complete Works of Coching Chu*] (Shanghai Science, Technology, and Education Press, 2004–), volume 4, 391–393.

60. Wang, "US-China Scientific Exchange."

61. Examples include Nixon's kitchen debate with Khrushchev, Eisenhower's giving Polaroid cameras as gifts to foreign leaders, and Johnson's present to the visiting president of South Korea—the Korean Institute of Science and Technology. On the last, see Kim Dong-Won and Stuart W. Leslie, "Winning Market or Winning Nobel Prizes? KAIST and the Challenges of Late Industrialization," in *Osiris* 13 (1998): 154–185. See also John Krige, *American Hegemony and the Postwar Reconstruction of Science in Europe* (MIT Press, 2006); John Krige and Kai-Henrik Barth, eds., *Global Power Knowledge: Science and Technology in International Affairs* (*Osiris* 21, 2006).

62. In the 1980s, for example, the FBI recruited Sylvia Lee, a staff member at the Los Alamos weapons laboratory, to keep track of Chinese scientists visiting the laboratory. In 1999, Lee's husband, Wen Ho Lee, was accused of nuclear espionage for China; he was later acquitted of all charges except for one of mishandling nuclear secrets. Dan Stober and Ian Hoffman, *A Convenient Spy: Wen Ho Lee and the Politics of Nuclear Espionage* (Simon and Schuster, 2001).

63. Neushul and Wang, "Between the Devil and the Deep Sea."

64. Wang, "US-China Scientific Exchange."

65. Wang, "US-China Scientific Exchange"; Yang Zhenning (C. N. Yang), "Deng Jiaxian," in *Yang Zhenning wenji* [*Writings of C. N. Yang*] (Central China Normal University Press, 1998), volume 2, 797–804; C. N. Yang, interview with Zuoyue Wang, July 21, 2011, Beijing.

66. "Yang Zhenning zhuanjuan 1971 nian (san zhi yi) [Special Files on C. N. Yang]," file 72-4-61, in Archives of the Chinese Academy of Sciences, Beijing.

67. On the history of Chinese students in the US since the late 1970s, see Zuoyue Wang, "'Xuehao shulihua … ': 1978 nian hou dalu liumei kexuejia yanjiu ['If You Learn Your Math, Physics, and Chemistry … ': A Historical Study of Chinese Scientists in the US since 1978]," in *Beimei huaqiao huaren xin shijiao* (Overseas Chinese Press, 2008). In turn, the infusion from China (and Asia in general) was in large part responsible for an up surge in the number of science PhDs, especially physics PhDs, awarded in the US in the 1980s. See the graphs of "Physics PhD's and all PhD's conferred in the US, 1900 through 2007" and "Citizenship of Physics PhD's 1967 through 2007" at http://www.aip.org. Thanks to David Kaiser for information regarding these trends.

68. Zuoyue Wang, "Chinese American Scientists and US-China Scientific Relations: From Richard Nixon to Wen Ho Lee," in *The Expanding Roles of Chinese Americans in US-China Relations: Transnational Networks and Trans-Pacific Interactions*, ed. Peter H. Koehn and Xiao-huang Yin (Sharpe, 2002).

69. Xinhua News Agency, "Sci-tech Cooperation Promotes Sino-US Relations," February 7, 2002 (http://www.china.org.cn/english/26753.htm).

70. Tamar Lewin, "China Surges Past India as Top Home of Foreign Students," *New York Times*, November 15, 2010.

71. Bai Chunli, "Keji chuangxin xuyao quanqiu de shiye he yishi," *Keji ribao*, September 4, 2007 (scitech.people.com.cn/GB/6211876.html). (Bai, who at the time he gave the speech quoted was vice president of the CAS, became president of that body in 2011.)

12 When *Structure* Met Sputnik: On the Cold War Origins of *The Structure of Scientific Revolutions*

George Reisch

One of the most dramatic moments of the Cold War occurred in October of 1957 when the Soviet Union launched the first artificial satellite. The day after the launch, a headline in the *New York Times* gave confirming details: SOVIET FIRES EARTH SATELLITE INTO SPACE; IT IS CIRCLING THE GLOBE AT 18,000 M.P.H.; SPHERE TRACKED IN 4 CROSSINGS OVER US. Eight days later, an article in the *Times* by John Finney cited the weight of the first Sputnik, 184 pounds, as "evidence of Soviet superiority in rocketry."[1]

Anyone reading Finney's article, titled "US Missile Experts Shaken By Sputnik," knew that most ordinary Americans were shaken too. Since the earliest days of the Cold War, the United States had been on guard against its suspicious Cold War adversary and its agents. Anti-communist politicians such as Senator Joseph McCarthy and academics such as Sidney Hook and James Burnham regularly warned that communism was a dangerous, powerful enemy. On the whole, however, Moscow's ideological successes—in China, North Korea, and Eastern Europe—remained geographically distant. Now, however, a mechanical agent of the enemy was directly overhead, circling the globe. Warnings that Moscow aimed to control the entire world seemed to have found an alarming technological confirmation.

In the language of Thomas Kuhn's monograph *The Structure of Scientific Revolutions*, one might say that Sputnik brought relations between the United States and the Soviet Union into a "new paradigm." That would be anachronistic, however—*Structure* was published in 1962. And it would be inapt, because *Structure* and its central theoretical construct, the "paradigm," concern science and its history, not Cold War politics. Yet Kuhn's famous conception of science and its history was intimately shaped by the concerns and debates through which America grappled with Sputnik. First conceived in 1953, when Kuhn was invited to write for the logical empiricist *International Encyclopedia of Unified Science*, *The Structure of Scientific Revolutions* evolved considerably in the course of the 1950s, through the Sputnik crisis of 1957 (which Kuhn once addressed in a public lecture), and into the early 1960s, when *Structure* and its theory of paradigms took their familiar, published form. The changes are too extensive to be detailed here, but one central thread in the development of *Structure* runs through the national debates surrounding Sputnik and Kuhn's relationships and encounters with two prominent Cold Warriors, James Bryant Conant and Hiram Bentley Glass. Tracing these events shows that Kuhn's book, though often taken to be timeless in its influence and importance, is nonetheless deeply rooted in the Cold War, one of the most distinctive and curious eras of American history.

Kuhn, Conant, and Western Liberty

One reason that *Structure* matured slowly was Kuhn's complicated relationship with Harvard University's president, James Bryant Conant. In the late 1940s, Conant recruited Kuhn, then a physicist, to study the history of science and help teach Conant's new general-education science course, Natural Science 4. Conant (a chemist) taught Kuhn how to think about the history of science and how to craft case studies. But Kuhn broke away from his mentor in some important respects. Conant and other liberal champions of science were mistaken, Kuhn concluded, to insist that science and its history were embodiments of liberalism, open-mindedness, and intellectual creativity. Professional scientists, Kuhn believed, were in fact relatively dogmatic and closed-minded. *Structure* is, among many other things, Kuhn's solution to the difficult problem of how to reconcile this dogmatism with the liberal conception of science and its history that he inherited from Conant.[2]

Kuhn was a professional scholar who isolated his political opinions from his intellectual work, but his critical revision of Conant's historiography had enormous political implications from Conant's point of view. Conant, after all, wasn't only Kuhn's mentor; he was the president of a leading university in the West and a prolific public intellectual who consistently defended freedom and intellectual liberty at the heart of both science and democracy. Conant was as alarmed by Sputnik as anyone else in the United States, perhaps more so. From his experience in the Manhattan Project, Conant knew well that scientific and technological feats could have immediate and destabilizing geopolitical effects.

For Conant and other liberal anti-communists, however, the Sputnik crisis had a silver lining. It led the US to accept the educational message they had been urging ever since the wartime alliance with the Soviet Union had turned into a anxious stalemate between two incompatible ideologies. In popular books such as *Education in a Divided World* and in many lectures and magazine pieces, Conant urged his fellow Americans to realize that public education, properly reformed and financially supported, could prove to be the country's best weapon in this new ideological struggle.[3] Public education could make good on American boasts about freedom, equality of opportunity, and economic growth while countering Soviet claims that the US was socially stratified and economically unjust. Better-educated, critically minded

Americans would also be less susceptible to communism's false, utopian promises, and the science gap revealed by Sputnik would be reduced as well-funded schools recognized and cultivated prime talent within the country's enormous and varied population.

Kuhn's close working relationship with Conant at Harvard ended in 1953 when Conant abruptly resigned the presidency of the university to accept the post of High Commissioner of Germany.[4] He and Kuhn remained in contact—sometimes closely, as when Conant helped Kuhn with his first book, *The Copernican Revolution*. Conant offered substantive comments and recommendations concerning the manuscript, then wrote a foreword that praised the book for promoting scientific understanding. This was specifically important, Conant wrote, "in Europe west of the Iron Curtain," where the cultural and political importance of science's inherent liberalism, he felt, had yet to be fully understood and embraced, even by anti-communist humanists and intellectuals.[5]

As far as Conant and any reader could tell, Kuhn, in his first book, had largely adopted Conant's understanding of science as an essentially anti-dogmatic, progressive, dynamic human institution that was functionally and symbolically opposed to authoritarianism in both society and science. From Conant's 1947 book *On Understanding Science*, which Kuhn had read in proofs and knew well, Kuhn adopted Conant's picture of science as a succession of "conceptual schemes."[6] As Kuhn described them, conceptual schemes "guide a scientist into the unknown, telling him where to look and what he may expect to find and this is perhaps the single most important function of the conceptual schemes in science."[7]

Though *The Copernican Revolution* supported the image of scientists as curious, open-minded explorers of "the unknown," Kuhn had long before come to reject this image. What Kuhn called his "Aristotle experience" had been pivotal in this. It had occurred in the summer of 1947, while he was studying the history of physics at Conant's direction. Kuhn later recalled:

I was sitting at my desk with the text of Aristotle's Physics open in front of me. ... Looking up, I gazed abstractedly out the window of my room—the visual image is one I still retain. Suddenly the fragments in my head sorted themselves out in a new way, and fell into place together. My jaw dropped, for all at once Aristotle seemed a very good physicist indeed, but of a sort I'd never dreamed possible.[8]

The experience convinced Kuhn that Conant's account of science evolving freely and progressively from one conceptual scheme to another was misleading. It seemed to reveal to Kuhn that systems of ideas had a powerful mental grip on the scientific mind. As a result, the history of science was punctuated, with periods of stasis broken only when one kind of mental grip was suddenly and completely replaced by another (not unlike the sudden gestalt shift Kuhn experienced that summer day and the revolutions he would later articulate in *Structure*).⁹ This powerful grip prevented most scientists from even wanting to explore "the unknown" with new and different conceptual schemes.

From "Sociology" to "Ideology"

From its first notes and outlines, *Structure* had little use for the stereotype of the critical, open-minded, and intellectually autonomous scientist. One illustrative moment was Kuhn's encounter with the Harvard philosopher of science Philipp Frank. Frank's Institute for the Unity of Science was then the official sponsor of the *International Encyclopedia of Unified Science*. In the early 1950s, shortly before encyclopedia editor, Charles Morris, at the University of Chicago approached Kuhn about contributing a monograph on history of science, Kuhn received from Frank an invitation to join a new committee within Frank's institute to promote sociology of science. Kuhn accepted the invitation eagerly, but he was critical of the explanatory document, titled "Research Project in the Sociology of Science," that Frank had written and included with the invitation.[10] Frank's conception of sociology of science bore the marks of Mannheimian sociology of knowledge, particularly the difference between logical and evidential reasons for beliefs and the larger sociological or "existential" factors that tend to interfere with logic and evidence or break ties when evidence supports different theories equally.[11] While typing his response, Kuhn respectfully but firmly suggested that Frank had misidentified the kinds of existential factors most relevant to science. The sociological forces affecting science, he argued, didn't come from outside scientific communities in the forms of political or religious pressures. They originated and operated *inside* those communities in ways that led Kuhn to ask "Would it not be appropriate to include in the committee's terms of reference an examination of those sociological factors which impinge upon an individual scientist not by virtue of his membership in a national community (say the United States), but by virtue of his membership in a narrower professional group (say the American Physical Society)?"[12] These sociological factors inside professional groups, Kuhn explained, didn't just surround and support scientific knowledge; they shaped and guided it fundamentally. They determined "the sort of problems which a scientist considers worth attacking, the sort of experiment which he employs to resolve his problem, the abstract aspects of the experiment which he considers relevant to the solution, and the logical and experimental standards which he demands as 'proof.'"[13] Kuhn agreed with Frank that logic and evidence operated within specific contexts that were framed and maintained by "existential" and sociological factors. But science had its own factors, and they were essential for understanding modern science.

Despite Kuhn's detailed response to Frank's invitation, it appears that nothing official came of it. At the top of his letter, Kuhn wrote "not sent." There is no further evidence that Kuhn participated officially in Frank's institute, which was approaching the end of its funding and its professional influence.[14] Still, the encounter appears to have been fruitful. Six months later, when Kuhn wrote to Charles Morris to finalize the title of his future monograph, he informed Morris that he planned to take a new approach. Morris had originally commissioned an essay in history of science, but Kuhn now envisioned a work in sociology. The monograph would still make use of historical case studies, he reassured Morris. But, Kuhn wrote, "my basic problem is sociological."

Kuhn's "basic problem" was how to reconcile the historical fact of total and revolutionary scientific change with the conservative, internal sociological dynamics that tended to fix practices, methods, and logical and experimental standards. Somehow, Kuhn believed, these contradictory features of science meshed and supported one another, but he didn't know how. There can be no doubt, however, that Kuhn was thinking about the problem in terms of social and political revolutions. Besides using the word 'revolutions' in his title, he invoked other political terms as he attempted to build a consistent theory of scientific change. For example, the "existential" and "sociological" factors that he earlier described to Frank now became "ideologies" and "ideological" factors that played similar, constitutive roles in creating and sustaining scientific knowledge. As an ideology, he explained to Morris,

a theory serves to direct the scientist's attention to certain sorts of problems as 'useful' and to certain sorts of measurements as 'important;' it dictates preferred techniques of interpretation, and it sets standards of precision in experiment and of rigor in reasoning. Above all, the theory, as ideology, is a source simultaneously of essential direction and of disasterous [sic] inhibition of the creative imagination.[15]

Once individual scientists were understood to work within these sociological or ideological dynamics, it became natural to characterize scientists as dogmatic—though Kuhn didn't use that word. Their "creative imagination" was inhibited and, Kuhn added, their range of scientific vision was reduced:

One of the most striking results of the 'ideological' portions of a professionally institutionalized theory is the relatively firm closure which it gives the field of scientific problems. Theories preserve themselves by restricting the attention of the profession to problems which can in principle be solved within the theory and by inhibiting the recognition of important incongruities in the application of the theory to nature. In some sense every theoretical orientation excludes the existence of totally unsolved problems.[16]

The typical scientists about whom Kuhn planned to write in his forthcoming monograph were nothing short of conceptually and intellectually blinkered. In the popular language of the day, they had been "brainwashed" by the powerful internal ideologies that controlled and sustained professional science. Clearly they were not intellectually "free" and unencumbered.

Sputnik and the American Public Mind

These early snapshots of *Structure* demonstrate how Kuhn's emerging conception of science had departed from Conant's liberal politics. Shortly after Kuhn's discussions with Frank and Morris, Conant would take up his new position in Germany. This is perhaps one reason why Conant and Kuhn never directly addressed or debated this disagreement at the time.[17] Another reason appears to be that Kuhn chose not to explicitly defend his new image of scientists ("normal" scientists, he would eventually call them) as dogmatic and closed-minded until he had solved his "basic problem" and discovered exactly how that dogmatism was consistent with, perhaps even theoretically required by, the historical advancement of science.

As Kuhn worked with Conant to complete *The Copernican Revolution* in the mid 1950s, the popular image of the scientist as the explorer of "the unknown" took precedent. When Sputnik made headlines in 1957, Kuhn seemed unsure of whether and how firmly to acknowledge the dogmatism he recognized in professional science. Then, at the University of California at Berkeley, Kuhn gave a talk addressing the Sputnik crisis titled "Sputnik & American Public Mind." The surviving notes suggest that it was a public evening lecture attended by a general audience eager to hear what a historian of science had to say about the sensational Soviet achievement and what it meant for American science and prestige.

Kuhn didn't sugar-coat the bad news. Sputnik had indeed punctured a "myth" of America's scientific and technological superiority over the Soviet Union and led the US to an "agonizing reappraisal" of its scientific prestige. Still, it was healthy for the US to face historical facts. "At least until about one generation ago," Kuhn explained, the US had been a laggard in "pure" or "basic science" relative to England and Europe. This was because American culture—the "American public mind"—didn't sufficiently value the kinds of formal, abstract thinking that enabled scientific progress. If the US wanted to best the USSR in science (as opposed to practical inventions, like the cotton gin or the light bulb, which had long been its forte), it would have to cultivate a stronger appreciation for "things of the mind."

Kuhn's diagnosis of the situation was awkward. Man-made satellites, he must have known, were technological achievements, not basic scientific achievements. They were much more like cotton gins or light bulbs than imagination-challenging advances in gravitation theory or nuclear physics. So it is difficult to see how a new fondness for abstraction in American culture might help close the satellite gap. Kuhn's nod to the politics of the satellite gap was awkward too. The US was soon to embrace the liberal consensus, defended by Conant and others, that scientific progress rested on intellectual freedom and open-mindedness. Within a year after Sputnik, for example, Congress passed the National Defense Education Act to fund education reforms, including the Biological Sciences Curriculum Study (BSCS) for biology and the Physical

Sciences Study Committee (PSSC) for physics. These reforms emphasized an understanding of scientific method as a tool for exploration and downplayed the importance of facts and their memorization. In this regard, they exalted science as free, self-critical, and creative exploration of 'the unknown'—as an institution that, once reformed, should thrive in the politically and intellectually free cultures of the West.

In his talk, Kuhn nodded approvingly at this consensus. But if one reads his lecture notes in the light of his unsolved "basic problem" about how to reconcile science's dogmatism with its historical dynamism, he seems to have been skeptical:

> Though there is little evidence on the point and that inconclusive, there is some reason to hope that the freedom from intellectual restraint implicit in at least the theory of democratic society may provide more fertile climate for basic science than autocracy.[18]

There was enough to be skeptical about: If the United States could close the satellite gap by squeezing better science out of its political and intellectual freedoms, then how had the Soviet Union, widely assumed to have little intellectual freedom under the dictates of dialectical materialism, managed to get ahead in space science? As for the evidence being "inconclusive," it was. Soviet science under dialectical materialist orthodoxy had a mixed record. The Sputnik satellites were successes, but Lysenko's biology was a tragic failure. Though Kuhn could join his audience in the patriotic "hope" that America's freedoms would somehow close the technology gap, he couldn't take a more definite, principled position until he had unraveled precisely how the conservative dogmatism of scientists functioned in the larger enterprise of science.

From Ideology to Paradigms

Kuhn's breakthrough came in 1960 when he reconceived the nature of the ideological conformity that held productive scientific communities together. It was not, after all, a conformity that produced visible, overt signs of agreement and consensus about propositions or definitions. Years later, explaining why it had taken him so long to write *Structure*, he recalled: "At the time I conceived of normal science as the result of a consensus among the members of a scientific community. Difficulties arose, however, when I tried to specify that consensus by enumerating the elements about which the members of a given community supposedly agreed."[19] The difficulty was that "in order to account for the way they did research and, especially, for the unanimity with which they ordinarily evaluated the research done by others, I had to attribute to them agreement about the defining characteristics of such quasi-theoretical terms as 'force' and 'mass', or 'mixture' and 'compound'. But experience, both as a scientist and as a historian, suggested that such definitions were seldom taught and that occasional attempts to produce them evoked pronounced disagreement. Apparently, the consensus I had been seeking did not exist, but I could find no way to write the chapter on normal science without it."[20]

The theory of paradigms allowed Kuhn to locate the source of this "unanimity" in a new place, far from the day-to-day sociology of scientific communities that exhibited less consensual agreement than Kuhn had expected. First, the theory pushed that source of agreement back in time to the training that preceded professional scientific life. Second, it found the shared ideology not in the form of propositions or specifiable beliefs but in practices—in the problem-solutions that all members of a community learned, repeated, and came to accept as second nature during their training. The point is clearest in the chapter of *Structure* titled "The Priority of Paradigms," which argues that paradigms come first in a scientist's training—both chronologically and (to borrow Frank's terminology) with an "existential" priority embedded in the rigors and conventions of scientific education.

Kuhn had solved his "basic problem": paradigms provided the basic sociological and ideological uniformity that professional science required, and conferred on working scientists a profound sense of confidence and dogmatism about current scientific beliefs. Yet, because paradigms were ultimately practices, not propositions or definitions or doctrines, scientists could and did disagree about substantive scientific matters, especially when asked to translate their tacit, paradigmatic commitments into explicit propositions, criteria, and definitions. This potential lack of consensus was, then, the seed from which scientific revolutions grew. When anomalies threatened the seeming truth and puzzle-solving power of a community's paradigm, these differences became paramount. Scientists in the same community would respond to the resulting crisis differently. At least one scientist, Kuhn's model of science held, would break through the reigning dogmatism surrounding the old paradigm and lead the rest to a revolutionary new paradigm.

"The Function of Dogma in Scientific Research"

With the theory of paradigms in hand, Kuhn stopped using the word 'ideology' as he had been using it up to that point in his notes and drafts. But he still believed that successful, professional scientists were dogmatic—and now, with *Structure* drafted and bringing Kuhn words of praise from several colleagues, Kuhn seemed to become more confident that dogmatism was an essential, important feature of science.[21] He then put dogmatism front and center. In "The Function of Dogma in Scientific Research," a paper he presented in 1961 at a conference in Oxford, he told his audience that the paper was "abstracted, in a drastically condensed form, from the first third of my forthcoming monograph, *The Structure of Scientific Revolutions*."

Like *Structure*, "The Function of Dogma" begins with a provocation, this one aimed directly at the reigning "image of the scientist as the uncommitted searcher after truth." Supposedly, the scientist is "the explorer of nature—the man who rejects prejudice at the threshold of his laboratory, who collects and examines the bare and objective facts, and whose allegiance is to such facts and them alone. ... To be scientific is, among other things, to be objective and open-minded."[22] The truth, Kuhn explained, is precisely the opposite:

> Though the scientific enterprise may be open-minded, the individual scientist is very often not. Whether his work is predominantly theoretical or experimental, he usually seems to know, before his research project is well under way, all but the most intimate details of the result which that project will achieve. If the result is quickly forthcoming, well and good. If not, he will struggle with his apparatus and with his equations until, if at all possible, they will yield results which conform to the sort of pattern which he has foreseen from the start.[23]

The source of this dogmatic confidence lay in paradigms, the "concrete problem-solutions" that science students learned and repeated through drills and that "the profession has come to accept as paradigms." The "unknown" that Kuhn and Conant celebrated in *The Copernican Revolution* had thus given way to confidence and dogmatism about the known.

Paradigms cultivated this dogmatism in three ways. First, paradigms are exclusive. A scientific community, "if it has a paradigm at all, can have only one." Second, paradigms are taken to ground true and unchanging representations of nature. "In receiving a paradigm," that is, "the scientific community commits itself, consciously or not, to the view that the fundamental problems there resolved have, in fact, been solved once and for all." Third, paradigms are the practical basis of most scientists' careers. "Given a paradigm," Kuhn wrote, scientists "strive with all their might and skill to bring it into closer and closer agreement with nature." During crises and revolutions, therefore, many scientists become defensive. "What they are defending," Kuhn wrote, "is, after all, neither more nor less than the basis of their professional way of life."[24]

What was the "function" of all this dogmatism? Though "a source of resistance and controversy," Kuhn explained, this dogmatism is also "instrumental in making the sciences the most consistently revolutionary of all human activities."[25] Dogmatism helps to ensure that scientists stay focused and that paradigm shifts aren't pursued capriciously. It helps to distinguish merely vexing puzzles, which will eventually be solved by dogmatic and determined scientists, from crisis-inducing anomalies that cannot be accommodated by the reigning paradigm and can lead only to a scientific revolution.

Conant, Glass, and the Anti-Communist Scientific Consensus

"The Function of Dogma" didn't go over very well at the conference.[26] The historian of science A. Rupert Hall worried that Kuhn's use of 'dogma' was, in effect, "an apology for weakness" on the part of insufficiently creative, free-thinking scientists.[27] The philosopher Stephen Toulmin argued that Kuhn had failed to distinguish the necessary use of prevailing ideas to "frame the questions" we put to nature from the need "to leave nature to answer [these] questions for herself, without prompting." It was true, he countered, that a scientist "relies upon preconceived ideas; but in a sense which emphatically does not involve any suggestion of dogmatism."[28]

Kuhn stood his ground against his British and European critics. The only commentator who persuaded him to moderate his claims about dogmatism was the American geneticist Hiram Bentley Glass. Like Conant, Glass was a natural scientist who had become a university administrator, a public intellectual, a historian of science, and an influential education reformer. From the late 1930s to the early 1960s, his writings in the *Bulletin of the Atomic Scientists* and his regular column in the *Baltimore Evening Sun* made Glass nearly as visible to the American public as Conant. Most important, with respect to Kuhn's claims about dogmatism in science, both Glass and Conant grappled publicly with the issue of intellectual freedom that had been raised by the controversies over communist teachers in the late 1940s and the early 1950s. Though they entered those debates from different institutional sides, Conant as president of Harvard and Glass as an officer of the American Association of University Professors, both endorsed the national consensus that intellectual freedom was essential both to science and to a healthy, democratic society. While Conant championed the liberal bona fides of Kuhn's *The Copernican Revolution* for their political relevance to the cultural Cold War in Europe, Glass used his position as editor of the *Quarterly Review of Biology* (from 1949 to 1965) to condemn Lysenkoism as an example of ideological interference in science.[29]

The liberalism that Conant and Glass shared was manifested in their historiography of science. In his book *On Understanding Science,* Conant defined science according to its propensity to change:

Science emerges from the other progressive activities of man to the extent that new concepts arise from experiments and observations, and the new concepts in turn lead to further experiments and observations. ... This dynamic quality of science viewed not as a practical undertaking but as development of conceptual schemes seems to me to be close to the heart of the best definition.[30]

"Almost by definition," he added, "science moves ahead."[31]

Conspicuously, Conant avoided using the word 'truth' in his definition, both because of his instrumentalist philosophical inclinations and because of the term's close associations with dogmatism and authoritarianism. Glass also avoided using the word in his conference paper as he surveyed the history of genetics from Bonnet and Maupertuis to emphasize the progressive, dialectical power of scientific ideas:

The modern view of the relation of heredity to development is not Bonnet's, but neither is it Maupertuis's. It has something of both, and something of neither. The two views, once held to be irreconcilable, have merged in a higher synthesis. As for our current views of heredity and of species, much the same may be said. [32]

Glass' aversion to the word 'truth' was evident in the conclusion of his survey: "It is in the dedication to a conceptual model which may seem to hold true, but cannot in fact describe nature in its fullness, that we find both the highest stimulus to current scientific investigation and the greatest barrier to ultimate knowledge."[33]

At the time of the conference, Glass was in his second year as a director of the Biological Sciences Curriculum Study, based at the University of Colorado at Boulder. His mission was to reform high school biology education in the wake of Sputnik and the National Defense Education Act. Though Glass admired some aspects of Kuhn's account (his own survey of genetics, advancing through different theoretical programs, he admitted, might well be characterized as something like a series of "paradigm shifts"), he couldn't accept Kuhn's descriptions of scientists, or science itself, as dogmatic. The adjective simply didn't fit the scientists Glass knew, scientists who "rather often discuss the validity of their basic assumptions." "I think more often today than when I was a younger scientist," he explained. Nor did it fit the increasing rate of scientific progress that Glass observed:

Within my own working life as a geneticist I have already seen two very fundamental overturns of prevailing conceptual models, or to use Mr. Kuhn's term, paradigms. The young scientist of today, therefore, must be trained to expect relatively frequent overturns of his basic ideas within his own field.[34]

On these matters of scientific education and training, Glass disagreed even more strongly—"sharply," he emphasized—with Kuhn's account. As a public intellectual who stood firmly behind the national consensus that science and intellectual freedom were inseparable, he could only have bridled at Kuhn's descriptions of students' undergoing "a relatively dogmatic initiation into a pre-established problem-solving tradition that the student is neither invited nor equipped to evaluate," an initiation that resulted in a "mindset," a "deep commitment," and "professional rigidity."[35] As "chairman of one of the science curriculum studies into which the National Science Foundation of the United States has poured some ten millions of dollars," Glass found this unacceptable. He spoke confidently "for my fellow biologists of this generation in America":

I have found complete unanimity among them in the belief that science must be taught—I do not say *has* been taught—as a variety of methods of investigation and inquiry rather than as a body of authoritative facts and principles. They also agree emphatically that students must be taught that scientific laws and principles are approximations derived from the data of experience and that they remain forever subject to alteration and correction or replacement in light of new evidence. I am appalled to think that, if Mr. Kuhn is right, we should go back to teaching paradigms and dogmas, not merely as temporary expedients to aid us more clearly to visualize the nature of our scientific problems, but rather as part of the regular, approved method of scientific advance.[36]

Of course, Kuhn hadn't made educational recommendations in "The Function of Dogma." But Glass clearly detected the unacceptable political implications for post-Sputnik education reforms in what Kuhn was saying. "If Mr. Kuhn is right," he argued, the liberal consensus enshrined within these reforms could well be wrong and possibly counterproductive. If the recent Soviet successes in space science had been the fruits of a relatively "autocratic" approach to education and planning (a possibility Kuhn hadn't ruled out in his earlier talk on Sputnik), the BSCS-style reforms championed by Glass would predictably fail to help close the technology gap with the Russians. It isn't surprising that Glass was "appalled."

In response, Kuhn reassured Glass that he was sympathetic to his reforms. "[T]he system [of education] they aim to change," he agreed, "is often no more than a parody of what scientific education should be." But Kuhn didn't back down. He wished Glass success, but predicted "that it may not be possible to carry the reform so far as he [Glass] would wish." However students were trained in science, and however frequently scientific revolutions occurred, Kuhn believed that paradigms and their dogmatic hold on the scientific mind were simply a feature of modern science: "I doubt that science will get on without them."[37]

But Kuhn conceded that he would cease to use the terms 'dogma' and 'dogmatism'. Glass himself had pointed the way for Kuhn at the beginning of his remarks, admitted that 'paradigm' was "a new word for me." Glass asked himself what, exactly, was the relationship between dogmas and paradigms, then urged Kuhn to accept his answer:

I now recognize that the paradigm looks backwards while moving forwards, whereas the dogma, a related creature with which I am more familiar, also looks backwards but stands its ground.[38]

In effect, Glass proposed a rapprochement between Conant and Kuhn. Whereas Conant had insisted that the essential quality of science was that it "moves ahead," Glass suggested that Kuhn's paradigms should be understood to have a similar motion, ever "moving forwards."

Kuhn embraced Glass' distinction between dogmas and paradigms, granting that it "both fits and furthers the purpose of my enquiry." He would "surrender ['dogma'] in favour of something like 'commitment to a paradigm'."[39] Understood as an essential, functional component of "normal science," 'dogma' and 'dogmatism' never appeared again in Kuhn's writings. As for the paper "The Function of Dogma in Scientific Research," after its original publication in the conference proceedings Kuhn forbade its inclusion in either of the two collections of his published essays.[40]

On the Invisibility of 'Dogma'

In fact, Kuhn's concession to Glass was superficial. Though he presented "The Function of Dogma" as a partial introduction to *The Structure of Scientific Revolutions*, *Structure* itself doesn't rely on the word 'dogma' or its variants in the same way. Kuhn had, in fact, already used phrases like "commitment to a paradigm" throughout the manuscript that he would soon finish writing and send to the University of Chicago Press. In the published version of *Structure*, one finds 'commitment' (on pages 11, 24, 25, 40), 'rigid' (on pages 19, 49, 166), "accepts without question" on page 47, "relatively inflexible [theoretical] box" on page 24, "take for granted" on page 19, 'assurance' on page 151, and "confidence in their paradigms" on page 165.[41] But one doesn't find 'dogma' or variants thereof, despite the fact that the crypto-dogmatism manifest in Kuhn's discussions plays the same pre-revolutionary function as the dogmatism described in "The Function of Dogma."[42]

Obviously only Kuhn himself knew why he used 'dogma' in this way in his presentation at the Oxford conference but not in the book that presentation was written to introduce. Had it occurred to him to use 'dogma' only after he had finished drafting *Structure*, the encounter with Glass may have dissuaded him from introducing the term in the final draft. Another possibility is suggested by Kuhn's careful treatment of the issues of scientific freedom and autocracy in his lecture on Sputnik. If 'dogma' was his first terminological choice for unveiling and articulating his new theory of science, he may have reasoned that a predominantly British audience, removed from the ongoing American preoccupation with dogmatism, authoritarianism, and other varieties of mental and ideological bondage, would be less distracted by the unsavory political connotations of the word; his American publisher and readers, on the other hand, would be better served by the array of euphemisms used in *Structure*. If so, Kuhn was exactly right, for it was the American Glass who recoiled most "sharply" at the Oxford conference.[43]

Whatever Kuhn's rationale, the uniqueness of "The Function of Dogma" and its subsequent obscurity relative to his later works should not lead us to conclude that his new image of science made only short-lived and incidental contact with the political and cultural preoccupations of Cold War America in 1961.[44] These selected episodes in the development of *Structure* indicate that matters of politics, sociology, and the geopolitical implications of how science was understood framed the intellectual matrix into which Conant invited Kuhn as he equipped himself for a career as a historian of science. Though some elements of this matrix—for example, the metaphor of political revolution and the belief (related to the national preoccupation with "mind-control" and "brainwashing") that ideas themselves can "grip" and control the human mind[45]—have long been visible to readers of *Structure*, they have not been recognized as Cold War markers. That recognition has been impeded not only by Kuhn's replacement of 'ideologies" by 'paradigms' circa 1960 and his retraction of the word 'dogma' in 1961, but also by the postwar style of intellectual professionalism that Kuhn's theory of paradigms helped to inaugurate.

As he put it in his unsent letter to Philipp Frank, Kuhn believed that large-scale sociological or "existential" pressures stemming from national or religious traditions no longer affected professional scientists in interesting or important ways. They may have in earlier centuries, Kuhn acknowledged, but "at this time and place" such external factors as "government, church, etc." seemed to have "relatively little impact upon decisions made by professional scientists about problems arising within their own sciences."[46] From this point of view, what divided Kuhn and Frank on this question was not only theoretical details of sociology but the ongoing professionalization and depoliticization of philosophy of science during the Cold War. By inviting Kuhn to join his new committee to promote sociology of science, Frank was reaching across an emerging professional divide between old-school logical empiricists (like himself and his co-editor, Charles Morris) who promoted philosophy of science as a broad unifying framework for modern life and younger scholars who, having witnessed Cold War persecutions of communist faculty members and the institution of loyalty oaths, adopted a professional posture that more firmly separated matters of scholarship from those of society and culture.[47]

Conant Reads *Structure*

In regard to this new professional, intellectual style, Conant was a member of the old guard, alongside Frank, Morris, and Glass: all of them believed that ideas about science and its history were importantly connected to matters of society and national welfare. Whereas Glass confronted Kuhn at the Oxford conference, Conant's engagement with Kuhn on these matters was more extended and complex. By the time *Structure* appeared, Conant and Kuhn had been colleagues for about fifteen years. One suspects that Conant, having supported and respected Kuhn for so long, was prepared to like *Structure* very much.

But when Kuhn sent Conant a copy of the manuscript in the spring of 1961 (only several weeks before the Oxford conference), Conant was plainly disappointed. Though he applauded the historicist, anti-positivist picture of science ("on the whole," he wrote, "I am very sympathetic to your unorthodox interpretation of science"), he urgently pointed out some inconsistencies and historiographic omissions, such as Kuhn's failure to address the role of "the practical arts" in the history of scientific change. The most pressing problem, however, was Kuhn's new theory of paradigms. The manuscript relied on 'paradigm' so much, Conant feared that Kuhn might come to be perceived and dismissed as "the man who grabbed onto the word 'paradigm' and used it as a magic verbal word to explain everything." "To my mind," Conant wrote, "the page on which you sum up your point without recourse to the word 'paradigm' is the clearest page in the whole document." Conant's detailed three-and-a-half-page critique leaves little doubt that he saw Kuhn's new theory of paradigms as unnecessary theoretical baggage that, ironically, obscured the dynamics of scientific revolutions. Kuhn's evident infatuation with 'paradigm' (a word "you seem to have fallen in love with!") had harmed *Structure*'s accuracy and credibility. ("A 'new world view' is implied by your treatment of *all* scientific revolutions but I query if this is not far too grandiose a characterization of most of the revolutions you treat as examples.") Though the manuscript didn't rely on 'dogma', Conant couldn't have missed the ways in which Kuhn's theory of paradigms clashed with his own liberalism. Like Glass, Conant rejected the image of intellectual conformity that he saw in the manuscript: "You tend to treat the scientific community far too much as a community with a single point of view."[48]

In the wake of Conant's criticisms of *Structure*, his objections to the dogmatic portrayal of normal scientists moved into the background of his relationship with Kuhn. Though it would be incorrect, one might even conclude that Conant changed his mind about all of his objections, even to the overuse of the word 'paradigm'. Weeks later, after Kuhn confessed that he was extremely disappointed by Conant's initial reaction, Conant softened his objections and reassured Kuhn that they concerned mainly "a matter of style and presentation."[49] Much relieved, Kuhn proposed dedicating the book to Conant.[50] A year and a half later, Conant found the printed book in his mailbox, read it, and seemed to like it even more:

I have just finished reading it and congratulate you most fervently. Needless to say I am grateful to you for the dedication and more than proud to have my name associated in this way with what is a truly important book.

You have not only presented a challenging and unorthodox interpretation of scientific history, but you have documented what you have written in a most impressive manner. ... Quite apart from the impact of your novel ideas, the setting forth of the scientific revolutions as you have is going to help many readers to understand science better.[51]

These congratulations weren't merely private and collegial. Conant was soon to write a new book of his own book in which, he told Kuhn, he would help promote *Structure*: "I refer to your book and I hope if nothing else this little volume may stimulate people to read yours."[52]

Had Conant himself adopted the new professionalism and decided to downplay or dismiss the unhealthy political implications he must have seen in *Structure*? Probably not—in *Two Modes of Thought* (a career-summarizing reflection on intellectual freedom and progress in science, law, business, and general education), Conant addressed the issues of dogmatism and intellectual uniformity in communities:

A free society requires today among its teachers, professors, and practitioners two types of individuals: the one prefers the empirical-inductive method of inquiry; the other the theoretical-deductive outlook. Both modes of thought have their dangers; both have their advantages. ... Above all, the continuation of intellectual freedom requires a tolerance of the activities of the proponents of the one mode by the other.[53]

The pluralistic title of Conant's book (*Two Modes of Thought: My Encounters with Science and Education*) rejects the claims in "The Function of Dogma" and *The Structure of Scientific Revolutions* that there is only *one* mode or method of scientific advance in modern science:

The more one studies the steps by which rapid advances have been made in the natural sciences, the more difficult it is to describe the ways in which wide generalizations and new concepts have originated. The one thing that does seem certain is that one must speak of the *ways*, for there is no single way. This is the reason why it is worse than nonsense to speak of *the* scientific method.[54]

In this respect, *Two Modes of Thought* takes up Conant's private criticism of *Structure*'s "far too grandiose" picture of revolutions. "By leaving out any reference to technology and advances in the practical arts (*including the practical art of experimentation and observation*)," Conant had written to Kuhn, "you distort the picture of science and get yourself into needless trouble about progress." In *Two Modes of Thought*, Conant offered a more balanced picture, insisting that any adequate history of "astronomy, physics, chemistry, or biology" must be ecumenical, for "the natural sciences as they stand today are the result of the careful use of the empirical-inductive method of inquiry together with the imaginative use of the theoretical-deductive." As if to counterbalance *Structure*'s one-sided, theoretical-deductive account, Conant devoted pages to "empirical-inductive" inquiry and practical arts, including ironmaking and a "modern example of the empirical-inductive approach," the trial-and-error invention of antiknock gasoline. As if to counterbalance *Structure*'s picture of dogmatic, "normal" science, Conant again presented his trademark definition of science as "an interconnected series of concepts and conceptual schemes that have developed as a result of experimentation and observation." This series, Conant asserted, led not to truth and finality—and certainly not to dogmatism—but only to "further experimentation and observation."

Though Conant called *Structure* "brilliant," his praise was specifically directed to this point about scientific truth. Conant recommended *Structure* to those who "wish to equate science with a quest for the ultimate structure of the universe." "I agree with Kuhn's conclusion in the final chapter of his brilliant book," he wrote, "that it is erroneous to measure scientific achievement by the extent to which we are brought closer to some ultimate goal."[55] The word 'paradigm' and other essential features of *Structure* do not appear in *Two Modes of Thought*.

Two Modes of History of Science

Conant also elaborated the geopolitical implications of the conceptual and methodological pluralism he endorsed—a pluralism that, he asserted, remained a source of the West's strength in its resistance to international communism and also helped to explain the origins of that nemesis: Bolshevism had rejected pluralism and embraced the dogma of dialectical materialism as a supremely powerful theory and "*Weltanschauung*" from which all legitimate answers could be deduced. As a result, the Soviet Union was doubly dogmatic, wedded to the "deductive theoretical" mode of thought and to one overarching theory—dialectical materialism. "I need not point out," Conant wrote, "the practical consequences of living in a totalitarian state with an *all-embracing* official dogma."[56] Yet he dwelled anyway on the dangers to science and society of intellectual conformity and the perils faced by "dissenters" who refused to conform:

> I am inclined to think social, political and intellectual regimentation is a necessary consequence of the belief that there is only one set of premises from which one may proceed to deduce conclusions about the way human beings behave and ought to behave. ... Even a superficial study of the books and magazine articles that have been published in the Soviet Zone and East Berlin in the last twenty years highlights the dangers inherent in a complete devotion to the theoretical-deductive mode of thought. If it assumed that the theoretical fabric is complete and all human problems can be solved by logical reference to this fabric, these dissenters are worse than heretics; they are unscientific![57]

Read in light of Conant's private complaint that *Structure* described scientific communities as having only "a single point of view" and the fact that Conant was soon to write *Two Modes of Thought* when he first read *Structure*, his argument appears to be a collegial but firm rebuttal of Kuhn's "new image" of science.

The route to scientific revolutions, Kuhn insisted, necessarily passed through the dogmatism of "normal science." No, Conant replied in *Two Modes of Thought*; scientific progress and creativity are fostered only by the interaction and conflict between different "modes of thought." A scientific paradigm, Kuhn explained, is tightly connected to a specific, all-embracing "world view." No, Conant objected; world views don't illuminate the workings of modern science. "Indeed," he emphasized, "the word *Weltanschauung* has come to epitomize the outlook of those who demand an all-embracing answer to the deep problems of human life and the nature of the cosmos."[58] A scientific dissenter, Kuhn explained, "who continues to resist after his whole profession has been converted" to a new paradigm, "has ipso facto ceased to be a scientist."[59] If that was so, Conant implied, modern professional science must be a totalitarian institution, not a free, liberal, tolerant one.

Conclusion

"Unless he has personally experienced a revolution in his own lifetime," Kuhn wrote in his chapter "The Invisibility of Revolutions," "the historical sense either of the working scientist or of the lay reader of the textbook literature extends only to the outcome of the most recent revolutions in the field." That historical sense, Kuhn added, is linear and progressive, "leading in a straight line to the discipline's present vantage."[60] On that view, the history of science studies after Kuhn runs in straight lines primarily from one scholarly text to another. Kuhn himself told us that *Structure* could revolutionize our understanding of science by critically responding to Karl Popper, to logical empiricism, and to presentistic "whig" historians of science.

As Glass' and Conant's reactions to *Structure* help show, however, *Structure* emerged not only in dialogue with other theoretical projects but also from within in a matrix of Cold War concerns—about the epic confrontation of liberalism and totalitarianism, about the nature and power of ideology, and the susceptibility of the open, liberal mind to dogmatism—all of which motivated and informed the historiography of science that Kuhn first learned under Conant's direction. Although Kuhn rejected specific elements of that matrix, such as the image of the open-minded scientist-explorer, and although his theoretical vocabulary of 'ideology' and 'dogma' fell by the wayside, *Structure* remains nonetheless shaped by these Cold War realities.

In *Two Modes of Thought*, Conant himself seemed to recognize that *Structure* was born of intellectual as well as political threads. He accepted *Structure*'s aversion to "truth," but not its historiography and the resulting picture of single-minded, dogmatic scientific communities. In Kuhn's words, that is, Conant experienced the Kuhnian "revolution in his lifetime," and his published remarks drew attention to these complexities and to the vital issues at stake. In only a few years, however, these complexities would begin to disappear. Kuhn's paradigms would eclipse Conant's conceptual schemes. New generations of scholars would treat *Structure* as a strictly historical and philosophical treatise, these several strands would be pulled taught into a single, unified line, and the revolution would be complete.

Notes

1. William J. Jorden, "Soviet Fires Earth Satellite into Space," *New York Times*, October 5, 1957; John W. Finney, "US Missile Experts Shaken by Sputnik," *New York Times*, October 13, 1957.

2. A controversial account of *Structure* by Steve Fuller holds that it was inspired by and faithful to the postwar agendas for science funding of James Conant and Vannevar Bush. Although there is much to support Fuller's original contention that *Structure* is an "exemplary document of the Cold War era," the intellectual and political relationship between Kuhn and Conant is more complex. The matter of dogmatism, explored here, is one area where Kuhn broke with Conant

and didn't, as Fuller writes, "simply [take] Conant's politics of science as uncontroversial—indeed, as a taken-for-granted worldview." See Steve Fuller, *Thomas Kuhn: A Philosophical History for Our Times* (University of Chicago Press, 2000), 5–6. Other issues on which *Structure* differed with Conant's views, briefly touched on here, include the nature of scientific progress and relationships between theoretical science and the practical arts and crafts.

3. See, for example, James B. Conant, "Wanted: American Radicals," *Atlantic Monthly*, May 1943: 41–45; Conant, "Science and the National Welfare," *Journal of Higher Education* 15, no. 8 (1944): 399–406, at 454; Conant, *Education in a Divided World: The Function of the Public Schools in our Unique Society* (Greenwood, 1948); Conant, *Modern Science and Modern Man* (Doubleday, 1952); Conant, *The Citadel of Learning* (Yale University Press, 1956). A helpful account of post-Sputnik education reforms involving H. Bentley Glass is John Rudolph, *Scientists in the Classroom* (Palgrave Macmillan, 2002).

4. For this and other episodes in Conant's career, see James Hershberg, *James B. Conant: Harvard to Hiroshima and the Making of the Nuclear Age* (Knopf, 1993).

5. James. B. Conant, foreword to Thomas Kuhn, *The Copernican Revolution* (Harvard University Press, 1957), xiii. Kuhn acknowledges Conant's assistance in Thomas Kuhn to James Conant, 29 June 1961, Box 25, Folder 53, Thomas S. Kuhn Papers, MIT.

6. Thomas Kuhn, *The Road Since Structure*, ed. James Conant and John Haugeland (University of Chicago Press, 2000), 275; James Conant, *On Understanding Science* (Yale University Press, 1947).

7. Kuhn, *Copernican Revolution*, 40.

8. Kuhn, *The Road Since Structure*, 16. Kuhn also discusses and dates his Aristotle experience on pages xi and xii of *The Essential Tension* (University of Chicago Press, 1977).

9. When asked in the 1990s about the relationship between *The Copernican Revolution* and *Structure*, Kuhn explained that *Structure*, and not his first book, was an effort to work out the implications of the Aristotle experience: "Oh, look, I had wanted to write *The Structure of Scientific Revolutions* ever since the Aristotle experience. That's why I had gotten into history of science—I didn't know quite what it was going to look like, but I knew the noncumulativeness; and I knew about what I took revolutions to be ... but that was what I really wanted to be doing." (*The Road Since Structure*, 292)

10. Frank to Kuhn, 2 December, 1952; Philipp Frank, "Research Project in the Sociology of Science, undated; Box 25, Folder 53, Thomas S. Kuhn Papers, MIT.

11. See Karl Mannheim, *Ideology and Utopia: An Introduction to the Sociology of Knowledge* (Harcourt Brace, 1954; originally published in 1929 by Friedrich Cohen as *Ideologie und Utopie*).

12. Kuhn, "Dear Professor Frank ... ," undated, Box 25, Folder 53, Thomas S. Kuhn Papers, MIT.

13. Ibid.

14. On the history of Frank's Institute, see George Reisch, *How the Cold War Transformed Philosophy of Science: To the Icy Slopes of Logic* (Cambridge University Press, 2005), chapter 15.

15. Kuhn to Morris, 31 July 1953, Box 25, Folder 53, Thomas S. Kuhn Papers, MIT.

16. Ibid.

17. However, when Conant first read *Structure* in manuscript form, in 1961, this disagreement did manifest itself, and it figured in Conant's strong reservations about the book's implications for understanding scientific progress. These matters are treated below, and briefly in Hershberg, *James B. Conant* (860, note 84).

18. "Sputnik & American Public Mind" is dated December 5, 1957. Kuhn's remarks are typed and paraphrased on notecards (Box 3, Folder 12, Thomas S. Kuhn Papers, MIT).

19. Kuhn, *Essential Tension*, xviii.

20. Ibid. xviii–xix.

21. As is evident from correspondence in Kuhn's files and information he supplied to the University of Chicago Press, the readers included Ernest Nagel, Bernard Barber, N. Russell Hanson, H. Pierre Noyes, Robert K. Merton, Paul Feyerabend, and Conant himself (Box 25, Folders 54–55: "UC Press Correspondence," Thomas S. Kuhn papers, MIT).

22. Thomas Kuhn, "The Function of Dogma in Scientific Research," in *Scientific Change: Historical Studies in the Intellectual, Social, and Technical Conditions for Scientific Discovery and Technical Invention, from Antiquity to the Present*, ed. A. C. Crombie (Basic Books, 1963), 347.

23. Ibid., 347–348.

24. Ibid., 352, 353, 360, 363.

25. Ibid., 349.

26. One participant who liked the paper was Michael Polanyi, whose conception of scientific knowledge as "tacit" and "personal" bears similarity to Kuhn's theory of paradigms. The intellectual relationship between Polanyi and Kuhn with respect to matters of influence and priority remains controversial—see, e.g., Martin Moleski, S.J., "Polanyi vs. Kuhn: Worldviews Apart," *Tradition & Discovery* 33, no. 2 (2006–2007), 8–24; Struan Jacobs, "Michael Polanyi and Thomas Kuhn: Priority and Credit," *Tradition & Discovery* 33, no. 2 (2006–2007): 25–36; Jacobs, "Thomas Kuhn's Memory," *Intellectual History Review* 19 (2009): 83–101.

27. "Commentary by A. Rupert Hall," in *Scientific Change*, ed. Crombie, quoted from 374.

28. "Discussion: S. E. Toulmin," in *Scientific Change*, ed. Crombie, quoted from 383.

29. No in-depth biography of Glass exists, but two useful resources are Audra Wolfe, "The Organization Man and the Archive: A Look at the Bentley Glass Papers," *Journal of the History of Biology* 44 (2011): 147–151, and Frank C. Erk, "H. Bentley Glass," *Proceedings of the American Philosophical Society* 153 (2009): 327–339. On Glass' activity as a public intellectual, see Bentley Glass, "Academic Freedom and Tenure in the Quest for National Security," *Bulletin of the Atomic Scientists* 12, no. 6 (1956): 221–223, at 226; Glass, "Liberal Education in a Scientific Age," *Bulletin*

of the Atomic Scientists 14, no. 11 (1958): 346–353; Glass, *Progress or Catastrophe: The Nature of Biological Science and Its Impact on Human Society* (Praeger, 1985).

30. Conant, *On Understanding Science*, 24.

31. Ibid., 25.

32. H. Bentley Glass, "The Establishment of Modern Genetical Theory as an Example of the Interaction of Different Models, Techniques, and Inferences," in *Scientific Change*, ed. Crombie, quoted from 541.

33. Glass, "The Establishment of Modern Genetical Theory," 541.

34. "Discussion: H. Bentley Glass," in *Scientific Change*, 381–382.

35. Kuhn, "The Function of Dogma in Scientific Research," in *Scientific Change*, 351, 363, 369, 350. See also Kuhn, "The Essential Tension: Tradition and Innovation in Scientific Research," in *The Essential Tension*, 225–239, esp. 229.

36. "Discussion: H. Bentley Glass," in *Scientific Change*, 382.

37. "Discussion: T. S. Kuhn," in *Scientific Change*, 386–395, quoted from 391.

38. "Discussion: H. Bentley Glass," in *Scientific Change*, 381.

39. "Discussion: T. S. Kuhn," in *Scientific Change*, 390, 392.

40. See Kuhn, *Road Since Structure*, 2, note 1. In Kuhn's papers, there exists a 1991 request from a publisher to reprint "The Function of Dogma in Scientific Research" that Kuhn declined while offering instead his early essay "The Essential Tension" (Box 20, Folder 23, Thomas S. Kuhn papers, MIT).

41. The page numbers refer to the second (1970) and third (1996) editions of *The Structure of Scientific Revolutions* (University of Chicago Press).

42. According to the Oxford paper, one function of dogma was to operate as "an immensely sensitive detector of the trouble spots" from which revolutions might be sparked ("The Function of Dogma in Scientific Research," 349). On page 166 of *Structure*, Kuhn reassured readers that scientists' intellectual "rigidity provides the community with a sensitive indicator that something has gone wrong" with the dominant paradigm.

43. Another occasion on which Kuhn seemed keenly aware of the need to manage politicized language concerned the title of his monograph. Besides *The Structure of Scientific Revolutions*, he told Morris, "all the alternatives [that he had come up with so far] make use of the word 'ideology', which, at least in the title, I should like to avoid" (Kuhn to Morris, 31 July 1953, Box 25, Folder 53, Thomas S. Kuhn Papers, MIT).

44. The current state-of-the-art philosophical reconstruction of *Structure* holds that Kuhn "retracts his use of 'dogma' entirely, in consideration of its inappropriate connotations." See Paul Hoyningen-Heune, *Reconstructing Scientific Revolutions: Thomas S. Kuhn's Philosophy of Science* (University of Chicago Press, 1993), 168, note 3.

45. Kuhn refers to the "hold," the "deep hold," and "the grip" of ideas on scientists' minds on pages 5, 45, and 88 of *Structure*. For an insightful discussion of "brainwashing" and Cold War liberalism, see Timothy Melley, "Brainwashed! Conspiracy Theory and Ideology in the Postwar United States," *New German Critique* 103, 35, no. 1 (2008): 145–164.

46. Kuhn, "Dear Professor Frank ... ," Box 25, Folder 53, Thomas S. Kuhn Papers, MIT.

47. For more on the depoliticization of philosophy of science, see Don Howard, "Two Left Turns Make a Right: On the Curious Political Career of North American Philosophy of Science at Midcentury," in *Logical Empiricism in North America*, ed. Gary L. Hardcastle and Alan W. Richardson (University of Minnesota Press); George Reisch, *How the Cold War Transformed Philosophy of Science*. On the depoliticization of philosophy more broadly, see John McCumber, *Time in the Ditch* (Northwestern University Press, 2001). On the depoliticization of the postwar academy in general, see Ellen Schrecker, *No Ivory Tower* (Oxford University Press, 1986). When Frank invited Kuhn to join his Institute in late 1952, Frank himself was on the FBI's radar and soon to be investigated on the basis of a rumor (which J. Edgar Hoover found credible) that he "came to the United States for the purpose of organizing high level Communist Party activities" (FBI memo quoted in Reisch, *How the Cold War Transformed Philosophy of Science*, 268).

48. Conant to Kuhn, 15 June 1961, Box 25, Folder 53, Thomas S. Kuhn Papers, MIT. (Words italicized in quoted passages were underscored in the original letters.)

49. Conant to Kuhn, 11 July 1961; Box 25, Folder 53, Thomas S. Kuhn Papers, MIT.

50. Kuhn to Conant, August 5, 1961, Box 25, Folder 53, Thomas S. Kuhn Papers, MIT.

51. Conant to Kuhn, 19 December, 1962, Box 4, "Structure of Scientific Revolutions Correspondence, C–D," Thomas S. Kuhn Papers, MIT.

52. Conant to Kuhn, 29 July 1963, Box 4, "Structure of Scientific Revolutions Correspondence, C–D," Thomas S. Kuhn Papers, MIT.

53. James B. Conant, *Two Modes of Thought: My Encounters with Science and Education* (Trident, 1964), xxxi.

54. Ibid., 18.

55. Ibid., 3–4, 8–12, 13.

56. Ibid., 86.

57. Ibid., 87.

58. Ibid., 82.

59. Kuhn, *Structure*, 159.

60. Ibid., 137, 167.

13 Big Science and "Big Science Studies" in the United States and the Soviet Union during the Cold War

Elena Aronova

In the 1960s, Big Science was identified as a new phenomenon, associated with changes in the organization of scientific research in the aftermath of World War II.[1] As the Big Science mode of research blossomed and expanded in the second half of the twentieth century, it became a widespread mode of scientific research in the natural sciences. More recently, in the social sciences, research has been increasingly organized around big projects involving big budgets and big interdisciplinary teams. Big Science has also become a favorite analytical category for historians of science who have used it for talking not only about Cold War science but also about its predecessors in earlier decades or even centuries. Indeed, the precedents of Big Science, understood as a mode of research characterized by large scale and extensive state involvement in its funding, could be found in the imperial sciences in the nineteenth century and in the great Victorian projects of knowledge that similarly had at their center the presence of big state as well as big business. As an intervention into this discussion, the historian Jon Agar has suggested distinguishing between the *phenomenon* of Big Science as a mode of organization of scientific research and the "*labeling* of 'Big Science' as something of concern, ... a product of the ... long 1960s."[2] Endorsing this useful distinction, this chapter aims to historicize Big Science as an analytical category, showing how the categories we typically use to characterize science during the Cold War were themselves products of Cold War science.

In the early 1960s, the phenomenon of Big Science became a focus of debates among social and natural scientists, political analysts, and scholars of diverse disciplinary and political backgrounds. On both sides of the "Iron Curtain," social analysts articulated awareness that the large-scale growth of science after World War II had significant implications, for better or for worse, for modern societies, identifying Big Science as a contemporary phenomenon requiring characterization and careful study, especially in terms of its social and political implications.

In the first section of this chapter I focus on the discussions in the United States concerning the political and social consequences of Big Science, which found its most prominent articulation in the early 1960s within the network of intellectuals associated with Congress for Cultural Freedom. In the second section I focus on the parallel discussions in the Soviet Union, where the discussion of the same phenomenon was articulated in terms of the theory of Scientific Technological Revolution (STR), promulgated in the 1960s and the 1970s. As I discuss in the third section, these discussions provided a context for reading Kuhn's *Structure of Scientific Revolutions* (1962), both in the United States and in the Soviet Union, in those decades.

I argue in this chapter that Big Science, as an analytical category, was deployed as a resource to debate, negotiate, and rationalize the concerns and anxieties of the Cold War on both sides of the Iron Curtain. Throughout the Cold War, both the United States and the Soviet Union advocated their ability to offer and display different visions of modern industrial (or rather "post-industrial") society, and Big Science played major role in these powerful imageries. In their different political settings, American and Soviet scholars conceptualized the phenomenon of Big Science, tying together their country's preferred model of social order with an image of the "right" scientific order and, at the same time, creating parallel imageries of societies built around modern technology. In the context of the Cold War and the expansion of the sciences into Big Science, these scientists and scholars also helped to invent a set of new subjects while reconceptualizing science as a social activity and in various ways exploring science-society-politics nexus. In both political settings, the discussion of social and political implications of Big Science contributed to the construction of the public space in which studies of science (or, rather, of Big Science) emerged as an important—and politically relevant—area of expertise.

The "*Minerva* debate": Big Science and "Big Science Studies" in Cold War America

In an editorial in the first issue of *Minerva*, the journal's founding editor, the University of Chicago sociologist Edward Shils, described the scope and the aims of the new journal:

The governmentalisation of science and scholarship is, in part, a product of intellectual development and its changed relationship to technology, which entails costs which can only be borne by government, and returns, in which governments, even in capitalistic societies, have a great and appropriate interest. The governmentalisation of science in the past decade and a half is also, in part, a result of the Cold War—as are also, to some extent, the political embarrassments and concerns of science. *Minerva* will be concerned with the indirect as well as the direct influence of the Cold War on the role of science and scholarship and on the performance of their true calling.[3]

The "governmentalisation" of science, as Shils emphasized, implied newly defined relations between American scientists and the federal government. In the aftermath of World War II, in Shils' words, scientists had "become politicians."[4] Politics was always part of science, but now, Shils argued, the relationship between politics and science had become more explicit and prominent. This intensified interdependency between science and politics was manifested on different levels, ranging from the development of scientific advisory representation in the government to "scientists' movements" such as the Pugwash Conferences. The aim of *Minerva*, Shils stated, was to create a forum for discussing, describing, documenting, and examining these recent changes and how they affected the relations between science and politics: "Without seeking to formulate a code to govern them, [*Minerva*] will try, through the description of what has happened and through the analysis of the conditions of good and poor relationships, to develop a better understanding of why things sometimes go wrong and to contribute to a better practice."[5]

During the 1960s, *Minerva* developed into a forum for discussions of the changing relations between science, politics, and national policy in the United States. The first and the most prominent topic of the newly founded journal, which became a continuing theme in the first decade of its publication, was the phenomenon of Big Science and its social and political implications. In 1961, the term "Big Science" was simultaneously coined by two physicists—Alvin Weinberg, director of the Oak Ridge National Laboratory and a member of the President's Science Advisory Committee, and Derek de Solla Price, a physicist turned historian.[6] In his influential 1961 essay in *Science* on the phenomenon that he called Big Science, Weinberg argued that the large-scale centralized scientific operations exemplified by gigantic National Laboratories such as Oak Ridge—which were linked to big facilities, big budgets, and big publicity—had dramatically changed the major features and societal role of science. While believing in the positive effect of Big Science and its advantages for American science, Weinberg argued that the funding for Big Science projects should be confined to National Laboratories, "to prevent the contagion" of Big Science from spreading to the universities. Big Science, in Weinberg's view, should be "segregated" from the more traditional modes of research in order to "prevent it from taking over Little Science."[7]

Weinberg's ambivalent position with regard to Big Science was widely shared by scientists of different disciplinary backgrounds who argued that large-scale, lavishly funded, team-operated research in Big Science threatened individual scientists' initiative and innovation.[8] Moreover, Big Science, with its huge expenditures, had placed new and unprecedentedly large demands on society. As Weinberg noted, however, the right question to ask was not whether Big Science was "ruining science." Since Big Science was "here to stay," the more appropriate thing to do, in Weinberg's view, was to systematically examine and characterize this phenomenon and its social and political implications. Arguing that "Big Science was an inevitable stage in the development of science," Weinberg warned scientists and the public about the broad social and long-term consequences of Big Science as a contemporary cultural phenomenon, comparing it to the construction of the pyramids of Egypt or the Palace of Versailles.

Like those cultural landmarks, Big Science was, in Weinberg's view, "a supreme outward expression of our culture's aspirations" that had "created many difficult problems, both philosophic and practical."[9] "These questions" addressing the cultural, social, and political implications of Big Science, Weinberg noted, "are so broad, and so difficult, that I cannot do more than raise them here."[10]

The questions that Weinberg sketched in his short 1961 *Science* essay were at the center of Shils' concerns when he founded *Minerva*. While planning the first issue of the journal, in May of 1962, Shils invited Weinberg to contribute an expanded version of his *Science* essay.[11] Weinberg apparently didn't respond, but he did respond to Shils' second written request several months later. With his letter Weinberg sent the text of the lecture he presented at the meeting of the local honorary engineering society at the University of Tennessee earlier that year.[12] Shils responded enthusiastically and suggested that it be published in *Minerva*.[13]

Weinberg's article—titled, at Shils' suggestion, "Criteria for Scientific Choice"—focused on the question of how funding for Big Science projects, requiring large federal patronage, should be allocated in a democratic society: Should scientists play a leading role in political decisions concerning their science, or should their role be limited to providing technical expertise to politicians?[14] As a director of an exemplary Big Science operation closely working with national-security and military agencies, Weinberg didn't question the power that scientists, especially physicists in the wake of the World War II, had at influencing decision-making and obtaining large state resources while staying largely free from political control. The problem for Weinberg was that all decisions pertaining to Big Science were inevitably political, based more on judgments of politicians and few elite scientists, mostly physicists, than on diversified scientific or technical expertise. How, Weinberg asked, should the government decide what to prioritize when it came to "different, often incommensurable, fields of science"? "We shall," he wrote, "have to choose between, for example, high-energy physics and oceanography or between molecular biology and science of metals ... whose only common characteristic is that they all derive support from the government."[15] The problem of such choices, Weinberg maintained, "arises in 'Big Science', not in 'Little Science.' ... It is only when science really does make serious demands on the resources of our society—when it becomes 'Big Science'—that the question of choice really arises."[16] Weinberg's solution was to "decentralize" these choices by resorting to the "systematic application of a set of criteria." Weinberg suggested three groups of criteria: "technological merit" (the balance between the costs of research and the prospective return in the form of industrial applications), "scientific merit" (measured as much by indirect repercussions as by direct promise), and "social merit" (promise of high payoffs for health, food production, defense or prestige). Molecular biology, for example, in Weinberg opinion, had all three merits—"scientific," "technological," and "social"; physics was currently overrated; space research was only "masquerading" as science: "If we do space-research because of prestige, then we should ask whether we get more prestige from a man on the moon than from successful control of the water logging problem in Pakistan's Indus Valley Basin. If we do space-research because of its military implications, we ought to say so. ..."[17]

Weinberg's article, although mostly concerned with the pragmatic issue of the allocation of funds for Big Science projects, simultaneously framed a set of more general themes. Big Science, Weinberg emphasized, was explicitly political. In Weinberg's view, the qualitative change in the relation between science and the state had occurred when science achieved the size and complexity of Big Science operations, all of which were embroiled in institutional, bureaucratic, and national, as well as international, politics. Moreover, as Weinberg's discussion implicitly suggested, it was the Cold War that had set up this new agenda for science, with its incentive to maintain the nation's superiority in all fields that might contribute, directly or indirectly, to the high-technology postwar economy. Weinberg concluded that scientists and the general public had no choice other than to accept the Big Science, but that they should ask questions about its political and social consequences and try to come up with reasonable answers.

Weinberg's essay opened the discussion of the social and political implications of Big Science on the pages of *Minerva*. Shils carefully planned the debate. Along with the paper from Weinberg, he asked Michael Polanyi, whom Shils deeply admired, to contribute a paper for the opening issue.[18] Polanyi was an enthusiastic supporter of Shils' journal, praising Shils for the "great achievement" of his "enterprise in bringing out the first issue of *Minerva*" and assuring his friend that he "subscribed to everything you say in the editorial introduction."[19] The essay Polanyi wrote for the opening issue of *Minerva* was his famous "The Republic of Science." [20]

In his essay, Polanyi articulated his long-held views on science and its governance, arguing that the social order of science constituted a coherent and self-governing system, with no central authority, internally coordinated "by mutual adjustment of independent initiatives." This system, Polanyi held, "works according to economic principles similar to those by which the production of material goods is regulated."[21] As Polanyi put it in his famous science-as-market analogy, "in the free cooperation of independent scientists we shall find a highly simplified model of a free society."[22] According to this view, the governmental support of science, though welcome, should not in any way imply the control of scientific enterprise, either in the form of planning or regulation, since any form of state intervention would undermine the order of science based on the self-governing autonomy of the scientific community. Science, Polanyi argued, should maintain its traditional "Little Science" mode of organization, as a decentralized network of independent self-coordinated initiatives, since only this mode of organization could provide science with its crucial strength.[23]

The long-cherished ideal of the self-governing autonomy of science was, however, in conflict with the Big Science mode of organization of science, as Weinberg's discussion made clear. With its centralization of research and decision making through centralization of facilities Big Science obviously didn't conform to the image of science as a spontaneous pursuit of knowledge by scientists-citizens in a self-governed republic of science, free of control and regulation. The theme was continued by scientists and scholars whom Shils invited to respond to the opinions expressed by Weinberg and Polanyi—"the *Minerva* debate on scientific choice" as one of the participants, philosopher Stephen Toulmin, called it.[24] In his review summarizing the "*Minerva* debate" Toulmin concluded that Polanyi's image of the "republic of science" was out of date:

> In real life, the republic of science cannot stand apart from the general commonwealth. Back in the 1930s, Polanyi's campaign to defend the autonomy of science against projects for a Nosey-Parkerish state centralism had a real point. By the 1960s, the need for academic science to be self-governing seems to be being conceded even in Russia and Polanyi's protestations are—surely—more insistent than they need be. As the social sciences too approach their coming-of-age, his distinction between the republic of science and the rest of the community becomes excessively disjunctive. The urgent question today is, rather, how the self-governing republic of science is to be integrated, not only into the broader academic confederation, but into the whole community of citizens.[25]

Polanyi responded to the critique without, however, changing his general position. In "The Growth of Science in Society," published in *Minerva* in 1967, he reflected on the relations between science and society in the age of Big Science. Polanyi argued that Big Science doesn't challenge the science-as-market model. The market-based economic model was still valid, even in seemingly non-market situations: "The marginal principle of economics offers the conceptual model for [scientific research]. ... Funds and appointments serving scientific research must be distributed in a way that promises the highest total increment to science. ... The two great principles of scientific growth [are] the granting of independence to mature scientists and the imposition of scientific values on their performances ... achieved by ... *self-coordination by mutual adjustment* and *discipline under mutual authority*.[26]

Weinberg did not find this view appealing. In 1968, reviewing a book by the science journalist Daniel Greenberg with the telling title *The Politics of Pure Science*, in which Greenberg contested the "traditional scientific view" that the internal system of science is a sufficient guarantee to require no outside surveillance, control or reinforcement, Weinberg wrote:

> This traditional view has been articulated eloquently in the pages of *Minerva* by Professor Polanyi. In Polanyi's "republic of science" countervailing "political" forces are kept in rational equilibrium by the internal workings of the value system of science, a value system the ultimate sanction of

which is nature itself. Yet, on this issue—the effectiveness of the republic of science in keeping the game honest—I find myself more in sympathy with the journalist Greenberg than with the scientist Polanyi.[27]

In 1991, Weinberg was even more explicit in his assessment of his own position, writing about his advocacy of Big Science as "being organized, more or less by an intrusive government":

> This "socialist" view of science contrasts with Polanyi's Republic of Science, in which myriad independent practitioners determine the course of science. The Republic of Science is a free market and decentralized. My scientific enterprise is more socialist and centralized. Actually, I would say that where Polanyi's democratic republic is a good model for Little Science, my socialist republic applies more to Big Science.[28]

And Weinberg added another twist to "*Minerva* debate." The choices rationalizing the decisions pertaining to Big Science, Weinberg argued, should be made not by politicians but by "some well-informed observers"—experts not in sciences per se but rather in meta-studies of science. "For this reason alone philosophic debate on the problems of scientific choice should lead to a more rational allocation of funds."[29] Big Science, in other words, required what might be called "Big Science studies"—an independent and decentralized expertise, which would provide a systematic study of Big Science mode of research and advise the government accordingly.[30] Weinberg jokingly remarked, in retrospect, that his papers in *Minerva* had launched his career as a "moonlight philosopher of scientific administration," and the "*Minerva* debate" had contributed to the recognition of "the importance of philosophic examination of the sanctions for public support of science," stimulating "something of a cottage industry in the philosophy of science policy."[31]

What Weinberg called a "cottage industry in the philosophy of science policy" had originated earlier, however. The founding of *Minerva* in 1962 had been preceded by three years of activities and workshops on science policy and science politics that Shils had organized under the auspices of the Congress of Cultural Freedom (CCF). The examination of the phenomenon later called Big Science was at the center of the intellectual agenda of the CCF. Both Shils and Polanyi were active members of the CCF and major drivers of the CCF Seminar Program and its "Study Groups," focused, among other areas of concerns, on the discussion of science as a social and political institution. The quest for "expertise" in meta-studies of science that Weinberg envisioned in his essays was at the center of the CCF's own aims and aspirations.

The Congress for Cultural Freedom and Its "Study Groups"
The Congress for Cultural Freedom is remembered as the spearhead of Cold War American cultural diplomacy, established with the aim to secure support for the Marshall Plan (Economic Recovery Program) in the sphere of culture and ideas by building

a transnational network of anti-Stalinist scholars and establishing strong bonds between European and American intellectuals. The CCF's goals were to be achieved through widely publicized endeavors such as the Festival of the Twentieth Century, as well as through CCF-affiliated journals such as *Encounter, Preuves, Der Monat*, and, later, *Minerva*.[32]

The CCF's first visionary was Arthur Koestler, a Hungarian émigré whose widely known novels denounced totalitarianism and voiced calls to arms for the defense of freedom and the fight against Stalinism. Koestler's "militant liberalism" wasn't embraced, however, by other visionaries of the CCF. Thus, Nicolas Nabokov, Secretary General of the CCF, referring to Koestler's original plan for the CCF, which he envisioned as a "combat unit" devoted to the cause of the defense of freedom, expressed the general opinion in 1952: "the Congress [should be established] in the minds of the European intellectuals as a positive, and not only as a political, organization."[33] After Stalin's death, it became even more obvious that the simple message of "anticommunism" wasn't enough to sustain intellectual appeal and gather significant audience among the European and American intellectual elite.[34]

The CCF lost no time establishing its "positive program," which emerged in the mid 1950s from the network of intellectuals associated with the CCF, most prominent among them Raymond Aron, Daniel Bell, and Edward Shils. The CCF's "positive program" was encapsulated in the rhetoric that later found its sharpest articulation in Bell's 1960 book *The End of Ideology*. Within the CCF, the "End of Ideology slogan" was adopted, in words of Michael Polanyi, as "an expression of [the CCF's] predominant aims [and] our official pronouncement."[35]

The "end of ideology" rhetoric signified an important shift in the CCF's self-perception, "from an instrument of struggle against totalitarianism to an international forum for debate."[36] In Polanyi's words, the "end of ideology" was a "distinctive, passionately sober approach to culture and politics" that encouraged factual and calm examination of political systems and contemporary societal phenomena, rather than mere denunciation of totalitarian regimes.[37] This rhetoric sought to move away from simplistic early Cold War dichotomist imageries of incommensurable Eastern and Western world systems and competition promulgating instead the images of convergence and compromise. It promoted the view that the common forces resulting from the dramatic advances of science and technology were leading both Western capitalism and the Soviet Bloc's socialist and collectivist systems to adopt similar methods of socioeconomic management. In words of Konstantin Jelenski, a Polish émigré writer who led the Eastern European Division at the CCF, there was a "growing realization that the realities of industrialism are perhaps a more important determining factor socially than political systems, whatever their ideological origin."[38]

The "end of ideology" wasn't merely a normative position, it sought to offer a substantial reformulation of the ideals and goals of classical liberalism. In Polanyi's words, the "end of ideology" was the means to "secure a post-Marxian basis for liberalism throughout the world."[39] These realities of the postwar world had shaken the belief in an unregulated free market system and *laissez-faire*. The "end of ideology" presented a renewed defense of capitalist "free society," with its central ideals of the "free market" and *laissez-faire* economics, seeking to attune it to the socio-political implications of changes in political economy in the postwar industrialized world, and to the subtleties and compromises of the Cold War with its postwar coalitions.

With this agenda, the "end of ideology" rhetoric emphasized the studies of science as the topic of central concern. With its emphasis on "sober," sophisticated, dispassionate socioeconomic analysis of modern industrial societies and their political systems, the "end of ideology" promoted the view that a dramatically increased role of science in the realm of public affairs and politics had delineated a new phase in the development of Western liberal democracies. Science, its history, and its politics were to be assessed, especially in terms of their implications for democracy, liberalism, and freedom. The CCF intellectuals sought to offer such an assessment, and they did this in a big way.

The CCF "methodology": The Study Groups and the Seminar Program

As Rebecca Lemov has pointed out, "methodological thought" became a preeminent concern during the Cold War.[40] Within the CCF this certainly was the case. With its "end of ideology" rhetoric calling for a non-ideological "sober" research based on facts, rational method, and science, rather than ideological conceptions and "messianic claims," the CCF encouraged the application of scientific methods to social problems. Among the models for the CCF's activities was the Committee for Economic Development (CED), which was seen by the CCF leaders as an exemplary organization that had developed sophisticated methods of collecting, processing and analyzing vast amounts of social data:

The Congress [for Cultural Freedom] would do well to adopt this CED technique or some appropriate variant of it whereby data which has been collected by specialized institutions on different topics of general importance would be submitted to a small study group of the Congress which would prepare on the basis of the detailed examination of the available data, recommendations for a special international policy statements, to be issued on the name of the Congress. Such statements ... could conceivably have a beneficial influence on the policies of other organizations, of foundations, and even of governments.[41]

The CED model of quantitative analysis of large sets of social data was, however, soon abandoned in favor of a qualitative analysis. The main form of the CCF's activities became the CCF Seminar Program, conceived as an innovative form of interdisciplinary contact, a method of "intellectual confrontation," in which critical discussion was methodically staged. One of the CCF officers claimed that "the seminar concept ...

provides the most adequate form for the treatment of many issues ... stimulating thinking and discussion about ... new ways of organizing intellectual confrontation."[42] To achieve this healthy "intellectual confrontation," the CCF's seminars and study groups meetings adopted the practice of commissioning two or more position papers presenting opposing views "on a critical contemporary problem" (for example, "an optimistic and pessimistic view of the quality of life in industrial society") and submitting them for discussion in small groups.[43]

The CCF Seminar Program consisted of international seminars and a large number of smaller local meetings. The themes were diverse. Science and its role in the postwar world was but one of many topics, but it always loomed large on the agenda.[44] The unstated goal of the Seminar Program was to develop the CCF into a clearing house for independent expertise on the pressing issues of the day—expertise that would eventually be used to advise government policy makers. The discussions that took place, which were carefully recorded and transcribed, were ultimately intended to result in a series of publications devoted to "major present preoccupations of the intellectual world."[45]

Edward Shils, one of the chief designers of the CCF's seminar program, also helped to introduce another form of the CCF activities: study groups. CCF study groups originated in the "Study Group on Science and Freedom," organized by Michael Polanyi as part of the 1953 Hamburg Congress and an outlet of his Society for Freedom in Science that he formed in the 1940s.[46] For Polanyi, the system of the organization of science as it had developed since the Scientific Revolution was a resource for liberal capitalist democracy.[47] The discussion of these ideas at 1953 Hamburg Congress resulted in Shils' proposal to set up a Study Group within the Congress that would sharpen the picture of the relations of science and the state in the contemporary world.[48]

At the 1955 conference in Milan, Shils suggested that the Study Group be extended into a "future Congress on Science and Freedom" that would focus on the "impact on the academic community of the great historical changes of the past two decades, and particularly of the greatly increased need for finance from outside sources, and the invasion of practical and technical tasks in the academic sphere, with all its resulting social, political and intellectual problems."[49] The envisioned Congress, eventually held in Hamburg in 1959, was devoted entirely to discussion of the dramatic changes in the role of technology and science that had occurred since World War II. The aspiration was to produce a comprehensive account of the "Technical Age" into which both Western and non-Western societies had entered.[50] However, the topic was obviously too grandiose and too important to be limited to one meeting. As a solution, Shils suggested dividing it among several permanent Study Groups. The CCF adopted Shils' proposal, and six Study Groups were inaugurated, each having its own program within the general theme outlined at the 1959 Hamburg Congress. The Study Groups continued to function thereafter, with different levels of success and productivity.

As Shils outlined the "division of labor," one Study Group, directed by Shils himself, would focus on the political aspects of the "Technical Age," characterized by the "decline of ideologies" in Western societies and the rise of technical and scientific expertise with their growing role in the realm of public affairs. Two other groups would be concerned with the features, both positive and negative, of modern industrial and mass society, with one group focusing on the features of society while the other focused on the individual's relation to culture and society in the "Technical Age." Then, two groups would focus on the role of the "vast class of intellectuals"—artists, scientists, and academics of all kind—in the contemporary world divided by the Cold War. Finally, the last group would study "the signs already visible of affirmation in the Communist world itself of those values which we hold dear. It will seek to pinpoint indications that such values do provide a point of convergence for various contrasting historical developments."[51] The plan was adopted by the Study Groups' "directors"— Edward Shils, Michael Polanyi, Daniel Bell, Raymond Aron, and Nicolas Nabokov. In subsequent years, the topics of Study Groups changed, although there always was a continuity with the original themes suggested by Shils. Shils, as a Study Group leader, was interested in science policy and science politics; Polanyi was concerned with the relationship of thought, mentalities, and contemporary politics; Raymond Aron led the discussions of the conditions of stability, both in democracies and ideological orthodoxies; Nicolas Nabokov led the discussions of the role of the arts in the postwar world.[52]

In 1960, Shils focused his Study Group on "Scientific Policy—the Cooperation of Government, Economy and the Universities in the Development and Application of Scientific Research."[53] That year and the next, he invited leading academics and scientists to take part in discussions, or to contribute a paper or a statement to be discussed. As one of the founders of the *Bulletin of the Atomic Scientists*, and as a co-founder and vice-director of the University of Chicago's Office of Enquiry into the Social Aspects of Atomic Energy, Shils had long been involved in the atomic scientists' movement. Now he capitalized on this network, trying to enlist atomic scientists in a more general and theoretical discussion of the interactions between science and society. "The kind of persons we wish for," he told J. Robert Oppenheimer in 1960, "are scientists concerned with problems of scientific policy, 'scientific administrators,' civil servants and politicians especially concerned with the development of science and other aspects of the application of scientific knowledge."[54]

By 1962, Shils was ready to move his Study Group's discussions into the public realm. The CCF Study Group on Science Policy was transformed into the CCF-affiliated journal *Minerva*. Establishing *Minerva*, Shils transferred some of the approaches and formats of the CCF Study Groups workshops and discussions into this forum. For the opening issue of *Minerva*, in line with the methodology of discussion that was adopted at the CCF, Shils commissioned two opposing position papers on Big Science: an "optimistic" one, by the physicist and science administrator Alvin Weinberg, and a "pessimistic" one, by the physical chemist and philosopher Michael Polanyi. Unlike the CCF seminar program, however, *Minerva* was established as a public forum with the intention of influencing not only the intellectual elite (the major "target" of the CCF's activities) but also the broader academic community and the makers of science policy.

In the 1960s the CCF sponsored a number of influential magazines and organized large and small international conferences and seminars on a wide range of topics, including science and its roles in the broader culture, in society, and in politics. By the mid 1960s, the CCF was regarded as a big success. Referring to the Congress' role in countering the appeal of the pro-communist Left and in creating a receptive atmosphere for the formation and support of a movement that often thought of itself as on the anti-communist Left, one of the CCF's best known members, the diplomat and the State Department's Soviet expert George Kennan, wrote in 1959 to Nicolas Nabokov: "I can think of no group of people who have done more to hold our world together in these last years than you and your colleagues. In this country [i.e. the United States] in particular, few will ever understand the dimensions and the significance of your accomplishments."[55]

"Who paid the piper": The CIA connection and the moral crusade of the CCF
In 1967, a series of newspaper publications revealed the close association of the CCF with the US Central Intelligence Agency. The CIA connection presented an ultimate—and very sensible—test for the theoretical discussion of the societal effects of the "governmentalisation of science" and the issue of "the indirect as well as the direct influence of the Cold War on the role of science and scholarship" (quoting Shils' 1962 editorial). During the 1960s, the revelations of the CIA's infamous involvement in operations in Iran and Guatemala in the 1950s, and in Cuba and Vietnam in the 1960s, made the question of "governmentalisation of science" as much a moral as an epistemological position for the CCF-associated intellectuals and for their critics.[56]

The revelation of the CIA funding behind the CCF and its activities caused public outrage. What can a "free thinker" say about his "freedom," asked the *Sunday Times* of London, "when he finds out that his free thought has been subsidized by a ruthlessly aggressive intelligence agency as part of the international Cold War?"[57] A report in the *New York Times* similarly pointed out that the intellectuals who were funded by the CIA (with or without their knowledge) were "being used for concealed government propaganda," which made a "mockery" of intellectual freedom. In the aftermath of the controversy, the historian Christopher Lasch commented that "the whole wretched business seemed inescapably to point to the conclusion that cultural freedom had been consistently confused with American propaganda, and that 'cultural freedom,' as defined by its leading defenders, was—to put it bluntly—a hoax."[58]

For most of the CCF associates, the revelation about the Congress' link to the CIA didn't come as a surprise. As Sidney Hook admitted retrospectively in his autobiography, "I have heard, like almost everyone else, that in some way the CIA was involved in funding the congress. Everyone mentioned it, even though no one had any hard evidence. ... In my own mind I had no doubt that the CIA was making some contribution to the financing of the Congress. ... Everyone involved in the activities of the Congress had heard rumors of covert CIA support."[59] Most of the CCF's members continued to be engaged in the organization, either knowing or suspecting its source of funding, assuming that as long as they are not dictated or controlled in their intellectual activity, they can claim their intellectual independence and integrity. Within few days after the revelations in the *New York Times*, several distinguished CCF associates—John K. Galbraith, George Kennan, J. Robert Oppenheimer, and Arthur Schlesinger Jr.—wrote a letter to the editors of *New York Times* stating: "on the basis of our own experiences with the Congress over the past 16 years—with its seminars, its artistic festivals, its magazines, its staff—we can say categorically that we have no question regarding the independence of its policy, the integrity of its officials, or the value of its contribution. In our experience the Congress ... has been an entirely free body, responsive only to the wishes of its members and collaborators."[60]

Neither *Minerva* nor its editor was directly attacked in the press, but Edward Shils was as disturbed as others when he saw that their valuable enterprise was being discredited. Like other CCF-associated intellectuals, Shils insisted that the CCF's magazine's editorial independence wasn't corrupted, emphasizing his own loyalty only to the "commitment to cultural freedom."[61] Deploying the "end of ideology" rhetoric and posture, Shils went so far as claiming that the CCF wasn't "political." In a letter to Crawford Goodwin, a professor of economics at Duke University and the program officer in charge of European and International Affairs at the Ford Foundation, he wrote: "it might be reasonably claimed that the Congress ... was not political. It sought to promote the understanding and solution of fundamental problems which concern serious intellectuals ... cutting across the boundaries of nationality, party, intellectual field and discipline. ... It created and fostered a sense of affinity among these intellectuals in a way which is, I think, unique in the history of the present century."[62]

The CCF's leaders should have felt that they were between a rock and a hard place. The complexities of the interrelation between politics and science that they had being discussing at length in the seminars and study groups appeared to represent now their own predicament. In some ways, the CCF intellectuals were reasoning according to the very logic of "Big Science" that they were disentangling. They were accepting, with Weinberg, that Big Science had changed not only science but also the way the relation between science and the state was understood. If Big Science was political then "Big Scientists" could not but be politicians in their claims on freedom and independence of science vis-à-vis state and politics.

The positions taken in the "*Minerva* debate" continued to resurface within the network of the CCF intellectuals, now on the "moral plane." The "intellectual confrontation" staged by Shils on the pages of *Minerva* resurfaced as a "moral confrontation" among the CCF leaders now facing a moral conundrum. Polanyi, once again, presented a "dissenting" view. Reasoning consistently with his theoretical defense of "Little Science," Polanyi was much concerned about a "little person": Michael Josselson, the CCF's Executive Director and a dedicated driving force behind the CCF's activities since its inception. After he was revealed to be a CIA officer, Josselson was ostracized by other CCF leaders, who unequivocally insisted on his immediate resignation. Polanyi, alluding to Arthur Koestler's famous novel, called the decision to "eliminate" Josselson from the CCF "another darkness at noon." For Polanyi this decision was, first of all, morally wrong. Moreover, it wasn't "rational." It didn't resolve the moral conundrum; it was only a way to claim unawareness about the CCF's connection to the CIA, a position that Polanyi found neither moral nor rational. In a letter, he asked Raymond Aron "What kind of figure are we going to cut? Are men like you or me ... going to declare that in 15 years we did not notice that we were being manipulated to serve sinister purposes? Are we going to proclaim our awakening, a new version of *The God That Failed*?"[63]

The fate of their "valuable enterprise" was, in Polanyi's eyes, inseparable from the fate of its "little" leader, even if it would imply to "lose face" and admit the strong connection between the CCF and the CIA. Appealing to the CCF's support "for saving Mike [Josselson], the Congress, and our honour," Polanyi wrote the following to Aron:

I would have served the C.I.A. (had I known of its existence) in the years following the war, with pleasure. We were faced with an ubiquitous madness, supported by an empire and organized on conspiratorial lines. ... In the years after 1950 we battled against a phalanx of Stalinist or Stalinisant intellectuals throughout Europe, for the vindication of free thought, which was despised and ridiculed by those who are now forcing us to dismiss Mike Josselson, because he had accepted the support of like-minded American officials, who appreciated the ideals he was fighting for.[64]

In the end, it was the forced resignation of Josselson rather than the revelation of CIA funding that prompted Polanyi to resign from the CCF. As he explained to Pierre Emmanuel, "I expressed the feeling that I could not remain connected with our organization, if we decided to eliminate Michael Josselson from it. ... I beg you, therefore, to accept my resignation from the community to which I have so long adhered."[65]

With funding secured by the grant from the Ford Foundation, the CCF quietly became the International Association for Cultural Freedom.[66] The IACF "inherited" the CCF's magazines and, aside from a few resignations, continued to rely on old networks until it quietly dissolved itself in 1979, never having enjoyed as much success as the CCF had enjoyed throughout the 1960s.[67]

The Broader Context of the *Minerva* Debate": The Theories of "Post-Industrial Society"
The "*Minerva* debate" articulated the awareness that the growth of science, for better or for worse, had significant implications for modern society. By the 1970s, this view had become common. Jacques Barzun, Spencer Klaw, Edward Shils, Daniel Bell, John K. Galbraith, and many other social theorists promoted the view that access to the power of the atom, the computer revolution, the exploration of the cosmos, and the greater cultural, social, economic, and political significance that science had come to hold in Western societies delineated a new type of social order—a "post-industrial society."

The theory of "post-industrial society" was the most articulate descriptor of a new postwar social structure in which advanced science and technology played a major role. Like the "end of ideology," the theory of "post-industrial society" was formulated within the CCF's network of intellectuals. It was most prominently articulated by Daniel Bell in his 1973 book *The Coming of Post-Industrial Society*.[68] However, the concept was born much earlier, within Bell's Study Group and the CCF seminars.

In his discussion of the changing role of the intellectuals, Bell attributed central importance to the technological developments that transform and increase the number of scientists and men of knowledge, seeing them as the dominant element within the new social structure in a society where the social functions of science had considerably expanded. The post-industrial society was a society in which the growing role of intellectual activities implied not only the mobilization of science for the sake of both knowledge and capital production, but also suggested that the distinction between economic and social processes were increasingly vanished. In post-industrial society, new forms of community emerged as alternatives to market-based norms of organization, and a "professional and technical class" of scientific workers displaced the "old working class."

The "post-industrial society" rhetoric resonated with the changing context of the Cold War. By the late 1960s, a partial détente with the Soviet Union and the "communist" countries in Europe made the anti-communist rhetoric of the 1950s obsolete, though of course didn't make anti-communism obsolete. The anti-communism of the late 1960s and the 1970s required a new rhetoric, and the "post-industrial society theory" supplied exactly what was needed, presenting the images of the emergence of a unitary "post-industrial society" in both mature capitalist and socialist "technostructures"—the so-called "convergence theories," widely promulgated throughout the 1970s. As Bell affirmed, behind the argument about convergence lay a recognition of the fact that, in Bell's words, "the market was rediscovered in the Communist socialist world, and the market was losing its importance in the Western economies."[69]

Economic performance and its prerequisites, scientific and technological advances, were the defining elements of modern life and a measure of success in the Cold War.

The "*Minerva* debate" provides a perfect illustration of how the "planning-versus-market" dichotomy didn't hold in the world of Big Science shaped by Cold War. The Western democracies and the Soviet Bloc were claimed to represent two incommensurable and opposing forms of economic modernity, epitomized in two ideal types—Soviet *planning* and American *free market*. Yet, in the debates on the pages of *Minerva* these two ideal types were redefined, re-negotiated, and reconciled to accommodate actual practices of social planning and control existing on both sides of the Iron Curtain. As a result, the "planning-versus-market" dichotomy was broken down into a number of different positions across the entire spectrum from planning to market.

Shils, eager to include the perspective from the Soviet side, published several papers in which the physicist Pyotr Kapitsa discussed the planning of science in the Soviet Union. The papers by Kapitsa published in *Minerva* were translations of articles that had been published in *Pravda* and in *Literaturnaia Gazeta* in 1966.[70] Kapitsa argued that Soviet science, which had been planned at the state level for several decades, had reached the point where a more effective approach to research should be implemented—one that would take the views of scientists into account. Emphasizing the value of the creative element in science, Kapitsa argued that "the plan" should support rather than restrict the freedom of scientific creativity and should be implemented by scientists "free[d] from petty controls."[71]

Kapitsa's deliberations came at the time when the prestige and authority of scientists had risen dramatically in the Soviet Union, and Soviet physicists used this power and prestige to renegotiate their relations with the party-state and to launch an active campaign for a liberal reform in science policy.[72] Referring to the advocates of the view that science should be left to the scientists in the Soviet Union, Kapitsa wrote: "The scientists who maintain that scientific work cannot be planned because it develops spontaneously are wrong"[73] Instead, Kapitsa advocated a more moderate position, arguing for a reform of the system of planning of science that would allow for more initiative on the part of scientific institutions and scientists themselves. In this way, Kapitsa contended, Soviet science could take advantage of the "freedom of science," a cherished value of Western scientists, as well as of the strength of Soviet institutional system of support. "We have money for research," Kapitsa noted. "[T]he state is not mean and we can get it more easily than, for instance, American scientists."[74]

Kapitsa's articles, in Shils' view, resonated with the consensus articulated during the "Minerva debate." Indeed, the reasoning of a representative of Soviet scientific community matched the tone of the discussion by American scientists and scholars trying to come to terms with Big Science's centralization and planning coming along with generous federal and military patronage. Sending a translation of one of Kapitsa's articles to his assistant, Shils wrote "This article is of first importance. It must get as much publicity as possible when appears."[75]

The publication of Kapitsa's essays in *Minerva* also marked a watershed in the way Soviet experience was presented in discussions of the interrelation of science and politics by American scientists after World War II. In the early 1950s, the "Lysenko affair" and the scandalous ban of genetics in the Soviet Union in August of 1948 became the most forceful arguments in favor of the "autonomy" of science and *laissez-faire*. Thus, in 1953, at the CCF's Science and Freedom conference in Hamburg, Polanyi forcefully argued against attempts by John D. Bernal and other left-wing British scientists to promote Soviet science policy as an appropriate model for the West. Polanyi backed his argument against planning of science by reminding listeners of the damage that had been done by Lysenko's misconceived effort to "run science for the public good."[76] By the early 1960s, however, in the context of the debate on Big Science on the pages of *Minerva*, Lysenko wasn't in the picture anymore. In the early 1960s the Soviet Union was seldom described as failing. Its economy tended to grow faster than that of the United States, and especially after the launch of Sputnik in 1957 the Soviet slogan of "catching up and overtaking" the West didn't appear to be patent nonsense.[77] In the 1960s and the 1970s, a more important lesson from the Lysenko affair was the acknowledgment that the state planning of science may produce different results depending on the nature of the state.

How were all these issues seen on the other side of the Iron Curtain? The theoretical discussions of the impact of the growth of science on society in the postwar world weren't restricted to the West. Simultaneously with the debate on "Big Science" in the United States, a parallel discussion of what was called the "Scientific-Technological Revolution" was taking place in the Soviet Union. Like their American counterparts, Soviet scientists and scholars responded to the anxieties and concerns of the Cold War, adapting and transforming them in highly specific and often peculiar ways.

The Soviet "Counterpart" of Big Science: The Theory of "Scientific-Technological Revolution"

In the late 1950s and the 1960s, in the wake of the sensational achievements of Soviet nuclear physics, the construction of the first nuclear power station and the first nuclear-powered ice breaker, and, above all, the Soviet space-exploration program, a new term entered the Soviet political and philosophical lexicon: "Scientific-Technological Revolution" (Nauchno-Tekhnicheskaia Revoliutsiia).[78] It came to denote the postwar scientific achievements and technological innovations that had taken place since World War II and that promised to change the socioeconomic conditions in the USSR and other countries.[79]

A comprehensive theory of Scientific-Technological Revolution (hereafter abbreviated to STR) was first developed during the 1960s by philosophers and social theorists in East Germany and Czechoslovakia, the most technologically advanced countries in the Soviet Bloc. In Czechoslovakia, the concept of the STR, developed by Czech philosophers, provided the Czech reform movement led by Alexander Dubček with its philosophical agenda. The program of Czech economic reforms ("socialism with a human face") was the result of teamwork by philosophers, economists, sociologists, psychologists, engineers, and natural scientists commissioned by Dubček's government and headed by the philosopher Radovan Richta, who assumed the directorship of the Institute of Philosophy in Prague in 1968. A programmatic collective monograph, *Civilization at the Crossroads: Social and Human Implications of the Scientific and Technological Revolution*, was published in Czech in 1967 and in English in 1969.[80] The central argument of Czech reformers was that the STR marked a new epoch "in the evolution of productive forces" and required the adaptation of the socialist economic system to the demands of modern industrialization and scientific-technological development. The STR was critically different from the "first industrial revolution" in many important aspects, Czech reformers argued, because in the socialist countries the qualitatively new possibilities of the STR were combined with an economic system inherited from the first industrial revolution. "These considerations," according to Richta's report, "underscore the vital need for radical economic reforms now being introduced in the socialist countries."[81] In practice, such an adaptation would require a more flexible and transparent economic system, and openness to the worldwide exchange of information and ideas.

The Soviet invasion of Czechoslovakia in August of 1968 and the end of Dubček's "new way toward socialism" had immediate consequences for philosophical discussion of the STR. Czech specialists who played leading roles in the debates of the 1960s, particularly those who were active in the reform movement, lost their formal positions, although some continued to work in less visible roles as researchers. Some theoretical innovations of Czech reformers, particularly concerning the role of social groups and the nature of social relations under socialism, were explicitly rejected.[82] The major effect of the events of 1968 on the theories of the STR was a shifting of the focus from the reformist cause (that is, the need for socialist societies to adjust to the requirements of the STR) to discussion and evaluation of the STR's anticipated or unanticipated social and political consequences.

The 1970s marked the beginning of what might be called the "era of the STR" in Soviet political discourse, when the STR became a central notion in the official statements of Soviet leaders. The greatly increased official commentary on the STR was part of the campaign to formulate national policies and to mobilize bureaucratic support for the major economic and political decisions of the late 1960s and the early 1970s. Nixon's visit to Moscow in 1972, followed by the accords on Soviet-American cooperation in space exploration, environmental matters, and trade, marked the beginning of détente, which was presented in official Soviet pronouncements as the end of the Cold War. Rather than ending the Cold War, however, détente introduced a new phase of it, marked by increased economic cooperation and trade.

The extension of foreign trade and the importation of Western technology in the Soviet Union had already increased greatly under Khrushchev, after the domestic liberalization and opening up of the Soviet economy to the outside world after Stalin's death. Although in the late 1950s and the 1960s the Soviet economy was growing faster than American one, with Soviet technological confidence boosted by the successful launching of Sputnik in 1957 and by other achievements in space exploration and in nuclear technologies, top Soviet officials acknowledged major weaknesses in domestic research and development. The economic historian Philip Hanson traced the beginning of the internal debate on the economic reforms within the Party Central Committee to 1958, when a Plenum of the Central Committee encouraged systematic attempts to make the planning of Soviet R&D more comprehensive.[83]

The oil crisis of 1973 created new opportunities for the exporting of oil and opened new possibilities for foreign purchases and restructuring of Soviet economy. The steep rise in oil prices radically improved the Soviet Union's terms of trade and gave an enormous boost to the foreign purchases, leading Brezhnev's government to adopt a long-term policy of technology transfer from the highly industrialized capitalist West.[84] During the period 1965–1972, two huge vehicle manufacturing plants, one for cars (the Fiat-Togliatti plant) and one for trucks (the Kama River plant), were constructed by purchasing equipment and general technical services from Italian and American companies.[85] The Kama River complex was parceled out to a number of different Western contractors, since no single Western company was prepared to take the risk of being the general contractor.[86] In both cases, the whole systems of technological know-how were transplanted from capitalist West to socialist East. Thus, during the construction of Fiat-Togliatti plant in what is now the province of Samara, Western contractors had overall responsibility for designing, equipping, installing, and commissioning the production facilities. In addition, they trained the Soviet workers, which resulted in significant cross-border flow of machines and people. Western specialists flew to the construction site and to Moscow, where the design bureau was located; Soviet citizens flew to Italy to learn how to operate the equipment that was being installed.

The emphasis of Brezhnev's government on the importation of whole systems of technology from the capitalist West made theories of the STR the main element in evaluating the effects (especially undesired and unanticipated ones) of the infusion of foreign technology into socially and politically different societies. In this context, Soviet STR theorists were expected to offer a comprehensive discussion of the relations between technology and society. They did as well as they could within the constraints of the official philosophical discourse. Questioning conventional interpretations of the exchange of technology as a form of cultural diffusion, they postulated analytical distinctions between the form and the content of modern technologies, between "technique" (gadgets and machines) and "technology" (social and economic relationships embedded in supposedly value-free machines), and between direct and indirect effects of technology on society. In a sharp contrast with usually highly abstract philosophizing, this discussion was tied to the real-life situations and processes. For example, in 1972 the philosopher Genrich Volkov contended that some technological innovations, for example computer technologies, increased labor productivity indirectly through the changes in production relations, while other technologies had a direct effect on labor process, "relatively independent of socio-economic operations." Thus, Volkov concluded, "an assembly line would require the same type of highly specialized, mechanical operations, no matter whether it is installed in a Detroit plant or in a plant in Sverdlovsk."[87]

These discussions revealed deep disagreements about the social and political implications of technology transfer. The authors of a 1972 volume titled *Nauchno-tekhnicheskaia revoliutsiia i social'nyi progress* contended that new technical hardware couldn't simply be grafted onto existing processes of labor, production, and management; rather, the existing processes required changes in order to accommodate new machines. Technical breakthroughs could be utilized effectively only if adjustments were made to the larger social systems. For example, the installation of computers would not produce "revolutionary" changes in the forms and organization of production, or in decision-making practices, unless they were accompanied by changes in the organization of the flow and content of technical and social information.[88] The underlying argument was that techniques could be considered value-free, but they were embedded in the value-laden technologies that enabled them to perform social and political functions.

The major producer of literature on the STR in the 1970s was the Institute for the History of Science and Technology (Institut Istorii Estestvoznaniia i Tekhniki) in Moscow. The IHST traces its origin to the Institute for the History of Science established under the directorship of Nikolai Bukharin in 1927, which was disbanded after the arrest of Bukharin and then that of another short-tenure director in 1938 during the Stalinist purges.[89] In 1945 the institute was reestablished with a new goal—to provide historical evidence to the Soviet nationalistic campaign in the wake of World War II. In the early years of the Cold War, the main function of the IHST was to assert the superiority of Soviet science.[90] In 1962, with the appointment of the new director, the philosopher Bonifatii M. Kedrov, the profile of the IHST changed dramatically.[91] Kedrov's program included a strong philosophical component and established strong connections between two Moscow institutes, the Institute for the History of Science and the Institute of Philosophy. One of the IHST's newly formed departments, dedicated to "general problems of the history of science," was staffed by philosophers from the Institute of Philosophy, who were expected to provide direction and theoretical framework to historians of specialized branches of science.

The "philosophical turn" of a historians' Institute served several goals. One of these goals concerned the new role for the history of science, as it was redefined in the 1960s. During the nationalistic campaigns of the late 1940s and the 1950s, the IHST dutifully produced or supervised numerous works that claimed priority for Russian science in almost any significant scientific discovery and trumpeted the superiority of Soviet science system over the Western one. At the same time, as a result of these militant nationalistic campaigns many historians retreated to descriptive and antiquarian modes of scholarship, or to exotic fields such as Arabic mathematics and ancient science.[92] In the 1960s, neither of these strategies of historians suited the new political agenda epitomized by the notion of the STR. The rhetoric of the STR had emphasized *recent* scientific developments and had encouraged study of the scientific and technological achievements of major *Western* industrial nations, rationalizing the new emphasis of Soviet economics on the transfer of Western technology and know-how. Kedrov's 1962 program emphasized that the IHST's structure and research focus should be changed to "correspond to the present day situation" and to ensure the production of up-to date "synthesizing and analytical work on the development of recent science and present-day science."[93] New departments—one dedicated to "the Scientific-Technological Revolution" and one to "General Problems of the History of Science"—were created to ensure the implementation of these changes.[94]

During the 1970s, the IHST produced or supervised a large number of collectively authored books on the STR.[95] One of the most important was *Man-Science-Technology: A Marxist Analysis of the Scientific-Technological Revolution* (1973), written by members of two Moscow institutes—the Institute of Philosophy and IHST—teamed with the Czechoslovak Institute of Philosophy and Sociology in Prague, where Radovan Richta and a number of other authors of the Czech reforms manifesto *Civilization at the Crossroads* continued to work as researchers after the Prague Spring.[96] Whereas *Civilization at the Crossroads* had contained only few references to Marx and none to Lenin, and had cited primarily Western works dating back to the 1940s (for example, works by John D. Bernal), *Man-Science-Technology* tied the notion of STR to the classics of Marxism, though it placed no particular emphasis on Marxism-Leninism.

The malleable discourse of the STR served various ends. On the political level, theories of the STR were deployed both by conservative Party leaders (to justify and rationalize the preservation of the status quo) and by those who sought to rationalize "revolutionary" transformations in many areas of Soviet life by reducing organizational "irrationality" and "optimizing" economic decision making.[97] On a theoretical level, the STR represented a general theory of social change, and hence an important modernization of Soviet Marxism.[98]

Whereas classic Marxist-Leninist theory of the 1930s had emphasized class conflict as the motor of social change, the basic premise of the STR was that the advancement of science and technology had become the principal driver of societal transformation. Unlike standard Soviet Marxism, this new outlook didn't take for granted that socialist society was the most advanced simply because of the public ownership of the means of production. Instead, the STR promoted a new image of competition between the two world systems based not on class conflict or military victory but on the Soviet Union's allegedly superior ability to develop, manage, and apply advances in science and technology.

For the Soviet philosophers, the theories of the STR presented a vantage point from which to create a new outlook and a modernized version of Marxism-Leninism, replacing the emphasis on the role of class struggle with the view that the ability to harvest the achievements of science and technology was a defining element of modern life. The theories of the STR extended Soviet Marxism-Leninism beyond the calcified form promulgated under Stalin. As the major evolving part of the official Marxist-Leninist theory in post-Stalinist Soviet Union, the STR discourse also restored, to a certain degree, the intellectual function to Soviet official philosophy, in the sense that political struggles were not solely over positions and personal power but also over ideas and the meaning of Marxism-Leninism.

The theories of the STR were conceived in dialogue with and as a response to the writings of American social analysts evaluating and forecasting the effects of large-scale growth of scientific research on society in the atomic and post-atomic age.[99] The theories of "post-industrial society" and "convergence theory" were thoroughly referenced, debated, and reviewed in the literature on the STR. In contrast to Daniel Bell and other theorists of "post-industrial society," Soviet analysts largely denied the emergence of a unitary "post-industrial society" in both mature capitalist and socialist systems.[100] Yet they presented an array of different opinions on social and political consequences of the STR. The opinions ranged from the assertion that the STR would result in a transition to "the STR society" and the convergence of the two systems into a unitary and international "industrial communism" to the view that the STR was connected not with convergence but with a divergence between the capitalist and Communist social systems and with the expectation of an "intensified ideological struggle."[101]

The bottom line of these diverse opinions was a cautious recognition that some of the *problems* of capitalist and socialist societies, in their drive toward an advanced stage of modernization, could be, indeed, common, but the Soviet theorists insisted that the capitalist and socialist societies differed in their *responses* to these problems. Georgi Shakhanazarov, a political scientist and a Central Committee department official, summarized the question in his 1979 book *Fiasko Futurologii*:

To be sure, in practice all or nearly all social phenomena bear the imprint of the prevailing system and class ideology. Even such seemingly nonsocial phenomena as *technology* and *technique* are no exception. They, too, can [acquire specific characteristics] in distinctive social conditions, giving rise to distinctive social consequences. ... Still, we must not overlook the fact that [these characteristics derive] not from the intrinsic nature of *technology* and *technique* but from the method of applying them in concrete social conditions. The gadgets and technologies are per se products of the human brain and human labor. ... Their spread and the increasing resemblance of production processes is, therefore, no argument in favor of any convergence of the social systems. ... Past experiences have shown that similar problems may be resolved in quite different ways, depending on the social conditions and the aims sought by the prevailing political forces.[102]

The theories of the STR was a Soviet counterpart to the American concept of Big Science. On the both sides of the Iron Curtain natural scientists and social theorists attempted to delineate a new epoch in human history based on the advancements of science and technology in the new, atomic or post-atomic, age. Both were articulated in response to the local economic and political situation and the needs of both states during the Cold War. In both political settings, reflection by natural scientists and social analysts on the social and political consequences of Big Science, as well as the articulation of the need for independent expertise on Big Science, was an important context for the nascent field of "science studies."[103] As the reception of Thomas Kuhn's work in the Soviet Union demonstrates, the Soviet "Big Science studiers" responded to the same anxieties and concerns of the Cold War as their Western counterparts, while adapting and transforming them in specific and often peculiar ways.

Reading Kuhn's *Structure* against the Backdrop of the Discussions on Big Science in America and the Soviet Union

Historical accounts of Science Studies (or Science and Technology Studies) usually highlight Thomas Kuhn's seminal book *The Structure of Scientific Revolutions* (1962) as challenging the received view of science and triggering the burgeoning of Science Studies as an academic field in subsequent decades. As David Hollinger has pointed out, however, many if not all of Kuhn's ideas can be found in the works of American intellectuals who were writing before or simultaneously with Kuhn.[104] Hollinger singled out Daniel Bell's *The End of Ideology*; Derek Price's *Little Science, Big Science*, Fritz Machlup's *The Production and Distribution of Knowledge in the United States*, Warren Hagestrom's *Scientific Community*, and Don K. Price's *The Scientific Estate* as constituting the immediate context of Kuhn's *Structure*.[105] Not only did those works emphasize the "communitarian aspect" of science; they also represented an important shift in the perception of science: from timeless and universal "science" without national or political affiliations to science as a concrete, historical, and interacting community.[106]

The "*Minerva* debate" on Big Science might be seen as part of this broader discussion that constituted the context of Kuhn's *Structure*. Although the *notion* of Big Science is not found in *Structure*, the *phenomenon* of Big Science can be easily discerned in the background. Kuhn, a physicist turned historian, depicted the world he knew best—the practices and the political economy of physical science that overcame revolutionary changes in the wake of World War II. Kuhn's scientists were team workers, "organization men" who followed instructions and defended their "paradigms." These were the scientists of Big Science, not "little science." However, these connotations were only implicit in Kuhn's work.

Edward Shils sought to make it explicit. In 1962, Shils asked Stephen Toulmin to review of Kuhn's *Structure* for *Minerva*. Indeed, the book came out in the midst of the discussions of the social and political implications of Big Science on the pages of *Minerva*. In his request, Shils made clear that he wanted to do some "translation work" for Kuhn's *Structure*, relating it to the discussion of science policy and science politics for which *Minerva* provided a forum in the early 1960s: "[T]he main point about Kuhn is that it should be centered around the implication of Kuhn's conception of scientific development for the planning and administration of science—not an easy task!"[107] Perhaps it was rather symptomatic that the review of Kuhn's *Structure* never appeared in *Minerva*: Kuhn's work was largely irrelevant to the "*Minerva* debate*.*"

In the Soviet Union, Kuhn's *Structure* was also read with an eye to the debates on the social and political implications of the growth of science and its role in society in the atomic and post-atomic age. The STR theory provided a specific context for reading Kuhn in the Soviet Union. During the 1960s, Kuhn's *Structure* didn't evoke any particular interest in the Soviet Union, although it was discussed and sympathetically reviewed almost immediately after its publication. In 1963, Henry Guerlac, a Cornell University historian of science, visited Leningrad's branch of the Institute for the History of Science and Technology and gave a lecture on "the development of the history of science in the USA."[108] In 1965, the IHST researcher Ludmila Markova published first Soviet review of *Structure*, offering a sympathetic summary. Markova emphasized *Structure*'s significance as a turning point for the history of science and noted its proximity to Marxist thought, characterizing Kuhn's book as a "dialectical interpretation" of the revolutions in science.[109] However, despite these interventions, it was not until the mid 1970s that Soviet reaction to Kuhn reached momentum with the publication of a Russian translation in 1975. The translation triggered a broad discussion of Kuhn's book among philosophers and scientists.

In many respects the Soviet discussion of Kuhn's work in the 1970s resembled the Western one.[110] Soviet philosophers and scientists were troubled, just as their Western counterparts were, by the "incommensurability" thesis. They argued that scientific change was a rational and logical choice, not a somewhat mysterious and irrational "gestalt switch." Kuhn's failure to discuss the sources of new knowledge was also criticized. From the perspective of Soviet critics, Kuhn's concept plausibly accounted for the survival of paradigms but didn't provide a sufficient explanation for the arrival of new paradigms.

Many Soviet reviewers pointed out that Kuhn's model of scientific change wasn't new—that Marx, Engels, and Lenin had recognized early on that the development of science occurs through periodic radical transformations, rather than through gradual grafting of new knowledge onto old. Most explicitly, critics pointed out, the view that the development of science occurs through leaps can be found in Engels' "law" of the transformation of quantitative changes into qualitative changes. Kedrov put it bluntly: "Kuhn put forth a view, long established in Marxism, that progress in science occurs through periodic shifts from the evolutionary to the revolutionary periods of development [of science], through the constant transition from one step to the next one, in the infinite progressive path to the absolute truth."[111] At the same time, Soviet critics noted that the apparent proximity between Kuhn's concept of "paradigm shifts" and the laws of dialectics was deceptive, since Kuhn's concept was largely concerned with how scientists achieve *agreement* as to what is true, rather than with how science *produces* truth—the objective knowledge of reality. One of Kuhn's critics put it this way:

> It is not difficult to find certain points where Kuhn's concept comes into contact with dialectical materialist theory of knowledge. These points of apparent proximity between the two include the implied interconnection and interdependence of theoretical and experimental practices in science, the protest against the absolutization of logical methods of studies of science, the assertion of the social conditioning of scientific research, etc. However, it would be a mistake to talk about any proximity between Kuhn's views and the basic tenets of the Marxist theory of knowledge. One of these major tenets lies in the answer to question about the relation between science and truth. This question is out of the scope of Kuhn's analysis, as the notion of "truth" does not play any role in his concept.[112]

The most prominent criticism, however, concerned Kuhn's focus on the universal features (structures) of scientific revolutions. As many Soviet critics emphasized, Kuhn's analysis, seeking for universal patterns in the development of science, didn't account for the unique features of individual revolutions. As the physicist Vitalii Ginzburg, put it, "Kuhn's scientific revolutions are as alike as two drops of water."[113] However, Ginzburg argued, the refutation of phlogiston theory is *qualitatively* different from the challenges to Newtonian mechanics by the theory of relativity, and the conflation of these two different phenomena leads to relativism, to which Ginzburg strongly objected.

Ginzburg's 1976 article opened a discussion of Kuhn's *Structure* on the pages of *Priroda*, the Soviet Union's leading popular-science magazine. The concluding article in this discussion was written by Kedrov, by this time the most prolific Soviet writer on scientific revolutions. Kedrov agreed with Ginzburg that "each scientific revolution must be studied separately as a unique and non-repetitive phenomenon."[114] Not only do revolutions differ across time, Kedrov reasoned; there are also significant differences between revolutions in different scientific disciplines, each having its own character, subject matter, and relation to other disciplines. As Kedrov pointed out, the problem with Kuhn's incommensurability thesis derived from his failure to distinguish qualitative differences between revolutions.

By this time, Kedrov had developed a comprehensive "typology" of revolutions in science in his several book-length accounts, arguing that revolutions in different centuries have different causes, effects and outcomes. One of his books was characteristically titled *Scientific Revolution: Substance. Typology. Structure. Mechanism. Criteria.*[115] Kedrov distinguished and characterized four "types" of scientific revolutions. The first "type" was the Copernican revolution characterized by Kuhn. Then there was the "Kantian Revolution" that forged the ideas of evolution. Late in the nineteenth century and early in the twentieth, the "New Revolution in the natural sciences" consolidated representations of nature based on mathematical abstractions and probability. Finally, there was the Scientific-Technological Revolution, a new phenomenon that could not be understood by reducing its causes, effects, and outcomes to the previous revolutions in science. Thus, Kedrov contended, "Although we could say, following Kuhn, that in each case there was a radical break with the system of interconnected concepts and views (what Kuhn calls "paradigms"), such a generalized (and hence abstract) approach can hardly be fruitful for the study of the revolutionary development of science."[116]

Overall, Kuhn's work was received in the Soviet Union with sympathetic interest but without any particular enthusiasm. One of the reasons was the existence of a competing discourse of scientific revolutions—the STR. The STR theorists, many of whom participated in promoting and institutionalizing the new field of *naukovedenie*—
—a Soviet version of "science studies"—were interested almost exclusively in the period since World War II. For them, history was happening here and now, in the post-atomic age, and promised a revolutionary transformation of the world. Politically primitive and counterrevolutionary, Kuhn, with his focus on normal science as a stabilizing social practice, was largely irrelevant for the theorists of the STR.

Conclusion

With the changes in the political economy of science that occurred during and after World War II—changes associated with science's dramatically increased economic dependence on public resources and the military patronage—the conventions for representing scientific enterprise had also changed on both sides of the Iron Curtain. By the early 1960s, scientists and social analysts had responded to the political challenges of the time by insisting on the image of science as a concrete, historical, and interacting community of scientist-citizens rather than timeless and universal "science" without national or political affiliations. The historian David Hollinger characterized the period as "a watershed in the history of discourse about science," which moved away from the traditional focus of philosophy of science on science method and intellectual history as foremost preoccupation of history of science and toward a focus on science's social, political, and economic contexts and aspects.[117]

The discussions of the social and political consequences of Big Science had contributed to this "paradigm shift" in the perception of science. In the United States, the loosely connected network of intellectuals associated with the Congress of Cultural Freedom contributed to a construction of a public space in which the relations between science and politics were debated, focusing the discussion on organization of science, science policy, and the planning of science. With the development of Science Studies as an academic discipline, the issues of science policy became marginalized, with science studies primarily focused on knowledge production—the central concern of Science Studies since the 1970s.[118]

The establishment of the journal *Science Studies* in 1970 marked the end of this earlier version of postwar "science studies"—"Big Science studies." Perhaps the founders of *Science Studies* didn't fully realize the extent to which the launch of this journal was a slap in the face to the CCF-associated "Big Science studiers." The founders of *Science Studies* (later renamed *Social Studies of Science*), Roy MacLeod and David Edge, both recorded the moment of the creation of the new journal in their recollections. As Edge recollected, John Maddox—then the head of Macmillan's journals division—had been enthusiastic about the idea of a new quarterly journal called *Science Studies*, even hoping that it would become a weekly. Edward Shils' reaction was furious, however. He "wanted to sabotage us, [saying] that we were committing academic suicide. ... He wanted to stop us. ... He did not stop us, but he kept treating us like we were ... I don't know ... just nuts."[119] MacLeod presented a more polite version of the same story in his published memoir: "Edward Shils cordially discouraged me from doing anything of the sort. There was simply not enough good material, he said, with the implication that anything 'good' he would publish in *Minerva* himself."[120] Shils gave his own account of this moment in a letter to Shepard Stone in 1970, immediately after the launch of *Science Studies*: "He [John Maddox] undertakes to establish another periodical on more or less the same subject as *Minerva* ... and he had one of his handy-girls prepare a promotional letter for this journal (it is called, I think, *Science Studies*) in which it is alleged that there is no competition between the magazines because *Minerva* deals only with relations between science and government. This is, of course, a caricature of the wide range of subjects treated by *Minerva*."[121]

In hindsight, the vision of "science studies" that Shils and his associates were advocating differed from academically linked Science Studies programs established at the universities in the United States and the United Kingdom in the 1960s and the 1970s. It was different, too, from "science studies" or *naukovedenie* in the Soviet Union. Yet, I would argue, these alternative approaches to studies of science, in which scientific achievements and growth appeared inseparable from issues of science policy, organization of science, science governance, ethics of science, and the planning of science, constituted an important part of the "pre-history" of Science Studies as we know it today, by establishing public forums for analysis and debate.

The case of the CCF and the "*Minerva* debate" can also illustrate the complexities and ambiguities of "cultural cold wars." The outcomes of discussions on science initiated under the auspices of the CCF were *shaped* by the organization's explicit political agenda but were not directly, let alone unequivocally, *determined* by the political demands. Although the studies of science during the Cold War encapsulated the political concerns and anxieties of the time, there was no single Cold War "party line." The CCF and its engagement in "science studies" presents a story akin to other social sciences that have received ample support during the Cold War: Area Studies, behavioral science, human relations, development studies, American Studies, and a host of other "studies" and interdisciplinary "clusters" that served national interest but not necessarily in predictable ways.[122] The CCF intellectuals' claims of being "independent" and "free" in their thinking would not survive the 1960s. The revelation of the CIA's sponsorship of the CCF shattered comfortable assumptions of scholars in the service of the state. Yet their quest for "middle ground," reconciliation, and compromise, as part and parcel of their conceptions of scholarship and service to the state, was effectively shifting the debate away from simplistic Cold War narratives of East-West competition.

As the term "Big Science" gained currency after the end of the Cold War, it gradually lost its political, connotations. This chapter, by exploring the context in which the concept of "Big Science" was framed originally—in Cold War America, as well as in the post-Stalinist Soviet Union—suggests that extending the notion of "Big Science" to later periods may be problematic. Although specific "characteristics" of Big Science as a mode of organization of science can be discerned as a set of traits applicable to different examples of large-scale scientific enterprises in the past, such extension draws attention away from the explicit Cold War connotations of this notion, and from what

Stephen Toulmin called the "political instrumentality" of Big Science shaped by the political economy of the Cold War.

Notes

1. See Peter Galison, "The Many Faces of Big Science," in *Big Science: The Growth of Large-Scale Research*, ed. Peter Galison and Bruce Hevly (Stanford University Press, 1992); James H. Capshew and Karen Rader, "Big Science: Price to the Present," *Osiris* 7 (1992): 3–25; Steven Shapin, *Scientific Life* (University of Chicago Press, 2008); Jon Agar, *Science in the Twentieth Century and Beyond* (Polity, 2012).

2. Agar, *Science in the Twentieth Century*, 330 (emphasis added). This distinction is also made in Capshew and Rader's "Big Science."

3. Edward Shils, "Minerva [editorial]," *Minerva* 1, no. 1 (1962): 5–17, at 16.

4. Ibid., 9.

5. Ibid., 12.

6. Alvin M. Weinberg, "Impact of Large-Scale Science on the United States," *Science* 134 (1961): 161–164; Derek de Solla Price, *Science Since Babylon* (Yale University Press, 1961).

7. Weinberg, "Impact of Large-Scale Science," 162. On Weinberg's championing of the view of Big Science as a pathological condition and his use of the metaphors of "disease," see Capshew and Rader, "Big Science."

8. On scientists' reservations and criticism of Big Science, see Shapin, *Scientific Life*, 80–89. On biologists' reactions on different modes of Big Science in the 1960s, see Elena Aronova, Karen Baker, and Naomi Oreskes, "Big Science and Bid Data in Biology: From the International Geophysical Year through the International Biological Program to the Long Term Ecological Research (LTER) Network, 1957–Present," *Historical Studies in the Natural Sciences* 40, no. 2 (2010): 183–224.

9. Weinberg, "Impact of Large-Scale Science," 161.

10. Ibid.

11. Edward Shils to Alvin Weinberg, November 2, 1962, Minerva Records, Special Collections Research Center, Regenstein Library, University of Chicago (hereafter cited as Minerva Records), box 1, folder 12.

12. Alvin Weinberg to Edward Shils, November 15, 1962, Minerva Records, box 1, folder 12.

13. Edward Shils to Alvin Weinberg, November 28, 1962, Minerva Records, box 1, folder 12.

14. Alvin Weinberg, "Criteria for Scientific Choice," *Minerva* 1/2 (1963): 159–171.

15. Ibid., 159.

16. Ibid., 171.

17. Ibid., 170.

18. Polanyi was the source of Shils' perhaps deepest intellectual inspiration, which, as Stephen Turner put it, "cannot be reduced to a model of 'influence.' ... [Rather,] it amounted in a way to a kind of dialectical partnership that stimulated Shils' thought and Polanyi's as well." Stephen P. Turner, "Obituary for Edward Shils," *Tradition and Discovery* 22, no. 2 (1996): 5–10. See also Louis H. Swartz, "Michael Polanyi and the Sociology of a Free Society," *The American Sociologist* 29 (1998): 59-70.

19. Michael Polanyi to Edward Shils, 15 November 1962, Minerva Records, box 1, folder 4. As Shils later admitted, Polanyi became less interested in the journal when it became more focused on "science-policy." Edward Shils, "A Great Citizen of the Republic of Science: Michael Polanyi, 1892–1976," *Minerva* 14, no. 4 (1976): 1–5.

20. Michael Polanyi, "The Republic of Science: Its Political and Economic Theory," *Minerva* 1, no. 1 (1962): 54–73.

21. Ibid. For a thorough discussion of Polanyi's political philosophy, see Mary Jo Nye, *Michael Polanyi and His Generation: Origins of the Social Construction of Science* (University of Chicago Press, 2011).

22. Polanyi, "The Republic of Science," 54.

23. As Mary Jo Nye had demonstrated, these views were rooted in Polanyi's career and the experiences he had in the 1920s and the early 1930s as a physical chemist in the Kaiser-Wilhelm-Gesellschaft Institutes in Berlin-Dahlem—the institutions that enjoyed the government support at the same time not limiting the scientists' autonomy. Mary Jo Nye, "Historical Sources of Science-as-Practice: Michael Polanyi's Berlin," *Historical Studies in the Physical and Biological Sciences* 37, no. 2 (2007): 409–434.

24. Stephen Toulmin, "The Complexity of Scientific Choice," *Minerva* 2 (1964): 343–359. For a more detailed discussion of the "*Minerva* debate," see Elena Aronova, "The Congress for Cultural Freedom, Minerva, and the Quest for Instituting 'Science Studies' in the Age of Cold War," *Minerva* 50 (2012): 307–337.

25. Toulmin, "The Complexity of Scientific Choice," 354.

26. Michael Polanyi, "The Growth of Science in Society," *Minerva* 5, no. 4 (1967): 533–545, at 543–544.

27. Alvin Weinberg, "Scientific Choice and the Scientific Muckrakers: Review Article," *Minerva* 7, no. 1–2 (1968): 52–63, at 61.

28. Alvin Weinberg, "Origins of Criteria for Scientific Choice," *Current Contents* 32 (1991): 18. For a discussion, see Aant Elzinga, "The Rise and Demise of the International Council for Science Policy Studies (ICSPS) as a Cold War Bridging Organization," *Minerva* 50 (2012): 277–305, at 296.

29. Weinberg, "Criteria for Scientific Choice," 160.

30. Ibid.

31. Alvin Weinberg, "Edward Shils and the 'Governmentalisation' of Science," *Minerva* 34 (1996): 39–43, at 42.

32. On the history of the CCF, see Frances Stonor Saunders, *The Cultural Cold War: The CIA and the World of Arts and Letters* (New Press, 2000); Volker R. Berghahn, *America and the Intellectual Cold Wars in Europe: Shepard Stone between Philanthropy, Academy and Diplomacy* (Princeton University Press, 2001); Giles Scott-Smith, *The Politics of Apolitical Culture: The Congress for Cultural Freedom, the CIA and Post-War American Hegemony* (Routledge, 2002); Peter Coleman, *The Liberal Conspiracy* (Free Press, 1989).

33. Cited in Coleman, *The Liberal Conspiracy*, 56.

34. On Koestler's "hard-line" in the early days of the CCF, see Berghahn, *America and the Intellectual Cold Wars*; Scott-Smith, *The Politics of Apolitical Culture*.

35. Michael Polanyi to V. B. Karnik, January 11, 1961, Michael Polanyi Papers, Special Collections Research Center, Regenstein Library, University of Chicago (hereafter cited as Polanyi Papers), b. 6, f. 1.

36. Scott-Smith, *The Politics of Apolitical Culture*, 139.

37. Michael Polanyi, "CCF, Memo," 19 November 1961, Polanyi Papers, b. 6, f. 6.

38. K. A. Jelenski, "Introduction," in *History and Hope: Progress and Freedom; The Berlin Conference of 1960*, ed. Jelenski (Routledge & Kegan Paul, 1962), 1.

39. Memo "Study Group of the Committee for Science and Freedom held in Paris 29 August to 1 September 1956," Polanyi Papers, b. 33, f. 10.

40. Rebecca Lemov, "'Hypothetical Machines': The Science Fiction Dreams of Cold War Social Science," *Isis* 101 (2010): 401–411.

41. "Memorandum" (n.d.), International Association for Cultural Freedom Papers, Special Collections Research Center, Regenstein Library, University of Chicago (hereafter cited as IACF Papers), Series III, box 99, folder 1.

42. Ibid.

43. Ibid.

44. "CCF Study Groups," Series III, bb. 27, 28, 29. IACF Papers.

45. "Memorandum," (n/d), IACF Papers, Series III, b. 99, f. 1.

46. "Committee on Science and Freedom. Report on the First Year's Activities: July 1954–August 1955." Committee on Science and Freedom Study Group, Agenda (Paris, 1956). IACF Papers, Series III, Box 12, folder 4. On Polanyi's Society for Freedom in Science, see Nye, *Michael Polanyi and His Generation*.

47. "Hamburg Congress on Science and Freedom, General Summary," IACF Papers, box 5, folder 7.

48. "Shils Comments on Polanyi," Science and Freedom Congress, Hamburg 1953, IACF Papers, box 5, folder 2.

49. Committee on Science and Freedom Study Group, Agenda (Paris, 1956). IACF Papers, Series III, b. 12, f. 4.

50. Edward Shils, "Outline of the General Conference of the Congress for Cultural Freedom," April 1959, International Association for Cultural Freedom/Congress for Cultural Freedom/ International Council on the Future of the University Records, Special Collections Research Center, Regenstein Library, University of Chicago (hereafter cited as IACF/ICFU Papers), b 1.

51. Ibid.

52. CCF Study Groups, IACF Papers, Series III, b. 27–29.

53. Edward Shils to A. K. Brohi, 11 April 1960, IACF/ICFU Papers, b. 1.

54. Edward Shils to J. Robert Oppenheimer, 12 July 1960, IACF/ICFU Papers, b. 1.

55. Cited in Coleman, *The Liberal Conspiracy*, 9.

56. On the CIA and its involvement in funding scientific research in various fields, see Ronald Doel and Allan A. Needell, "Science, Scientists, and the CIA: Balancing International Ideals, National Needs, and Professional Opportunities," in *Eternal Vigilance? 50 Years of the CIA*, ed. Rhodri Jeffreys-Jones and Christopher Andrew (Frank Cass, 1997). As Doel and Needell show, scientists were generally willing to aid the CIA in providing scientific intelligence on atomic, biological, and chemical weapons as well as basic science fields in the 1950s. Discussing the ways different scientists involved in intelligence gathering had tried to negotiate, reconcile and/or in various ways struggled with fundamental differences between the "ideals and values" of intelligence gathering and those of science, Doel and Needell argued that these experiences "have profoundly altered the image and practice of science in postwar America."

57. Cited in Christopher Lasch, *The Agony of the American Left* (Andre Deutsch, 1970).

58. Lasch, *The Agony of the American Left*, 104–105.

59. Sidney Hook, *Out of Step: An Unquiet Life in the 20th Century* (Harper & Row, 1987), 451.

60. "Copy of the text of the letter sent on May 4th to the Editor of the *New York Times* by John Kenneth Galbraith, George Kennan, Arthur Schlesinger, Jr." [1967], IACF/ICFU Papers, b. 1.

61. Edward Shils to Crawford, Goodwin, n/d. IACF/ICFU Papers, b. 1, f. 12.

62. Ibid. As a countermeasure against the accusations thrown by the press, Shils suggested a project aimed at producing a well-documented history of the CCF, which would imply the organization of the archive of the CCF records documenting its activities and the oral history interviews with the "intellectual figures who played leading roles in the history of the Congress," such as Michael Polanyi, Raymond Aron, Arthur Koestler, Willy Brandt, George Kennan, Daniel Bell, as well as with the members of the CCF executive staff: Michael Josselson, Nicolas Nabokov,

Francois Bondy, Konstantin (Cot) Jelenski, Pierre Emmanuel, and Melvin Lasky. IACF/ICFU Papers, b. 1, f. 12.

63. Michael Polanyi to Pierre Emmanuel, 9 April 1968, Polanyi Papers, b. 6, f. 13.

64. Michael Polanyi to Raymond Aron, May 9, 1967, Polanyi Papers, b. 6, f. 10.

65. Michael Polanyi to Pierre Emmanuel, 9 April 1968, Polanyi Papers, b. 6, f. 13.

66. "Press Release, Monday, October 2, 1967," Polanyi Papers, b. 6, f. 10.

67. On the CCF after its reorganization in 1967, see Aronova, "The Congress for Cultural Freedom."

68. Howard Brick, "Optimism of the Mind: Imagining Postindustrial Society in the 1960s and 1970s," *American Quarterly* 44, no. 3 (1992): 348–380.

69. Transcript of Proceedings of the conference "Post-Industrial Society and Cultural Diversity," June 12–14, 1970. IACF Papers, Series III, b. 80, f. 3.

70. Pyotr Kapitsa, "Slovo o progresse," *Pravda*, January 20 1966; Kapitsa, "Uchenyi i plan," *Literaturnaya Gazeta*, March 5, 1966; Kapitsa, "Scientific Policy in the USSR: The Scientist and the Plans," *Minerva* 4, no. 4 (1966): 555–560.

71. Edward Shils, "Kapitsa's views on how science should be planned," *Minerva* Records, box 8, folder 10.

72. Slava Gerovitch, *From Newspeak to Cyberspeak: A History of Soviet Cybernetics* (MIT Press, 2002).

73. Kapitsa, "Scientific Policy in the USSR," 556.

74. Ibid., 559.

75. Edward Shils to Marion Bieber, n/d [1966], Minerva Records, box 7, folder 14.

76. Michael Polanyi, opening address, 1953 Congress "Science and Freedom," Hamburg, IACF Papers, box 5, folder 7.

77. For an overview of the Soviet economy, see Philip Hanson, *The Rise and Fall of the Soviet Economy and Economic History of the USSR from 1945* (Longman, 2003).

78. Although the literal translation of "tekhnicheskaia" should be "technical" (in contrast to "tekhnologicheskaia" that stands for "technological" in Russian) we are using a standard rendering in English literature on Scientific-Technological Revolution.

This section is based on the previously published article: Elena Aronova, "The Politics and Contexts of Soviet Science Studies (*Naukovedenie*): Soviet Philosophy of Science at the Crossroads," *Studies in East European Thought* 63, no. 3 (2011): 175–202.

79. In the 1970s and the 1980s the literature on the STR was thoroughly reviewed by American Sovietologists and political analysts. See Erik P. Hoffmann and Robbin F. Laird, *"The*

Scientific-Technological Revolution" and Soviet Foreign Policy (Pergamon, 1982); Hoffmann and Laird, *The Politics of Economic Modernization in the Soviet Union* (Cornell University Press, 1982); Hoffmann and Laird, *Technocratic Socialism: The Soviet Union in the Advanced Industrial Era* (Duke University Press, 1985); Cyril Black, "The Scientific-Technological Revolution: Economic to Scientific Determinism?" (occasional paper presented at Kennan Institute for Advanced Russian Studies, 1979); Arnold Buchholz, "The Role of the Scientific-Technological Revolution in Marxism-Leninism," *Studies in Soviet Thought* 20 (1979): 145–164; Arnold Buchholz, "The Scientific-Technological Revolution (STR) and Soviet Ideology," *Studies in Soviet Thought* 30 (1985): 337–346; Erik P. Hoffmann, "Soviet views of 'Scientific-Technological Revolution,'" *World Politics* 30 (July 1978): 615–644; Friedrich Rapp, "Soviet-Marxist Philosophy of Technology," *Studies in Soviet Thought* 29 (1985): 139–150.

80. The title of the volume is reminiscent of *Science at the Crossroads*, Papers from the Second International Congress of the History of Science and Technology, ed. Nikolai I. Bukharin et al. (Frank Cass, 1971; originally published in 1931), of which Boris Hessen's paper on Newton is the most famous legacy. (I thank an anonymous reviewer for pointing out this connection.) Although *Civilization at the Crossroads* didn't make explicit references to that volume, it appears to be an homage to Bukharin, since Richta's conception contained the theoretical redefinitions of Marxist doctrine along the lines of Bukharin's theoretical position on the role of productive forces and the function of science (both natural and social sciences) in society. On Richta's theoretical views, see Vitezslav Sommer, "From 'Active Superstructure' to Culture of Socialist Experts: Czechoslovak Social Sciences and the Construction of 'Scientific' in the 1950s and 1960s," paper presented at workshop on Politics and Contexts of Science Studies during the Cold War and Beyond, Greifswald, 2012. On Bukharin's views, see Helena Sheehan, *Marxism and the Philosophy of Science* (Humanities Press, 1985).

81. Radovan Richta et al., *Civilization at the Crossroads: Social and Human Implications of the Scientific and Technological Revolution* (Australian Left Review Publications, 1969), 19.

82. Vladimir Kusin, *The Intellectual Origins of the Prague Spring* (Cambridge University Press, 1971).

83. Hanson, *The Rise and Fall of the Soviet Economy*, 92.

84. Roy Boyd and Tony Caporale, "Scarcity, Resource Price Uncertainty and Economic Growth," *Land Economics* 72, no. 3 (1992): 326–335; anonymous, "International Issues: Henry Kissinger's Views on Possible US Military Intervention Against the Oil Producers," *Time* 105 (January 20, 1975): 34–35; Antony E. Reinsh, Igor Lavrovsky, and Jennifer I. Considine, *Oil in the Former Soviet Union: Historical Perspective Long-term outlook* (Canadian Energy Research Institute, 1992).

85. On the history of the Fiat-Togliatti plant, see Lewis H. Siegelbaum, *Cars for Comrades: The Life of the Soviet Automobile* (Cornell University Press, 2008).

86. Ibid. In the American debates over this foreign policy development, the gains made available to the Soviet Union from easier access to Western machinery and know-how became the center of controversies over détente and Nixon-Kissinger strategy. See "International Development, Trade Policies, 1969–1972," http://www.state.gov/r/pa/ho/frus/nixon/iv/15710.htm.

87. Genrikh N. Volkov, *Man and the Challenge of Technology* (Novosti Press Agency Publishing House, 1972), 37.

88. Bonifatii M. Kedrov, Semen R. Mukulinsky, and Ilia T. Frolov, *Nauchno-technocheskaia revolutciia i sotcial'nyi progress* (Politizdat, 1972).

89. For the detailed account, see Aronova, "The Politics and Contexts of Soviet Science Studies." On the institutionalization of history of science in the Soviet Union in the 1920s and the 1930s, see M. S. Bastarkova, "Iz istorii razvitiia istoriko-nauchnykh issledovanii," *Voprosy Istorii Estestvoznaniia I Tekhniki* 61–63 (1978): 34–47; Semen S. Ilizarov, *Institut Istorii Estestvoznaniia I Tekhniki im. S. I. Vavilova, 1953–1993* (Nauka, 1993); Aleksandr N. Dmitriev, "Institut Istorii Nauki I Tekhniki v 1932–1936 gg," *Voprosy Istorii Estestvoznaniia I Tekhniki* 1 (2002): 3–36; Natalia L. Gindilis, "Predystoriia otechestvennogo naukovedeniia," *Voprosy Istorii Estestvoznaniia I Tekhniki* 2 (2009): 160–178.

90. As Slava Gerovich has argued in his account of the strategies of Soviet historians of science in dealing with the Cold War, Soviet historiography of science followed the political and social evolution of Soviet society. See Gerovitch, "Writing History in the Present Tense: Cold War-Era Discursive Strategies of Soviet Historians of Science and Technology," in *Universities and Empire: Money and Politics in the Social Sciences during the Cold War*, ed. Christopher Simpson (New Press, 1998).

91. Kedrov, an important figure in Soviet history and philosophy of science, had been writing on a range of issues and had been involved in major Soviet political debates over science since the 1920s; for the detailed discussion of his work and career, see Aronova, "The Politics and Contexts of Soviet Science Studies." Kedrov's most original work in the history of science was his meticulous hour-to-hour reconstruction of Mendeleev's discovery of the periodic law. (Kedrov called it "a microanatomy of scientific discovery.") As Michael Gordin pointed out, in that work Kedrov emphasized the role of textbooks in the construction of the periodic system—a line of reasoning quite resonant with the present day science studies approaches: see Michael Gordin, "The organic roots of Mendeleev's periodic law," *Historical Studies in the Physical Sciences* 32, no. 1 (2002): 263–290.

92. Many historians and literary critics retreated to a descriptive and factological style, or to "textology" (the publication of original texts with only minimal commentary and with no interpretation). For a discussion of Pushkin scholars during Pushkin Centennials in the late 1930s, when Pushkin was redefined and mythologized as a Soviet hero, see Karen Petrone, *Life Has Become More Joyous, Comrades: Celebrations in the Time of Stalin* (Indiana University Press, 2000), 113–149.

93. Prilozhenie "O napravlenii nauchnykh issledovanii i strukture Instituta istorii estestvoznaniia I tekhniki AN SSSR" k postanovleniiu Presidiuma AN SSSR ot 12 oktiabria 1962 (Archive of the Institute for the History of Science and Technology, Moscow, Russia, hereafter cited as IHST Papers).

94. Prilozhenie "O napravlenii nauchnykh issledovanii i strukture Instituta istorii estestvoznaniia I tekhniki AN SSSR" k postanovleniiu Presidiuma AN SSSR ot 12 oktiabria 1962 (IHST Papers).

95. The "collective volumes" on the STR published by IHST researchers or under the supervision of IHST included the following: A.A. Kuzin, N.N. Stokova, S.V. Shukhardin et al., *Sovremennaia nauchno-technicheskaia revolutciia. Istoricheskoe issledovanie* (Nauka, 1967); *Sovremennaia nauchno-technicheskaia revolutciia* (1967); *Nauchno-technocheskaia revolutciia i sotcial'nyi progress* (1972); *Chelovek—Nauka—Technika (opyt marksistskogo analisa nauchno-tecknicheskoi revolutcii)* (1973); *Nauchno-technicheskaia revolutciia i izmenenie struktury nauchnykh kadrov SSSR* (1973); *Nauchno-tekhnicheskaia revolutciia i obshchestvo* (1973); *Partiia i sovremennaia nauchno-technicheskaia revolutciia v SSSR* (1974). The names of only a few authors, usually the supervisors, were listed on the title pages, and some had no authors' names at all. The 1973 volume *Chelovek—Nauka—Technika* was the collaboration with Czech philosophers who were involved in the reform movement and were displaced or disgraced after the crushing of the Prague Spring.

96. Despite the suppression of the Prague reform movement, the Soviet and Czechoslovak academies continued to collaborate, but the volume was published without any names listed as the main authors, and was presented as the "combined product of a collective body of authors, members of three institutes." See *Chelovek—Nauka—Technika (opyt marksistskogo analisa nauchno-tecknicheskoi revolutcii)* (Politizdat, 1973), 3.

97. Hoffmann and Laird, *The Politics of Economic Modernization*.

98. Black, "The Scientific-Technological Revolution."

99. Similar discussions were taking place in both West and East European countries. On Big Science in Germany, see Gerhard A. Ritter, *Großforschung und Staat in Deutschland. Ein historischer Überblick* (Beck, 1992); Margit Szöllösi-Janze, *Geschichte der Arbeitsgemeinschaft der Grossforschungseinrichtungen: 1958–1980* (Campus, 1990).

100. Svetlana I. Ikonninkova, *Teoriia "postindustrial'nogo obshchestva": budushchee chelovechestva I ego burguaznye tolkovateli* (Mysl', 1975).

101. Buchholz, "The role of the scientific-technological revolution."

102. Georgii Kh. Shakhnazarov, *Fiasco futurologii* (Politizdat, 1979), 44–45.

103. Aronova, "The Politics and Contexts of Soviet Science Studies."

104. David Hollinger, "Science as a Weapon in *Kulterkampfe* in the United States During and After World War II," *Isis* 86, no. 3 (1995): 440–454, at 453.

105. Ibid.

106. David Hollinger, *Science, Jews, and Secular Culture: Studies in Mid-Twentieth-Century American Intellectual History* (Princeton University Press, 1996), 101.

107. Edward Shils to Stephen Toulmin, February19, 1964, Minerva Records, box 3, folder 10.

108. Otchet o rabote Leningradskogo otdeleniia Instituta istorii estestvoznaniia i tekhoniki AN SSSR v 1963 g. (IHST Papers, Protokoly zasedaniy direktcii).

109. Ludmila A. Markova, "Problema revolutcii v istorii nauki," *Voprosy Filosofii* 9 (1965): 178–182.

110. On the reception and the responses to Kuhn's Structure in the United States, see John H. Zammito, *A Nice Derangement of Epistemes: Post-Positivism in the Study of Science from Quine to Latour* (University of Chicago Press, 2004).

111. Bonifatii M. Kedrov, *Lenin i metodologicheskie voprosy istorii nauki* (Znanie, 1969).

112. V. M. Legostaev, "Filosophskaia interpretatciia konceptcii nauchnogo razvitiia Tomasa Kuna," *Voprosy Filosopfii* 11 (1972): 129–136, at 136.

113. Ginzburg considered Kuhn's book "trivial" in some respects though interesting and largely correct (from the point of view of a physicist). "Kuhn's scientific revolutions," he wrote, "are as alike as two drops of water, and, moreover, practically any change in science, however minor it is, can be called a revolution ... within a particular scientific community which can be as little as 25 individuals. ... If we would take the same approach to social changes, we should call any coup d'état or even a reorganization of an office consisting of 25 staff members, a revolution." Vitalii L. Ginzburg, "Kak razvivaetsia nauka?" *Priroda* 6 (1976): 73–85, at 78.

114. Bonifatii M. Kedrov, "O revolutcionnom kharaktere razvitiia estestvoznaniia," *Priroda* 10 (1976): 68–71.

115. Bonifatii M. Kedrov, *Nauchnye revolutcii: Sushchnost'. Tipologiia. Struktura. Mekhanism. Kriterii* (Znanie, 1980). See also Bonifatii M. Kedrov, *Lenin i revolutciia v estestvoznanii v 20 v. Filosofskie i estestvennye nauki* (Moskva, 1969) and Bonifatii M. Kedrov, *Lenin i nauchnye revolutcii. Estestvennye nauki. Fizika* (Znanie, 1980).

116. Kedrov, "O revolutcionnom kharaktere razvitiia estestvoznaniia."

117. Hollinger, *Science, Jews, and Secular Culture*, 101.

118. On the relation between Science Studies and Science Policy Studies, see Aant Elzinga, "Changing Policy Agenda in Science and Technology," in *Handbook of Science and Technology Studies*, ed. Sheila Jasanoff, Gerald E. Markle, James C. Petersen, and Trevor J. Pinch (London, 1995); Elzinga, "The Science-Society Contract in Historical Transformation: With Special Reference to 'Epistemic Drift,'" *Social Science Information* 36, no. 3 (1997): 411–445.

119. David Edge Oral History interview, transcript, British Society for the History of Science Oral History Project "The history of science in Britain, 1945–1965," Brotherton Library, University of Leeds, BSHS 10/8/7, on 52.

120. Roy MacLeod, "David Edge. In Memoriam," *Social Studies of Science* 33, no. 2 (2002): 181–183.

121. Edward Shils to Shepard Stone, 6 October 1970, IACF/ICFU Papers, b. 1, f. 16.

122. David C. Engerman, "Social Science in the Cold War," *Isis* 101 (2010): 393–400.

Concluding Remarks

John Krige

What have we learned from the chapters in this volume, which describe the performance of different sciences and technologies in a variety of national contexts during the Cold War? Obviously there is much that is familiar here. All the authors draw on the rich literature on Cold War science and technology to situate their particular arguments and to highlight their originality. At the same time, if we look at this collection as a whole, rather than treat each of its contributions as distinct elements, certain commonalities come into relief. In highlighting these features, I seek both to valorize the interest of the collection as a whole and to contribute to ongoing debates on science and technology during the Cold War.

The exchange between Paul Forman and Daniel Kevles on the effect of military patronage on scientific practice after World War II provided a common baseline for our contributors.[1] Their case studies confirm the deep engagement of postwar American science with the defense agencies and the US Atomic Energy Commission, to which these scholars originally drew our attention. At the same time, the findings here emphasize that directed research wasn't incompatible with doing outstanding science, that state patronage wasn't synonymous with state control, and that scientists were not, as Forman argued, "far more exploited by, than exploiting the new forms and terms of their social integration."[2] On the other hand, to say that "physics is what physicists do," as Kevles put it, fails to address the specificity of the state-science relationship in the Cold War (or any other historical period), and disarms the historian of any critical tools.[3] In this volume we seek to move beyond this stark dichotomy, and to throw new light on the practice of physics and other sciences, so as to build a picture of the intercalation of science and technology with the state after 1945 that neither sees it as betraying an ideal type nor uncritically normalizes its engagement with structures of power. Instead we see scientists pursuing a diversity of research agendas, from basic to applied, with varying degrees of relevance to weaponry, while their leaders actively worked with the state apparatus to construct a pluralistic institutional framework that, as Kevles put it, left "civilian scientists semi-autonomously tied to the military"—and we acknowledge that the prefix 'semi' leaves a fair bit of room for interpretation, not to say moral assessment.[4]

The studies presented here, along with other more recent work, help us to think more clearly through the question of what, exactly, the expectations of Cold War patrons were and how those expectations shaped, adjusted, modified, supported or discouraged certain scientific enterprises and activities. Certainly military patronage transformed the practice of science—and not just physics—in Cold War America, imposing a regime of knowledge production that was far more project-oriented, team-based, bureaucratized, and subject to the restrictions of national security than it had been. Many scientists (notably Merle Tuve and Norbert Wiener) balked at the adaptations required. And at least some Cold War habits lived on after that epoch was formally over. Science didn't return to its pre-Cold War ground state. All the same, along with adaptation came institutional and intellectual entrepreneurship, the details of which are laid out elegantly in this volume. Scientists exploited the contract system to create spaces for what they wanted to do, including fundamental research, within the limits imposed on them by funding agencies and administrators. The Cold War was at once a constraint and an opportunity.

This double aspect—constraint and opportunity—was possible because the experimental techniques and technologies that were funded by the military (including the Atomic Energy Commission) were plastic, they were mobile between research questions directed to very different ends, basic and applied, civilian and military. An acoustic tracking system built for the Navy to detect Soviet submarines could be used as a thermometer to measure ocean temperatures, and so to assess whether the planet was actually warming up (Oreskes). Radar apparatus designed, built, and paid for to enable the US Air Force to track Soviet ballistic missiles was used to confirm Einstein's theory of general relativity to an astonishing degree of accuracy by bouncing radar signals off Venus, measuring the time required for the echoes to return to Earth, and showing that they were slowed down by the gravitational pull of the sun (Wilson and Kaiser). Cheap radioactive tracers provided by AEC reactors built for the Manhattan Project, when combined with electrophoresis, centrifugation, or chromatography, could be used to identify cancerous growths, follow the movements of isotopes produced by nuclear waste and atmospheric tests, or trace pathways and metabolisms in ecosystems (Creager). Physicists and geochemists, exploiting instruments that had become standardized during World War II, transformed "traditional" geology departments by combining the quest for uranium with measurement of lead deposits at different depths in an ice pack, or with an investigation of the abundance of deuterium in nature (Shindell). Military funding in the early days of the Cold War was generous; donors were willing to trade a degree of control to secure the allegiance of the best researchers available; "semi-autonomous" scientists evolved strategies to please patrons and to pursue personal interests. He who paid the piper didn't so much call the tune as provide the instruments, the hardware, and the logistical support with which scientists could play many tunes, some of which called for deep conceptual understanding, some of which applied known truths, some of which were music to the military ear, and some of which were far from pitch-perfect.

For some scientists, this fusion of constraint and opportunity was an unfortunate if inevitable compromise. For others it was a new social paradigm required by the need to have access to increasingly complex and costly experimental equipment if one wanted to do cutting-edge research. The latter consciously worked within the framework of Cold War America, adopting its cultural norms and fashioning their identities accordingly. The military needed the scientists to help fight the war of tomorrow. Scientists wanted the resources that only the military could provide. The military secured their allegiance through a contract system that left some space for personal creativity. The concessions, while problematic for some, were unproblematic for the many who were imbued with the competitive determination to secure American scientific pre-eminence and so to contribute to US leadership of the "free world." (Krige)

As scientists became integrated into the apparatus of the national-security state, they also became adept at keeping classified knowledge, divulged only within a restricted circle, distinct from publicly available fundamental science, which could be shared openly. The studies presented in this volume emphasize that these two domains of knowledge production were mutually reinforcing. They shared a common base of tools, techniques and skills, and knowledge circulated back and forth between them. They were also interdependent: military patrons quickly realized that the freedom to do basic research was often a condition for the brightest and the best devoting their talents to doing applied and classified work. The coupling also enhanced scientists' legitimacy and their public image. Radioisotopes that were distributed widely as part of the Atoms for Peace program helped justify Congress' decision to entrust nuclear weapons to a civilian agency, and projected a benign public image of the AEC, at least for the first decade after the war (Creager). Planetary radar astronomy could coexist with missile and satellite tracking thanks to the modular electronics system of the Haystack radar at MIT's Lincoln Laboratory (Wilson and Kaiser). The very possibility of tapping into the international pool of scientific knowledge required unclassified research whose results could be shared with foreign partners (Schmid, Siddiqi, Krige). The national-security state didn't simply tolerate the co-production of classified and unclassified research; it understood them to be two sides of the same coin, reinforcing each other to sustain the quest for scientific and technological pre-eminence and industrial development.[5] The tension produced by this interlacing wasn't easily managed. Traumatized by the wartime use of the atom, the physical chemist Harold Urey tried to move into an area that he thought would be independent, only to find himself drawn back into the AEC fold. Only the AEC had the means to support the ambitious new initiative in isotope geochemistry that he envisaged (Shindell).

The studies in this volume suggest that, whereas American scientists for the most part managed their relations with their military patrons with ease, the interface with the public proved more difficult. The public's honeymoon with "the peaceful atom" ended when the health hazards of nuclear testing became evident, and eventually led to widespread opposition to nuclear power. The scientific and medical benefits of radioisotopes persisted, but they lost their value as an instrument of political legitimization of the vast nuclear complex (Creager). In the 1990s, Scripps Institution oceanographers who had blithely ignored the impact of the propagation of underwater sound on marine life during the Cold War were unprepared for the public outcry that greeted their proposal to place an acoustic thermometer in the ocean. The expertise the oceanographers had developed under Navy patronage went along with a critical public perception of their activities that thwarted their new goals (Oreskes)—oceanography was *not* anymore what oceanographers at Scripps did! Although the Cold War opened vast areas of scientific investigation (both materially and conceptually), it integrated scientists into a system of patronage that generated a culture of unaccountability, and that protected them from public scrutiny at least as effectively as did the pursuit of "pure" science in an academic milieu detached from the demands of the modern state.

Several of the chapters in this volume remind us that the late 1960s were a turning point in the American military's enthusiasm for science. The Defense Department's Project Hindsight and Congress' Mansfield Amendment demanded that closer attention be given to the previously assumed strong coupling between undirected fundamental research and the production of military technologies. The trauma of the Vietnam War, and the role of science and technology in developing advanced weapons systems for it, led to widespread public discontent and demands for the demilitarization of research, above all in academia. MIT's Lincoln Laboratory found itself at the heart of a debate over its military-academic mission, and the director of Haystack quickly took steps to shift funding for operations away from the Department of Defense onto the National Science Foundation and the National Aeronautics and Space Administration, which planned to use the array to make radar maps of the lunar surface (Wilson and Kaiser). NASA itself was forced to redefine its relevance to national goals once the competition with the Soviet Union that had marked its birth and early development were replaced by the more cooperative climate of détente. To sustain congressional support, NASA responded to the 1978 National Climate Program Act by planning for an extensive Earth observation program (Conway). Drawing on remote sensing technologies that had been developed for Earth scientists in NASA's Applications Program, and reaching out to planetary scientists who had studied the atmospheres of Mars and Venus, NASA eventually became one of the largest funders of these disciplines, with a budget for Earth-science research across all geoscience disciplines that was more than double that of the corresponding NSF Directorate—if one includes the cost of the satellites (Conway). Oceanographers at the Scripps Institution sought to maintain their funding stream by devising a program to study global warming that could be supported by a new program, funded by the Defense Advanced Research Projects Agency, specifically set up to investigate environmental problems after the end of the Cold War (Oreskes). If competition between the United States and the Soviet Union provided the dominant rationale for science funding in the early days of the Cold War, then new rationales were required to secure federal support for science from the 1970s on.

This collection addresses both the transformation of scientific practice during the Cold War and the re-evaluation of the social function of science and technology which that transformation inspired. These reflections merged with a more general discussion of the role of the state in promoting science, and with insistent calls for the definition of policies and criteria for the rational management of "big science" (Aranova, Schmid, Sidiqqi). Unexpected intellectual and institutional alignments attest to the depth of the transformation that was under way. The Central Intelligence Agency clandestinely funded *Minerva*, a new journal of science policy whose editor, Edward Shils, was both a staunch anti-communist and an opponent of Michael Polanyi's free-market conception of a healthy Republic of Science (Aranova). Thomas Kuhn was soon embarrassed by his enthusiasm for strict adherence to a paradigm as being essential to the problem-solving success of a scientific community, with its implicit critique of science as a critically engaged open society. Under pressure from his mentors, Kuhn rapidly withdrew his celebration of the value of "dogma" in scientific research, and was careful never again to describe the scientific community in language that was commonly invoked to deride Soviet control over the freedom of scientific expression (Reisch).

By bringing together chapters that deal with postwar science in very different countries and political systems, this collection emphasizes that the Cold War as an analytical category must itself be interrogated. None of its "defining characteristics" as regards the practice of science and technology should be reified, and none of them were invariable over time and geographical space during the latter half of the twentieth century. To begin with, as we all know, the extraordinary explosion of science in postwar America owed much to developments that occurred during World War II, if not before. The successful mobilization of science for that war produced the political will and the practical means to construct a dynamic research system afterwards. The mass spectrographs that were crucial to the emergence of geochemistry (Shindell), the reactors that produced radioactive tracers for biology, medicine, and ecology (Creager), the radar technologies that provided the backbone of early warning systems and of exotic tests of Einstein's relativity theory (Wilson and Kaiser), the rockets and missiles that served as scientific research tools, as delivery systems, and as platforms for spectacular techno-ideological displays of national prowess (Krige, Siddiqi), and even the pool of skilled manpower that Mao could draw on for his modernization programs in the 1950s (Wang) were all first developed before 1945, if not in the decades before.

A major historiographical question, of course, is how these social relations differed or overlapped in different countries or political systems. Though we have not managed to achieve a global reach in this volume, the chapters help us at least to address this question in terms of the "big three" Cold War powers—the United States, the Soviet Union, and China—and of Europe. Superpower competition drove support for science and technology in both the United States and the Soviet Union for two decades or more after 1945, but that competition was expressed in national goals that were adapted to the historical specificity of local conditions. The US emerged from World War II as the leading scientific and technological power, and had every intention of maintaining its pre-eminence. The Soviet Union, which lagged behind the US when the war ended, was just as determined to win the ideological and political struggle for the soul of mankind, aiming to catch up with the US and then overtake it. Entrepreneurial scientists and engineers in the Soviet Union's state-driven system (which was far more competitive than is usually recognized by Western scholars) used American supremacy as bogeyman whenever they could to win support for their pet projects, mirroring the strategies adopted in the West (Aronova, Schmid, Siddiqi).

In other countries the state defined its scientific and technological mission somewhat differently. France saw itself as a major regional power with growing global influence, but was concerned predominantly with modernization and reconstruction—and by the fear of being reduced to a "colony" of the United States (Krige). The ideological confrontation that marked the relationship between Washington and Moscow in the 1960s was of little concern to the French technocratic elite or to France's president, Charles de Gaulle. In China, Cold War science focused to a great extent on fostering self-reliance. The Communist Party positioned itself vis-à-vis both of the superpowers, seeking its own path to modernization and development by drawing on a strong sense of national loyalty. Scientists returning from the United States were as welcome as was technical assistance from the Soviet Union. Mao used what help he could get from both the East and the West to accelerate the modernization of a largely peasant society, much of it still trapped in tradition and a "pre-scientific" mentality (Schmalzer, Wang).

As we saw earlier, the 1970s ruptured the dominant dynamic of superpower rivalry and were witness to a major reconfiguration of international relations (closely intertwined with domestic developments and, some argue, with the onset of "globalization").[6] As was mentioned above, détente forced American science to seek new patrons and new rationales for funding (Conway, Oreskes, Wilson and Kaiser). The liberalization of trade with the Soviet Union and a huge influx of petrodollars after the increase in the price of oil that occurred in 1973 enabled the Brezhnev regime to adopt a long-term policy of technology transfer from the highly industrialized capitalist West that paved the way for the renewed confrontation that marked a second wave of US-USSR tension in the 1980s. It also sparked a spirited domestic debate on the implications of the "scientific technological revolution" for a Soviet social and political system that was increasingly dependent on Western imports (Aranova). In China, the Sovietization of science, technology and education of the 1950s and the early 1960s was gradually supplanted by its "Americanization" after President Richard Nixon, in a dramatic reversal of United States policy, encouraged the transnational circulation of Chinese scientists and engineers between the two countries. The market-oriented policies of Deng Xiaoping accelerated a process that began in the late 1970s. Today, notwithstanding brief periods of decline, there are almost 128,000 Chinese students studying in the United States (Wang). In sum, these chapters insist that it is perilous to think of the Cold War in monolithic terms, be it chronologically, geographically, or as a social system that shaped the practice of science.

Forman's and Kevles' analyses of the transformation of science-state relationships during the Cold War weren't simply intellectual exercises; they were also "political" interventions. Forman emphasized what he called the "false consciousness" of scientists—and of historians of science—who had "pretended a fundamental character to their work that it scarcely had."[7] Kevles insisted that military patronage was to a great extent compatible with traditional values of scientific autonomy, and imperative to laying the foundations of new exotic weapons systems. Both were intent on defining what should count as "science" in Cold War America, the one insisting that physics had strayed from its "true path" and the other that scientific practice had been reconfigured (yet again) by historical context.

The comparative approach we have taken in this volume allows us to see that similar boundary work was performed in other countries and social systems whenever and wherever the state sought to bend science and technology to national need.

Consider the Soviet Union in the 1950s. After Stalin's death, nuclear specialists there used their success in weapons development to take control of their careers from party ideologues. Like many of their Western counterparts, they insisted on characterizing what they did as "fundamental science." However, they also went further in response to the particular political imperatives of the Soviet Union, solidifying a series of research institutes that were independent of direct political control. Exploiting their new relative autonomy, they traveled extensively to learn what others were doing. They promoted two different reactor designs, one "international" and the other "Soviet," to be implemented in distinct engineering bureaus (Schmid). In the 1960s it was the turn of the rocket engineers to appropriate the label of scientist for themselves, and to compete ferociously for resources to develop different types of propulsion systems in a discursive field that conflated fundamental and applied, civilian and military, spectacular display and utter secrecy (Siddiqi). In Maoist China, science was above all an empirical method so that the mindful application of manure could count as scientific farming. It was also an instrument, at least in political rhetoric, insofar as *all* science was applied. Its exploitation was intertwined with the celebration of native techniques, the mobilization of the masses, the loyalty of scientists to the party-state and the affirmation of self-reliance, meaning autonomy both from foreign dependence and from the central government (Schmaltzer). Wherever we look, then, we see that the boundary between fundamental and applied science was widely contested as soon as the state emerged as major patron of research.

As one surveys these comparative studies, there seems to be one secure generalization that applies broadly: that researchers in all countries—and intellectuals who sided with them against "oppressive" regimes, whether capitalist, communist, or somewhere in between—developed strategies, appropriate to their local contexts, disciplines, and constraints, to carve out a space to sustain and instrumentalize traditional values of free inquiry and international exchange, even as they built devices that strengthened the power of ruling elites. They drew on universal values of science to ensure scientific autonomy as best they could, although, since practice was bounded by the "civil-defense-industrial complex," the semi-autonomy they had in practice came at the expense of the very values that were so loudly proclaimed.

Two contributions to this collection specifically interrogate the transnational flow of knowledge from different perspectives, one emphasizing the circulation of trained scientists and engineers between China and abroad (Wang) and one exploring the cross-border movement of science, technology, and skills between NASA and France (Krige). This approach demystifies the Cold War emphasis on scientific and technological achievement as a purely national affair and as a marker of national prowess. Machines have been the measure of men for several centuries, as Michael Adas has written, but superpower rivalry for global influence invested them with even greater significance.[8] Indeed, during the Cold War all of the major countries manipulated scientific and technological success to enhance national pride and to justify major investments in research and development by the state and by private industry, thereby effacing the network of international relationships in which their national research efforts were embedded. China's search for self-reliance was intended to engage the rural masses and enlist their practical knowledge in the transformation of the country. All the same, its revolutionary appeal was necessarily complemented by the need to mobilize an educated scientific elite, with foreign help, to build a modern industrial and military system. France, in its efforts to enter the domain of rocketry in the 1960s, relied on what it could learn from NASA, but also on émigré German engineers and a launch base in one of its colonial possessions.

Once we suspend the national frame to focus on the transnational flow of people and ideas, we find that no major Cold War scientific or technological development was uniquely indigenous; all were hybrid bundles of local and internationally acquired information and skills.[9] That circulation couldn't be taken for granted, however. Moscow withdrew its technical support from Beijing as soon as China began to emerge as a rival power. The United States opened its doors to Chinese scientists in the 1970s when Nixon redefined foreign policy in the region (Wang). France could take advantage of American help in space science and technology because a strong, scientific and technologically integrated Europe was an important element of American foreign policy in the 1950s and the 1960s (Krige). Intellectually dissolving national borders, as these chapters show, also requires softening, if not dissolving, academic boundaries between the history of science and technology, on the one hand, and the history of foreign policy and international society on the other.

Though we are confident that this collection breaks new ground in our understanding of the place of science in the Cold War, it is evident that much more remains to be done. Several directions for further research have emerged. One question that is mostly unaddressed and clearly left unresolved by our studies is how scientists, as individuals, understood and negotiated their relationships with the national-security state.[10] Paul Forman suggested that they were unable to face up to the distortion of their calling by the demands of their military patrons. Recently Joseph Masco has revisited this question, arguing that scientists were haunted—and thus motivated—by the dangers of nuclear destruction by a ruthless adversary, a fear that was ably managed as a tool to secure allegiance to national goals.[11] Both Forman and Masco suggest that scientists sacrificed substantial intellectual autonomy on the altar of the military-industrial-academic complex. For Forman this has led to the subversion of disciplinary rigor and personal integrity, and to the corruption of the critical faculty.[12] Bounded knowledge is incompatible with independent expertise; science is no longer subversive. The cultural construction of practices that tie research to grants, contracts and commercialization, and that spawn regulatory regimes that place severe restraints on free inquiry, needs to be understood and studied more closely.

There are several other lacunas in this volume. Quite obviously, we have not devoted any attention to science in the so-called Third World, notably India and Africa. The important work being done on these regions by a few historians of science and technology, and by diplomatic historians interested in "modernization," treats the Cold War as a global phenomenon and highlights the knowledge/power nexus that structured North-South relations.[13] It is also apparent, not only from this volume but also from wider debates in the historiographical community, that the 1970s are understudied and poorly understood.[14] That decade was not only one of détente. It was also a period of growing interdependence between states, of a move from a bipolar to a multipolar international system, of a concomitant decline in the autonomy of the nation-state (especially the United States), and of the emergence of non-state actors that had considerable capacity to force issues such as human rights and environmental degradation onto the political agenda. In short, the 1970s saw the onset of a process of "globalization" that was made possible by major scientific and technological breakthroughs, above all in communications and computing, whose implications for science during the Cold War are not addressed here.

We need to continue to learn from the transnational turn in history and to break the national frames of our analyses. The studies presented here, though grounded in national contexts, speak to the need to situate Cold War science, propelled as it may have been by national priorities, in the context of interdependence and interconnectivity that globalization involves. A transnational approach also helps us to challenge American exceptionalism, and to see the United States as one actor among others in a world system. And, perhaps crucially, the transnational approach recognizes that national research systems have always been embedded in international networks through which knowledge in all its forms has circulated, and selectively appropriated at multiple nodes (Krige, Wang). Of course science has always been transnational in the sense that it has crossed, and even defied, national borders. "Universalism" is one of its defining features—or so Robert Merton claimed. However, the centrality of science and technology to the postwar state has often led those of us who study the postwar period to focus exclusively on the national framework, perhaps for logistical or linguistic reasons and perhaps because our actors defined their projects in deeply nationalistic ways. Those self-definitions, as the studies here show, while a crucial part of the story, are not its entirety.

This points to another research question. American science and technology were certainly not "self-sufficient" before the Cold War, and they certainly are not so today. As the four major US weapons laboratories pointed out in 1999, the Department of Energy's laboratories now "conduct only 1 to 2 percent of the world's research and development," and their effectiveness depends "substantially on the capacity to access and apply the 98 to 99 percent of the work that is performed elsewhere."[15] The relative decline of American power in the late twentieth century is reflected in the embedding of its research system in a global network of knowledge production and circulation, such that it no longer can be fully understood through the lens of a national framework—if it ever could. Was the Cold War an exceptional period of scientific nationalism and self-sufficiency? Or was that idea itself part and parcel of Cold War ideology? There is still much work to be done.

Notes

1. Paul Forman, "Behind Quantum Electronics: National Security as a Basis for Physical Research in the United States, 1940–1960," *Historical Studies in the Physical Sciences* 18, no. 1 (1987): 149–229; Daniel Kevles, "Cold War and Hot Physics: Science, Security and the National Security State, 1945–1956," *Historical Studies in the Physical Sciences* 20, no. 2 (1990): 239–264.

2. Forman, "Beyond Quantum Electronics," 200, 229.

3. Kevles, "Cold War, Hot Physics," 263, 264.

4. Ibid., 249.

5. Alex Wellerstein, "Patenting the Bomb: Nuclear Weapons, Intellectual Property, and Technological Control," *Isis* 99, no. 1 (2008): 57–87.

6. Niall Ferguson, Charles Maier, Erez Manela, and David Sargent, eds., *The Shock of the Global: The 1970s in Perspective* (Harvard University Press, 2010).

7. Forman, "Beyond Quantum Electronics," 228.

8. Michael Adas, *Machines as the Measure of Men: Science, Technology and Ideologies of Western Dominance* (Cornell University Press, 1989); Adas, *Dominance by Design: Technological Imperatives and America's Civilizing Mission* (Harvard University Press, 2006).

9. Michael D. Gordin, *Red Cloud at Dawn: Truman, Stalin and the End of Atomic Monopoly* (Picador, 2009); John Krige, "Hybrid Knowledge: The Transnational Coproduction of the Gas Centrifuge for Uranium Enrichment in the 1960s," *British Journal for the History of Science* 45, no. 3 (2012): 337–357.

10. One exception is Matthew Shindell's study of Harold Urey.

11. Joseph Masco: "'Survival Is Your Business': Engineering Ruins and Affect in Nuclear America," *Cultural Anthropology* 23, no. 2 (2008): 361–398.

12. Paul Forman, "On the Historical Forms of Knowledge Production and Curation: Modernity Entailed Disciplinarity, Postmodernity Entails Antidisciplinarity," *Osiris* 27, no. 1 (2012): 56–97.

13. Gabrielle Hecht, ed., *Entangled Geographies: Empire and Technopolitics in the Global Cold War* (MIT Press, 2011); Hecht, *Being Nuclear: Africans and the Global Uranium Trade* (MIT Press, 2012); Nick Cullather, *The Hungry World: America's Cold War Battle Against Poverty in Asia* (Harvard University Press, 2010); David Engerman, "American Knowledge and Global Power," *Diplomatic History* 31, no. 4 (2007): 599–622.

14. Ferguson et al., *The Shock of the Global*.

15. *Balancing Scientific Openness and National Security Controls at the Nuclear Weapons Laboratories* (National Academies Press, 1999), 11.

About the Authors

Elena Aronova is a Research Scholar at the Max Planck Institute for the History of Science in Berlin. Her work focuses on the history of earth science, environmental science, and evolutionary science. Her forthcoming book *Science and Cultural Cold War: Thinking Science on the Opposite Sides of the Iron Curtain* examines the Cold War geopolitics of science and its role in the emergence of "science studies" on both sides of the Iron Curtain.

Erik Conway is a historian of science and technology at the Jet Propulsion Laboratory California Institute of Technology, where he studies and documents the history of space exploration, space science, earth science, and technological change. In 2009 he received the NASA History Award for "path-breaking contributions to space history ranging from aeronautics to Earth and space sciences," and in 2011 he received the Watson Davis and Helen Miles Davis Prize of the History of Science Society for *Merchants of Doubt: How a Handful of Scientists Obscured the Truth about Issues from Tobacco Smoke to Global Warming* (Bloomsbury, 2010), co-authored with Naomi Oreskes. His next book, a history of robotic Mars exploration, will be published by the Johns Hopkins University Press.

Angela N. H. Creager is the Philip and Beulah Rollins Professor of History at Princeton University. She is the author of *The Life of a Virus: Tobacco Mosaic Virus as an Experimental Model, 1930–1965* and *Life Atomic: A History of Radioisotopes in Science and Medicine*, both published by the University of Chicago Press.

David Kaiser is Germeshausen Professor of the History of Science and Department Head of the Program in Science, Technology, and Society at the Massachusetts Institute of Technology, where he is also a Senior Lecturer in the Department of Physics. His books include *Drawing Theories Apart: The Dispersion of Feynman Diagrams in Postwar Physics* (University of Chicago Press, 2005), which received the Pfizer Prize from the History of Science Society, and *How the Hippies Saved Physics: Science, Counterculture, and the Quantum Revival* (Norton, 2011), a *Physics World* "Book of the Year."

John Krige is the Kranzberg Professor in the School of History, Technology, and Society at the Georgia Institute of Technology. He is the author of *American Hegemony and the Postwar Reconstruction of Science in Europe* (MIT Press, 2006), and a co-author (with Angelina Long Callahan and Ashok Maharaj) of *NASA in the World: Fifty Years of International Collaboration in Space* (Palgrave Macmillan, 2013)

Naomi Oreskes is a Professor of the History of Science and an Affiliated Professor of Earth and Planetary Sciences at Harvard University. Her books include *The Rejection of Continental Drift: Theory and Method in American Earth Science* (Oxford University Press, 1999), *Plate Tectonics: An Insider's History of the Modern Theory of the Earth* (Westview, 2003), and (with Erik Conway) *Merchants of Doubt* (Bloomsbury, 2010). *Merchants of Doubt* was a finalist for a *Los Angeles Times* Book Prize. Her book with Conway, *The Collapse of Western Civilization: A View from the Future*, was published in 2014 by Columbia University Press.

George Reisch is the author of *How the Cold War Transformed Philosophy of Science* (Cambridge University Press, 2005). He is currently working on a book on the Cold War context of Thomas Kuhn's *The Structure of Scientific Revolutions*. He is the series editor for the Open Court Press Popular Culture and Philosophy series, a co-editor (with Brandon Forbes) of *Radiohead and Philosophy: Fitter, Happier, More Deductive* (Open Court, 2009), and a co-editor (with Gary Hardcastle) of *Monty Python and Philosophy: Nudge Nudge, Think Think!* (Open Court, 2006). He is managing editor and webmaster of the long-running philosophy journal *The Monist*.

Sigrid Schmalzer is an Associate Professor of History at the University of Massachusetts, Amherst, where she teaches Chinese history and the history of science. She focuses on the political, social, and cultural significance of science in post-1949 China, especially the meaning of science in socialism and that of socialism in science. Her first book, *The People's Peking Man: Popular Science and Human Identity in Twentieth-Century China*, (University of Chicago Press, 2008), received the Allan Sharlin Memorial Award of the Social Science History Association. She is currently completing her second book project, *Red Revolution, Green Revolution: Encounters with "Scientific Farming" in Socialist China*.

Sonja D. Schmid is an Assistant Professor in the Department of Science and Technology in Society at Virginia Tech. Her work focuses on the history of nuclear policy, risk and governance, and organizational learning. Her forthcoming book *Producing Power: The Pre-Chernobyl Origins of the Soviet Nuclear Industry* (MIT Press) investigates the mutual shaping of national energy policies, technological choices, and nonproliferation concerns during the Cold War.

Matthew Shindell is a postdoctoral fellow in Harvard University's Department of the History of Science. His PhD dissertation (University of California, San Diego, 2011) is a socio-biographical study of the life and career of the American chemist Harold C. Urey. He is currently a collaborator in a historical and social study of consensus and expert assessment being conducted under the auspices of the National Academy of Sciences' National Research Council.

Asif A. Siddiqi is a Professor of History at Fordham University and holds the Charles A. Lindbergh Chair in Aerospace History at the Smithsonian Institution's National Air and Space Museum. He specializes in the history of modern science and technology, Russian history, and postcolonial technoscience. He is the author of *The Red Rockets' Glare: Spaceflght and the Soviet Imagination, 1857–1957* (Cambridge University Press, 2010).

Zuoyue Wang is a Professor of History at the California State Polytechnic University at Pomona, where he focuses on the history of modern science and technology in the United States and China. He is the author of *In Sputnik's Shadow: The President's Science Advisory Committee and Cold War America* (Rutgers University Press, 2008). He is currently working on a transnational history of American-educated Chinese scientists during the Cold War.

Benjamin Wilson is a doctoral candidate in MIT's Program in History, Anthropology, and Science, Technology, and Society. He is writing a dissertation on the history of the community of nuclear arms control experts in the United States.

Index

Abraham, Itty, 228
Acausality, 17, 18
Acoustic Tomography of Ocean Climate, 142–172
Acupuncture, 80, 86, 95–97
Advanced Research Projects Agency, 16, 148, 164, 435
Aebersold, Paul, 35, 46
Agriculture, 37, 39, 78, 87–90, 94–98, 438
Air Force Cambridge Research Laboratory, 227
Air Force Office of Scientific Research, 20
Aleksandrov, Anatolii, 323
Allison, Samuel, 111, 112
American Petroleum Institute, 118
Anderson, Paul, 158
Anger, Hal, 48
Anti-nuclear movement, 59
Apollo program, 251, 256, 257, 302
Applied Fisheries Laboratory, 51, 52
Applied science, 5, 6, 75, 76, 85, 94, 191, 192, 317, 321–327, 330–334
Aquatic Biological Laboratory, 52
Aquatic ecosystems, 50–52
Area studies, 2
Arms race, 1, 12–16, 23, 33, 228
Armstrong, Neil, 189
Arnon, Daniel, 37
Aron, Raymond, 400, 403
Aronova, Elena, 79
Arthur, J. S., 298
Ash, Michael, 285, 286, 297, 298

Atmospheric chemistry, 108
Atomic bomb, 11–16, 31, 113, 319, 322, 332, 357
Atomic Bomb Casualty Commission, 56
Atomic energy, 31–33, 51, 57–59
Atomic Energy Act, 31, 57
Atomic Energy Commission, 16, 19, 45, 108, 431, 433
    distribution of radioisotopes by, 31–33
    reorganization of, 59, 60
    research funding from, 54–56, 110, 121–143
    Urey and, 120, 121
Atomic scientists, 109–132
Atoms for Peace initiative, 57, 433
Auerbach, Stanley, 52–54
Autonomy of science, 6, 409, 439
Autoradiography, 57

Ball, George, 240
Ballistic Missile Early Warning System, 277–279
Barker, Horace, 43
Barnes, Richard, 241, 244
Barstow, Robbins, 154
Barth, Charles, 254
Basic science, 5, 6, 21, 22, 75, 76, 80, 94, 131, 191, 317, 321–334
Beijing University, 86
Bell, Daniel, 400, 403, 407
Benioff, Hugo, 122, 129
Benson, Andrew, 43

Bergmann, Peter, 297
Beria, Lavrentii, 319
Bigeleisen, Jacob, 115
Big science, 263, 393–429, 435
  in Soviet Union, 190–193, 213, 214,
    409–415
  in United States, 394–409
Bikini Atoll, 52
Biochemistry, 33–39, 50, 56–58, 435
Biological Sciences Curriculum Study, 376
Bismuth, 33, 34
Blackett, P. M. S., 11, 12
Blamont, Jacques, 227, 228, 243–245
Blumgart, Herrmann, 33, 34
Bohr, Niels, 11
Boundary work, 243
Bowen, Norman Levi, 116
Bowie, William, 170
Bowles, Ann, 147–151, 168, 169
Boxer, Barbara, 155
Boxer fellowship program, 345–347
Boxer Rebellion, 345
Brain tumors, 47, 48
Bretherton, Francis, 261
Brezhnev, Leonid, 196, 198, 203, 211, 411
Bright, Charles, 229
Bringer, Karl, 227
Brooks, Harvey, 21
Brown, Harrison, 110, 111, 125–132
Brownell, Gordon, 48
Brush, Stephen, 115, 132
Buffer states, 15
Bundy, McGeorge, 237
Bush, George H. W., 251
Bush, Vannevar, 123, 124
Buwalda, John, 122, 129

California Coastal Commission, 160–166
California Institute of Technology (Caltech),
  110, 111, 122–132, 282
Calvin-Benson cycle, 33
Calvin, Melvin, 43–45
Calvin photosynthetic cycle, 43–45

Campbell, Ian, 126
Canada, 32
Cancer, 32, 37, 42, 45, 46, 59
Capitalism, 15, 401
Carbon, 33, 43, 45
Carbon dioxide, 142–144
Carson, Rachel, 56
Cassen, Benedict, 48
Causse, Jean-Pierre, 234
Central Intelligence Agency, 165, 232, 233,
  404–407, 420, 435
Chaikoff, Israel, 37
Challenger Space Shuttle, 262–264
Chargaff, Erwin, 19
Chelomei, Vladimir, 196, 210–212
Chen Mengxiong, 356, 357
Chen Ning Yang, 360
Chernobyl accident, 324, 333
Chiang Kai-shek, 79
Chievitz, Otto, 37
China, 4–6, 77–79, 343, 436
  diplomacy of, 93
  in Mao era, 78–90, 96–98
  Soviet Union and, 77, 81–86, 89, 90,
    349–357
  United States and, 77, 81, 90, 97, 358–362
Chinese American scientists, 359–363
Chinese Communist Party, 79–83
Chinese medicine, 80, 86, 94–97
Chinese science, 75–106, 437, 438
  Americanization of, 344, 355
  basic vs. applied, 75, 76, 85, 94
  revolutionary roots of, 79, 80
  self-reliant, 76–106, 357, 358, 362, 438
  Sovietization of, 349–357, 362
  transnational, 343–369
Chinese students, 345–347, 351–355,
  359–361, 437
Chlorofluorocarbons, 108, 253
Chromatography, 44, 45
Churchill, Winston, 193
Clark, Christopher, 150, 151, 154–157
Clark, Cindy, 156

Clauser, Milton, 301
Clayton, Robert, 130
Clements, Frederic, 50
Clewett, Dave, 165
Climate change, 141–187, 251, 255, 256, 435
Cohn, Mildred, 35
Cohn, Waldo, 37
Cole, Derek, 163
Columbia Space Shuttle, 259
Committee for Economic Development, 401
Committee on Isotope Separation, 120, 121
Committee on Space Research, 230–234
Compton, Arthur Holly, 111, 112
Conant, James, 113, 372–388
Congress for Cultural Freedom, 394, 399–407, 420
Conventional warfare, 11–13
Convergence theories, 407
Cook, S. F., 37
Cooper, Frederick, 229
Copernican Revolution (Kuhn), 373, 376, 379
Cosmochemistry, 115
Costa, Daniel, 151
Cronkite, Walter, 189
Cultural Revolution, 78–82, 86, 89, 90, 97, 343, 358, 360, 361
Cybernetics, 57
Cyclotrons, 35, 37, 41, 42
Czechoslovakia, 15, 410, 413

Dazhai production brigade, 87, 88
DDT, 56
Defend Diaoyu movement, 359
de Gaulle, Charles, 232, 236, 237, 240, 244
Dempster, Arthur, 111, 112
Deng Xiaoping, 77, 78, 98, 361, 437
Department of Defense, 19, 171, 293, 294, 301, 434
Deser, Stanley, 283
de Solla Price, Derek, 395
Dialectical materialism, 387
Dietrich, Bill, 164
Distant Early Warning Line, 276

Doel, Ronald, 108, 118
Dogmatism, 372, 378–388, 435
Donaldson, Lauren, 51, 52
DuBridge, Lee, 110, 111, 122–126
Dustin, Daniel, 278
Dyson, Freeman, 45

Earle, Sylvia, 155, 165, 166
Earth Observing System, 251, 252
Earth sciences, 132, 251, 252, 434
  climate change and, 143
  funding of, 108–110
  NASA and, 251–256, 259–266, 434
  physical laboratory methods in, 108, 109
Earth System Science, 261–264
Eastern Europe, 15, 240, 328
Ecology, 33, 49–59, 435
Eddington, Arthur, 273, 274
Edwards, Paul, 20, 57, 77
Einstein, Albert, 18, 114, 273–275, 280, 297, 298, 302, 303, 432
Eisenhower, Dwight, 13, 22, 57, 251, 282, 323
Embden-Meyerhof pathway, 34
Emiliani, Cesare, 117
Endangered Species Act, 147
End of Ideology (Bell), 400, 401
Environmental Defense Fund, 161
Environmental toxins, 56, 57
Epstein, Samuel, 115–117, 129, 130
Ericson, David, 117
European integration, 240, 241
European Launcher Development Organization, 237–243
Evans, Robley, 39
Ewing, Maurice, 117, 129
Experimental Design Bureau-1 (OKB-1), 196, 199

Faanes, Craig, 158
Farmer, Crofton, 254
Federal Contract Research Centers, 301
Feinstein, Diane, 155

Fermi, Enrico, 111
Fletcher, James, 256
Food chain, 52, 54, 55, 58
Forman, Paul, 2, 6, 17–23, 75, 141, 142, 190, 252, 439
Forman thesis, 17, 18
Forrester, Jay, 20
Foster, Richard, 52
France, 4, 436
  Soviet Union and, 241
  space science in, 227–250
Frank, Philipp, 374, 375
Frieman, Edward, 152, 155
Frutkin, Arnold, 231, 232, 236–239, 244
Fujita, Rodney, 161–163

Gaddis, John, 14
Gadgeteering, 19, 75, 141, 252
Gagosian, Robert, 157
Gaia hypothesis, 261, 262
Galbraith, John, 405
Galison, Peter, 56
Gambia, 93
Gamow, George, 273
Gauge theory, 18
Geiger, Roger, 113
Geiger counters, 47–49, 57
Geochemistry, 107–139, 435
  at California Institute of Technology, 122–132
  funding of research in, 108–113, 118–130
  isotope, 109–139
  at University of Chicago, 109–126
Geochronology, 126
Geological Society of America, 118, 119
Geology, 116, 117, 128, 129, 132, 143, 432
Geophysics, 108, 109
Geopolitics, 5–7, 32, 77, 230–234, 356
Germany, 11–14, 17, 18, 237, 238, 274
Geyer, Michael, 229
Glass, Hiram, 380–382, 388
Global economy, 14–16

Global Habitability, 260, 261
Globalization, 229, 243, 436, 440
Glushko, Valentin, 189, 190, 201–217
Goldsmith, Julian, 117
Goldstein, Richard, 297, 298
Gordin, Michael, 13
Gore, Albert, Jr., 148
Governmentalization, 394, 395
Government-funded research, 16–22, 109–113, 119–130, 145–148, 170–172, 192, 293, 294, 302, 396, 397, 434, 435
Graham, Loren, 190, 213
Grechko, Andrei, 212
Green, Paul, 280–282
Greenberg, David, 37
Greenhouse gases, 142–144
Green revolution, 78
Greenwood, Ron, 255
Greiff, Lotti, 114, 115
Groves, Leslie, 41, 113
Guggenheim Foundation, 129
Gutenberg, Beno, 122, 129

Halliday, Fred, 77
Hamblin, Jacob, 113
Hamilton, Alice, 107, 108
Hamilton, Joseph, 35, 39–42, 46
Hassid, William, 43
Heard Island Feasibility Test, 146, 147
Heilbron, John, 113
Hertz, Saul, 39
Hevesy, Georg von, 33, 37
Hickenlooper, Bourke, 57
Hill, Robin, 43
Holloway, David, 193, 197, 198
Hook, Sidney, 405
Hubbs, Carl, 150
Human experimentation, 33, 34, 39, 42
Hutchins, Robert, 112
Hutchinson, G. Evelyn, 33, 50
Hu Xiansu, 350, 351
Hyde, David, 153
Hydrogen bomb, 12

Index

Iangel', Mikhail, 196–198, 209–212
Incommensurability thesis, 417
India, 5
Institute for the History of Science and Technology, 412, 413
Institute of Atomic Energy, 325–327
Institute of Nuclear Studies, 108–128
Insulin, 84, 85
Intellectual work, context of, 1–3, 7, 17–24
Intelligent Design movement, 79
International Association for Cultural Freedom, 406, 407
International collaboration, 227–250, 322, 323, 330, 331, 362
International Conference on the Peaceful Applications of Atomic Energy, 323
International Geophysical Year, 356
Ioffe, Abram, 319
Isotopes, 5, 34, 35, 48, 113, 114, 120, 121, 125. *Also see* Radioisotopes

Japan, 11, 12
Jet Propulsion Laboratory, 258, 282, 285, 297, 298
Johnson, Howard, 300–303
Johnson, Lyndon, 231, 251
Joliot-Curie, Frédéric, 35, 236
Joliot-Curie, Irène, 35
Josselson, Michael, 406

Kamen, Martin, 43
Kapitsa, Pyotr, 408, 409
Kay, Lily, 56
Keeling, Charles David, 142, 143
Keldysh, Mstislav, 195, 196, 205, 210, 321
Keldysh Commission, 205–209, 213
Kennan, George, 404, 405
Kennedy, John, 251
Kevles, Daniel, 1, 2, 6, 21–23, 431, 437
Khariton, Iulkii, 194
Khrushchev, Nikita, 193–196, 199, 205, 211, 349, 351
Kiessling, W., 34

King, F. H., 97
Kingston, Robert, 281
Knowledge flows, 233–245, 358, 359, 438, 439
Koestler, Arthur, 400
Kojevnikov, Alexei, 190, 191, 318, 333, 334
Korean War, 349, 351, 356
Korolev, Sergie, 189, 190, 194–217
Kovda, V. A., 350
Krebs cycle, 34
Krebs, Hans, 34
Krige, John, 94
Kuenzler, Edward, 54, 55
Kuhn, Thomas, 5, 371–394, 415–418, 435
Kurchatov, Igor, 194, 319, 320, 323
Kuznetsov, Nikolai, 202, 203, 213, 214

Lamont Geological Laboratory, 117
Landsat, 253–255, 260
Landsat Data Continuity Mission, 253
Laurance, Donald, 255
Lawrence, Ernest, 35–39, 43
Lawrence, John, 35–39, 46
Lead pollution, 107, 108
League of Nations, 14
LeFeber, Walter, 13–15
Leningrad Physico-Technical Institute (LFTI), 319
Leninism, 78, 79
Leskov, Sergei, 189
Le Tianyu, 80, 81
Liang Sili, 357
Libby, Willard, 111, 129
Liberalism, 372, 373, 380, 385, 388, 400, 401
Liberia, 93
Life sciences, 32–37, 56–58
Lincoln Laboratory, 274–303, 433, 434
Lindee, Susan, 56
Lindeman, Raymond, 50
Liu Shaoqi, 84–86
Livingood, J. J., 39
Los Alamos Laboratory, 41
Lovelock, James, 261, 262

Lowen, Rebecca, 112
Lowenstam, Heinz, 117, 130
Lunar Polar Orbiter mission, 258
Lysenko, Trofim Denisovich, 81
Lysenko affair, 318, 409
Lysenkoism, 81, 350

MacDonald, Gordon, 143
Machel, Samora, 93
Mack, Pamela, 255
Maier, Charles, 243
Malinovskii, Rodion, 212
Manhattan Engineer District, 41, 51
Manhattan Project, 31, 41, 42, 58, 109–113, 125, 169, 236, 322, 432
Mansfield Amendment, 301, 434
Mao Zedong, 75–85, 89, 90, 96, 349–351, 361, 436
March Eighth Agricultural Science Group, 89
Marine Mammal Protection Act, 147–152, 163
Marine Mammal Research Program, 157–168
Marine mammals, 147–166
Marshall Plan Economic Recovery Program, 399, 400
Marxism-Leninism, 413, 414
Masco, Joseph, 439
Maser, 281
Massachusetts Institute of Technology (MIT), 274–288, 303, 434
Mass science, 84–90
Mass spectrometers, 109, 115, 121, 122, 126, 130
Material culture, 56–60
Mayeda, Toshiko, 117
Mayer, Joseph, 111, 114
Mayer, Maria Goeppert, 111, 115
May Fourth Movement, 349
McCarthyism, 79, 359
McCray, W. Patrick, 122, 123
McCrea, John, 116
McElroy, Michael, 254
McKinney, Charles, 116, 129, 130

McNamara, Robert, 241, 242, 300
Medical diagnostics, 45–49, 58
Metabolism, 33–37, 42–45, 49, 50
Meyer, James, 294, 295
Meyerhof, Otto, 34
Michurinism, 81
Military-Industrial Commission (VPK), 197, 198, 215
Military-industrial complex, 13, 19, 191, 192
Military patronage, 141, 142, 169–172, 190, 408, 419, 431–434
Miller, George, 155
Millikan, Robert, 122, 129
Minerva debate, 394–399, 403–409, 416, 420, 435
Ministry of Medium Machine Building (Sredmash), 326
Missile defense, 275–279
Missile engineers, 193–217
Missiles
  Blue Streak, 240
  intercontinental ballistic, 12, 25, 195, 200–207, 211–215, 276
  Silkworm, 85, 96
Mission to Planet Earth, 251, 262–264
Modernization theory, 78, 79
Moon missions, 189, 196, 251
Morgan, Karl, 52
Mozambique, 93
Muhleman, Duane, 297
Munk, Walter, 145–147, 153, 154, 159–162, 170
Mutual Assured Destruction, 13

Nabokov, Nicolas, 403, 404
National Advisory Committee on Aeronautics, 252, 255
National Aeronautics and Space Act, 258
National Aeronautics and Space Administration, 122, 230–235, 241–245, 251–272, 294–296, 319, 434
National Climate Program Act, 255, 259, 434
National Defense of Education Act, 376, 381

National Institutes of Health, 16
Nationalism, 5, 344, 347, 359, 362, 363
National Marine Fisheries Service, 149–159, 162, 168, 169
National Oceanic and Atmospheric Administration, 166, 252, 255, 256, 260
National Science Foundation, 16, 19, 119, 122–124, 302, 434
National security, 16, 19, 22, 23, 57, 190, 228, 236, 237, 432, 433
National Security Action Memoranda
  NSAM 294, 241, 242
  NSAM 354, 239, 240
  NSAM 357, 242
Nation-states
  goals of, 4–7
  and science, 16–24
Nettersheim (missile scientist), 227
Newhouse, Walter, 116, 117
Nier, Alfred, 115
Nimbus program, 252–255
Ninkovich, Frank, 230
Nixon, Richard, 256, 343, 358–360, 437–439
North Atlantic Treaty Organization, 235, 240
Northeast Radio Observatory Corporation, 301, 302
Nuclear energy, 1, 6, 32, 57–59, 318, 323–327
Nuclear fission, 13, 41, 318
Nuclear medicine, 4, 45–49
Nuclear physics, 317–342
Nuclear power plants, 326
Nuclear reactors, 59, 60, 327–331
Nuclear weapons, 1, 4, 11–16, 32
  China's development of, 85, 86
  control of, 114
  France's development of, 236, 237, 244
  proliferation of, 236–237, 243–244
  Soviet Union's development of, 191, 192, 319, 323, 3233
  testing of, 33, 50–52, 55, 434

Oak Ridge National Laboratory, 31, 45, 52–56, 59, 112, 395

Ocean acoustic tomography, 144–172
Oceanography, 1, 141–187, 434, 435
Odum, Eugene, 50–55
Office of Naval Research, 16, 20, 110, 113, 119–120, 124, 145, 146, 151, 169–172
Office of Scientific Research and Development, 21, 112
Office of Space Science and Applications, 252–256, 260
Olson, Jerry, 54
On Understanding Science (Conant), 373
Oppenheimer, J. Robert, 41, 42, 147, 405
Oreskes, Naomi, 94, 108, 266
Orwell, George, 11, 12
Ousley, Gilbert, 234
Overhage, Carl, 282
Oxygen thermometer, 129, 130
Ozone depletion, 253–255

Paleontology, 132
Party Rectification Movement, 80, 81
Patronage, 3, 7, 18–21, 108–110, 141, 142, 169–172, 190–192, 302, 396, 408
Patterson, Clair, 107, 108, 130–132
Pauling, Linus, 125
Pencil-beam scanning system, 278
Periodization, 5, 14, 343, 344
Pest control, 95–98
Pesticides and insecticides, 56, 87, 95, 98
Petroleum industry, 118, 119
Pettengill, Gordon, 278, 299
Photosynthesis, 43–45
Physical Sciences Study Committee, 376, 377
Physics, 2, 17–24, 56, 57
Physiological localization, 46
Pilz, Wolfgang, 227
Pitzer, Kenneth, 120
Planetary Ephemeris Program, 285
Planetary exploration, 256–259, 265
Pluralism, 387
Plutonium, 41, 42, 57
Point Loma, 152, 153
Poland, 15

Polanyi, Michael, 397, 398, 402–406, 435
Politics, 13, 14, 394–399, 420, 437
Post-industrial society, 407–409, 414
Potter, John, 152
Press, Frank, 129
Price, Robert, 280–282
Project Hindsight, 294, 434
Project Lincoln, 276
Project Meteor, 275
Project Whirlwind, 20
Public education, 372–377, 381, 382
Pugwash Conferences, 395
Purcell, Edward, 294
Pure science, 5, 6, 21, 22

Qian Xuesen, 96, 353, 354
Quantum electronics, 20
Quantum mechanics, 17, 18

Radar and radar facilities, 275, 278–303, 432, 433
Radford, William, 294
Radiation Laboratory, MIT, 35–43, 275
Radioactive waste, 31, 50–58
Radioisotopes, 31–73, 434
  artificial, 35–37, 53, 54
  availability of, 31–33, 57–60, 433
  debates over use of, 56–60
  export of, 57
  Manhattan Project and, 41, 42
  nuclear medicine and, 45–49
  as tracers, 32–45, 53–57, 432
  uses of, in ecology, 49–57
Radium Institute, 318, 319
Ramberg, Hans, 117
Rankama, Kalervo, 117
Rasmussen, Nicolas, 31
Reagan, Ronald, 260
Reisch, George, 79
Relativity, general theory of, 5, 273–275, 280–303, 432, 435
Research contracts, 109–113, 119–124, 128–132, 432

Research institutes, 5, 190–191, 318, 319, 325–327
Research Laboratory of Electronics, MIT, 275, 279, 280
Revelle, Roger, 119, 142–144, 162
Richta, Radovan, 410
Richter, Charles, 122, 129
Ride, Sally, 262
Ridgway, Sam, 152, 153
Rittenberg, David, 35
Roberts, A., 39
Rocket propellants, 199–210
Rocketry, 6
Rockets
  Antares, 216, 217
  Diamant, 237
  Europa, 237, 240
  N-1, 189, 199–201, 205–217
  R-5M, 194, 195
Romania, 15
Roosevelt, Franklin, 14, 15, 16, 169
Roosevelt, Theodore, 345
Rose, Naomi, 171
Rostow, Walt, 78, 239, 242
Rowland, Frank Sherwood, 108
Ruben, Sam, 43

Sakharov, Andrei, 194
Savannah River Ecology Laboratory, 54
Schlesinger, Arthur, Jr., 405
Schoenheimer, Rudolph, 35
Science Advisory Committee, 22
Science Applications, Inc., 148
Science-as-market analogy, 397, 398
Science Studies (journal), 419, 420
Scientific farming, 78, 88–90
Scientific knowledge, 13, 14, 17
Scientific Technological Revolution, 394, 409–415, 437
Scott, K. G., 37
Scripps Institution of Oceanography, 144–148, 151, 155–157, 166–168, 434, 435
Seaborg, Glenn, 39

Seasat A, 258
Sebring, Paul, 294, 295, 299–302
Seielstand, David, 164, 165
Seismological Laboratory, Caltech, 122, 129
Semi-Automatic Ground Environment, 276, 277
Shakhanazarov, Georgi, 414, 415
Shanghai Communiqué, 358
Shapiro, Irwin, 5, 274–300
Sharp, Robert, 125, 128, 129
Shelford, Victor, 50
Shen Dianzhong, 89, 90
Sherwin, Martin, 13
Shils, Edward, 394–409, 416, 419, 420, 435
Shute, Nevil, 13
Siddiqi, Asif, 228, 243, 244
Sidorenko, Viktor, 325
Sierra Leone, 93
Silent Spring (Carson), 56
Silver, Lee, 130
Sino-Japanese war, 344
Smil, Vaclav, 97, 98
Smyth, Henry DeWolf, 31
Sociology of science, 374, 375
Solar Mesosphere Explorer, 254
Solid-state physics, 20
Solley, Mayo, 39
Sound Surveillance System, 145, 148
Soviet Academy of Sciences, 190–196, 318–320, 332
Soviet Union, 4–6, 436
 and China, 77, 80–86, 89, 90, 349–357
 defense industry of, 197–199
 distribution of radioisotopes by, 32
 fall of, 343
 France and, 241
 learning in, 77–81
 philosophy of science in, 79
 post-Stalin reforms in, 321–324
 science in, 81, 189–225, 317–342, 409–415, 437
 space program of, 189–225
 and United States, 14, 15

Space Act, 254
Space exploration, 1, 129, 189–250, 256–258
Space Shuttle program, 253, 257–259, 262
Spiesberger, John, 146
Sputnik, 192, 195–197, 210, 251, 371, 372, 376, 377, 409–411
Sredmash, 326
Stalin, Joseph, 15, 16, 318–321, 351
Stavis, Benedict, 78
Stephens, Bill, 239
Stevens, Albert, 42
Stock, Chester, 122–125, 132
Stone, Robert, 40
Strategic Environmental Research and Development Program, 148
Stratospheric science, 252–256
Stroke, George, 279, 280
Structure of Scientific Revolutions (Kuhn), 371–394, 415–418
Struxness, Edward, 52
Studds, Gerry, 155
Substitute Alloy Materials Laboratory, Columbia University, 113, 120
Suess, Hans, 114, 142, 143
Superconducting Super Collider, 263
Sweden, 95
Sweet, William, 48

Taiwan, 356
Tansley, Arthur, 50
Technical knowledge, 13–17
Technological development, 22, 78, 79
Technology transfer, 233–245, 358, 359, 411–413, 436, 437
Teller, Edward, 111
Third World countries, 93, 94, 439
Thode, Harry, 115
Tiananmen Square protests, 343, 361, 362
Tilford, Shelby, 254, 255
Tilton, George, 130
Time-delay test, 279–303
Toulmin, Stephen, 398
Transnational science, 343–369

Trophic-dynamics, 50
Truman, Harry, 14–16, 22, 31, 57, 169
Tu Guangchi, 355
Two Modes of Thought (Conant), 385–387

Ukrainian Physico-Technical Institute (UFTI), 319
UN Framework Convention on Climate Change, 163
United Kingdom, 14, 32, 240, 241
United Nations, 90, 97
United States, 4, 5, 436
  alliance with Soviet Union, 14, 15
  China and, 77, 81, 90, 97, 358–362
  collaboration in space science, 227–250
  economy of, 14–16
  environmentalism in, 95
  foreign policy of, 230, 244, 245, 439
  military of, 6, 18–22, 56–59, 108–110, 169–172, 293, 294
University of Chicago, 108–126, 132
Upper Atmosphere Research Satellite, 254
Urey, Harold, 5, 34, 35, 109–126, 132
US Geological Survey, 253

Vavilov, Sergei, 318
Vietnam War, 77, 359, 434
von Arx, William, 141, 142, 145

Wallace, Henry, 15
Wasserburg, Gerald, 129
Weather satellites, 252–256, 259–260
Webb, James, 241–244
Weilgart, Linda, 152–154, 159, 171
Weinberg, Alvin, 395–399
Weiss, Herbert, 290, 291
Western Europe, and United States, 227–250
Westwick, Peter, 77
Weyl, Hermann, 18
White, Robert, 143, 154
Whitehead, Hal, 152–154
Wiener, Norbert, 56
Wilson, Roscoe, 282

Wilson, Woodrow, 15
Wilson cloud chamber, 34
Wood, Harry, 122
Woods Hole Oceanographic Institution, 144–146, 156, 157
Woodwell, George, 56
World Meteorological Organization, 144
World War I, 14–18
World War II, 11–14, 19, 191
Wunsch, Carl, 145

Xinhai Revolution, 349
Xu Teli, 80

Yang Guanghua, 355
Ye Qisun, 347
Yugoslavia, 15

Zachariasen, William, 111, 112
Zaveniagin, Avramii, 194
Zeldovich, Iakov, 194
Zeng Chengkui, 351, 359
Zhang Li, 355
Zhou Enlai, 349, 358
Zhukov, Georgii, 194
Zuckerman, Solly, 241
Zweig, David, 78

Printed in the United States
by Baker & Taylor Publisher Services